Repair, Rejuvenation and Enhancement of Concrete

Proceedings of the International Seminar
held at the University of Dundee, Scotland, UK
on 5-6 September 2002

Edited by

Ravindra K. Dhir
Director, Concrete Technology Unit
University of Dundee

M. Roderick Jones
Senior Lecturer, Concrete Technology Unit
University of Dundee

and

Li Zheng
Research/Teaching Fellow, Concrete Technology Unit
University of Dundee

 ThomasTelford

Published by Thomas Telford Publishing, Thomas Telford Ltd, 1 Heron Quay, London E14 4JD.
www.thomastelford.com

Distributors for Thomas Telford books are
USA: ASCE Press, 1801 Alexander Bell Drive, Reston, VA 20191-4400, USA
Japan: Maruzen Co. Ltd, Book Department, 3–10 Nihonbashi 2-chome, Chuo-ku, Tokyo 103
Australia: DA Books and Journals, 648 Whitehorse Road, Mitcham 3132, Victoria

First published 2002

The full list of titles from the 2002 International Congress 'Challenges of Concrete Construction' and available
from Thomas Telford is as follows

- Innovations and developments in concrete materials and construction
- Sustainable concrete construction
- Concrete for extreme conditions
- Composite materials in concrete construction
- Concrete floors and slabs
- Repair, rejuvenation and enhancement of concrete

A catalogue record for this book is available from the British Library

ISBN: 0 7277 3175 0

Printed and bound in Great Britain by MPG Books, Bodmin, Cornwall

PREFACE

Concrete is a global material that underwrites commercial well-being and social development. Notwithstanding concrete's uniqueness, it faces challenges from new materials, environmental concerns and economic factors, as well as ever more demanding design requirements. Indeed, the pressure for change and improvement of performance is relentless and necessary.

The Concrete Technology Unit (CTU) of the University of Dundee organised this Congress to address these issues, continuing its established series of events, namely, Creating with Concrete in 1999, Concrete in the Service of Mankind in 1996, Economic and Durable Concrete Construction Through Excellence in 1993 and Protection of Concrete in 1990.

The event was organised in collaboration with three of the world's most recognised institutions: the Institution of Civil Engineers, the American Concrete Institute and the Japan Society of Civil Engineers. Under the theme of Challenges of Concrete Construction, the Congress consisted of three Seminars: (i) Composite Materials in Concrete Construction, (ii) Concrete Floors and Slabs, (iii) Repair, Rejuvenation and Enhancement of Concrete, and three Conferences: (i) Innovations and Developments in Concrete Materials and Construction, (ii) Sustainable Concrete Construction, (iii) Concrete for Extreme Conditions. In all, a total of 350 papers were presented from 58 countries.

The Opening Addresses were given by Mr Jack McConnell MSP, First Minister of the Scottish Executive, Sir Alan Langlands, Principal and Vice-Chancellor of the University of Dundee, Mr John Letford, Lord Provost, City of Dundee, Professor Adrian Long, Senior Vice-President of the Institution of Civil Engineers, Dr Taketo Uomoto, Director of the Japan Society of Civil Engineers and Dr Terence Holland, President of the American Concrete Institute. The Congress had six Opening and six Closing Papers dealing with the main themes of the Seminars and Conferences. Opening Papers were presented by Professor Gerard Van Erp, University of Southern Queensland, Australia Dr Peter Seidler, Astradur Industrieboden, Germany and Professor Kyosti Tuttii, Skanska Teknik AB, Sweden, Professor Surendra Shah, Northwestern University, USA, Dr Philip Nixon, Building Research Establishment, UK and Mr Hans de Vries, Ministry of Transport, the Netherlands. Closing Papers were presented by Dr Gier Horrigmoe, NORUT Technology Ltd, Norway, Professor Andrew Beeby, University of Leeds, UK, Professor Peter Robery, FaberMaunsell, UK, Professor Heiki Kukko, VTT Building and Transport, Finland, Dr Mette Glavind, Danish Technological Institute, Denmark and Professor Yoshihiro Masuda, Utsunomiya University, Japan. The Congress was closed by Professor Peter Hewlett, Chief Executive of the British Board of Agrément, UK.

The support of 23 International Professional Institutions and 32 Sponsoring Organisations was a major contribution to the success of the Congress. An extensive Trade Fair formed an integral part of the event. The work of the Congress was an immense undertaking and all of those involved are gratefully acknowledged, in particular, the members of the Organising Committee for managing the event from start to finish; members of the International Advisory and National Technical Committees for advising on the selection and reviewing of papers; the Authors and the Chairmen of Technical Sessions for their invaluable contributions to the proceedings.

All of the proceedings have been prepared directly from the camera-ready manuscripts submitted by the authors and editing has been restricted to minor changes where it was considered absolutely necessary.

Dundee   Ravindra K Dhir
September 2002   Chairman, Congress Organising Committee

INTRODUCTION

All concrete structures deteriorate as the effects of structural and environmental loading take place over time. Sometimes unforeseen circumstances can occur that produce more rapid degradation than was initially envisaged. Whatever the particular type of exposure or deterioration mechanism involved, all these processes lead to the same point - an unacceptable loss of performance. This may be structural or aesthetic, but the result is that there is a need to repair, rejuvenate or enhance concrete.

To enable this, what has happened to the materials involved must be clearly identified, understood and quantified. This is often a difficult task, as there are no 'black boxes' that can record the history of the particular element. The forensics of deterioration can, therefore, be both problematic and costly.

Having said this, our understanding of the mechanisms of concrete deterioration has greatly improved over recent years and research has begun to identify, if not quantify, the underlying fundamental chemistry and materials science involved. There are new high technology analytical methods that can probe the nature of concrete down to its nano-scale. Undoubtedly these techniques will provide new insights to the effects of degradation processes and, thereby, enhance repair and rehabilitation methods.

Allied to this there has been a growing realisation that there is need for partnership between the scientific and engineering research communities. Without this, research can lack direction and practice can stagnate. Innovation and development are rarely straightforward but there is much to be learned from fast moving, technology-driven industries such as IT and aerospace engineering, where continuous innovation is embraced as a key business process. Even into the 21st century this still seems a paradigm away from our industry.

However, even the concrete construction industry does not stand still and is being required to provide ever higher performance for the client, both technically and economically. In turn, this will bring new challenges for the repair industry. Sustainable construction has turned the spotlight on the way we manage our building and infrastructure and far from being seen as a route to taxation of materials and wastes, it should be seen as an opportunity to re-engineer the maintenance methods and end-of-life disposal. The lessons of other industries are that the adoption of these strategies sharpens work practices and enhances productivity, while protecting the environment.

All of these point towards the fact that the processes of repair, rejuvenation and enhancement of concrete structures must be intimately woven into initial design and construction and should not be regarded as a separate engineering system that simply turns up when something has gone wrong. It is these systems that provided the basis for this Seminar.

The Proceedings of this Seminar; '*Repair, Rejuvenation and Enhancement of Concrete*' dealt with these subjects and the issues raised, under three clearly defined themes: (i) Degradation of Concrete in Structures, (ii) Assessment and Repair Techniques and (iii) Enhancement of Existing Structures. Each theme started with a Keynote Paper presented by the foremost exponents in their respective fields. There were a total of 48 papers presented during this International Seminar, which have been compiled to form these Proceeding

Dundee   R K Dhir
September 2002   M R Jones
                                                                                                                  L Zheng

iv

ORGANISING COMMITTEE

Concrete Technology Unit

INTERNATIONAL ADVISORY COMMITTEE

INTERNATIONAL ADVISORY COMMITTEE
(CONTINUED)

Professor R W Lindberg, *Professor*
Tampere University of Technology, Finland

Dr H-U Litzner, *Senior Chief Executive*
German Concrete Society (DBV), Germany

Mr J E McDonald, *Research Civil Engineer*
US Army Corp of Engineers, USA

Professor S Mirza, *Professor*
McGill University, Canada

Mr J V Paiva, *Head, Building Department*
LNEC, Portugal

Dr T Philippou, *Manager*
Heracles General Cement Company, Greece

Mr S A Reddi, *Deputy Managing Director*
Gammon India Limited, India

Professor F Saje, *Professor of Concrete Structures*
University of Ljubljana, Slovenia

Professor S P Shah, *Director ACBM*
Northwestern University, USA

Professor H E R Sommer, *Consultant*
Austria

Dr S Tangtermsirikul, *Head of School of Civil Engineering*
Thammasat University, Thailand

Professor K Tuutti, *Director*
Skånska Teknik AB, Sweden

Professor T Uomoto, *Head of Uomoto Concrete Laboratory*
University of Tokyo, Japan

Dr O H Wallevik, *Head of Concrete Division*
Icelandic Buidling Research Institute, Iceland

Professor T H Wee, *Associate Professor*
National University of Singapore, Singapore

Professor M A Yeginobali, *Director, R&D Institute*
Turkish Cement Manufacturers Association, Turkey

Professor H M Z Al-Abideen, *Deputy Minister*
Ministry of Public Works and Housing, Saudi Arabia

NATIONAL TECHNICAL COMMITTEE

Professor S A Austin
Professor of Structural Engineering, Loughborough University

Professor A W Beeby
Professor of Structural Design, The University of Leeds

Professor T Broyd
Research & Innovations Director, W S Atkins

Professor J H Bungey
Head of Department, The University of Liverpool

Mr N Clarke
Publications Manager, The Concrete Society

Mr G Cooper
Development Director, Fosroc International Ltd

Mr S J Crompton
Divisional Manager, RMC Readymix Ltd

Dr S B Desai OBE
Visiting Professor, University of Surrey

Professor R K Dhir OBE (Chairman)
Director, Concrete Technology Unit, University of Dundee

Dr K Elliott
Senior Lecturer, University of Nottingham

Dr S Garvin
Director (Scotland), Building Research Establishment

Professor F P Glasser
Professor, University of Aberdeen

Mr P G Goring
Technical Director, John Doyle Construction

Professor T A Harrison
BRMCA Consultant, Quarry Products Association

Professor P C Hewlett
Chief Executive, British Board of Agrement

Mr A Johnson
Divisional Director and Manager, Mott MacDonald Special Services

Dr M R Jones
Senior Lecturer, Concrete Technology Unit, University of Dundee

Professor R J Kettle
Head of Department, Aston University

Mr P Livesey
National Technical Services Manager, Castle Cement Limited

NATIONAL TECHNICAL COMMITTEE
(CONTINUED)

Professor A E Long
Dean of Faculty of Engineering, Queens University Belfast

Mr N Loudon
Senior Technical Adviser, Highways Agency

Mr A C Mack
Operations Director, Bovis Lend Lease (Scotland) Ltd

Professor P S Mangat
Head, Civil Engineering & Construction, Sheffield Hallam University

Mr G G T Masterton
Director, Babtie Group

Professor G C Mays
Deputy Principal, Cranfield University

Professor W J McCarter
Professor, Heriot-Watt University

Dr J Moore
Director, Standards & Technical, British Cement Association

Dr P J Nixon
Director, Centre for Concrete Construction, Building Research Establishment

Mr M Peden
Partner, W.A. Fairhurst & Partners

Dr W F Price
Senior Research Manager, British Cement Assocation

Professor P C Robery
Divisional Director - Midlands, Maunsell Ltd

Mr J M Ross
Chairman, Blyth & Blyth Consulting Engineers

Dr R H Scott
Reader in Engineering, University of Durham

Dr I Sims
Director (Materials), STATS Limited

Professor G Somerville OBE
Independent Consultant

Mr P Titman
Quality and Environment Manager, Edmund Nuttall Ltd

Dr P R Vassie
Research Fellow, Transport Research Laboratory

Professor S Wild
Head, Building Materials Research Unit, University of Glamorgan

COLLABORATING INSTITUTIONS

Institution of Civil Engineers, UK

American Concrete Institute

Japan Society of Civil Engineers

SPONSORING ORGANISATIONS WITH EXHIBITION

Aggregate Industries plc

Babtie Group

British Board of Agrement

British Cement Association

Building Research Establishment

Caledonian Slag Cement

Castle Cement Limited

CEMBUREAU, Belgium

Danish Technological Institute, Denmark

Degussa Construction Chemicals, Italy

Dundee City Council

Elkem Materials Ltd

FaberMaunsell

Fosroc International Ltd

Heidelberg Cement, Germany

Heracles General Cement Company, Greece

Institution of Civil Engineers

John Doyle Construction

Lafarge Cement UK

Makers UK Ltd

MBT Admixtures

Netzsch Instruments, Germany

North East Slag Cement

Palladian Publications

RMC Readymix Ltd

Rugby Cement

ScotAsh Ltd

SPONSORING ORGANISATIONS WITH EXHIBITION (CONTINUED)

Scottish Enterprise Tayside

Thomas Telford Publishing

UK Quality Ash Association

Waste Recycling Action Programme (WRAP)

Wexham Developments

Zwick Testing Machines Ltd

SUPPORTING INSTITUTIONS

American Society of Civil Engineers

Austrian Concrete Society

Belgische Betongroepering, Belgium

Concrete Institute of Australia

Concrete Society of Southern Africa, South Africa

Concrete Society, UK

Concrete Association of Finland

Czech Concrete Society

Entreprises Generales de France, France

European Concrete Societies Network

Fédération de l'Industrie du Béton, France

German Society for Concrete and Construction

Hong Kong Institution of Engineers

Indian Concrete Institute

Institute of Concrete Technology, UK

Instituto Brasileiro Do Concreto, Brazil

Irish Concrete Society

Japan Concrete Institute

Netherlands Concrete Society

New Zealand Concrete Society

Norwegian Concrete Association

Singapore Concrete Institute

Swedish Concrete Association

CONTENTS

Preface   iii

Introduction   iv

Organising Committee   v

International Advisory Committee   vi

National Technical Committee   viii

Collaborating Institutions   x

Sponsoring Organisations With Exhibition   x

Supporting Institutions   xi

Opening Paper
Repair, rejuvenation and enhancement of concrete - a fast growing market   1
K Tuutti, Skånska Teknik AB, Sweden

THEME 1: DEGRADATION OF CONCRETE IN STRUCTURES

Keynote Paper
Diagnosing and avoiding the causes of concrete degradation   11
I Sims, STATS Limited, United Kingdom

Experimental study of the wear of a concrete surface under ice friction   25
B Fiorio

Maintenance management system for existing concrete marine structures   37
H Yokota, T Tanabe and A Moriwake

Appraisal of thermophysical properties of a concrete dam   49
M M Safarov, Z V Kobuliev, O H Amirov, M S Muhamadiev and M A Zaripova

Corrosion processes in the concrete of the dams on the river Angara   55
M A Sadovich and A A Sokolvskaya

Maintenance and refurbishment of concrete in water retaining structures   65
P J Edwards

A technological model for predicting rebar corrosion produced by   75
covercrete carbonation
A Giovambattista, L Eperjesi and E Ferreyra Hirschi

Influence of fire on the mechanical behaviour of concrete specimens   83
S Kumar

Mechanism of damage for the alkali-silica reaction: relationships between   93
swelling and reaction degree
M J Riche, M E Garcia-Diaz , M D Bulteel, M J M Siwak and M C Vernet

Reconstruction of runway 9R-27L at Hartsfield Atlanta International Airport the 33 day wonder - a case history
S R Kuchikulla, D Molloy, F O Hayes, A Saxena and D S Saxena
103

Durability of concrete for transport construction
L A Fedner, J V Nikiforov, S N Efimov and A B Samohvalov
111

Ultra thin whitetoppings: a 2-D finite element approach
W De Corte, D De Leersnyder and E De Winne
119

Thaumasite in concrete structures: some UK case studies
D Wimpenny and D Slater
127

Analysis of a large database of reinforced concrete bridge inspection records
C Mc Parland, A Abu-tair, A Nadjai and J F Lyness
139

Performance evaluation of bridge balustrades in South Africa
C E Ackerman
149

Methodology for characterization of ASR in aggregates in Chihuahua, Mexico
C C Olague, P Castro and G Wenglas
159

Chloride penetration into Swedish road bridges exposed to splash from de-icing salts
A Lindvall
169

Performance, repair needs and renovation of plain concrete pavement of an international airport
S Z Bosunia and A M Hoque
179

THEME 2: ASSESSMENT AND REPAIR TECHNIQUES

Keynote Paper
Documentation of Electrochemical Maintenance Methods
O Vennesland, Norwegian University of Science and Technology, Norway
191

Accelerated examination of long-standing reinforced concrete viaduct
S M Skorobogatov, B P Pasynkov, A V Chernyavskiy, A M Mukhametshin, A V Kurshpel and A K Yagofarov
199

Use of intermittent vibration measurements for determining the integrity of concrete structures
N R Short, O T Owolawi, M G Wood, J E T Penny and J A Purkiss
207

Statistical nonlinear analysis of concrete structures
M Vorechovsky, R Pukl, V Vesely, V Cervenka and R Rusina
217

The effect of pore size on ion migration in concrete during electrochemical chloride extraction
M Siegwart, B J McFarland, J F Lyness and W Cousins
227

Influence of specimen size on measured direct tensile strength of concrete   237
A K H Kwan, P K K Lee and W Zheng

Bond deterioration of reinforcing steel in concrete due to corrosion   247
L Amleh, M S Mirza and B B N Ahwazi

Experimental investigation of the failure patterns and mechanical   257
properties for plain concrete
L Zheng and S Wang

In situ mechanical determination of the concrete elastic modulus   267
G Bocca and M Crotti

Microwave non-destructive testing of fibre concrete using free space   277
microwave measurements
H M A Al-Mattarneh, D K Ghodgaonkar and W M bin W A Majid

Diagnostic assessment and repair of honeycomb concrete in Algiers   289
new airport reinforced concrete building
S Kenai and R Bahar

Non-destructive testing of the microstructural development in hardening   297
cement-based materials by ultrasonic pulse velocity measurements
G Ye, K van Bruegel and A L A Fraaij

Electrochemical systems for repair of reinforced concrete structures   307
N Davison, A C Roberts and J M Taylor

Tees viaduct chloride extraction trial   317
D A Kimberley

Properties of pre-tensioned prestressed concrete members supplied with   329
cathodic protection for ten years
T Aoyama, H Seki, M Abe and K Igawa

The influence of rebar orientation on electrochemical chloride extraction   339
D W Law and A N Fried

THEME 3: ENHANCEMENT OF EXISTING STRUCTURES

Keynote Paper
Structural Challenge of Historic Structures - A Case Study on Renewing
the Reichstags Building for German Parliament in Berlin   349
M Maier, Leonhardt, Andrä and Partners, Germany

Products and systems for the protection and repair of concrete   361
structures: the current position of European standards
G C Mays

Strengthening of concrete structures with externally bonded   371
reinforcement - practical applications in Belgium
S Ignoul, K Brosens and D Van Gemert

FRP for bridge deck strengthening   381
B Sadka and A F Daly

Fibre reinforced mortar for the rehabilitation of historic   391
RC buildings - The Lirick Theatre in Assisi
M Mezzi, R Radicchia, G Mantegazza and A Gatti

Accidental impact loading of concrete structures in the marine environment   403
T M Browne and P M Watry

The principle of information entropy in fracture mechanics of   415
buildings - elements of theory of catastrophes
S M Skorobogatov

Application of reliability theory in service life prediction of initiation time   423
J Andrade and D Dal Molin

Failure probability assessment for additional design load of an existing structure   433
D Bandyopadhyay and S Saraswati

A laboratory and on-site carbonation data correlation in estimating the life   443
span of concrete structures
M F Nuruddin, A B M Diah, K S Ali and H M Saman

Repair of concrete beam-column joints using fibrous composites   453
M Shannag, S Barakat and M Kareem

Carbonation induced corrosion of reinforcement   465
G Bouquet

Closing Paper
Concrete maintenance and repair: the lessons and the future   477
P C Robery, FaberMaunsell, United Kingdom

Late Paper
Motorways: cement or asphalt   489
T Chrzan

Congress Closing Paper
Concrete: Vade Mecum   495
P C Hewlett, British Board of Agrément, UK

Index of Authors   **511**

Subject Index   **513**

OPENING PAPER

REPAIR, REJUVENATION AND ENHANCEMENT OF CONCRETE – A FAST GROWING MARKET

K Tuutti

Skånska Teknik AB

Sweden

ABSTRACT. Fixed assets within European countries that approach the end of service life will probably soon exceed the total stock exchange value. Research efforts during the past decades will serve the society with knowledge for assessment of old structures and a tool for development of new materials and systems for repair, rejuvenation and enhancement of concrete.

Keywords: Fixed assets, Capital destruction, Service life, Deterioration mechanisms, Assessment methods, Strengthening systems, Research, New materials.

Professor K Tuutti, is Research Director at Skånska Teknik AB, a global project development and construction company with about 80.000 employees. The main efforts are focused on establishment of knowledge networks, centres of excellence, etc. as adviser tools. He is also active as professor for the Department of Building Materials at Lund Institute of Technology. He specializes in mechanisms that impact the durability of materials and air quality properties in buildings. He has been profoundly committed in the procedure of establishment of several Swedish National research programmes.

INTRODUCTION

Concrete is a material that has been widely used during thousands of years for countless applications worldwide. The material is, generally speaking, a cultural heritage that has enveloped our societies in industrial countries. In the past 50 years, since the second world war, has shown a great growth in investments in buildings, bridges, roads, underground infrastructure for supply systems, etc, that embody a total capital exceeding several times the value of all stocks quoted on national exchange lists, see Figure 1.

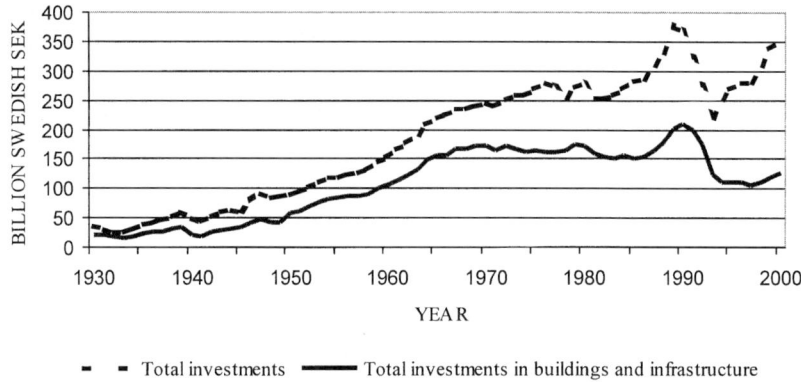

- - Total investments   ──── Total investments in buildings and infrastructure

Figure 1   Swedish investments in buildings and infrastructure during the past 70 years compared with all other investments quoted as actual exchange rate [1]

The main investments have been made in the immediate past 35 years with an annual constant volume. A comparison with the actual value of the Swedish stock market indicates that the employed capital in the building sector is about 4 times higher. Almost the same situation exists in all industrial countries in the northern hemisphere. Politically interesting tasks will be the robustness and functionality of existing structures together with possibilities of rejuvenation and enhancement of different structural parts and materials. Structures built in the beginning of the 50´s are now in the end part of their service life and a lot of decisions and actions are needed, such as possibilities for replacement, demolition or repair and capital investments, are the future demographics with an increasing population of people beyond the age of normal workforces a trap or a possibility? - Is the growth of big cities sustainable with all infrastructures for supply of water, wastewater, electricity, logistics for goods and people, business and living conditions optimal?

PREDICTED CONSTRUCTION MARKET IN INDUSTRIALIZED COUNTRIES

We will certainly face a more complicated and interesting life than we are able to image today. Mankind's life was easy compared to future efforts in maintaining the serviceability in the industrial countries. Even such a small city as Stockholm with a population of 1 million inhabitants is locked in the existing structure. Future supply systems will be obliged to resort to underground systems and existing systems must be maintained, repaired and reinvested with an intellectual capacity not needed before, see Figure 2.

Figure 2   Two photos of the central part of Stockholm at the end of 1800 and hundred years later illustrating the difficulty of a paradigm change of the city structure, Wikström and Lundgren [2]

Investments in research have continuously increased the past decades with focus on durability, service life estimations, repair, etc. which have been reported in almost all concrete conferences. The society will benefit from such visionary activities processed essentially by universities.

The repair and maintenance market will increase rapidly the next decade, caused by the amount of structures that will achieve the end of their service lives and the complicatedness and expensive methods that must be used to ensure accurate serviceability. The existing demography in European countries, evident population entering retirement age, will not simplify matters. The total population in developed countries will decline and the supply of young people will diminish. Probably a paradigm shift would occur where the next society is lacking valuable knowledge or people for transferring existing knowledge.

Information technology as interactive knowledge stations would not overtake the whole role as education institutions. Construction managers, material specialists, researchers, etc. would be raised to the category high technology and the portion of busy older people in this market will certainly increase.

New and smart materials are also entering the construction sector. Extensive research has been undertaken into the use of composite materials, as carbon fibre strengthening systems. An exiting vision is to refine wooden fibres, which have a strengthening potential 10 times carbon fibres, for use in carbonated cement matrixes that could on one hand be a binder for the fibre and on the other barrier for biological processes in fibres.

Smart materials are defined, by CMB at Chalmers University of Technology and University of Reading, as those that recognize their environment and changes and can adapt to meet these changes. New materials in repair and strengthening systems would both increase the load bearing capacity and act as sensors for future assessment of needed actions. Concrete could also be defined as a passive smart material due to its sometimes self-repairing properties, crack-sealing tendencies in different environments, but new strengthening materials for concrete will be a key factor for investments in research.

THE MATERIAL CONCRETE

At first attempt concrete seems to be low technology product consisting of cement, aggregate, water and sometimes additives and admixtures. In early stage of this evolution, concrete was a low technology product but also robust from mixing to fulfill functional requirements. Today even the cement chemistry will be a science within itself with complicated inorganic structures and reactivity conditions. Several additives have entered concrete science such as fly-ash, silica fume, glass filler and different compounds that the societies pushed over from the waste market for solidification into a concrete matrix. Engineers may be unaffected by such changes as long as the concrete strength at 28 days performs the requirements. Conversely, scientists that should understand the materials properties and especially the aging and durability are confused in all possible mechanisms explaining the complexity of such processes. The material industry and the society expects much from ongoing research and we may consider if there is a balance between investments and expectations that would make materials available for use as compounds in concrete. On the other hand scientists have implemented new products with extraordinary properties in strength and durability by the use of several of newly developed compounds and by understanding the mechanism of optimal positioning of aggregate particles combined with fibres and additives. This type of knowledge develops intermittently and would be a future tool also in understanding the performance of existing structures.

DEFINITION OF SERVICE LIFE

Newly constructed huge structures, such as the Öresund-bridge, have had service life requirements in the design phase. The first attempt was to ensure the load bearing structural parts for about 100 years that would be complicated to repair and costly to maintain. The philosophy was not to expect the bridge to collapse after this time of service life, but some years grace for inspection and assessment of needed protection systems or repair work. The most important knowledge for this type of predictions stems from the material science.

Old structures in the end of their service life are often seriously damaged and eventually near a collapse. The first step in these assessments is to analyse the actual load bearing capacity and if there would be big risks for a sudden and eventual brittle fracture. After this procedure the material experts would be valuable for prediction of the future deterioration as a function of time or the impact of different repair systems on the deterioration processes. These types of processes need collaboration with both structural engineering and material science. This type of collaboration, historically, has not been as frequent as we would request today.

Generally the knowledge in this complicated area is not as precise as we would want. We are today able to predict the general trends in different processes but the variation in both the materials and the environments are sizeable which will affect the accuracy of such predictions. However, these prediction methods used for existing structures combined with sampling will increase the accuracy.

The definition of the service life could be stated as the time a structure in a specific environment will ensure the main properties, security against collapse and acceptable aesthetic conditions. However, owners of important structures would probably also implement economical assessments, interruptions in serviceability and changes in image if unexpected disturbance would occur.

THE MOST ACCURATE DETERIORATION MECHANISMS

In literature different deterioration mechanisms for concrete have been reported. Practically the dominating types of destruction are:

- Reinforcement corrosion.
- Frost attack in cold regions.
- Alkali silica reactions.
- Chemical attacks.

Reinforcement corrosion is an electrochemical process depending on the micro environmental conditions close to the metal phase. Corrosion of reinforcement is initiated as a result of carbonation or by the presence of chlorides in the concrete. The oxidation process will increase the volume and create splitting forces on the concrete matrix. Normally we would see signs of the ongoing process by rust streams and cracked concrete covers. Corrosion of reinforcement in concrete structures is of a more global interest than other mechanisms and naturally also more thoroughly studied. Diffusion processes, mathematical modelling and the relevance of different material parameters are well known. However, one important parameter, the corrosion threshold value, must be further studied before we can expect a general breakthrough in this subject. Theoretical calculations and modelling of the time of initiation in a chloride rich environment demonstrate the lack of knowledge for the important parameter, the threshold value of chloride ion concentration, which changes the passive stage to an active corrosion stage, see Figure 3.

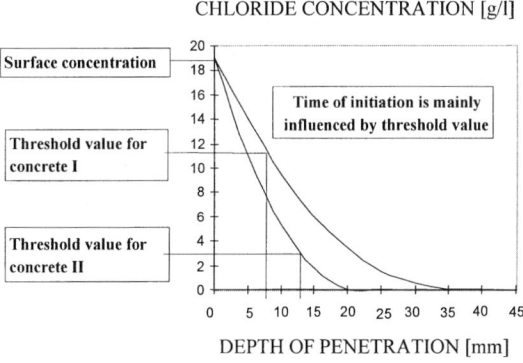

Figure 3   Schematic sketch of initiation time for two different concrete's as a function of depth. High threshold values have been measured for concretes with low water cement ratios and constant humid conditions. Practical experiences indicate also that the penetration of chloride ions is not the most important parameter.

The frost attack is a physical process by which the volume transforms water into ice. Splitting forces will appear in the pores if the expansion is prevented. The term critical degree of saturation and standardised frost resistance test procedures are milestones in the frost resistance domain. The severities of frost attack depend on both the moisture loads in the structure and the presence of salt. Salt can be supplied to structures both as salts used for thawing purposes and by contact with seawater. Old structures may also have been produced with a salt addition to the concrete mix.

Alkali silica reactions will take part in alkali soluble materials such as glass, some types of natural aggregates etc. These reaction products also increase in volume with general cracking of the concrete matrix.

Chemical attacks could be of different mode of action either expansive by formation of new chemical structures or soluble on the cement paste.

Principal prerequisite for all of these deterioration mechanisms is the environmental conditions that could create a flux of aggressive substances into the concrete. Some of these substances are not normally comprehended as dangerous such as water but essential for all of these mechanisms. Today it is possible to predict the moisture conditions in different materials also if fluctuations occur, see Figure 4.

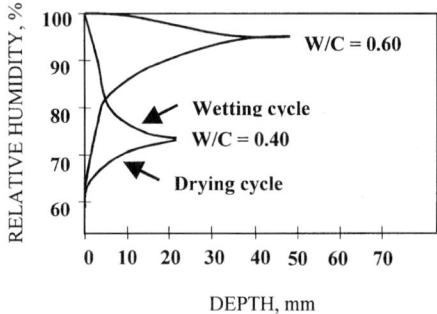

Figure 4   Mathematic modelling of the moisture variation in two different concretes, W/C 0.40 respectively 0.60, with an environmental cycling of 1 day capillary suction and 1 month drying in RH = 60%, Arfvidsson and Hedenblad [3]. Consequently a high performance concrete would create low and constant moisture conditions in the reinforcement zone.

ASSESSMENT OF CONCRETE STRUCTURES

Knowledge in structural engineering and in material science is the key component for assessment of service lives, recommended repair systems, choice of materials for repair and execution of repair work. Is the object close to a collapse, what type of processes have caused the destruction, would deterioration continue even after repair, would we predict different impact on properties from different repair systems or materials, would we expect a more durable structure after repair are some interesting tasks upon entering the market for repair, rejuvenation and enhancement of concrete. We may consider that many questions will be raised but all of these could not be answered.

Several research programs have been processed during the past twenty years and experts could today use a lot of detail information presented in the literature. The complicated nature of the information will hold back a general use of these methods. However, logical thinking and simplification of these theories would facilitate the use particularly in the first assessment that should aim in elucidation of the urgency of intervention. In the European project, 309021 CONTECVET, manuals for assessment of concrete structures have been developed and evaluated by owners of concrete structures. Numerous practical cases have been analysed in order to express the usefulness of such assessment methods. The structure of simplified assessment is presented in Figure 5.

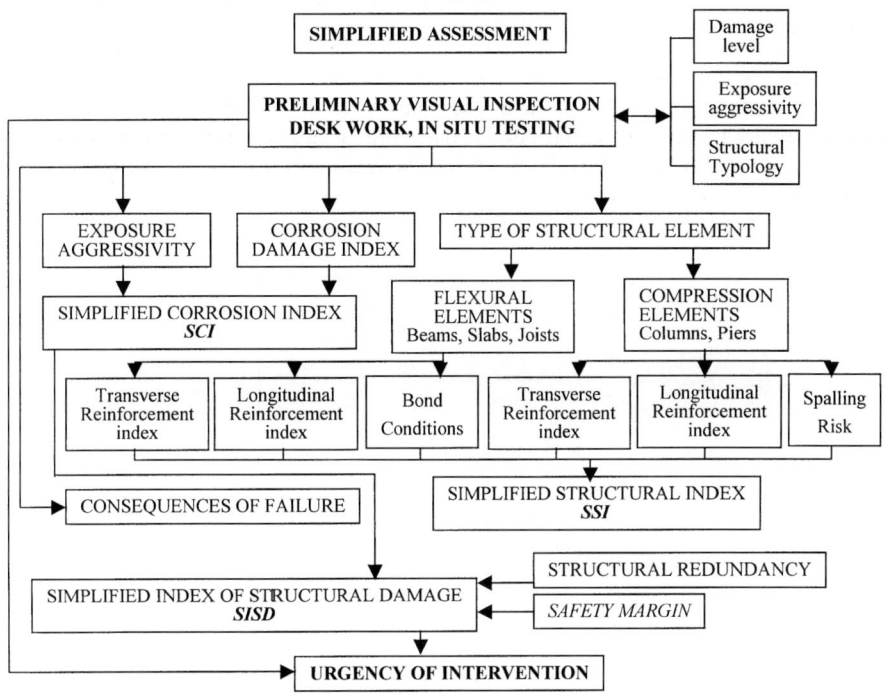

Figure 5   Simplified assessment structure for concrete structures [4]

The European Commission is still financing continued research projects in this field but predictions of the amount of structures that should be examined in the near future indicate lack of time for improved knowledge and implementation into the society and the industry.

REPAIR STRATEGY

Structures that are in a phase of great demand of repair indicate that the material properties have changed by the impact of the environment. Time of service verify the suitability of a material or a combination of materials for a specific use. An important issue will be the possibilities to restore the structure to its former condition or improve properties as load bearing capacity. However, it is not obvious that a repair would improve the durability properties. Several practical cases have shown a faster deterioration after a repair process.

The strategy of a repair design procedure is to use experts on both material science and design probably people working in universities. Generally it is an obvious fact that repair of concrete is much more difficult compared to steel, timber, brick, etc. The problem is connected to the concrete pore structure that is more or less filled with water. An impermeable sealing would protect the material against aggressive substances but could also saturate parts of the material, especially areas close to the impenetrable layer, which accelerate deterioration processes that are influenced on high water content. Therefore, the use of permeable materials, such as concrete itself, will be of advantage for repair systems.

Choice of repair method must be adjusted to cause of damage or deterioration mechanism, range of damage and damage position. New polymer-based repair materials are introduced on the market, more or less impermeable, and must act together with the old concrete. The role of water is significant some processes will increase in rate with increasing water content other processes will act oppositely. Impermeable repair layers will raise negative properties in cold regions where frost attack is considerably destructive, see Figure 6.

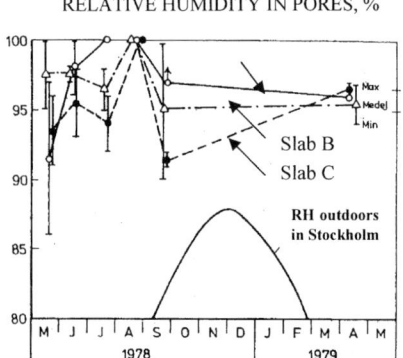
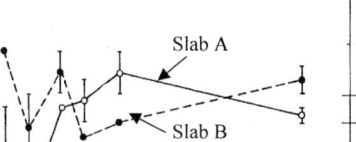

Figure 6   The Figure on the left shows measured relative humidity during a year in repaired concretes where the applied materials are impermeable. The Figure on the right illustrates the moisture conditions in repaired concretes where the repair material is permeable, as concrete with water cement ratio 0,45. An impermeable surface, even if it will prevent water suction into the material, could increase the water content in the old concrete behind a dense surface layer, Fagerlund and Svensson [5]

Replacement of damaged concrete is a common method for concrete repair. This repair methodology must be used for heavily damage concrete and is recommended for chloride-induced corrosion. Water-jet techniques are very efficient processes for removal of hard concrete and will also induce some leaching of chloride ions from the surrounding concrete areas. These methods will not cause any harm to the reinforcement compared to hammer-drill methods. Extensive repair work is recommended to use water-jet techniques, especially if the work will be carried out on horizontal joists.

Repair of concrete structures include sometimes the shifting of one material to another material. The old fashion ingredients in reinforced concrete would not give the optimum properties in the most aggressive environments. Black steel could be replaced by carbon-fibre systems or stainless steels of different chemical compositions.

Polymers will give a better surface protection against acids than high performance concretes. Consequently the environment surrounding concrete has a major influence on the type of deterioration mechanism that will create damage and also affecting the rate on destruction processes.

Generally concrete repaired with concrete of higher quality than the original are a safe method of restoring structures. The thickness of a new layer on an old concrete would have significant impact on robustness of the repair system. The problematic task is to generate full and durable adhesion in the joint between old concrete and the repair material. Anchorage in existing reinforcement and studs in the old structure would generate advantageous properties for the repaired structure. Divergences in volume changes by time between the old structure and the new applied materials are of great importance in the choice of the repair system. Sometimes completely new structures have been constructed around heavily damaged sections. Such drastic repair systems will take over all load-bearing functions and the properties of the old concrete will be of no importance. However, this type of surrounded poor concrete would probably increase in volume for all time, which must be called to attention to in the design of the repair system.

Sometimes a surface treatment is an optimal repair action. Today we are able to use varies materials that will improve the durability and the aesthetics of repaired structures. These methods can be divided into

- Impregnation.
- Sealing.
- Coating.
- Painting.

Impregnations of concrete surfaces, which are able to penetrate deep, have shown positive impact on the slowdown of destruction processes.

SEVERELY EXPOSED STRUCTURES

Concrete standards or codes have always classified structures into different environmental classes. However, codes of today are more restrictive due to our experience and the existing know-how. The environment class concept is a way of describing the environmental load to which structures are expected to be exposed.

This background gives three typical structures that are in great need of repair:

- Heavily exposed civil engineering structures, such as bridges, quays, etc.
- Old protruding surfaces on buildings, such as balconies, facades, etc.
- Parking decks with relatively low concrete quality.

The deterioration mechanism for civil engineering structures and parking decks is the presence of chloride ions. Generally chlorides raise conditions that are difficult to repair and depend on the process of steel corrosion. Only two possibilities seem to be conceivable, namely removal of chloride-contaminated concrete or a cathodic protection of the reinforcement. New types of inhibitors had also entered the market, but the methods are still lacking experience of a long exposure time. Frost attacks that peel off parts of the surface reduces automatically also the chloride concentration but likewise requires a replacement of concrete. In dry environments some trials have been made with corrosion control by a reduction of the internal relative humidity.

Replacement of cracked areas for concrete structures seems not to be effective for chloride initiated corrosion processes. Normally we are able to identify fractions of ongoing corrosion by ocular inspection. Repair actions for deteriorated areas will cause new concrete spalling after some years. The reason is that the propagation time from active corrosion to a visible crack pattern takes about 10 years. Facades are quite different due to a lower concrete quality. These structures are exposed to a dryer environment and we will normally not detect any frost problems. However, spalling of concrete is increased by time, depending on a slow steel corrosion process. Corrosion in such structures is induced by carbonation, a pH reduction in the concrete pore water. Repair actions with the purpose to reduce the corrosion process by a reduced relative humidity or carbonation preventing treatments have been relatively successful. During the past two decades the Swedish efforts has been focused on the repair of balconies. These structures had before repair varying conditions caused by a low concrete quality in origin. The successful strategy has been to replace the edge parts with a high quality concrete. Nordic experiences indicate that structures containing chloride ions in such concentrations that the corrosion process is initiated should be classified as heavily damaged concrete. A concrete façade could contain such critical chloride content if chloride content additives were used during the casting procedure, which were a common method 20 to 40 years ago.

CONCLUSIONS

We are entering a new stage in the construction business where the repair and rejuvenation of old concrete structures would probably be the most dominating discipline in the next decade. The demography, few young engineers and numerous retired experts, in industrialised countries will have a major impact on the market structure. A lack of knowledge would increase the general interest in the high technology repair market to fill the void. Universities and institutions controlling the knowledge and diffusion of knowledge will be a key actor supporting the industry.

New materials and smart materials will be used in different applications for both strengthening and rejuvenation of concrete structures.

Material science and structural engineering experts will cooperate in research projects and pilot studies for contribution to a better understanding of mechanisms in different deterioration processes and the structural impact.

REFERENCES

1. Statistics from the Swedish Construction Federation, Personal Communication, Stockholm, 2001.

2. WIKSTRÖM, J, LUNDGREN, J, Perspektiv på Stockholm, publishing company Max Ström, Stockholm, 1991, ISBN 91-89204-05-0.

3. ARFVIDSSON, J, HEDENBLAD, G, Calculation of moisture variation in concrete surfaces, Building Materials, Lund Institute of Technology, Lund, 1992.

4. EC Innovation Programme IN30902I CONTECVET, A validated users manual for assessing the residual service life of concrete structures, chaired by George Somerville, British Cement Association.

5. FAGERLUND, G, SVENSSON, O, Durability of repair systems for concrete balconies, Swedish Cement and Research Institute, Fo 2:80, Stockholm, 1980.

THEME ONE:

DEGRADATION OF CONCRETE IN STRUCTURES

DIAGNOSING AND AVOIDING THE CAUSES OF CONCRETE DEGRADATION

I Sims

STATS Limited

United Kingdom

ABSTRACT. Concrete can be exceptionally durable, but is sometimes regarded as being inherently defective. Durability is now challenged by complex mixtures, demanding applications and increasing use of marginal materials. Most causes of degradation can be prevented and experience from dependable diagnosis of previous failures is invaluable. The essential steps to successful, reliable diagnosis are summarised and the importance of continuity from desk study, through site inspection to laboratory findings is stressed. Petrography is identified as an indispensable procedure in the investigation of concrete deterioration. A selection of both established and relatively newly recognised threats to concrete durability is reviewed and some particular issues are highlighted for each mechanism. It is suggested that the occurrence of the thaumasite might have been underestimated in recent surveys.

Keywords: Alkali-reactivity, Bicarbonation, Carbonation, Diagnosis, Durability, Ettringite, Freeze-thaw, Petrography, Sulfate, Thaumasite, Weathering.

Dr I Sims, graduated in geology from London University in 1972, then undertook PhD research into some mechanisms of concrete deterioration, including seawater attack, reinforcement corrosion, alkali-silica reactivity and the conversion of high-alumina cement. In 1975, he joined Sandberg, specialist consulting materials engineers in London, where he established and developed their geological department, gained experience in a wide range of construction and building materials and applications all over the world and rose to become a Senior Associate in their Consultancy Group. Since 1996, he has been a main board Director of STATS Limited, specialist engineering, materials and environmental consultants based in St Albans, where he is responsible for materials consultancy and expert witness services. He has co-authored books, published many practical papers and articles and served on various national and international committees and working parties.

INTRODUCTION

Concrete has the potential to be an exceptionally durable construction material and historic structures survive to demonstrate this truth. Yet concrete in recent times has gained an unfortunate reputation for being almost inherently defective, including its alleged susceptibility to 'concrete cancer', a journalistic epithet now taken to cover a full range of major and minor defects. This need not and should not be the case. Although research continues to help engineers and scientists to understand the detailed behaviour of concrete, and to improve its performance in various conditions, the main causes of in-service degradation are actually well established and usually avoidable. Most of these mechanisms have been exhaustively investigated and ought to have been eradicated, but nevertheless remain a potent threat to durability on a world scale; examples include alkali-reactivity and chloride-induced corrosion.

However, concrete is a beguilingly complex material, which is exposed to a wide range of conditions. This complexity has steadily increased over time, with developments in cements, the ever-widening use of additions and admixtures, sometimes in bewildering cocktails, plus the growing pressures for the greater use of waste, recycled and 'marginal' aggregate constituents. Also, in some special cases, demanding applications and particularly exposure conditions can present new challenges. In these ways, new circumstances can be generated for concrete, giving rise to fresh durability threats or the reappearance of previously controlled problems.

An assurance of concrete durability in structures is not adequately achieved by prognostic extrapolation from theoretical studies or laboratory experimentation alone. It is important to learn from previous experience, so that defects should be reliably diagnosed and the findings used to guide preventative practice. The purpose of this contribution is to stimulate debate by the presentation of comments, including some perhaps controversial observations, based upon practical experiences of concrete investigation and endeavours to assure concrete durability.

DIAGNOSIS

Any investigation into the cause(s) of concrete degradation should be conducted systematically and without prejudgement. An apparently obvious cause is not always the only factor and, sometimes, may eventually be found to be irrelevant. A concrete motorway bridge in continental Europe, for example, was dismantled in the misguided belief that it was irreparably damaged by alkali-silica reaction (ASR). Investigation, carried out during demolition, certainly found evidence of ASR, but it was a secondary feature, brought about by water percolation and alkali concentration facilitated by cracking of an under-reinforced and consequently over-loaded deck [1].

An effective diagnosis requires continuity between the various aspects of an investigation, including: i) a desk study into the design and specification of both the structure and the concrete, ii) site inspection, iii) sampling and iv) laboratory examination, analysis and testing. This ideal may be achieved with some research-based investigations, but regrettably it is quite uncommon in commercial cases. All too often, no one carries out task i), a site engineer attempts step ii), the sampling stage is entrusted to a contractor and the laboratory work is conducted in ignorance of the site circumstances and with little or no information about the sampling locations. Co-operation and teamwork are essential to successful and dependable diagnosis. In particular, it is imperative that the laboratory specialist is involved with the site inspection and especially with the selection of sampling locations.

Interpretation is then a matter of integrating all of the desk study, site and laboratory findings, so that the primary and any contributory causal factors can be identified. Unequivocal outcomes are the exception and further or corroborative work is often desirable. Recipients of investigation reports should be cautious over definitive or overly dogmatic conclusions, as diagnosis is often a subjective interpretation of objective but possibly incomplete findings. By the same measure, investigators should not lose sight of the limitations of the evidence, which is often pronounced in funding-sensitive commercial investigations.

It is a truism that the value of the laboratory investigation phase is dependent on the relevance of the samples taken from the structure; hence the need for the laboratory scientist to be involved with the sampling and to have first-hand appreciation of the context from which the samples were derived. Most engineers and scientists are familiar with practices for obtaining statistically representative samples of materials, but sampling for investigative purposes is different. In that case it is important to select sampling locations that represent, not the bulk material, but the features or defects under investigation, or variations in circumstances that might have some bearing on their causation. Comparative samples might be taken, for example, from variously affected and unaffected locations, or from positions exposed in service to different weathering conditions. Examples of such selective sampling regimes are described for ASR[2] and for the thaumasite form of sulfate attack (TSA)[3].

These concrete samples, often drilled cores, will need to be viewed in the laboratory with a full understanding of their site context. If this cannot be achieved by an intimate involvement of the laboratory specialist on site, which is the preference, then each sample must be accompanied by unambiguous notes, sketches and good quality photographs. Any degradation during sampling, particularly any sample loss, must be recorded. In any case, all samples must be clearly marked to show identity and orientation. It is good practice for concrete samples to be wrapped immediately to prevent moisture loss (for example using Cling-film) and to be further wrapped or be otherwise protected from physical damage in transit to the laboratory.

Once in the laboratory, the samples will be subjected to an appropriate programme of examination, analysis and testing. There is a wide variety of traditional, specialised and innovative techniques for the laboratory investigation of concrete[4] and the experienced materials scientist will select the methods that are most applicable in a particular case. However, it is strongly recommended that any laboratory assessment of hardened concrete should include, and preferably commence with, petrographical examination.

Adapted from similar procedures that are routinely used in the systematic description of natural rocks, petrography is indispensable in the investigation of concrete deterioration[5]. It is a direct-observational technique, which ranges from the initial visual inspection of samples in their as-received state, through optical microscopy using various types of specimen and different levels of magnification, to scanning and other forms of electron microscopy. One examination, by an experienced petrographer using good quality equipment, can provide a breadth of factual information on the sample concerned that will be invaluable for diagnosis. This information will routinely embrace, for example: data on constituents and composition, including additions, water/cement ratio and air content (differentiating entrapped and entrained); indications of quality, such as the thoroughness of mixing, any segregation or bleeding, any excessive micro-cracking and the adequacy of compaction; an assessment of condition, including degrees of carbonation and any signs of leaching or weathering; and the identification of any evidence for the operation of decay mechanisms, such as ASR or TSA.

It is well recognised that petrography is an indispensable procedure for the diagnosis of ASR[2] and TSA[6,7], but it is not always appreciated that petrography is an equally powerful technique for the overall investigation of hardened concrete. The relatively simple early stages, for example using an unaided eye or only low-power stereo-microscopy, are essential and should not be excluded in the mistaken belief that only the comparatively sophisticated examinations at higher magnification are important. Just as it is necessary for the laboratory samples to be viewed in the context of their locations on the structure, so it is vital for small areas examined in detail under a microscope to be viewed in the context of their positions on the samples.

It is particularly poor practice to exclude optical microscopy altogether, believing that electron microscopy will provide all the answers. On the contrary, there would then be a grave risk that the small-scale evidence of electron microscopy would be over-emphasised and its significance misinterpreted. In one case of a concrete water-retaining structure outside the UK, cracking was reported by the owner, cores were drilled by a contractor and submitted to a research laboratory, which immediately subjected small portions to examination by scanning electron microscopy (SEM). A thorough search using SEM eventually identified some alkali-silicate gel deposits and the cracking was pronounced to be caused by ASR. After more than two years of expensive legal dispute, it was finally accepted that this diagnosis was entirely incorrect. The concrete mix was high in ground granulated blastfurnace slag (ggbs), then poor supervision on site had permitted excess water to be added prior to placing and for curing to be completely inadequate. As a result, the concrete surfaces exhibited severe drying shrinkage cracking, the pattern of which was characteristic of its actual cause and inconsistent with the mis-diagnosed cause.

After initial visual examination of all the samples, variously using the unaided eye and low-power stereo-microscopy, selected sample portions should be subjected to examination in large-area thin-section under a high-power petrological microscope. Auxiliary examinations can be carried out using finely-ground concrete surfaces or, in reflected-light mode, using highly-polished specimens. Increasingly, conventional petrography is augmented by fluorescence microscopy, which is useful for studying microporosity and/or micro-crack patterns. When necessary, observations in thin-section can be quantified by point-counting. STATS employs a preparation technique that, when appropriate, enables particular areas of thin-sections to be subsequently investigated using SEM and its attached micro-analyser. This has proved to be particular helpful with investigations for TSA.

SOME ESTABLISHED THREATS
Weathering

Most external concrete surfaces are subjected to weathering, although obviously the degree of exposure and the severity of the weathering agents varies greatly. In research, most interest has centred on the ability of concrete to resist the most dramatic threats to durability, such as freezing and thawing cycles (especially when exacerbated by de-icing chemicals), seawater action and, in some climatic regions, salt weathering. However, in the UK and similar temperate climatic regions, the most common weathering threats are posed by rain, rain driven by wind and subsequent drying conditions.

It is easy to overlook the weathering effects caused by gradual dissolution and leaching arising from regular exposure to slightly acidic rain-wash, also the physical degradation that can result from wetting & drying fatigue.

Advanced leaching of concrete produces a distinctive texture of matrix scouring and secondary re-deposition in thin-section (Figure 1), but can be difficult to detect in the earlier stages. Although many additions are now available that potentially help concrete surfaces to be more weather-resistant, it remains the case that the greatest benefit is derived from conventional concrete material that is designed, mixed, placed and finished to the highest standards. Of course, it follows that a successful combination of high quality traditional concreting practice and the appropriate use of additions and/or admixtures can achieve an optimum level of durability potential.

Concrete surfaces that reach critical water saturation are vulnerable to damage caused by freeze-thaw cycles. This is a minor threat in most parts of the UK, but assumes great important in other parts of the world, such as Scandinavia, Central Europe and the interior and northern areas of North America. Freeze-thaw cycles have the capacity to be very destructive and are often the direct cause of damage that has been initiated by other mechanisms. In Scotland, for example, the condition of concrete that had cracked as the result of using shrinkable aggregates was typically greatly worsened by subsequent freeze-thaw damage. Also in Canada, where many structures have been affected to some extent by forms of alkali-reactivity and chloride-induced reinforcement corrosion, it is commonly freeze-thaw action that has caused the main damage. Indeed, in terms of diagnosis of such cases, it can be difficult to unravel whether inevitable concrete degradation has been simply accelerated by freeze-thaw action, or instead that originally minor damage has been transformed into major damage by freeze-thaw, or even that potential resistance to freeze-thaw was critically compromised by the initial mechanism.

Air-entrainment is often specified as a means of providing resistance to freeze-thaw action, although controls over cement content, water/cement ratio and appropriate additions might be similarly effective, at least in UK conditions[8]. However, air-entrainment is typically monitored by measuring total air content in the fresh concrete, rather than separately measuring the entrained air-void component or actually checking the air-void system in the finally hardened concrete. Thus, failures have occurred when the content of entrained air-voids is less than intended, or, more critically, because the entrained air content in the exposed surface zone is lower than required for resistance whilst that in the main part of the concrete is within specification[9]. A concrete airfield runway in the UK exhibited surface delamination caused by freeze-thaw action, despite the mix having yielded appropriately high total air contents and initial testing of hardened concrete cores indicating satisfactorily high contents of entrained air-voids. More refined application of the microscopical test method[10], demonstrated a crucial loss of entrained air from the surface-most few millimetres, apparently as a result of the methods used for placing and finishing.

Normal Carbonation

Concrete undergoes a gradual process of carbonation during exposure to the atmosphere[11]. In practical terms, the main effect of carbonation is significantly to reduce the alkalinity of the concrete from around pH=13 for the original non-carbonated material to around pH=8 once fully carbonated. Whilst a concrete alkalinity of pH=13 protects steel reinforcement from corrosion, an alkalinity of pH=8 does not, so that carbonation of the full cover depth in reinforced concrete exposes the embedded steel to a distinct risk of corrosion. For this reason, it is common for routine condition surveys of concrete structures to include measurements of the depth of carbonation beneath the exposed surfaces.

Such measurements of carbonation depth are usually carried out using a phenolphthalein indicator solution[12], which is sensitive to alkalinity rather than carbonation. Whilst this method is simple and easy to apply in quantity, it is unable to detect partial carbonation and can thus provide a misleading level of reassurance. Thin-section microscopy sometimes reveals that a significant zone of partial carbonation (Figure 2) separates the fully carbonated surface-most zone or 'carbonation front', which is detected by phenolphthalein treatment (by not being stained), from the non-carbonated region. Both the partially carbonated and non-carbonated areas are stained purple by the phenolphthalein solution. This potential for overlooking partially carbonated concrete, and thus under-estimating the risk of corrosion, may be particularly serious, as research has indicated that the rate of corrosion increases steeply as the carbonation front approaches the steel[13].

Thin-section microscopy also enables the carbonation pattern to be precisely mapped, including features such as cracks and bleed channels that facilitate localised deeper penetration by carbonation, which can sometimes compromise the protection to steel being offered by an otherwise non-carbonated concrete cover.

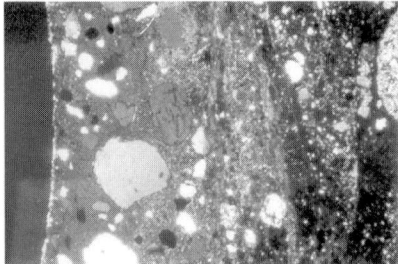

Figure 1 Leaching & re-deposition in concrete

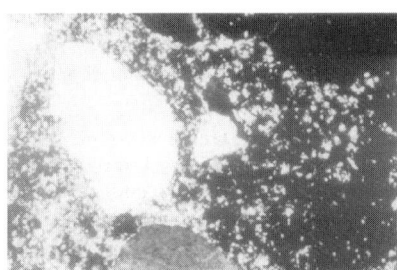

Figure 2 Zones of carbonated, partially carbonated & non-carbonated concrete (left to right)

Alkali-Reactivity

Concrete specialists have been struggling to overcome the potential threat from alkali-reactivity for more than 60 years[14]. As there is a number of related but different reactions, and because the risk is dependent on a range of factors that include all the constituents, the way in which they are used together in the concrete and the environmental conditions in service, each country has tended to seek and sometimes identify local solutions. In the UK, for example, preventative measures were developed, appear to have been effective and have now evolved to offer a flexible range of options[15,16].

The UK preventative approach only covers the most common type of reaction, ASR. Affected UK structures, which were mainly investigated during the 1970s and 1980s, presented a fairly repetitive situation, implicating particular combinations of aggregate materials[17]. A wide range of cements and additions is also available for use within the UK, with few serious transportation problems around a relatively small (but highly populated) geographical area. Accordingly, it was possible to devise a scheme that effectively accepts aggregate as being potentially reactive (described in the UK as having a 'normal' reactivity) and then minimises the risk of ASR damage by controlling the concrete alkali content through a choice of options.

This UK system means that many as-designed concrete mixes will be assessed as meeting the preventative criteria without further precautions. It also enables the amount of project-specific aggregate assessment and testing to be restricted to new materials with no performance history, structures that will be exposed in service to unusually severe conditions or structures that are deemed exceptionally 'sensitive' in that the design life is prolonged and/or the consequences of failure would be unacceptable.

In recent times, for example, great efforts were made in an endeavour to ensure that the concretes used for the Sizewell B nuclear power station and the Channel Tunnel would not be damaged by ASR. It may be fair to say that the UK solution has led to a degree of complacency over ASR potential, so that insufficient assessment of aggregates and concrete mixtures is currently being undertaken.

Such a comfortable situation does not exist world-wide. Many regions and countries are still confronting a real threat to concrete durability from alkali-reactivity and UK engineers have to make due allowances for this factor when working overseas[18,19]. In some countries, there is a wider diversity of aggregate materials, including some varieties that would be recognised as having a 'high' reactivity according to the UK scheme, also there can be fewer available options for controlling concrete alkali content and in-service exposure conditions may be more extreme. Thus, world-wide, there remains an urgent need to find ways of assessing aggregate combinations that are both fully reliable and practicable in terms of test duration.

In 1988, a RILEM technical committee (TC 106, now TC ARP) was charged with developing such methods for universal application and has made good progress in respect of ASR[20,21]. An integrated assessment scheme has been established, which first classifies the aggregate combination by petrography, then provides a rapid screening test for materials deemed potentially reactive and a longer term concrete expansion test for ultimate reassurance (Figure 3). TC ARP is currently embarking on an international trial of an accelerated concrete expansion test that has provided some encouraging initial results. This new method holds out the prospect of being usable as a project-specific performance test for ASR.

It has long been recognised that some carbonate aggregates can similarly participate in an expansive reaction with alkalis within hardened concrete. In particular some materials in regions of Canada have been identified as exhibiting alkali-carbonate reactivity (ACR). This type of alkali-reactivity has usually been regarded as being a comparatively rare regional occurrence and much less is known about its mechanism(s) and the best approaches to minimising the risk of damage.

Recently the importance of ACR on a world scale has been highlighted by reports from China[22], where large quantities of potentially reactive dolomitic rocks and aggregates have been identified and damage to structures investigated. There is thus now an international imperative for improving our understanding of ACR and RILEM TC ARP has started to respond to this challenge.

Geological materials are infinitely variable and thus have the capacity to pose materials scientists with complex problems. In this way, it cannot be supposed that rocks and aggregates will simply exhibit potential for either ASR or ACR, or neither. Cases have been identified in China that exhibit potentially reactive silica and potentially reactive carbonate in the same material.

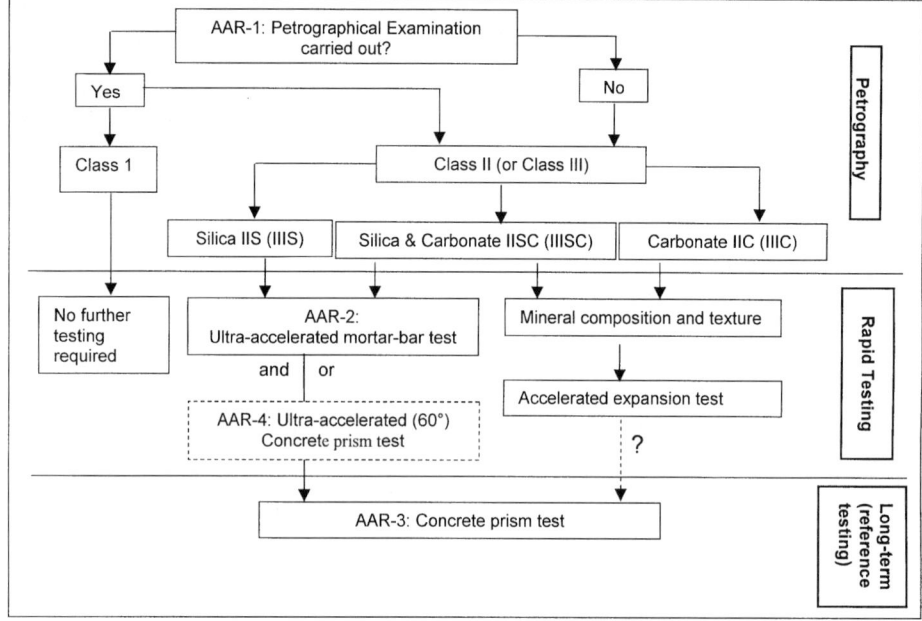

Figure 3   RILEM TC-ARP integrated aggregate assessment scheme

SOME NEW THREATS

Aggressive Carbonation

Carbonation of concrete by interaction of cement hydrates with carbon dioxide in the atmosphere is reasonably well understood and typically poses an indirect threat to the durability of reinforced structures, rather than endangering the integrity of the concrete material itself. However, in some circumstances, ground water may contain dissolved carbon dioxide and can attack buried concrete in a very destructive manner[23]. Absorption of carbon dioxide by ground water is not unusual and may reduce pH, except that buffering by bicarbonate already present increases the amount of absorbed carbon dioxide that is required to reduce pH[24]. However, in some mountain and high moorland areas, where the water is relatively free of dissolved salts, small amounts of dissolved carbon dioxide can cause the water to become acidly aggressive towards concrete. Such areas, of course, can be associated with major concrete structures, including dams, related structures and tunnels. The author has recently encountered examples in which desert ground waters containing dissolved carbon dioxide may have been responsible for unexpectedly high degrees of carbonation of deeply buried concrete.

Another form of 'aggressive carbonation' has been recognised more recently, in which the integrity of the cementing matrix of concrete can be seriously impaired. Thaulow & Jakobsen[25] termed this 'bicarbonation' and described a distinctive petrographic texture in which the matrix is replaced by coarse 'popcorn' crystals of calcite and silica gel (Figure 4). It would seem that low quality, porous concrete is more vulnerable to this bicarbonation and Thaulow et al[26] have concluded that the *"process occurs when the availability of calcium hydroxide in the concrete is too low to maintain the pore liquid at pH sufficiently high to have high concentration of carbonate ions compared to bicarbonate ions".* An aggravating factor was thought to be *"the presence of competing demands for calcium hydroxide either from pozzolanic activity, leaching, or sulfate attack".* French & Crammond[27] have reported a similar process, termed by them 'sub-aqueous carbonation', affecting UK concretes buried in wet ground rich in Ca, HCO₃ and SO₃, in which the formation of coarse carbonate and thaumasite (see later) occurred together.

These new findings indicate that carbonation is not always a benign form of alteration to concrete material and petrographers must learn to recognise and interpret the characteristic texture of bicarbonation. The most effective prevention will be recognition of the threat in advance, so that concrete can be designed to be suitably impermeable and, if necessary, surfaces can be protected by impervious or sacrificial tanking.

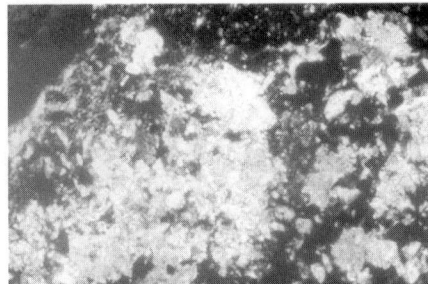 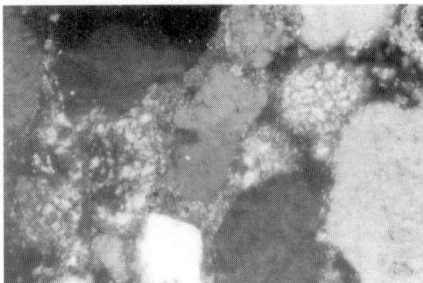

Figure 4 - Coarse calcite texture indicative of sub-aqueous carbonation

Figure 5 - Mortar matrix altered to ettringite and thaumasite

Delayed Ettringite Formation

Comparatively recently, it has been discovered that the normal cement hydration reactions can be modified when the concrete is exposed to early prolonged high temperature and subsequent wet conditions. In these circumstances, which can prevail for example with the manufacture of precast concrete units, the normal formation of calcium sulfo-aluminate hydrate (or ettringite) is initially suppressed and damage can later occur to the hardened concrete by the seemingly expansive process of 'delayed ettringite formation'[28,29]. Steam cured concrete railway sleepers and other precast units were found to be especially at risk, but in situ concrete can also be affected[30].

Ettringite is a normal product of the hydration of Portland cement, but is usually sub-microscopic in its primary occurrence. Coarser ettringite can occur in concrete as the result of leaching and re-crystallisation (commonly infilling or lining voids and cracks), or conventional sulfate attack on the calcium aluminates, when the distinctive formations of acicular crystallites are visible by optical thin-section microscopy[5]. Delayed ettringite formation (DEF) also leads to microscopically visible formations of ettringite and typically these include characteristic ettringite rims around the periphery of aggregate particles. However, diagnosis is not always straightforward; it has been shown that ettringite rims are not invariably present[31] and, moreover, secondary infilling by ettringite of existing peripheral cracks around aggregate particles can be misinterpreted as evidence of DEF.

Hobbs[32,33] has claimed that cracking caused by DEF has "in many instances" been wrongly attributed to ASR. It is certainly undeniable that similar crack patterns can appear on the concrete construction, because both DEF and ASR involve the cumulative and non-uniform effect of many internal centres of expansive reaction. It is also the case that evidence of DEF and ASR may co-exist in the same concrete, so that judgement must be exercised regarding the primary causation of damage. However, such misdiagnosis is most likely to have arisen from inadequate investigation, particularly when assumptions were made from macroscopic crack patterns and geographical location, without carrying out competent petrography. Hobbs[34] gave an example of a cement-rich thick-section concrete wing-wall in an area of the UK that had experienced ASR in many structures: petrography identified both DEF and ASR, but the intensity of the latter was considered insufficient to explain the considerable cracking.

Widespread misdiagnosis of DEF as ASR is unlikely, where petrography was carried out and its interpretation undertaken by experienced specialists. However, prior to the first published recognition of DEF, there would be an expectation of some misdiagnosis, especially as the occasional coexistence of ASR gel and ettringite had long been recognised as a feature, possibly associated with water migration and the local concentration of leached alkalis and sulfates. Sims[1] has reported an example of an investigation in the 1980s in which ASR was correctly discounted as the previously alleged cause of cracking of a large concrete anchor block, but instead the cracking was explained as being initiated by thermal contraction and plastic settlement, developed by materials movements. Later review of the petrography had suggested that DEF might have actually been the cause of the cracking.

Thaumasite Form of Sulfate Attack

Erlin & Stark[35] first reported on an unusual form of sulfate attack in concrete in which the product was the complex thaumasite (calcium carbo-sulfo-silicate hydrate) rather than the more familiar ettringite[36]. Petrographers would thereafter occasionally report on the occurrence of thaumasite in concrete. Research found that the thaumasite form of sulfate attack (TSA), as it became known, could be induced quite easily in the laboratory, but reported cases[37] were often regarded as exceptional. In the UK, the discovery of buried concrete columns that were seriously degraded by TSA on some motorway bridges raised alarm and an urgent government-sponsored investigation was undertaken under the chairmanship of Clark[6,7].

Clark's committee reported that TSA was a rare occurrence that required all of four 'primary risk factors' to be present in addition to the cement in the concrete: a source of sulfates (usually from adjacent ground), wet conditions (especially mobile water), a source of carbonate (usually in the aggregate) and low temperatures (<15°C). Agencies around the UK responsible for major concrete highway structures, mainly bridges, were then invited to review their structures against these risk factors and to investigate further a sample of those assessed as being at some risk of TSA. This exercise seems to have led to a perhaps surprisingly limited number of investigations, beyond those already included within the programme overseen by Clark, which were principally from the same region. In the author's experience, even those investigations that were carried out were decidedly limited in scope.

The selection of structures for investigation was based on the assumption that the presence of all four primary risk factors was a pre-requirement. However, investigation by STATS of other structures, including some from locations outside the UK, has indicated that TSA can occur in the absence of some of the recognised risk factors. Several cases have involved concrete or mortar in which the aggregate contains no carbonate constituent[38]. One of these examples was from a structure on the coast of the Mediterranean, where low temperatures are uncommon, and, moreover, the samples comprised mortar from the high-rise superstructure and not from the buried foundations. Jointing mortar that was being investigated for premature deterioration, initially thought possibly to be caused by salt weathering, was found by petrography and micro-analysis to be variably altered to patches of ettringite and thaumasite (Figure 5), but never mixtures of the two.

Owing to the UK process of selection against primary risk factors that might not always apply, there must be a concern that affected structures remain undiscovered and thus not investigated. Routine condition surveys of concrete structures, whilst sometimes including petrography, rarely include sampling and examination of buried concrete, so that future failures might remain undetected.

Petrographical examination, with allied analyses, is the main diagnostic procedure for recognising TSA[3] and advanced cases are readily identifiable. However, the prognostic interpretation of less advanced examples can be problematical. As well as TSA, Clark[6] describes 'thaumasite formation' (TF) to cover cases in which the reaction product is only found infilling pre-existing voids and cracks but not causing deterioration. In practice, when assessing samples from structures that are being routinely investigated and have so far given no cause for concern, it can be difficult to distinguish between TF and the incipient stages of TSA. The identification of thaumasite in concrete, even in small and apparently innocuous quantities, will naturally give rise to some alarm, but it is still important for the investigator not to misinterpret the early signs of a mechanism that could potentially proceed to cause serious degradation. Sims & Huntley have suggested some guidance[38].

CONCLUSIONS

- Concrete can be an exceptionally durable material and previous experience can be harnessed to prevent degradation in service.

- Dependable diagnosis is achieved by continuity of investigation, from desk study, through site inspection to the interpretation of laboratory findings. Selection of sample locations is a critical factor.

- Petrography is indispensable in the investigation of concrete deterioration.

- Established and avoidable threats to concrete durability include weathering and freeze-thaw action, atmospheric carbonation and alkali-reactivity. Partial carbonation is an indication of reduced durability that is frequently overlooked. Progress is now being made with international measures for assessing and avoiding the threat from various forms of alkali-reactivity.

- Relatively newly recognised threats to concrete durability include aggressive carbonation, delayed ettringite formation (DEF) and the thaumasite form of sulfate attack (TSA). Some examples of DEF may have previously been misdiagnosed. UK guidance on TSA may have underestimated the frequency of occurrence and some examples do not confirm the presence of all risk factors.

ACKNOWLEDGMENTS

The author is grateful for the understanding of his family during weekends taken in preparing this paper and for the expert assistance of Miss Siân Kitchen in editing the document. He also wishes to thank Mr Niels Thaulow for supplying information and Dr Bill French for his advice and for permission to reproduce the photomicrograph in Figure 4.

REFERENCES

1   SIMS, I, Phantom, opportunistic, historical and real AAR - Getting diagnosis right. In: Shayan, A (Ed), Alkali-aggregate reaction in concrete, Proceedings of the 10th International Conference, Melbourne, Australia, 1996, 175-82.

2   PALMER, D (Chairman), The diagnosis of alkali-silica reaction - Report of a Working Party. British Cement Association, 45.042, Slough, UK, 1992, 44pp.

3   SIMS, I, HARTSHORN, S A, Recognising thaumasite. Concrete Engineering International, 1998, 2, (8), 44-48.

4   RAMACHANDRAN, V S, BEAUDOIN, J J, Handbook of analytical techniques in concrete science and technology. William Andrew Publishing/Noyes Publications, Norwich/Park Ridge, USA, 2001, 964pp.

5   ST JOHN, D A, POOLE, A B, SIMS, I, Concrete petrography - a handbook of investigative techniques. Arnold (Hodder), London, UK, 1998, 474pp.

6   CLARK, L A (Chairman), The thaumasite form of sulfate attack: Risks, diagnosis, remedial works and guidance on new construction - Report of the Thaumasite Expert Group. Department of the Environment, Transport & the Regions, London, UK, 1999, 180pp.

7   CLARKE, L A (Chairman), Thaumasite Expert Group one-year review. Department of the Environment, Transport & the Regions, London, UK, 2000, 22pp (http://www.construction.detr.gov.uk/thaumasite/oneyear/index.htm).

8   HOBBS, D W, MARSH, B K, MATTHEWS, J D, Minimum requirements for concrete to resist freeze-thaw attack. In: Hobbs, D W (Ed), Minimum requirements for durable concrete - Carbonation- and chloride-induced corrosion, freeze-thaw attack and chemical attack, British Cement Association, 45.043, Crowthorne, UK, 1998, 91-129.

9   SANDBERG, A, COLLIS, L, Toil and trouble on concrete bubbles. Consulting Engineer, 1982, November, 32-5.

10 ASTM, Standard test method for microscopical determination of parameters of the air-void system in hardened concrete. American Society for Testing and Materials, C457, Philadelphia, USA, 1998, 14pp.

11 SIMS, I, The assessment of concrete for carbonation. Concrete, 1994, 28, (6), 33-8.

12 ROBERTS, M H, Carbonation of concrete made with dense natural aggregates. Building Research Establishment, IP 6/81, Watford, UK, 1981, 4pp.

13 PARROTT, L J, Some effects of cement and curing upon carbonation and reinforcement corrosion in concrete. Materials & Structures, 1996, 29, (187), 164-73.

14 SWAMY, R N (Ed), The alkali-silica reaction in concrete. Blackie, Glasgow, UK, 336pp.

15 BRE, Alkali-silica reaction in concrete, Parts 1 to 4. Building Research Establishment (CRC Ltd), Digest 330, Watford, UK, 1999, 8+8+8+4pp.

16 HAWKINS, M R (Chairman), Alkali-silica reaction: minimising the risk of damage to concrete - Guidance notes and model clauses for specifications - Report of a Concrete Society Working Party. The Concrete Society, TR30, 3rd Ed, Slough (Crowthorne), UK, 1999, 72pp.

17 SIMS, I, Alkali-silica reaction - UK experience. In: Swamy, R N (Ed), The alkali-silica reaction in concrete, Blackie, Glasgow, UK, 1992, 122-87.

18 SIMS, I, Alkali-reactivity - solving the problem worldwide. Concrete, 2000, 34, (10), 64-6.

19 BÉRUBÉ, M-A, FOURNIER, B, DURAND, B (Eds), Alkali-aggregate reaction in concrete - Proceedings, 11th International Conference, Québec City, Canada. CRIB, Laval & Sherbrooke Universities, Sainte-Foy, Québec, Canada, 2000, 1406pp.

20 NIXON, P J, SIMS, I, Universally accepted testing procedures for AAR - the progress of RILEM Technical Committee 106. In: Bérubé, M-A, Fournier, B, Durand, B (Eds), Alkali-aggregate reaction in concrete, Proceedings, 11th International Conference, Québec City, Canada. CRIB, Laval & Sherbrooke Universities, Sainte-Foy, Québec, Canada, 2000, 435-44.

21 SIMS, I, NIXON, P J, Alkali-reactivity - a new international scheme for assessing aggregates. Concrete, 2001, 35, (1), 36-9.

22 TANG, M-S, DENG, M, XU, Z, LAN, X, HAN, S, Alkali-aggregate reactions in China. In: Shayan, A (Ed), Alkali-aggregate reaction in concrete, Proceedings of the 10th International Conference, Melbourne, Australia, 1996, 195-201.

23 COWIE, J, GLASSER, F P, The reaction between cement and natural waters containing dissolved carbon dioxide. Advances in Cement Research, 1991/1992, 4, (15), 119-34.

24 TERZAGHI, R D, Concrete deterioration due to carbonic acid. Journal of the Boston Society of Civil Engineers, 1949, 36, 136-60.

25 THAULOW, N, JAKOBSEN, U H, Deterioration of concrete diagnosed by optical microscopy. In: Proceedings of the 6th Euroseminar on Microscopy Applied to Building Materials, Reykjavik, Iceland, 1997, 282-296.

26 THAULOW, N, LEE, R J, WAGNER, K, Effect of calcium hydroxide on the form, extent, and significance of carbonation. In: Skalny, J, Gebauer, J, Odler, I (Eds), Materials Science of Concrete: Calcium Hydroxide in Concrete. The American Ceramic Society, Special Volume, Westerville, USA, 2001, 191-202.

27 FRENCH, W J, CRAMMOND, N J, Sub-aqueous carbonation and the formation of thaumasite in concrete. 20th Cement and Concrete Science Conference, The Institute of Materials, London, UK, 2000.

28 HEINZ, D, LUDWIG, U, Mechanisms of secondary ettringite formation in mortars and concretes subjected to heat treatment. In: Scanlon, J M (Ed), Concrete durability: Katharine and Bryant Mather International Conference. American Concrete Institute, SP-100, Detroit, USA, 1987, 2059-71 (Paper SP100-105).

29 LAWRENCE, C D, DALZIEL, J A, HOBBS, D W, Sulphate attack arising from delayed ettringite formation. British Cement Association, ITN 12, Slough, UK, 1990, 43pp.

30 QUILLIN, K, Delayed ettringite formation: in-situ concrete. Building Research Establishment (CRC Ltd), Information Paper 11/01, Watford, UK, 2001, 8pp.

31 POOLE, A B, PATEL, H H, SHIEKH, V, Alkali silica and ettringite expansions in 'steam cured' concretes. In: Shayan, A (Ed), Alkali-aggregate reaction in concrete, Proceedings of the 10th International Conference, Melbourne, Australia, 1996, 943-8.

32 HOBBS, D W, World wide durability problems with concrete and trends in prevention. In: Proceedings of Concrete Meets the Challenge, Sun City, South Africa. Concrete Society of South Africa, 1994 (ISBN 0-9583831-3-8).

33 HOBBS, D W, Concrete deterioration: causes, diagnosis, and minimising risk. International Materials Reviews, 2001, 46, (3), 117-144.

34 HOBBS, D W, Diagnosis of the cause of cracking in four structures in which ASR is occurring. In: Shayan, A (Ed), Alkali-aggregate reaction in concrete, Proceedings of the 10th International Conference, Melbourne, Australia, 1996, 209-218.

35 ERLIN, B, STARK, D, Identification and occurrence of thaumasite in concrete. Highway Research Record, 1966, (113), 108-13.

36 HARTSHORN, S A, SIMS, I, Thaumasite - a brief guide for engineers. Concrete, 1998, 32, (8), 24-7.

37 CRAMMOND, N J, HALLIWELL, M A, The thaumasite form of sulfate attack in concretes containing a source of carbonate ions - a microstructural overview. In: Malhotra, V M (Ed), Advances in Concrete Technology, 2nd Symposium. American Concrete Institute, SP 154, Detroit, USA, 1995, 357-80 (SP154-19).

38 SIMS, I, HUNTLEY, S A, The thaumasite form of sulfate attack - breaking the rules. (Submitted for 1st International Conference on Thaumasite in Cementitious Materials, Building Research Establishment, Watford, UK, June 2002).

EXPERIMENTAL STUDY OF THE WEAR OF A CONCRETE SURFACE UNDER ICE FRICTION

B Fiorio

Universite de Cergy Pontoise

France

ABSTRACT. Marine structures constructed in polar and sub-polar regions are exposed to ice interactions controlled by the drift of ice floes. In the case of concrete structures, friction at the crushed ice-structure contact contributes to the wear of concrete. To study the small scale effects of the friction-induced wear of concrete, we have performed cyclic friction tests between concrete plates and ice. The contact conditions have been typical of what can be expected in ice-structure interaction conditions. Optical observations of ice and concrete surfaces and concrete surface topographic measurements were also performed.

Our experimental results underline the wear process: cement paste particles are pulled out of the concrete surface, which leads to the decrease of the embedding of small aggregates in cement and finally to there ejection out of the concrete surface. Calculation of the wear rate has been made for the different concrete plates. The evolution of the wear rate with time is made of two stages. A large wear rate is observed for the initial stage. It corresponds to the wear of a superficial layer of cement paste. In the second stage, the wear rate reduces to a lowest level where it is independent of the average roughness of the plate.

Keywords: Concrete, Ice, Structure, Interaction, Contact, Friction, Wear, Roughness.

Dr B Fiorio, was working at the Laboratoire de Glaciologie et Géophysique de l'Environnement in Grenoble, France, when he began working on ice-concrete friction and wear. He now teaches civil engineering at the University Institute of Civil Engineering Science of the University of Cergy-Pontoise, France. He carries on his research work in the Civil Engineering Laboratory of the same university. His main subjects of interest are lightweight concretes and new structural fibres reinforced concretes.

INTRODUCTION

Offshore structures constructed in Polar Regions are exposed to the drift of ice floes. This generates dynamic loads on structures and wear of concrete. The design of offshore structures must obviously take into account the large forces generated at the structure-ice contact. It must also take into account the effect of the wear of the concrete to guarantee acceptable mechanical strength to the structure during its lifetime.

These phenomena depend among other things on the flow of the crushed-ice layer that is usually observed at the structure-ice interface [1], [2]. Owing to the pressure exerted by ice on the structure, this crushed ice is extruded out of the contact zone. Extrusion tests [3] and numerical simulations [4], [5] have shown that this flow is mainly controlled by the structure-ice friction conditions observed at its limits. In the case of concrete structure, the friction conditions are the concrete on ice friction itself and the concrete wear, which supply the interface with concrete wear particles.

To characterise more accurately concrete wear at small scale (i.e. the centimetre scale), friction tests were performed to study the influence of the essential parameters involved, and also to attain a better understanding of the underlying physical mechanisms active at this scale. As concrete-ice friction is a connected phenomenon, we have made a joint study of this last phenomenon and of concrete wear.

The present paper describes the experimental work done and the results obtained.

EXPERIMENTAL SET-UP

Concrete and Ice Samples

The tests were performed with a micro-concrete, thought to be representative of the composition of a structural concrete near its surface where large size aggregates are under-represented. The micro-concrete was made of a PORTLAND CPA 45 cement, a fine sand (0.2 to 0.6 mm granulometry) and a coarse sand (3 to 5 mm granulometry). The mixing was made with the following proportion: $W/C = 0.6$ and $S_1/C = S_2/C = 1.25$, where W, C, S_1 and S_2 are respectively the weight of water, cement, fine and coarse sand. This micro-concrete was used to cast 175 mm x 150 mm plates whose initial average roughness was controlled by moulding. Two types of plate have been casted: a rough plate (initial average roughness of 0.28 mm) and a smooth plate (initial average roughness of 0.11 mm). Figure 1 gives examples of the typical surface geometry of the two plates.

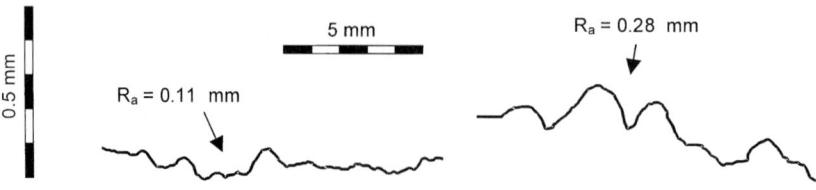

Figure 1   Typical geometry of the surface of the concrete plates

The ice was columnar freshwater ice prepared in the laboratory by growing deionized water from a germ of small-grained granular ice, in order to obtain columnar grains approximately 8 mm in diameter. Ice sample were cylindrical (60 mm diameter and 90 mm height).

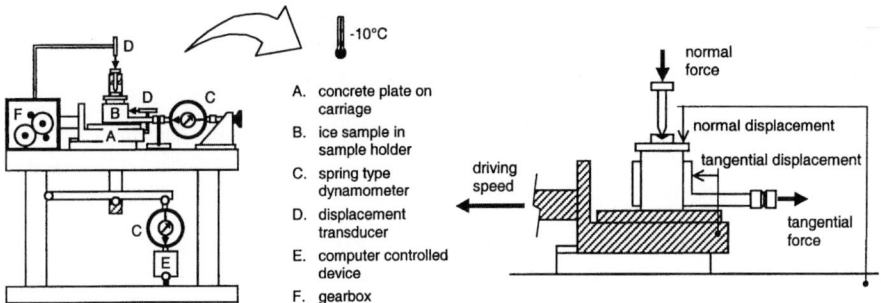

A. concrete plate on carriage
B. ice sample in sample holder
C. spring type dynamometer
D. displacement transducer
E. computer controlled device
F. gearbox

Figure 2   Friction apparatus

The micro-concrete and the ice used were rather simple materials compared to the wide complexity of the concrete-ice contact conditions observed in offshore conditions [6] [7], but this simplification was needed to perform friction and wear tests in well defined conditions, and to bring out the main physical mechanisms involved.

Friction Apparatus

The friction tests were performed with a direct-shear-box machine, designed for soil mechanics studies and modified to perform interfacial shear. The apparatus (see Figure 2) was placed in a cold room at -10°C ±0.5°C. The concrete plate was fixed on a carriage which moved horizontally by alternate translation along a 30 mm stroke. The carriage driving speed was constant for each test (0.1, 0.5, 1, 5 or 10 mm/min, that is, respectively, 1.67×10^{-6}, 8.35×10^{-6}, 1.67×10^{-5}, 8.35×10^{-5} or 1.67×10^{-4} m.s^{-1}). The cylindrical ice sample was placed over the concrete plate and held in place with a cylindrical sample-holder maintained at 1mm from the concrete surface. The normal stress was imposed on the ice-concrete interface through a lever, which was operated by a computer-controlled electromechanical device. Normal stress remained constant during the test and was chosen in the range 25-800 kPa. Two spring-type dynamometers were used to record the interfacial normal and tangential forces, denoted by F_n and F_t, respectively. The corresponding normal and tangential stresses were calculated respectively as F_n/A and F_t/A, where A is the nominal area of contact ($A = 28.3$ cm^2), and were obtained within ± 0.16 kPa and ± 0.49 kPa, respectively.   Two LVDT's displacement transducers were used to record the normal and tangential displacements of ice over the concrete plate (with an accuracy of ± 0.01 mm and ± 0.1 mm, respectively).

Wear Measurements

To study the wear of the concrete under ice friction, a series of friction tests were performed on the same concrete plate and recorded test after test the evolution of the concrete surface geometry. To this aim, the recording of the topography of six reference profiles on the top of the concrete surface, each 26 mm long, was performed before and after each friction test (see

Figure 3 the position of the six profiles in the friction track). The maximum height h_M and the mean height h_m of the profile were calculated for each profile as defined in Figure 4. Then the maximum abrasion amount w_M and the mean abrasion amount w_m were calculated as:

$$w_M = \frac{1}{6}\sum_{i=1}^{6}\left(h_{Mf,i} - h_{M0,i}\right) \quad ; \quad w_m = \frac{1}{6}\sum_{i=1}^{6}\left(h_{mf,i} - h_{m0,i}\right) \tag{1}$$

where subscript 0 refers to the initial state of the concrete plate (before the first friction test) and subscript f refers to the state of the concrete plate after a given test. Subscript i refers to one of the six profiles ($i = 1 ; 2 ; 3 ; 4 ; 5 ; 6$).

The average roughness (CLA) of the concrete plate was obtained as the mean value of the average roughness $R_{a,i}$ of the six reference profiles. This last value is the arithmetic average of the mean deviation of each profile topography:

$$R_{a,i} = \frac{1}{\lambda}\int_{0}^{\lambda}|y(x)|dx \tag{2}$$

where the mean deviation $y(x)$ described the profile (see Figure 4) and λ is the length of the profile.

After each test, a series of observations of the concrete and ice contact surfaces was performed, by the use of optical microscopy techniques. Theses observations gave information on the physical mechanisms involved in friction and wear.

The whole experimental set-up is described in more details by [8] and [9].

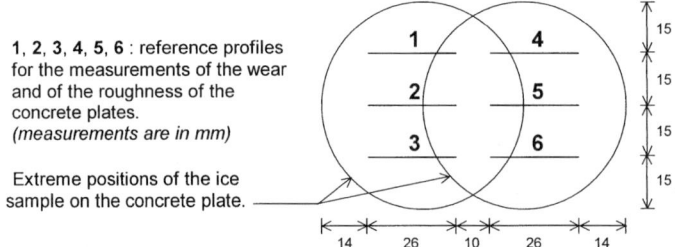

Figure 3   Position of the six reference profiles on the concrete plates for wear and roughness measurements

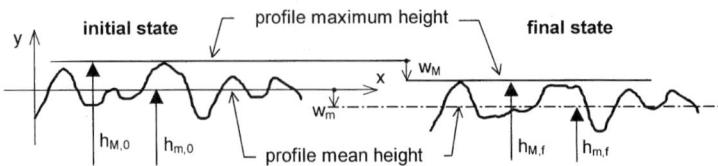

Figure 4   Maximum (h_M) and mean (h_m) height of a profile and calculation of the maximum (w_M) and mean (w_m) abrasion amount

RESULTS

Wear Measurements of the Concrete Plates

Table 1 gives the results of the wear measurements performed on the smooth concrete plate. σ_n, S and R_a define the friction condition at the concrete-ice interface (respectively, the normal nominal stress on the interface, the sliding speed and the average roughness measured after completion of the test). d_s is the cumulative sliding distance (distance covered by the concrete on ice from test #1 to the considered test). w_m and w_M are the mean and maximum abrasion amount described in the previous paragraph.

Table 2 gives the results of the wear measurements performed on the rough concrete plate (same presentation as Table 1).

Observations of the Frictional Face of Concrete and Ice Samples

Optical observations performed after each friction test on ice and concrete samples have shown the formation of a layer of finely crushed ice and concrete wear particles at the interface (Figure 5). It was composed of cement and fine sand particles that were circulating inside the interface. A part of them was ejected out of the contact zone during the friction test.

Table 1   Evolution of the maximum (w_M) and mean (w_m) abrasion amount of the smooth plate recorded during the series of friction tests; σ_n, S and R_a define the friction conditions at the concrete-ice interface

FRICTION TEST	σ_n (kPa)	S (mm/min)	R_a (mm)	d_s (m)	w_m (mm)	w_M (mm)
			0.113	0.00	0.000	0.000
SP # 1	200	0.1	0.114	0.42	0.047	0.048
SP # 2	200	1	0.102	1.49	0.092	0.153
SP # 3	200	10	0.098	2.09	0.129	0.174
SP # 4	400	0.1	0.098	2.68	0.180	0.260
SP # 5	800	0.5	0.084	3.28	0.190	0.253
SP # 6	800	1	0.072	4.17	0.278	0.361
SP # 7	50	10	0.073	4.77	0.289	0.373
SP # 8	800	10	0.069	5.36	0.284	0.364
SP # 9	200	0.5	0.069	5.96	0.308	0.383
SP # 10	100	0.1	0.079	6.47	0.346	0.448
SP # 11	200	0.5	0.073	6.50	0.331	0.404
SP # 12[†]	332	10	0.074	7.51	0.308	0.402
SP # 13	200	0.1	0.063	8.11	0.341	0.432
SP # 14[†]	332	1	0.067	9.12	0.355	0.446
SP # 15	100	0.1	0.076	9.72	0.393	0.494
SP # 16[†]	332	5	0.083	10.73	0.390	0.448
SP # 17	800	10	0.061	10.91	0.434	0.475
SP # 18	200	5	0.079	11.50	0.411	0.514
SP # 19[†]	332	0.5	0.066	13.11	0.416	0.485
SP # 20	50	1	0.083	14.07	0.441	0.545
SP # 21	800	0.1	0.069	14.75	0.452	0.520
SP # 22	800	5	0.070	15.35	0.445	0.575
SP # 23	800	0.1	0.069	15.94	0.442	0.542

[†] Friction test performed under variable normal stress. The given value of σ_n is the average value measured for the friction test.

Table 2   Evolution of the maximum (w_M) and mean (w_m) abrasion amount of the rough plate recorded during the series of friction tests; σ_n, S and R_a define the friction conditions at the concrete-ice interface

FRICTION TEST	σ_n (kPa)	S (mm/min)	R_a (mm)	d_s (m)	w_m (mm)	w_M (mm)
			0.281	0.00	0.000	0.000
RP # 1	200	10	0.243	0.60	0.167	0.215
RP # 2	200	0.5	0.272	1.19	0.153	0.207
RP # 3[†]	224	10	0.119	2.74	0.535	0.659
RP # 4	200	10	0.132	3.34	0.684	0.766
RP # 5[†]	224	5	0.132	4.89	1.000	1.109
RP # 6	200	1	0.141	5.96	1.079	1.231
RP # 7[†]	224	1	0.171	7.51	1.134	1.216
RP # 8	200	0.1	0.168	8.11	1.120	1.179
RP # 9	200	5	0.149	8.70	1.165	1.239
RP # 10	50	1	0.175	9.77	1.149	1.210
RP # 11	50	0.1	0.173	10.10	1.133	1.182
RP # 12[†]	260	0.5	0.192	10.94	1.145	1.181
RP # 13	400	10	0.192	10.98	1.189	1.210
RP # 14	400	0.1	0.192	11.03	1.266	1.280
RP # 15	400	1	0.185	11.07	1.344	1.350
RP # 16	50	10	0.177	11.37	1.197	1.211

[†] Friction test performed under variable normal stress. The given value of σ_n is the average value measured for the friction test.

ice surface after friction

wear particles : ice, cement
and small aggregates

wear particles after melting and refreezing

agglomerates of cement particles
(orientation is the result of refreezing)

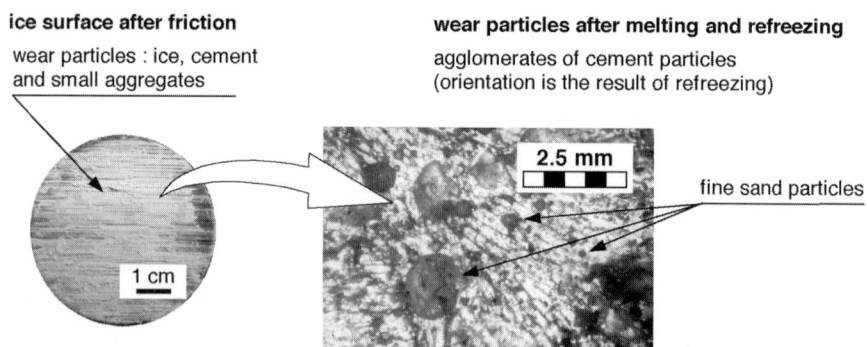

2.5 mm

fine sand particles

1 cm

Figure 5   Wear particles at the concrete-ice interface

qualitative observations of the appearance of the wear of the concrete plates surface have shown the abrasion of the cement paste and of the fine sand particles. This phenomenon was resulting in a few friction tests to the uncoating of the coarse sand particles, which then became more and more prominent on top on the concrete plates (see Figure 6). At the end of the series of friction tests, we have observed the formation of grooves in the sliding direction.

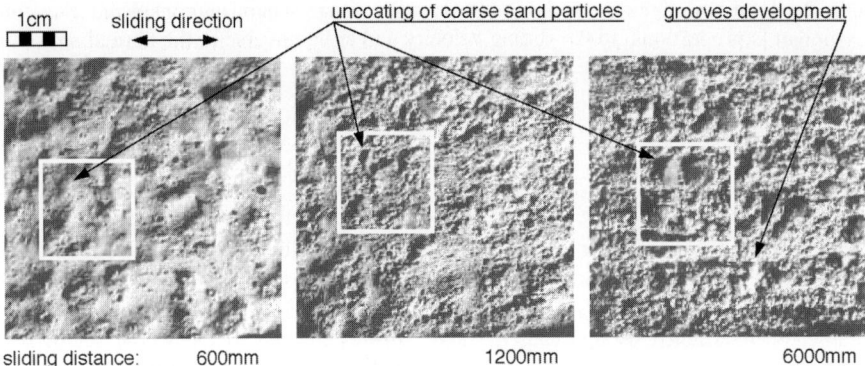

Figure 6   Progressive uncoating of the coarse sand particles

Concrete on Ice Friction

The level of friction during the friction tests was characterised by the value of the friction coefficient f calculated as the ratio $f = <\tau> / <\sigma_n>$, where $<\tau>$ and $<\sigma_n>$ are, respectively, the average values of the tangential and normal stresses recorded over the stabilized regime of friction. Transitory regime of friction due to the change of direction of the ice sample on the concrete plate was not taken into account for the calculation of the friction coefficient. The values of the friction coefficient are given in Table 3. Table 3 also gives the average value of the nominal tangential stress $<\tau>$.

Table 3   Friction coefficient f measured during the friction tests on the smooth concrete plate (SP tests) and on the rough concrete plate (RP tests)

FRICTION TEST	f	$<\tau>$MPa	FRICTION TEST	f	$<\tau>$MPa	FRICTION TEST	f	$<\tau>$MPa
SP # 1	0.52	0.10	SP # 14	0.61	0.20	RP # 4	0.67	0.13
SP # 2	0.63	0.13	SP # 15	0.47	0.05	RP # 5	0.71	0.16
SP # 3	0.65	0.20	SP # 16	0.69	0.23	RP # 6	0.57	0.11
SP # 4	0.48	0.19	SP # 17	0.66	0.26	RP # 7	0.64	0.14
SP # 5	0.52	0.42	SP # 18	0.56	0.11	RP # 8	0.48	0.10
SP # 6	0.55	0.44	SP # 19	0.61	0.20	RP # 9	0.56	0.11
SP # 7	0.58	0.03	SP # 20	0.58	0.03	RP # 10	0.58	0.03
SP # 8	0.66	0.53	SP # 21	0.58	0.46	RP # 11	0.60	0.03
SP # 9	0.57	0.11	SP # 22	0.65	0.52	RP # 12	0.64	0.17
SP # 10	0.46	0.05	SP # 23	0.58	0.46	RP # 13	0.72	0.29
SP # 11	0.58	0.12	RP # 1	0.70	0.14	RP # 14	0.52	0.21
SP # 12	0.73	0.24	RP # 2	0.61	0.12	RP # 15	0.61	0.24
SP # 13	0.53	0.11	RP # 3	0.74	0.17	RP # 16	0.63	0.03

In the range of normal stress, sliding velocity and average roughness considered, the friction coefficient is proportional to the sliding velocity and to the inverse of the normal stress. It is an increasing function of the average roughness and can be described by the following empirical law:

$$f = \left(1 + \frac{S}{s_1} + \frac{\sigma}{\sigma_n}\left(1 + \frac{S}{s_2}\right)\right)\left(\frac{R_a}{r}\right)^\alpha$$

[3]

where σ_n, S and R_a denote the normal stress, the sliding velocity and the average roughness, respectively, s_1 and s_2 are two constants with the dimension of a velocity, σ is a constant stress, r is a constant length, and α is a dimensionless constant. The numerical values of the parameters determined by applying the least squares method to the whole set of results (Tables 1 and 2) are : $s_1 = 7.62 \times 10^{-4}$ m.s^{-1}; $s_2 = 4.45 \times 10^{-4}$ m.s^{-1}; $\sigma = 15.06$ kPa ; $r = 236.78$ mm ; $\alpha = 0.0784$.

As an example, Figure 7 shows the evolution of the friction coefficient, calculated with relations [4], in the range of normal stress and sliding speed of our tests, drawn for an average roughness of 0.1 mm (corresponding to the smooth concrete plate). For other values of the average roughness, the general shape of the friction curves is preserved. For a detailed study of concrete on ice friction, one can refers to [8] and [9].

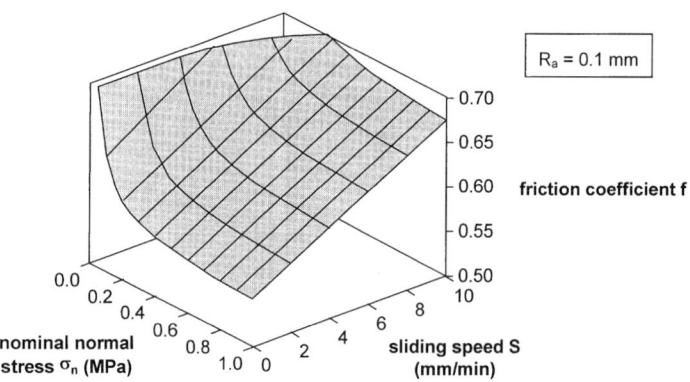

Figure 7   Calculated friction coefficient for concrete-ice friction on the smooth plate

ANALYSIS

Concrete Plate Degradation Mechanisms

To characterize the wear of the concrete surface, we have studied the evolution of the topography of the six reference profiles that were spaced out over the frictional area of the concrete plates. Figure 8 gives the example of the topographic evolution observed for profile 5 on the smooth plate. Wear appear as the result of two phenomena: a general wear, gradual and uniformly distributed over the frictional area, and a catastrophic wear, much faster but space and time localised.

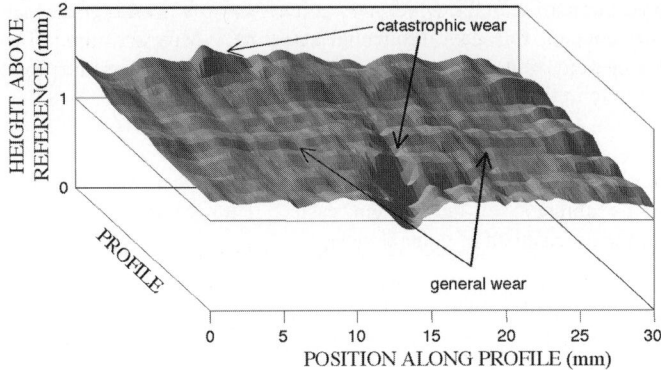

Figure 8   Smooth plate profile 5 topographic evolution due to ice friction
Profile age is proportional to the number of friction tests performed on the concrete plate

General wear can be characterized by the relation between the maximum and the mean abrasion amount and the sliding distance of ice on concrete (Figure 9). It occurs in two stages. The initial stage, where the superficial layer of cement paste is abraded, in particular at the top of concrete asperities (maximum abrasion amount can be twice the mean abrasion amount) is characterized by high and roughness dependent maximum and mean abrasion rate. The permanent stage is characterized by lower maximum and mean abrasion rate which do not depend on the roughness of the concrete plate. The lower abrasion rates obtain in this case are the consequence of the presence of protruding coarse sand particles above the cement paste level. General wear corresponds to the mechanical wear of the cement paste. Results regarding friction coefficient show indeed that nominal tangential stress was of the order of a few hundreds of kPa.

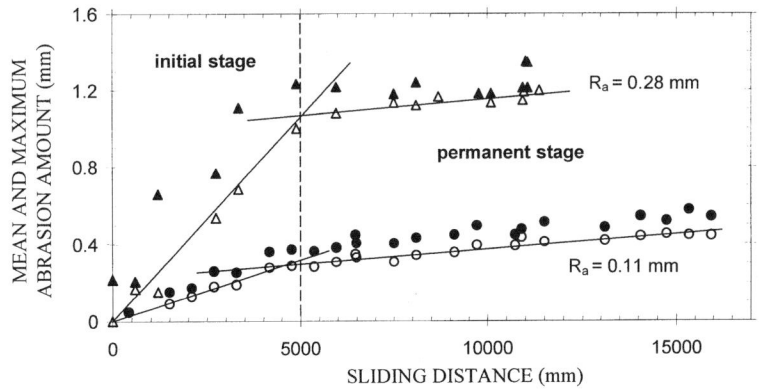

Figure 9   Evolution of the mean (white symbols) and maximum (black symbols)
abrasion amount of concrete during the wear test

Reduce to the real area of contact, which was at least ten to a hundred times smaller than the nominal area of contact, this gave tangential stress of 1 Mpa or more. Tensile stress in concrete was therefore able to reach the tensile strength of this material, explaining the presence of concrete in the wear particles.

Catastrophic wear corresponds to the pulling out of coarse sand particles from the surface layer of the plate. This happens when the particle-cement paste bond has been sufficiently embrittled by the abrasion of the cement paste. Thus catastrophic wear appears as a consequence of the mechanism of general wear.

With the exception of catastrophic wear they do not mention, we find here the same results as [10] with values of the mean abrasion rates 150 times higher (20 mm/km instead of 0.15 mm/km). As the results of [10] were obtained with concrete (maximum size of aggregates of 25 mm) instead of micro-concrete, this suggests an important effect of the high sized aggregates. These last could play an important part in concrete abrasion as a buffer layer between ice and cement paste. By limiting ice stress on the cement paste, this buffer layer slows down the general abrasion and reduces the mean abrasion rate. Differences in the mean abrasion rates can be attributed to differences in the protection of the cement paste provided by the aggregates. This last is all the more important than the aggregate size is high.

CONCLUSION

We have shown that the wear of concrete, in the case of friction on ice, was the combination of two different mechanisms: general wear and catastrophic wear. This work also shows how wear is related to the composition of the concrete, in particular to the size of the aggregates. More should now be done to make clear the role of the contact conditions.

REFERENCES

1.   JORDAAN, I. J., TIMCO, G. W., Dynamics of the ice-crushing process, Journal of Glaciology, Vol 34, No118, 1988, p 318-326.

2.   TUHKURI, J., Experimental observations of the brittle failure process of ice and ice-structure contact, Cold Regions Science and Technology, Vol 23, 1995, p 265-278.

3.   SAYED, M., FREDERKING, R. M. W., Two-dimensional extrusion of crushed ice, Part 1, Experimental, Cold Regions Science and Technology, Vol 21, 1992, p 37-47.

4.   SAVAGE, S. B., SAYED, M., FREDERKING, R. M. W., Two-dimensional extrusion of crushed ice, Part 2, Analysis, Cold Regions Science and Technology, Vol 21, 1992, p 37-47.

5.   SINGH, S. K., JORDAAN, I. J., XIAO, J., SPENCER, P. A., The flow properties of crushed ice, Journal of Offshore Mechanics and Arctic Engineering, Vol 117, 1995, p 276-282.

6. CAMMAERT, A. B., MUGGERIDGE, D. B., Ice interaction with offshore structures, Van Norstrand Reinhold, New York, 1988, p 432.

7. FORLAND, K. A., TATINCLAUX, J. C. P., Kinetic friction coefficient of ice, CRREL Report 85-6, USACE, 1985, p 40.

8. FIORIO, B., MEYSSONNIER, J., BOULON, M., Experimental study of the friction of ice over concrete at the centimetre scale, Proceedings of the 7th International Offshore and Polar Engineering Conference, Honolulu, USA, 1997, Vol 2, p 466-472.

9. FIORIO, B., Etude expérimentale du frottement glace-structure à l'échelle centimétrique, Thèse de troisième cycle de l'université Joseph Fourier - Grenoble 1, Grenoble, France, 2000, p 190.

10. ITO, Y., HARA, F., SAEKI, H., TACHIBANA, H., The mechanism of the abrasion of concrete structures due to the movement of ice sheets, Proceedings of the International Conference on Concrete under Severe Conditions, Sapporo, Japan, 1995, p 465-474.

MAINTENANCE MANAGEMENT SYSTEM FOR EXISTING CONCRETE MARINE STRUCTURES

H Yokota

Port and Airport
Research Institute

T Tanabe

National Institute for Land and
Infrastructure Management

A Moriwake

Toa Corporation

Japan

ABSTRACT. Since concrete structures in marine areas are exposed to extremely severe environments for materials, it is important to establish a comprehensive maintenance management method for keeping marine structures in good condition during their design lives. As a part of the method, a decision-making support system for maintenance management of existing berthing structures is developed. This system enables to predict the process of future deterioration, to implement maintenance plans for minor repair, major repair, strengthening etc and to estimate rising costs due to maintenance work. This paper presents the outline of the system focusing on deterioration models to concrete and rebars and verification results of the system by applying to an existing open-piled marginal wharf.

Keywords: Coastal structures, Maintenance management system, Chloride content, Rebar corrosion, Diffusion coefficient, Degree of deterioration, Prediction, Inspection.

H Yokota is Head of Structural Mechanics Division in the Port and Airport Research Institute, Yokosuka, Japan. He specialises on structural behaviours and design of concrete structures. His recent research interests include the effect of deterioration of materials to overall performance of concrete structures.

T Tanabe is Head of Port Facilities Division in the National Institute for Land and Infrastructure Management, Yokosuka, Japan. He specialises on Design Standard for Port and Harbour Facilities in Japan. His recent research interests include the maintenance code for concrete structures.

A Moriwake is Chief of Materials Laboratory in the Technical Research Institute, Toa Corporation, Yokohama, Japan. He received his Dr Eng from Tokyo Institute of Technology in 1996. He has been closely associated with durability of concrete structures.

INTRODUCTION

Reinforced concrete structures in marine areas are exposed to extremely severe environments for materials. Consequently, some reinforced concrete marine and coastal structures suffer from heavy deterioration, which may result in lack of their structural capacities against requirements. The typical structural deterioration there is caused by chloride ions in seawater that induce corrosion of rebars embedded in concrete. Many studies have been conducted to date about the mechanism of chloride-induced deterioration and useful results are available to tackle the problems.

Periodic maintenance is the only measure to be taken in reinforced concrete marine structures after commencement in service for avoiding heavy deterioration and consequent loss of structural capacities. In addition, effective and rational maintenance can be achieved with understanding both the process and progress of deterioration and the relationship between materials deterioration and change in structural capacities. For this purpose, it is important to establish a comprehensive maintenance management system that provides engineers with quantitative information on deterioration as well as several alternatives of maintenance work.

A management system for the rational maintenance to existing berthing facilities has been developed in this study. The system enables to predict the process of future deterioration, to implement maintenance plans for minor repair, major repair, strengthening etc and to estimate rising costs due to maintenance work. In particular, to support engineers in making decisions of measures taken against predicted future deterioration, some alternatives will be offered based on a combination of maintenance cost minimisation and quality maximisation approach.

This paper presents the outline of the developed maintenance management system, focusing on calculation models to predict corrosion of rebars and its effect on structural capacities. By applying to an existing open piled marginal wharf for 28 years in service, accuracy of the system is verified and suitable values of parameters for the prediction are discussed.

SYSTEM DESCRIPTION

The management system in this study [1] evaluates current structural performances particularly load-carrying capacity and predicts the future progress of deterioration in target structural members and structures. Then, it proposes the most probable measure to maintain or to restore structural performances depending on estimated life-cycle costs. The system focuses on deterioration of materials caused by environmental actions in the ocean, but does not cover structural overall deformation such as settling, tilting, and sliding. The system consists of the following five stages:

- Data input: Basic data are supplied to the system regarding information on materials and structures as well as the results of inspection on current conditions of structural members.

- Evaluation of current deterioration: The estimated results of materials deterioration by the system are adjusted to those of inspection followed by modification of calculation parameters.

- Prediction of deterioration: The future deterioration process of structural members is predicted by using material degradation models with the modified parameters.

- Proposal of measures: Several methods of measures including their execution timing and frequencies are proposed from the viewpoints of costs and design lives.

- Decision of the most probable method: The most appropriate measure is specified by engineers among alternatives of proposals on the basis of benefit versus cost judgement.

PREDICTION ABOUT PROGRESS OF DETERIORATION

Parameters Required for Prediction

There are three fundamental important parameters for the prediction: chloride content on the surface of concrete, C_o, an apparent diffusion coefficient, D_{ap}, and the depth of concrete cover to a rebar embedded, c. Using these three parameters, it is possible to calculate the chloride content in concrete at a certain position and time. C_o is a parameter to representatively express the environmental condition around structural members.   As deterioration of a structure widely differs due to diversity of environmental conditions, the quantitative evaluation to environmental conditions, that is, setting a proper value to C_o is very important. The values of C_o widely scatter depending on structural conditions, materials used, and environments.

The measured values of C_o in various open-piled wharves nationwide in Japan are plotted in Figure 1, where p denotes for the non-exceedance probability. The figure shows a much variation and a coefficient of variation of about 50%, which implies that we should take proper safety margin to set the value of C_o. The design values of C_o are specified to be 13kg/m^3 (0.56% vs the density of concrete) for structures in splash zones and 9kg/m^3 (0.39%) for those along coastlines [2]. When C_o in a target structural member is not measured, those specified values can be used as the first trial.

D_{ap} represents the permeability of chloride in concrete, which has the close relation to durability of structures. The value of D_{ap} can be calculated by fitting Fick's second law of diffusion to the measured chloride distribution in concrete as presented in Eq (1).

$$C_{cl}(t,x)=C_o\left[1-erf\left(\frac{x}{2\sqrt{D_{ap}\,t}}\right)\right]$$ (1)

Where, C_{cl} is chloride content, t is duration, x is the depth from the surface of concrete, and *erf* is an error function.

Figure 2 shows the measured values of D_{ap} in the same wharves as those depicted in Figure 1. This also shows much variation. When the exact value of D_{ap} is not available, D_{ap} can be calculated with the type of cement used and water to cement ratio of concrete. An equation for ordinary Portland cement concrete is as follows:

$$\log D_{ap} = \left[4.5(W/C)^2 +0.14(W/C)-8.47\right]+ \log\left(3.15\times10^7\right)$$ (2)

The depths of concrete cover also vary but not so much as the other two parameters. It is possible to use the designed value for the prediction.

Progress of Deterioration

Once fixing the three parameters described above, it is possible to calculate the chloride content at the position of rebars with Eq (1) and consequently predict the progress of deterioration.   Figure 3 shows the description of prediction about the progress of deterioration.

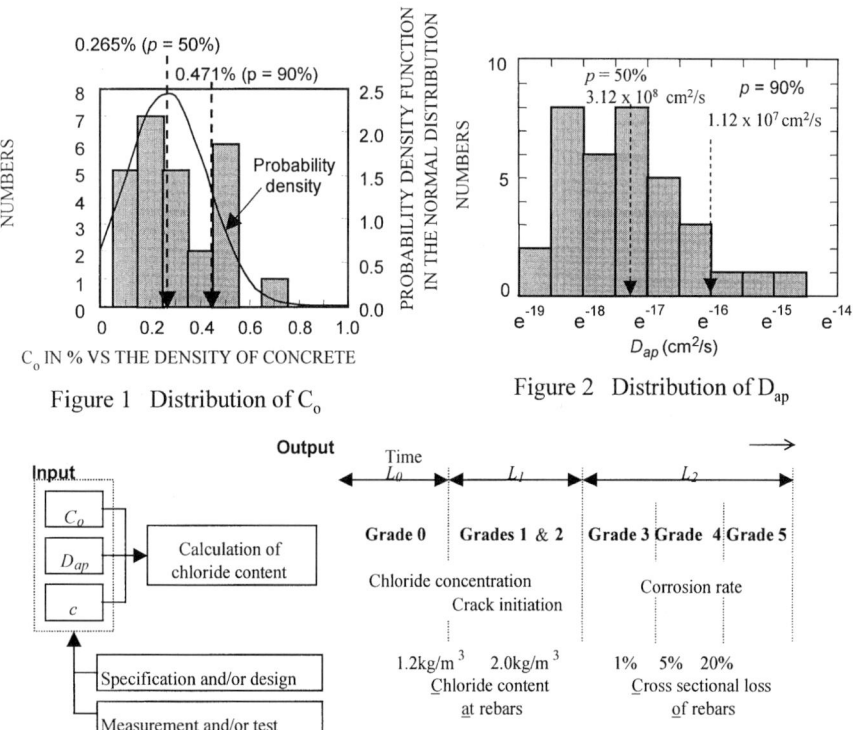

Figure 1   Distribution of C_o

Figure 2   Distribution of D_{ap}

Figure 3   Prediction about the progress of deterioration

When the chloride content reaches a threshold value at the position of rebars, corrosion of rebars will start and progress rapidly. The threshold value to initiate corrosion of rebars is specified to be generally 1.2 kg/m^3 [2]. After starting corrosion of rebars, the cross sectional area of rebars decreases due to corrosion and may reach the ultimate stage where structural capacities will be totally lost. The corrosion rate of rebars can be estimated by the following equation:

$$V_{red} = \frac{4\,aw}{\phi\gamma_{Fe}}e^{\frac{\alpha}{a}t}$$

(3)

Where, V_{red} is the volume loss of a rebar due to corrosion, γ_{Fe} is the density of rebars, a is a coefficient to represent the relationship between corrosion loss and crack width, w is an initial crack width due to corrosion, ϕ is a diameter of rebar, α is a coefficient to represent the relationship between corrosion rate and crack width, and t is elapsed time. V_{red} corresponds to the cross sectional loss of a rebar.   The values of a, w, and α are assumed to be 1500 mg/cm^2, 0.005 cm, and 220 mg/cm^3/y respectively in this study.

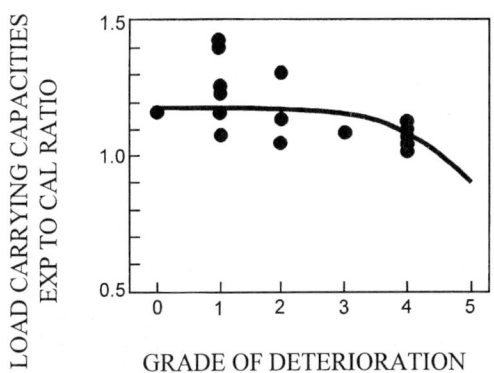

Figure 4   Load carrying capacities versus the grade of deterioration

Evaluation and Expression of Deterioration

The predicted progress of deterioration represented by the chloride content and the cross sectional loss of rebars will be linked to the grades of deterioration of structural members. The grade of deterioration is generally evaluated based on the results of visual inspection to the appearance of structural members.   In the inspection manual on port structures [3], the grade of deterioration is defined as the following six grades; Grades 0 to 5 from the viewpoint of rebar corrosion as follows:

Grade 0: No deterioration

Grade 1: Spotted rust stain and/or partially spread small cracks found
Grade 2: Partial rust stain, partial delamination of concrete, and/or many small cracks
Grade 3: Much rust stain, partial spalling and/or many cracks
Grade 4: Much delamination or spalling and/or many wide cracks
Grade 5: Widely spread delamination or spalling of cover concrete

These grades were not quantitatively defined, thus the prediction of deterioration in this study was tried to link them as shown in Figure 3. Since no cracks or unsatisfied features appear on the surface of concrete, the chloride content is considered smaller than 1.2 k g/m^3 in Grade 0. As corrosion of rebars progresses, cracks in concrete cover may form due to expansion of rebars. The time interval from the start of rebar corrosion to initiation of a crack is rather short in reinforced concrete superstructures of open-piled wharves because oxygen will be easily supplied there. Moriwake et al [4] pointed out that the threshold value to initiate a crack in concrete cover is about 2.0 kg/m^3.

The experimental results [4] also concluded that there is a relationship between corrosion current density, I_{corr} in μA/cm^2 and chloride content, C_{cl} in kg/m^3 as $I_{corr} = 0.025C_{cl}^{0.15}$.

Although the increase in chloride concentration does not directly accelerate rebar corrosion, corrosion current tends to be large because the corrosion widely spreads over the surface of a rebar. Therefore, when the chloride content reaches 2.0 kg/m^3, the stage of Grade 2 was considered to be finished.

After initiation of cracks in the cover concrete, load carrying capacity is a primary performance to be considered. Deterioration of the load carrying capacity became dominant after the cross sectional loss of rebars is larger than about 10% [5]. On the basis of the result of load tests on reinforced concrete beams exposed to marine environments for about 20 years [6], it was made clear that load carrying capacity was not so much deteriorated before reaching Grade 5 as shown in Figure 4.

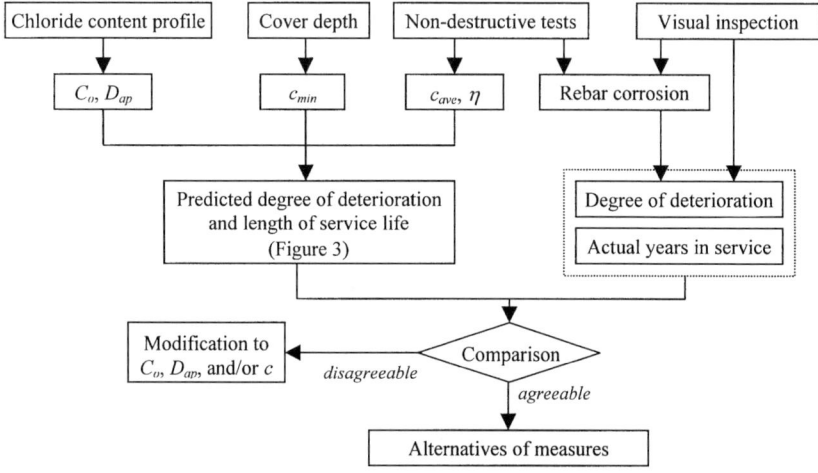

Figure 5   Modification of calculation parameters

Therefore, after Grade 3, the limit value of each Grade is expressed in terms of cross sectional loss of rebars: up to 1% for Grade 3 and 5% for Grade 4. Reinforced concrete superstructures in open-piled wharves generally have a safety margin for ultimate load carrying capacity of about 20%, thus Grade 5 remains until 20% in this study.

Adjusting the Predicted Degrees of Deterioration to the Inspected One

As described before, although some codes and specifications provide us with their standard values, each of them is merely one possible value within a wide range of scattering. Also even if using actual measured values, it may give us incorrect prediction. This is because the system involves some assumptions of the threshold values. Therefore, it is required to adjust the predicted results to the inspected results. The flow of adjustment is shown in Figure 5. The adjustment is to be done with modification to the three parameters C_o, D_{ap}, and c.

The system calculates the lengths of time during which a target structural component remains in respective grades of deterioration. Through the comparison between the actual year in service and the predicted length of time to reach the inspected grade of deterioration, the calculation parameters are modified for predicting future progress of deterioration. The predicted length of time is made by the following way depending on the *inspected grade* of deterioration, where L_0, L_1, and L_2 are defined as presented in Figure 3.

- **Grade 0**: On the basis of the authors' past study, corrosion spread over more than 50% of the total surfaces of rebars in some specimens even if the inspected result was Grade 1. Therefore, there may have corrosion in Grade 0 and the current condition is assumed to be at the end of Grade 0. Nevertheless, when the service life so far is shorter than a half of L_0, there should have no modification. When some detailed inspection is applied such as non-destructive tests, the current condition is assumed to have experienced two-thirds of the period of L_0.

- **Grades 1 and 2**: There are few differences in rebar corrosion between the two grades. The current condition is $L_0+0.5L_1$ in case of Grade 1 and L_0+L_1 in case of Grade 2.

- **Grade 3**: The process of deterioration can be subdivided by the degree of rebar corrosion because partial cross sectional loss of rebar is often observed. However, the criterion of subdivision has not been made clear. In this paper, the status of this grade is defined as that the cross sectional loss of rebar is 1.0% as described before. Therefore, the length of time at the current condition is $L_0+L_1+L_{11}$, where L_{11} is the period of Grade 3.

- **Grade 4**: As the same criterion as Grade 3, the limit of cross sectional loss of rebars is set to be 5%. Thus, the length of time at the current condition is $L_0+L_1+L_{11}+L_{12}$, where L_{12} is the period of Grade 4.

- **Grade 5**: The maximum cross sectional loss is 20%. The length of time is $L_0+L_1+L_2$.

When non-destructive techniques are applied to investigation, some useful information may be collected. In this study, only defects in the cover concrete will be evaluated with the results of non-destructive tests; that is, the reduction factor, η, to depth of cover concrete ranges from 0.5 to 1.0 depending of the degree of defects.

APPLICATION OF THE SYSTEM TO AN EXISTING OPEN-PILED WHARF

Description of the Wharf

The management system was applied to a prototype of structure to verify its accuracy. The target structure is a reinforced concrete deck in an existing open piled marginal wharf at Port T, Japan. The structure was built in 1970 to 1973, which has been in service for 28 years as a container-handling berth. The standard section of the wharf is shown in Figure 6.

Status of Deterioration

The results of visual inspection to the grade of deterioration of beams and slabs are shown in Figures 7 and 8 respectively. Approximately 30% of all the beams were categorized into Grade 0, while about a half of slabs was judged to be slightly deteriorated as Grade 1.

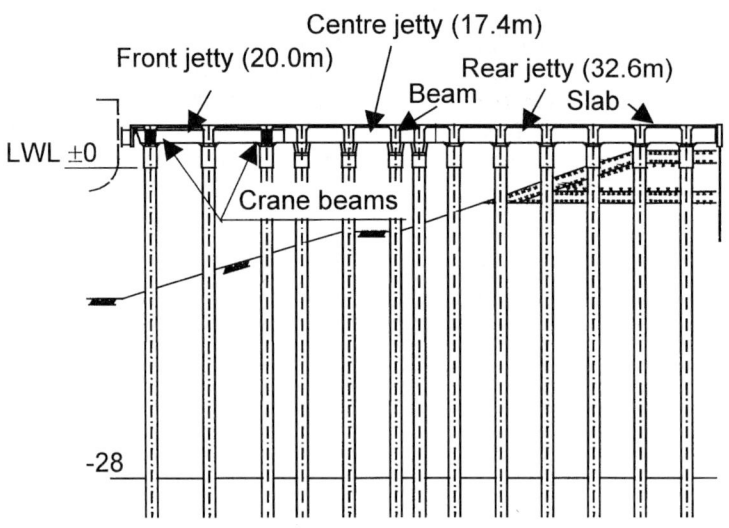

Figure 6   Section of the target wharf

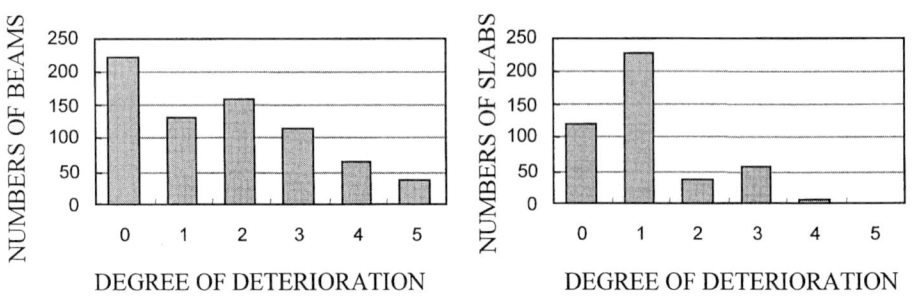

Figure 7   Deterioration of beams   Figure 8   Deterioration of slabs

Table 1 lists the measured results of C_o, D_{ap}, and c. The respective values with the non-exceedance probability, p=70%, 80%, and 90% were calculated on the assumption that C_o and c follow the normal distribution and D_{ap} does the log-normal distribution.

The specified design values of D_{ap} is 3.19×10^{-8} cm^2/s that was obtained by Eq (2) for ordinary Portland cement with water to cement ratio of $W/C = 0.45$.

Table 1   List of measured parameters for prediction

	BEAMS			SLABS	
C_o (kg/m³) $n=12$	Average	7.89	C_o (kg/m³) $n=4$	Average	2.08
	Standard dev.	5.75		Standard dev.	0.59
	$p=70\%$	10.88		$p=70\%$	2.38
	$p=80\%$	12.72		$p=80\%$	2.57
	$p=90\%$	15.26		$p=90\%$	2.84
	Specified	13.0		Specified	9.0
D_{ap} (cm²/s) $n=12$	Average	2.19×10^{-8}	D_{ap} (cm²/s) $n=4$	Average	1.93×10^{-8}
	Standard dev.	1.17×10^{-8}		Standard dev.	3.65×10^{-8}
	$p=70\%$	2.80×10^{-8}		$p=70\%$	2.12×10^{-8}
	$p=80\%$	3.17×10^{-8}		$p=80\%$	2.24×10^{-8}
	$p=90\%$	3.68×10^{-8}		$p=90\%$	2.40×10^{-8}
	Specified	3.19×10^{-8}		Specified	3.19×10^{-8}
c (mm) $n=13$	Average	58.83	c (mm) $n=10$	Average	58.26
	Standard dev.	20.47		Standard dev.	10.48
	$p=70\%$	48.19		$p=70\%$	52.75
	$p=80\%$	41.64		$p=80\%$	49.20
	$p=90\%$	32.60		$p=90\%$	44.77
	Minimum	20.00		Minimum	43.00
	Designed	60.00		Designed	43.00

n: number of data measured

Results of Prediction

While the progresses of degrees of beams and slabs were calculated by the system, but the results of beams are described here. The estimated years when the degree of deterioration reaches Grade 2 are shown in Figure 9 under the condition of designed concrete cover and the average diffusion coefficient.   When the representative degree of deterioration of the structure was defined as the average of all the data: Grade 2, the estimated years were shorter than 28 years, being on the safe side.

Figure 9 also shows the result after changing the representative degree of deterioration of all the beams from Grade 2 to 3. The broken line and solid and dashed line are the results from Grade 2 to 3 and from Grade 3 to 4 respectively. The estimated year approached the true value of 28 years. The most appropriate result was obtained with parameters of the average surface chloride content, the average apparent diffusion coefficient, and the designed value of concrete cover depth.

When the respective specified or designed values were chosen as these three estimation parameters, the estimated year was 4 to 18 years shorter than the actual years in service. Furthermore, the representative value of deterioration of the whole structure showed better prediction when it was defined as the one higher grade than the average of all the inspected grades.

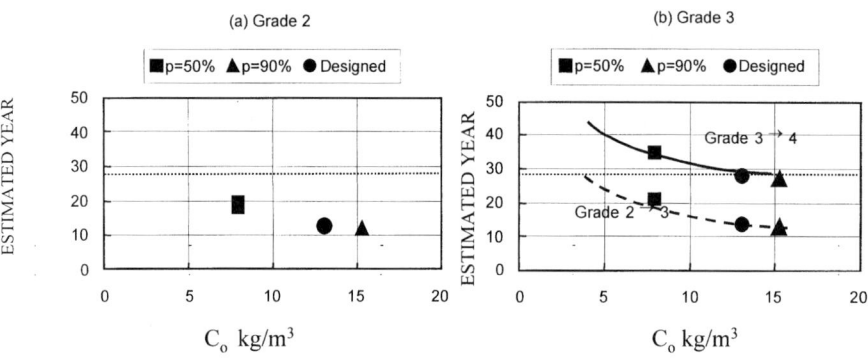

Figure 9   Estimated years to reach the target deterioration
in case of D_{ap}: p = 50% and c: designed

Table 2   List of input parameters and results of prediction

MEMBER	BEAM	SLAB	REMARK
C_o	7.8 kg/m^3	5.4 kg/m^3	60% of the specified value
D_{ap}	$2.07^{\times}10^{-8}$cm^2/s	$2.07^{\times}10^{-8}$cm^2/s	65% of the specified value
c	59 mm	58 mm	Average value
Rebar dia.	16 mm	16 mm	Designed value
Predicted years	Grade 0: 0 to 14	Grade 0: 0 to 18	
to remain in	Grade 1: 15 to 18	Grade 1: 19 to 26	
each Grade	Grade 2: 19 to 21	Grade 2: 27 to 33	
	Grade 3: 22 to 37	**Grade 3: 34 to 49**	
	Grade 4: 38 to 52	Grade 4: 50 to 64	
	Grade 5: 53-	Grade 5: 65-	
Actual year	28	28	

In conclusion, Table 2 summarizes the predicted years to remain each grade of deterioration unless measured results of those parameters are available. The predicted grades expressed in thick letters of each structure refer to those judged by the inspection. When the calculation parameters were set to the respective value in Table 2, the results by the system fit to those by the inspection. Further studies should be required to provide with more accurately predicted results.

CONCLUSIONS

The following conclusions were drawn from the presented study here:

1. The maintenance management system including estimation of the progress of deterioration was developed with reasonable accuracy.

2. The respective average of measure results of C_o, D_{ap}, and c were suitable to make future progress of deterioration to a target structure.

3. Unless measured data are available for use in prediction, the following modification made it possible to accurately predict the degree of deterioration:

   • Representative degree of deterioration: 80% probability (about 1 degree higher than the average degree of deterioration);

   • Surface chloride content: approximately 60% of the specified value;

   • Diffusion coefficient: approximately 60 to 65% of the specified value; and

   • Concrete cover: the average value of measured results.

ACKNOWLEDGEMENTS

The authors would like to extend their sincere appreciation to Dr H Hamada and Dr M Iwanami, Port and Airport Research Institute and Mr S Kodama, National Institute for Land and Infrastructure Management for their cooperation in this study.

REFERENCES

1. KODAMA, S, TANABE, T, YOKOTA, H, HAMADA, H, IWANAMI, M, and HIBI, T. Development of Maintenance Management System for Existing Open-Piled Piers, Technical Note of the Port and Harbour Research Institute, No 1001, June 2001.

2. JAPAN SOCIETY OF CIVIL ENGINEERS. Standard Specification for Design and Construction of Concrete Structures, Maintenance, 2001.

3. PORT AND HARBOUR RESEARCH INSTITUTE ed Manual on Maintenance and Rehabilitation of Port Structures, Coastal Development Institute of Technology, 1999.

4. MORIWAKE, A. Study on Durability and Maintenance of Reinforced Concrete Deck in Open-Piled Pier against Chloride Induced Deterioration, PhD Thesis submitted to Tokyo Institute of Technology, May 1996.

5. IWANAMI, M, YOKOTA, H, and AKIMOTO, T. Load Carrying Capacity of Deteriorated Reinforced Concrete Beams Due to Electrolytic Corrosion and Nondestructive Evaluation of Corrosion of Rebars, Proc the 3rd Regional Symposium on Infrastructure Development in Civil Engineering, Tokyo, December 2000, pp 621–628.

6. YOKOTA, H, FUKUTE, T, HAMADA, H, and MIKAMI, A. Structural assessment of deteriorated reinforced concrete beams under marine environments for more than 20 years, Concrete Durability and Repair Technology, Proc the International Conference, Dundee, 8-10 September 1999, pp 251–257.

APPRAISAL OF THERMOPHYSICAL PROPERTIES OF A CONCRETE DAM

M M Safarov   Z V Kobuliev
O H Amirov   M A Zaripova
Tajik Technical University
M S Muhamadiev
Baipazine Electrical Power Station
Tajikistan

ABSTRACT. This paper describes a study of thermophysical properties of concrete of dam hydroelectric power station. It is known, that heat conductivity the platinum's of hydroelectric power stations cannot be defined by an experimental way. Therefore we have put a task to estimate heat conductivity of a part of a dam of hydroelectric power station consisting from concrete. During construction of a dam of power station test from a concrete part frequently undertake. The tests have cylindrical, prismatic, cubic form with the which approximately identical with conditions of dams ic the constancy of a moisture, temperature etc is observed. Theoretical and experimental methods study heat conductivity concrete in the room temperature. Common of relative error of measuring heat conductivity under confidential probability $\alpha = 0,95$ range 4.2%. These materials, when flowing out of the reactor vessel, may interact with the concrete of the reactor building thus introducing decomposition products of concrete into the original mixture. These decomposition products are mainly; SiO_2, FeO, MgO, CaO and Al_2O_3 in different amounts depending on the nature of the concretes which are considered.

Keywords: Pratinum's of hydroelectric power station, Thermophysical properties, Theory, experimental, Models.

M M Safarov is a Chartered Thermophysical Properties Matter and Senior Lecturer in the Tajik Technical University in the Department of Thermal Engineering. His research focuses mainly on technology.

Z V Kobuliev is currently undertaking research into the environmental studies.

H Amirov is currently undertaking research into the environmental studies and composition materials at the Tajik Technical University .

M A Zaripova is currently undertaking research into the development of performance of concreter in the Civil engineer at the Tajik Technical University.

M S Muhamadiev is currently undertaking research into the development of performance specifications for carbonation resistance of concrete in the Baipazine Electrical Power Station at the Nurek.

INTRODUCTION

Severe accidents of nuclear reactors involve many situations such as pools of molten core material, melt spreading, melt/concrete interactions, etc. The word "corium" designates mixtures of materials issued from the molten core at high temperature; these mixtures involve mainly: UO_2, ZrO_2, Zr and, with minor amounts, Ni, Cr, Ag, In, Cd. The practical application of scientific achievements requires knowledge of the different materials and products which are subjected to storage, technological treatment, and use. Among them thermophysical properties and their quantitative characteristics are of great importance, as thermal treatment is widely used in the national economy, specifically in the industry.   Development, improvement, and intensification of the thermal treatment processes are based on the main principles of modern technology. The researched objects undertake by job lots and in definite quantity. In test laboratories and building managements are defined mechanical and physical properties.   It is necessary to note, that at change external of the factors (temperature, moisture, and pressure etc) change various parameters determining properties of concrete parts of dams. With change external of the factors, also vary heatphysical of property, is especial heat conductivity, specific capacity of researched tests. Many technical problems of hardening and melting of metals, freezing and thawing soil, thermochemical of destruction and others are connected to the analysis of tasks heat conductivity at change of a modular condition of substance. The most simple models of such processes are considered Layli and Klapeiron (1831), and also Stephan (1891). For an estimation of factor heat conductivity an ingot test by us are used melting of the decision on L S Leibenzon. In all statements the absence convection (for example, convection of water is supposed at hardening an ingot).

SURVEY

In the reference [1, 2] bring results experimental stud's thermophysical properties (heat conductivity, specific heat capacity, density and temperature conductivity) of concrete at the rooms temperature (Table 1). From Table 1 one can see of heat conductivity and temperature conductivity dependence of density concrete with increase of density.

Table 1   Thermophysical properties difference concrete at the rooms temperature

NO	MATERIALS	HUMIDITY MASS W, %	DENSITY ρ,kg/ m^3	T, K	λ, W/m K	Cp, J/kgK	a, 10^7, m^2/s
1	Concrete stone road-metal	8	2000	293	1,3	840	7.7
2	Concrete dry	0	1600	293	0,84	840	9.0
3	Gland concrete	8	2200	293	1,55	840	9.2
4	Slag concrete	13	1500	293	0,70	800	5.8
5[*]	Concrete stone road metal	10	1700	293	1,1	840	8.0

* Our results

EXPERIMENTAL

Heat Conductivity

Measurements of heat conductivity were conducted by means of flat wall [3]. Temperature measurements are made with the use of a Chromel-Alumel thermocouple. The deferential thermocouple registers the deference in temperatures of the investigated layers. To the side flat of boundless wall thickness δ support up of constant temperature t_{f_1} and t_{f_2}; $t_{f_1} > t_{f_2}$.

Temperature wall to the X define of formula:

$$t = t_{f_1} - \frac{t_{f_1} - t_{f_2}}{\sigma} x \qquad , \qquad (1)$$

if, λ not dependence temperature.

Model and Theory

The account is carried out on model as the cylinder having porous structure, and in pores is astringent. Processes of carry of heat through such structure we shall consider stage by stage. At the first stage we shall estimate heat conductivity of a porous grain consisting from sand, assuming, that in pores there is water. We use known model mutual penetrating of a material (model G N Dulnev etc), which components form mutual penetrate run through a lattice (Figure 1). At the first stage we shall estimate heat conductivity of a material of the cylinder, ie mutual penetrating of a porous material filling by road-metal [4-6]:

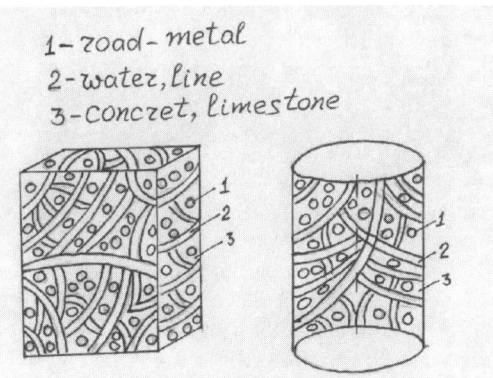

Figure 1   Model for calculation heat conductivity composition materials (model Dulnev)

$$\lambda^1 = \lambda_c \left[c^2 M + \upsilon (1-c)^2 + 2\upsilon c (1-c)/(\upsilon c + 1 - c) \right], \qquad (2)$$

Here,

$$c = 0,5 + A \cos(\ 2\pi - \arccos\ \frac{\varphi}{3})$$

$$\upsilon = \frac{\lambda_p}{\lambda_c};$$

$$\begin{cases} m_2 \le 0,5; A = -1; \varphi = 1 - 2m_2 \\ m_2 \ge 0,5; A = 1; \varphi = 2m_2 - 1 \end{cases}$$

m -parameter, characterise crack materials.

This cylinder considers, that it is second components a material, one road- metal, second components a material lime etc.

$$\lambda^{11} = \lambda^1 \left[1 - m_r / \left(\left(1 - \upsilon^1 \right)^{-1} - \left(1 - m_r \right) / 3 \right) \right], \tag{3}$$

$$\upsilon^1 = \frac{\lambda_r}{\lambda^1}, \qquad\qquad m_r\text{- volume concentration road-metal.}$$

At a finishing third stage carried out account heat conductivity "λ' " of granular system, which grains have heat conductivity, "λ''", and between them there is a gas-fill.

$$\lambda = \lambda'' \left(\left(y_1^2 / \left(0,5h_m + (1 - 0,5h_m) \phi \right) + \left(D/y_3^2 + A \left((1 - 0,5h_m - B + 0,5h_m / \upsilon_c \right) \right. \right. \right.$$
$$+ 2\upsilon_r \left(D - F + \omega \ln \left((\omega - D)/(\omega - F) \right) \right) / (1 - \upsilon_r) \right)^{-1} \right)^{-1} + \upsilon_{2Cn} E)) / y_1^2; \tag{4}$$

$$\upsilon_{c.3.} = \frac{\lambda_{c.3.}}{\lambda_1}; \upsilon_{2cn} = \frac{\lambda_{2cn}}{\lambda_1}; h_m = 0; \upsilon_g = \frac{\lambda_g}{\lambda_1}$$

Degradation of Concrete in Structures

On the offered above technique were carried out account heat conductivity porous ingots having the cylindrical from. The results of model account compared to experemental data received Kobuliev Z V (Table 1). It is possible to make of comparison a conclusion, that on the offered technique it is possible to calculate heat conductivity of the described above structures an error, commensurable with an error of given initial data.

Thermal Diffusivity

Thermal conductivity (λ) values can be calculated from measurements of the thermal diffusivity, density, and specific heat as [7-9]:

$$\lambda = a\rho c_p, W \bullet m^{-1} K^{-1} \tag{5}$$

Thus,

$$a = \frac{\lambda}{\rho c_p}, m^2 / s \qquad , \tag{6}$$

Results calculation thermal diffusivity concert at the room temperature shows to the Table 1.

Practical Implications

The novel test method developed allows the relative performance of existing and new materials to be compared under strict laboratory conditions in a relatively short-term period. The method is a step towards the development of an explicit design procedure for concrete durability and can allow designers to work alongside concrete suppliers to optimise material characteristics in a given environmental.

CONCLUSIONS

An infrared thermal imaging system has been developed to measure heat conductivity and thermal diffusivity distribution in dam hydroelectric power station. The system provides accurate measurements of heat conductivity, specific heat, thermal diffusivity values for standard samples. Measured heat conductivity, specific heat, thermal diffusivity images indicated variations in sample density and humidity.

ACKNOWLEDGMENTS

The authors acknowledge the financial support of the Baipazine Power Stations. The authors gratefully acknowledge the assistance of the Development "Automatizeition systems" Mr A A Naimov for the material presented in the figures and for helpful comments based on their analysis of considerable thermophysical data.

REFERENCES

1. MISNAR, A, 1968.   Heat conductivity solid materials, liquids, gaseous and this composition. M Mir. 464 p.

2. PEKHOVICH, A I, JIDKIKH, V M, 1974.   Calculation heat regime solid materials. M- L Energy. 304 p.

3. JURNEVA, V N, 1976. Heat technical reference T 2.

4. DULNEV, G N, ZARICHNIYAK, U P, 1974.   Heat conductivity of solutions and composition materials. L.

5. DULNEV, G N, 1976. Coefficient transfer to the linary systems. L.

6. SAFAROV, M M, 1993. Thermophysical properties simple ethers and water solutions hydrazine in the dependence temperatures end pressures. Dissertation d t s Dushanbe. 450 p.

7. TAYLOR, R E, and MAGLIC, K D, 1984. Pulse Method for Thermal Diffusivity Measurement, Compendium of Thermophysical Property. Measurement Methods. Vol 4, Plenum Publishing Corp, N Y, pp 305-336.

8. TAYLOR R E, 1975. Critical Evaluation of the Flash Method for Measuring Thermal Diffusivity, Rev Int Hates Temper et. Refract, pp 141- 145.

9. TAYLOR R E, Lawrence R, Holland, and Roger K, Crouch, 1985, Thermal Diffusivity Measurements on Some Molten Semiconductors, High Temp- High Press, 17. Pp 47-52.

CORROSION PROCESSES IN CONCRETE OF THE DAMS ON THE RIVER ANGARA

M A Sadovich

A A Sokolovskaya

Bratsk State Technical University

Russia

ABSTRACT. During the whole operation life of Bratsk and Ust-Ilimsk hydroelectric plants chemical analysis of water filtering through the pressure front of the concrete dams was under systematic monitoring by sampling in places of concentrated outcome in the inspection galleries of a dam. Long-term cumulative observations made up a basis for study of the indicated dams concrete corrosion caused by the Angara water in the process of its filtration through the cracks formed in the period of the dam erection, through construction joints and other defects. Mathematical-statistical analysis allowed the classification the processes of corrosion from the point of view of their time dynamics and divide them correspondingly into stationary (stable) and non-stationary. On the basis of a calculation of the cement stone component removal along with the ideas of the process dynamics the pressure front filtration development forecast has been made and their danger level has been estimated. The conception developed is of rather a general character and can be used when analyzing the state of concrete dams of the Angara cascade and other similar cases.

Keywords: Concrete corrosion, Concrete endurance, Water filtration, Filtration seepage, Dam, Pressure front, Leaching, Component removal.

M A Sadovich, Head of the Construction Technology Department of Bratsk State Technical University. The scope of his scientific interests includes the technology of monolithic concrete, study of hydraulic concrete, diverse study in the field of construction materials, technology of cold-weather concrete placement etc., scientific consultant of the construction firm "Bratskgesstroi", member of the University Academic Council.

A A Sokolovskaya, a postgraduate researcher of Bratsk State Technical University. Author of several scientific articles on corrosion of hydraulic concrete.

INTRODUCTION

Eastern Siberia is one of the major suppliers of electric power in the United Power Grid of Russia. At the same time a fair amount of electricity is generated by hydropower plants of the Angara cascade. Under these conditions performance reliability of hydropower plant concrete dam exposed to the effect of the Angara waters is of special significance.

It is obvious that the interest to the reliability of concrete dams will steadily rise in proportion to aging and change of service conditions forming under the influence of constantly varying environment. In full measure it refers to high concrete dams of Bratsk and Ust-Ilimsk hydro electric plants on the Angara river that have been running for more than thirty years.

Concrete endurance in water medium is one of the major longevity problems of concrete hydraulic structures. Depending on real dam running conditions, the effect of filtrated water on the concrete can have a destructive pattern that becomes apparent in the increase of seepage and, as an after-effect, in the reliability degradation of the construction. The process of corrosion running in the body of dam can lead to serious damages of the pressure front that will entail in huge economic outlay for its reconditioning. In this connection the state of concrete of sustained operation constructions requires strict and skilled supervision.

The conception of concrete corrosion process of Bratsk and Ust-Ilimsk dams was developed on the basis of quality and quantity analysis of the cement stone basic components removal depending on the amount of filtration seepage which in its part was considered as one of the main parameters of the process.

ANALYSIS OF AGGRESSIVENESS OF THE ANGARA WATER TOWARDS CONCRETE

Aggressive water attack is understood as its ability to destroy different building materials by acting upon them with dissolved salts and gases or washing away their component parts. The aggressive water attack on concrete constructions, particularly the Angara dams, permanent and potentially hazardous structures under heavy water head gain specific importance.

Evaluation research of water chemistry of Bratsk and Ust-Ilimsk reservoirs conducted by the Chemistry Laboratory of Ust-Ilimsk HPP in 1977-1999 and Hydrochemistry Laboratory of Irkutsk State University in 1977-1998 allowed to form notions on the water chemistry of the indicated reservoirs.

Long-term data analysis of water chemistry of the Angara artificial reservoirs showed that predominant anion is ion HCO_3^- (its water content is 70-160 mg/l) and the predominant cation is Ca^{2+} (its water content is 15-30 mg/l). General water mineralization of all high Angara river reservoirs doesn't exceed 250 mg/l. Free CO_2 content in reservoir water varies within the limits from 2,5 mg/l in summer time to 22 mg/l in winter period of the year. According to hydrogen index pH, Bratsk and Ust-Ilimsk reservoir waters are alkaline, however, gradual decline of pH from 8,8 to 7,7 is being observed during the whole observation period. Water aggressiveness towards concrete is estimated according to a number of indices (Table 1).

Table 1   Indices of non-aggressiveness and corresponding content
of aggressive components in the water of the Angara river

INDEX OF AGGRESSIVENESS	INDEX OF NON-AGGRESSIVENESS OF FLOW MEDIUM FOR PRESSURE STRUCTURES	CONTENT IN THE WATER OF THE ANGARA RIVER
Bicarbonate hardness, mg/l	Over 64,05	64,1 ÷ 122,0
Hydrogen index pH	Over 5	7,5
Content of aggressive carbon dioxide, mg/l	Less than 10	4.0 ÷ 12.0
Summary content of chlorides, sulfates, nitrates g/l	Less than 10	20 mg/l

Correlation of the Angara cascade reservoir water chemistry with the reduced criteria of non-aggressiveness shows that combination of leaching and carbon dioxide corrosion can take place in the dam concrete. A cross section of the dam is shown in Figure 1.

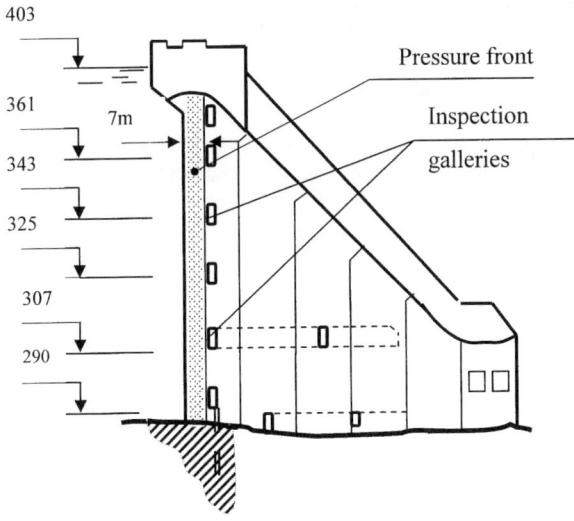

Figure 1   Cross section of a gravitation dam

ANALYSIS OF FULL-SCALE TEST RESULTS

In the process of condition control of the concrete pressure front of Bratsk and Ust-Ilimsk dams the sampling of water filtrated through the concrete in seepage sites of inspection galleries was carried out .

Filtrate sampling methods met the following requirements: sampling time and site registration; sampling, storing and handling were to eliminate changes in the content of the defined components.

In compliance with universally recognized concepts leaching corrosion of cement stone is connected with dissolution and removal of main structure forming component, which provides concrete high strength characteristics and environment alkalinity necessary for steady existence of cement stone minerals [1,2,3].

Actually, at the filtration chemistry analysis the ions of calcium were discovered not to be removed but precipitate in the process of concrete filtration approximately in the half of the survey cases, i.e. calcium ion concentration in the filtrate is lower than in the reservoir water [4]. Simultaneously an abrupt increase of sodium and potassium ions concentration is observed in comparison with the reservoir water (Figure 2).

The presence of removal of the mentioned compounds from the concrete depends on the filtration seepage. At minor seepage (up to 0,005 l/min) there is a decrease of ions Ca^{2+} content in comparison with the reservoir (precipitation). Here an abrupt increase of ions N^+ and K^+ content and corresponding decrease of free CO_2 up to its complete absence in the filtrate is marked. At seepage rates in the interval from 0,005 to 0,02 l/min both precipitation and removal of calcium ions is observed. At higher seepage rates (over 0,02 l/min) the removal of ions Ca^{2+} and Na^+, K^+ takes place.

The calculation of removed substances testifies that sodium rather than calcium prevail compounds in the filtrate (Table 2).

Phenomenon of predominant removal of sodium compounds from the concrete is not a fact of common language. Nevertheless, it is quite explicable. To build a pressure front of Bratsk and Ust-Ilimsk dams Portland slag cement of Krasnoyarsk plant was used. It contains up to 50% of Magnitogorsk blust-furnace slag. The cement content of alkaline oxides Na_2O+K_2O is defined by their presence in clinker (up to 1%) and in mineral additives.

From literature sources it is known that there are about 2% of sodium and potassium oxides in Magnitogorsk slag. Hence, at the content of alkaline oxides in cement up to 50% of blast-furnace slag, the content of alkaline oxides in cement makes up 1-1,5%. Since sodium and potassium oxides are chemically more active than calcium hydroxide, the process of their dilution in the water filtrated is imposed on the general pattern of cement stone corrosion.

It is important to note, that the degree of saturation of the filtrate with ions of sodium and potassium and corresponding rise of filtrate alkalinity depend on filtration rate. So, at filtration rates up to 0.02 l/min filtrate pH rises to 10-12 in comparison with 7-8 in the reservoir. At higher rate this abrupt rise of alkalinity is not marked.

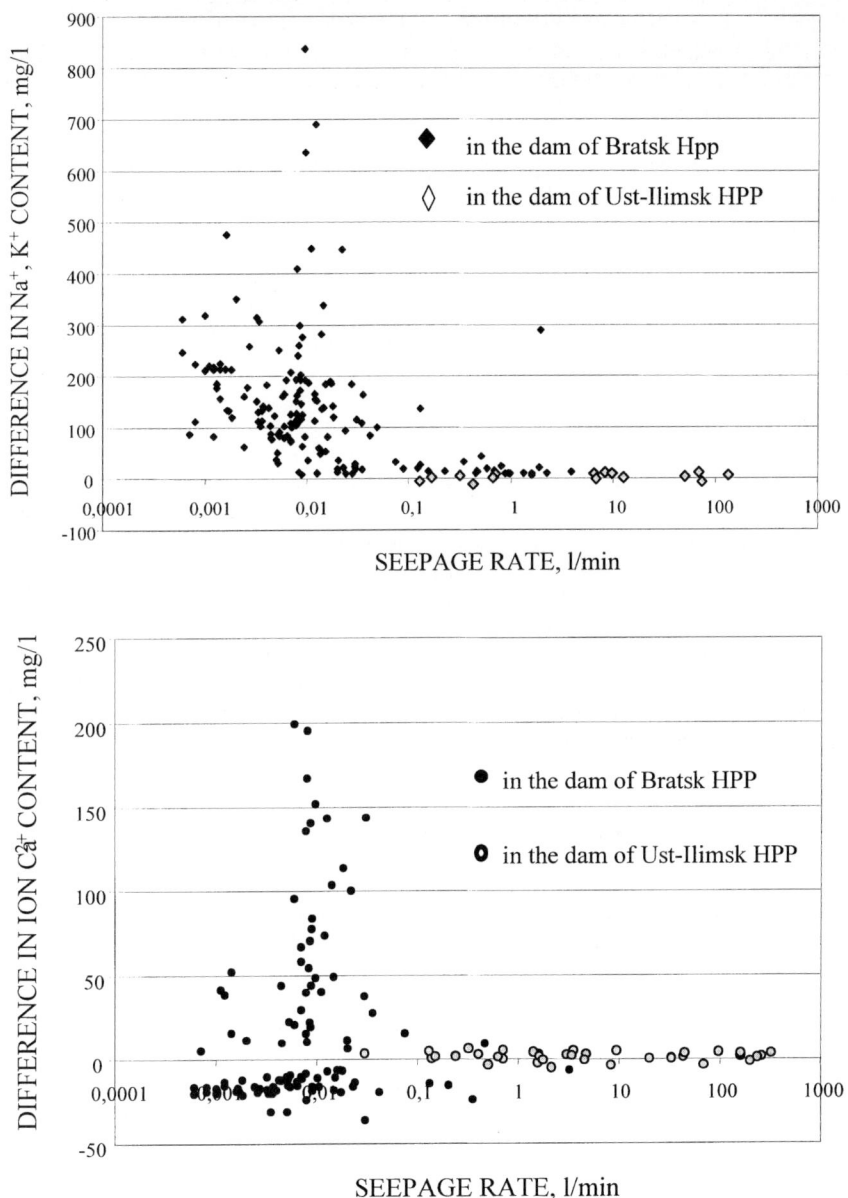

Figure 2   Relation between the amount of filtration flow and the differencein ions Ca^{2+} and Na^{+} content in the filtrate in comparison with reservoir water

Table 2   Average values of annual amount of dam concrete components removed precipitated by filtrated water counting on CaO and Na_2O

SITE OF SAMPLING			AVERAGE ANNUAL REMOVAL, KG PER YEAR	
			CaO	Na_2O
Mark 307 section 40 drain 5			0.529	6.15
Mark 307	s 46	d 5	1.29	0.468
Mark 307	s 55	d 4a	0.72	1.26
Mark 307	s 59	d 6a	-0.04	0.46
Mark 325	s 36	d 3	n/o	15.73
Mark 325	s 39	d 4a	n/o	12.46
Mark 325	s 41	d 4	3.393	7.52
Mark 325	s 45	d 5	0.323	1.99
Mark 343	s 31	d 5	1.107	2.26
Mark 343	s 34	d 3	-0.12	0.58
Mark 343	s 35	d 5	0.13	0.432
Mark 343	s 37	d 5	-0.05	0.4
Mark 343	s 39	d 6	0.07	0.398
Mark 343	s 54	d 4	-0.066	0.52
Mark 361	s 52	d 4	-0.011	6.6
Mark 361	s 58	d 5	0.157	2.43
Mark 361	s 60	d 2	-0.153	1.21
Mark 361	s 61	d 5	-0.342	3.0
Mark 361	s 63	d 4	-0.024	0.34

Notations:

n/o - is not observed due to minor difference in ion concentration in the filtrate in comparison with the reservoir water.

+ - removal of components with a filtrate.

- - precipitation of components on ways of filtration.

As it has already been noted, the Angara water can be a source of carbon-dioxide corrosion due to presence of aggressive carbonic acid (Table 1). It is necessary to point that recently there is a trend towards rise of water dissolved carbon dioxide content caused by biochemical processes and other factors exerting influence on the reservoir water chemistry.

Existence of pressure front concrete carbon dioxide corrosion is verified by the rise of concentration of carbonate ions CO_3^{2-} and HCO_3^- in the filtrate in comparison with the reservoir water. Simultaneously an abrupt decrease of free CO_2 is marked, up to its complete absence in the filtrate that confirms its conversion in carbonate compounds. The search of correlation links between ions Ca^{2+} and $Na^+ + K^+$ and carbonate ions allowed to find out the presence of a close correlation link only between the content of ions $Na^+ + K^+$ and carbonate CO_3^{2-} in the filtrate (correlation coefficient is 0,91-0,95). In this connection, availability of calcium bicarbonate $Ca(HCO_3)_2$ becomes improbable. The process of carbon dioxide corrosion runs in such a way that first of all easily dissolved compounds of Na_2CO_3 and $NaHCO_3$ type are formed. The pointed peculiarity allows to give explanation to such a phenomenon as calcium sedimentation on the ways of filtration. As far as formation of bicarbonate calcium in the reservoir water becomes possible only with availability of a definite amount of free CO_2, its conversion into carbonate compounds Na and K distorts this natural balance and the reaction runs towards decomposition of a bicarbonate into a precipitating carbonate and carbon acid. Decomposition of a bicarbonate releases new portion of carbon dioxide which is able to converse into carbonate compounds of sodium and potassium.

$$Ca(HCO_3)_2 \leftrightarrows CaCO_3 + CO_2 + H_2O$$

Low soluble $CaCO_3$ is precipitated on the filtration ways or is water removed on the surface of the concrete where it precipitates as saline sediments. This is supported by chemical analysis of saline sediments that showed that $CaCO_3$ makes up 93 % of their content.

However, the impact of carbonate compounds of Na and K is not limited by decomposition of bicarbonate calcium. Emergence of new carbonate compounds Na^+ and K^+ in the filtrate sufficiently increases its hardness in comparison with reservoir water which inevitably results in the decrease of solubility of $Ca(OH)_2$ in the cement stone, i.e. attenuation process of lime leaching can take place.

The suggested pattern of a corrosion process considering the effect of sodium and potassium oxides can not be regarded ignoring the filtration seepage, the amount of which defines the degree of saturation of the filtrate by the products of corrosion. Precipitation of CaCO3 is possible only at sufficient saturation of the filtrate by carbonate compounds of sodium and potassium that occurs to a different degree by filtration rates up to 0.02 l/min.

Further increase of filtration seepage and corresponding fall of concentration of the soluble components of the cement stone changes the pattern of the corrosion process in such a way that dissolution and removal of both alkaline oxides and calcium hydrocarbonate is determined only by the diffusion of the mentioned compounds from concrete. At discharges of more than 0.5 l/min the addition of concentration of the controlled ions in the filtrate in comparison with the reservoir water is so small that it creates illusion for absence of corrosion. It is evident, that these most dangerous sites of filtration in the dam require a more accurate monitoring for the ion concentration both in the filtrate and in the reservoir water so that the true pattern of the corrosion process would not elude the researcher.

Thus, the suggested pattern of a corrosion process in concrete affected by the water under study takes into account the effect of the cement present easy soluble sodium and potassium alkali and describes the state of the processes depending on filtration processes that in their turn determine the concentration of soluble compounds in the filtrate. The stated approaches allow to explain such a little known phenomenon as decrease of calcium ion concentration in the filtrate in comparison with reservoir water.

An important characteristic of concrete condition on the ways of filtration is the time dynamics of filtration seepage. The carried out analysis of change of measured filtration discharges presented as time series allowed to evolve filtration sites that are specified by the stationary process state with a trend towards attenuation and the sites with relatively large outlay in the form of non-stationary process without any trend towards attenuation. Principal dependencies of amount of the cement stone removed components on the observable filtration discharges were discovered.

At filtration rates up to 0,005 l/min the process of cement stone removal from the concrete has a stationary attenuating pattern, and the removal amount at separate inspection sites doesn't exceed 1 kg/year.

At filtration rates from 0,005 to 0,02 l/min the process of attenuation is mainly stationary, in some cases with a pronounced trend towards attenuation. The removal of the cement stone components is up to 6 kg/year.

At filtration rates over 0,02 l/min the process, as a rule, is non-stationary with peaks in separate years at some inspection sites equal to 30 kg/year.

EXPERIMENTAL VERIFICATION OF Na AND K OXIDES EFFECT

To verify the proposed assumptions on the effect of diffusion of alkaline ions Na^+ and K^+ on the state of bicarbonate in the reservoir water and free CO_2 an experiment aimed at simulation of processes running in the concrete filtrated water at small filtration seepage was conducted. The essence of the experiment was in stepwise saturation of the reservoir water of known composition with ions of sodium and monitoring for the quantity content of calcium ions and free CO_2 in the solution. To this end 5 samples of reservoir water of 300 mg each were taken and after water chemistry analysis the alkali NaOH of 0.1 gram-molecule/l concentration in volumes 15, 30, 45, 60 and 90 ml was added. The data on the content of sodium and potassium ions determined after 6 hours of aging are given in table 3.

It shows that increase of concentration of sodium ions is accompanied by the fall of calcium ions content in comparison with water reservoir that corresponds to settling-out in the form of CaCO3. At this stage the reaction most likely runs according to the scheme:

$$Ca^{2+}+2HCO_3^- +2Na^+ +2OH^- \Rightarrow CaCO_3 \Downarrow +2Na^+ +CO_3^{2-}+2H_2O$$

Availability of free CO2 in the solution after adding of NaOH is not discovered.

The second part of the experiment was verification of the assumption that saturation of the filtrate with sodium compounds affects the lime solubility.

A hanging of $Ca(OH)_2$ weighing 50 mg (table 3) was put into each of 5 samples and in reservoir water. Determination of calcium ions content in the solution two hours after introduction of the lime carbonate showed that in pure reservoir water a part of $Ca(OH)_2$ has dissolved (increment of Ca^{2+} concentration took place), whereas in samples saturated with sodium ions the lime carbonate not only remained insoluble but its availability in the solution contributed to further settling-out of $CaCO_3$ (decrease of Ca^{2+} content).

Table 3 The results of experimental verification of Na and K oxides effect

№ sample	pH	Ion content in res. water mg/l Ca^{2+} Na^+	Volume of added alkali NaOH per 300 ml of water, ml	Ion concentration after adding alkali, mg/l Ca^{2+} Na^+		pH after adding NaOH	CO_2 free content, mg/l	Additive $Ca(OH)_2$ per 200 мл of solution, mg	Concentration Ca^{2+}, mg/l	Appearance
1			15	43.2	211.6	9.6			21.6	
2			30	40.08	340.7	9.76			19.04	
3	7.5	50.1 7.98	45	36.2	452.3	9.8	Is not discovered	50	18.5	Turbid whitish solution, after settling has a flocculent white sediment
4			60	30.06	591.8	9.9			17.03	
5			90	25.5	855.9	9.96			13.1	
6			0	50.1	7.98	7.5	7.7	50	157.3	Clear solution

At this stage the following reactions can take place, each provoking the settling-out of $CaCO_3$.

$$1.\; Ca^{2+}+2HCO_3^-+Ca^{2+}+2OH^-\Rightarrow 2CaCO_3\!\Downarrow+2H_2O$$
$$2.\; 2Na^++CO_3^-+Ca^{2+}+2OH^-\Rightarrow CaCO_3\!\Downarrow+2Na^++OH^-$$
$$3.\; 2Na^++2HCO_3^-+Ca^{2+}+2OH^-\Rightarrow CaCO_3\!\Downarrow+2Na^++CO_3^{2-}+2H_2O$$

The experiment has confirmed the stated assumptions about the effect of diffusion of alkaline ions Na+ and K+ on the state of bicarbonate in the reservoir water, free CO_2 and lime carbonate under conditions corresponding to small filtration discharges.

CONCLUSIONS

The evaluation of the concrete state of dam pressure front over more than thirty years of operation allows to ascertain the availability of set ways of filtration. Some of them are characterized by a trend toward attenuation of filtration seepage and the other keep a non-stationary status and pose a definite threat of development. The sites of emergence of filtration are timed to crack and vertical drains and so most liable filtration lines go along vertical cracks of the concrete body and the horizontal construction joints through which drains can also be injected. Due to absence of particular data on filtration ways and corresponding area of concrete surface washed by water, the evaluation of filtration effect on the concrete status remains to be conducted according to comparative characteristics for evidently safe and relatively adverse sites of filtration.

On the whole over the period of the dam operation cement stone components removal at specifically inspected sites counting on Na_2O makes up 30-50 kg and at some sites it is up to 100 kg. The removal of CaO is significantly less, up to 30 kg. For the concrete of the pressure front that has hardened for some years the content of $Ca(OH)_2$ makes up 15-20 kg per cubic meter or respectively 1- 1, 5% of the cement body which conforms to 2-3 kg /m^3 (with slag Portland cement discharge of 230 kg /m^3 of concrete). To remove the stated amount of Na_2O with the filtrate it is necessary for the process of leaching to span 10-20 cubic meters of concrete but as long as the solution starts from the surface washed by the filtrated water one could assume that this surface makes up 100-200 cubic meters with an active zone depth of 10 cm which is completely real. At that, rather a big supply of $Ca(OH)_2$ still remains in the concrete.

In spite of the fact that the removal of Na_2O is not dangerous for the concrete the nature of the phenomena stated is destructive since it contributes to the development of capillary porosity of the cement stone and the fall of pH environment.

At the same time there are seepages of filtration water at which the calculated amount of the cement stone components removed with a filtrate is significantly higher than the above analyzed. At some sites the removal makes up 15 kg per year and over without any trend toward the fall of filtration seepage. From the point of view of concrete state such filtration sites are most dangerous as the removal over the whole period of operation is estimated in hundreds of kilograms.

One could assume that the concrete contacting the filtration flow has already lost rather a tangible amount of $Ca(OH)_2$ and the diffusion of Na_2O and K_2O comes out of the depth layers of concrete.

Thus, as a result of the analyses of the available data on the filtration in the concrete of the Angara dams pressure front both comparatively safe and most adverse sections were discovered. Their state monitoring is to be specified and continued.

REFERENCES

1.   V.M. MOSKVIN, F.M. IVANOV, S.N. ALEXEEV, Ye. A. GUZEEV. Corrosion of concrete and reinforced concrete, methods of their protection.- Moscow, Stroiizdat, 1980, 536 pp.

2.   S.N. ALEXEEV, F.M. IVANOV, N. MODRY, P. SCHIESSEL. Longevity of reinforced concrete in corrosive media. – Moscow, Stroiizdat, 1990,320 pp.

3.   V.A. KIND. Corrosion of cements and concrete in hydraulic structures. – Moscow Cosenegroizdat, 1955, 190 pp.

4.   M.A. SADOVICH, T.F. SHLYAKHTINA, Z.I. SOLOV'YOVA. Peculiarities of corrosion process of Bratsk HPP concrete dam pressure front. "Hydraulic engineering construction", № 3, 2000, (pp 14-17).

MAINTENANCE AND REFURBISHMENT OF CONCRETE IN WATER RETAINING STRUCTURES

P J Edwards
Montgomery Watson Harza
United Kingdom

ABSTRACT. This Paper describes the various types of water retaining structure encountered in water supply and wastewater treatment within the water industry. It goes on to describe in some detail the various causes of chemical and physical influences which degrade the reinforced concrete of which many are constructed. The Paper describes proven refurbishment methods for repairing these vital structures to give additional useful design life and concludes with a preferred type of contract strategy for carrying them out in practice.

Keywords: Water retaining structures, Durability breakdown, Refurbishment strategy, Framework agreement.

Philip Edwards is a Project Engineer and Senior Chartered Civil Engineer with MWH engaged on the execution of the Asset Management Plan for the water company United Utilities PLC. He specialises in the design of new structures and the refurbishment of existing structures.

INTRODUCTION – THE NEED FOR MAINTENANCE AND REFURBISHMENT

Maintenance of water assets is becoming a major industry (of a total capital programme of £18 billion over 5 years in England and Wales, £7 billion will be spent on maintenance alone) and is vital for the non-interruption of water transmission, the safeguarding of the integrity of structures and the guarantee of water quality.

Any client who owns water assets will wish to squeeze extra life out of them instead of investing in costly new-build projects. This is where 'maintenance and refurbishment' to promote additional design life becomes an attractive proposition.

Water assets such as aqueducts, pipelines, bridges and water retaining structures within water supply and wastewater treatment systems will suffer deterioration with age. This deterioration takes many different forms such as corrosion, erosion and degradation of the cast iron, steel, masonry or concrete members of which the asset under study may be constructed.

This Paper will discuss the initial investigation and inspection into the condition of water retaining structures solely and the causes of deterioration, including the various mechanical and chemical influences which degrade the reinforced concrete of which they are constructed. These influences include carbonation, chloride infestation and alkali aggregate reaction with respect to the material concrete.

The Paper will also discuss the problems of excessive leakage from water retaining structures and the ingress of pollutant water through all components of the structure with resultant effects on water quality. Currently government departments are extremely concerned about both of these aspects from the points of view of loss of a valuable water resource and public health issues respectively. The proposed repair and refurbishment regime required to give additional design life to the asset will then be discussed.

Finally, the means of procuring and implementing the refurbishment work, including the use of 'framework' contracts with selected contractors skilled in the field of rehabilitation, will be investigated. This form of contract enables substantial cost savings to be made during design and construction.

DEFINITION OF WATER RETAINING STRUCTURES

The water retaining assets of a typical water company are many and varied and may best be described in the larger sense as liquid retaining structures. They range from service reservoirs and clear water tanks in the water supply processes to storm tanks and sedimentation tanks in the wastewater treatment processes.

Generally in this country water retaining structures are designed to British Standard BS 8007 by methods based on limit state philosophy in which the maximum acceptable serviceability crack widths in reinforced concrete design are chosen to produce water retaining surfaces in given exposure categories.

The various types of liquid retaining structures owned by a typical water company are listed below.

Water Supply

- Service reservoirs for the storage of potable water in a water supply network.

- Balancing tanks for equalising pressure within a water supply network.

- Clear water tanks for the storage of potable water at a water treatment works.

- Intake structures at a water treatment works.

- Settlement tanks at a water treatment works.

- Filtration tanks at a water treatment works.

- Sludge consolidation tanks at a water treatment works.

- Washwater tanks at a water treatment works.

Wastewater Treatment

- Inlet structures at a wastewater treatment works.

- Stormwater tanks at a wastewater treatment works.

- Filtration tanks at a water treatment works.

- Sedimentation tanks at a wastewater treatment works.

- Sludge consolidation tanks at a wastewater treatment works.

- Storage tanks within a wastewater network.

PROCESSES OF DETERIORATION IN CONCRETE IN WATER RETAINING STRUCTURES

Durability in reinforced concrete may be defined as the ability of the material concrete to resist deterioration of its essential protective properties through chemical and physical influences resulting in steel reinforcement corrosion and subsequent loss of structural integrity ultimately leading to failure.

In the main, the performance of concrete in the cast state is monitored by strength considerations alone and the ultimate compressive strength is the primary control property monitored through the crushing of standard test cubes.

It is therefore ironical that 'durability' which is a non-tested property of concrete, not necessarily related to strength, may ultimately determine the useful life of a structure since it governs concrete and hence steel reinforcement decay.

This decay process is initiated by a number of chemical and physical influences, either affecting the zone between the surface and the reinforcement, or through the whole section of a given reinforced concrete structural member.

Decay due to chemical influences such as carbonation, chloride attack and acid attack generally takes place within the surface zone and is determined by the porosity or permeability of this zone. This permeability, which is generally measurable, relates to a number of factors such as:

- Mix design (proportions of cement and cement replacement materials, such as ground granulated blast furnace slag and pulverised fuel ash).

- The ratio of water to cement.

- The cover to the reinforcement.

- General site practices with respect to placing, compaction and curing.

It is this permeability, defined as a measure of the voids or capillaries within the surface zone, which enables moisture to transmit external agents, such as gases and chemicals, thus promoting the decay process. Some of the reactions occurring due to ingress within this zone are understood, such as carbonation, acid and sulphate attack; whilst others such as chloride attack and alkali aggregate reaction, have only partially been investigated and the remedial works necessary to counteract them are in an experimental stage.

Decay due to physical influences such as cracking, impact and abrasion damage is more obvious than chemical or electrochemical processes. Generally, the result is damage beyond the surface zone and this can intrude deeply into the section of the structural member resulting in the loss of structural capacity.

The following physical influences on concrete durability have likewise been recognised and recorded: thermal cracking, settlement shrinkage, structural movement, joint deterioration, freeze-thaw action and mechanical damage.

These processes of durability breakdown are detailed below.

Durability Breakdown Due to Chemical Influences

Carbonation

Carbonation is the progressive breakdown of the alkaline environment surrounding steel reinforcement due to the ingress of acidic carbon dioxide gas from the atmosphere. The carbon dioxide is carried into the concrete surface by moisture through cracks and pores formed as a result of durability breakdown.

The pH of the alkaline environment is reduced from 12 to around 9 as the carbonation front reaches the steel reinforcement. The process is known as depassivation or dealkalisation. At a pH of 9 the steel will corrode in a uniform manner to iron hydroxide and iron oxide with resultant expansion which will eventually cause the surface of the concrete to crack and spall.

A comparable chemical reaction to carbonation occurs with the acidic gas sulphur dioxide. where the process is known as 'sulphonation'.

In each case, the process is accelerated by the presence of industrial pollution, moisture, temperature change and breakdown of durability of the surface concrete which is also known as the 'covercrete'.

Carbonation will not attack structures which are totally submerged since the optimum moisture content for this reaction is in the range 60-70%. However, internal surfaces of structures subject to wetting and drying are at risk and also any external concrete surface open to the atmosphere, particularly in an aggressive industrial environment.

Chloride attack

Durability breakdown due to chloride attack is an electrochemical process during which the steel reinforcement and surrounding concrete are subject to electrolytic cell action. Anodes are set up at various discrete locations on the steel surface and the subsequent oxidation with resultant material loss occurs in the form of reinforcement 'pitting'. This corrosive action is non-expansive and is accompanied by severe attenuation of reinforcement at the anode locations and is unlike the carbonation process where corrosion is uniform along the length of the reinforcement. The action can continue undetected without visible signs at the surface because the concrete does not spall under the non-expansive chloride attack, however the loss of structural capacity may be severe.

The inclusion of chlorides within the concrete can occur in two different ways. With the case of 'combined' or 'built-in' chlorides, the use of unwashed marine aggregates or calcium chloride as a setting agent results in chlorides being bound into the chemical structure of the concrete. Alternatively, the chloride infestation can result from migration through the covercrete from the environment as in the case of de-icing salts and sea spray in marine or coastal structures.

In general, the chloride attack is exacerbated by the presence of a carbonation front but will continue in the presence of alkaline conditions in the vicinity of the reinforcement.

For structures in the water industry, water towers at a coastal location are susceptible to wind-blown salts and sea spray. In addition, coastal structures and areas where de-icing salts are used are particularly at risk from chloride infestation.

Alkali aggregate reaction

This can occur when the alkaline pore fluid in concrete reacts with the siliceous minerals in some aggregates to form a calcium alkali silicate gel. This gel imbibes moisture giving a volume expansion which can disrupt the concrete. The most common form of alkali reactivity is alkali silica reaction (ASR). Aggregates such as greywacke, chert and flint can be susceptible to ASR.

There is an equivalent reaction involving certain reactive carbonate aggregates known as alkali carbonate reaction. The presence of alkalis within the pores of the concrete will generally passivate the steel reinforcement against corrosion but the expansive gel associated with these aggregates may, under certain circumstances, result in non-uniform expansion resulting in severe map-cracking at the concrete surface and subsequent loss of durability.

Thaumasite

Precautions are required to minimise the risk of alkali reactivity in connection with the formation of ettringite and thaumasite in the use of certain aggregates. This can occur when naturally occurring sulphates of sodium (Glauber's Salt), calcium (gypsum or selenite), magnesium (Epsom Salt), iron (pyrite), potassium and aluminium are present in some soils and groundwater. Soluble sulphates may also be present in some mining spoil, industrial wastes or contaminated ground.

The sulphates can react with the concrete to produce ettringite (a calcium aluminate sulphate hydrate) giving a volume expansion which can disrupt the concrete. Where the concrete is exposed to cold, very wet conditions and a source of carbonate is present; there can be a different reaction with the concrete involving the formation of thaumasite (a calcium silicate carbonate sulphate hydrate) and this may lead to a softening of the concrete.

Demineralised water

It has been discovered that the storage of certain types of water (generally from upland sources), where sparingly soluble carbonate ions and soluble bicarbonate ions are not in equilibrium, can result in calcium ions being leached from the surface of the concrete. This results in a highly friable non-durable surface to the concrete.

Acid attack

The surface of the concrete can be readily attacked by acids in pipework and containment structures associated with water supply and wastewater disposal.

A particularly severe problem occurs in sewerage systems where slow flows and humid conditions in the sewage can lead to the bacteriological production of acidic hydrogen sulphide which deposits on the walls of pipes as sulphuric acid. This can corrode the walls of a pipe within a short timescale.

Sulphate attack

The action of sulphates (particularly sodium sulphate) carried in groundwater on concrete is fairly well understood and the use of sulphate-resisting Portland cement is widespread in the water industry.

Sulphates react with calcium salts in the concrete to produce alkaline gypsum and ettringite. The reaction is expansive and results in the breakdown of the concrete surface and severe loss of durability.

When sulphate attack, resulting in highly alkaline decay products, is combined with the potentially expansive aggregate, an extreme form of alkali aggregate reaction occurs.

Durability Breakdown Due to Physical Influences

Thermal cracking

The expansion of the concrete during the exothermal hydration reaction can result in severe cracking through the section. This is generally allowed for by the provision of horizontal tensile reinforcement.

Some researchers have concluded that the heat of hydration of cements has been progressively raised over recent years due to the alteration in the composition of modern cements required to give a rapid early strength gain.

Settlement shrinkage

There are a number of influences on concrete in its plastic state prior to complete cure as follows.

- Plastic Shrinkage: differential shrinkage between surface and core.

- Drying Shrinkage: surface drying immediately after set.

- Plastic Settlement: settlement and void formation over reinforcement.

- Segregation: constituents of concrete separated under vibration or by being dropped.

- Shutter Movement: loss of fines.

- Cold Joints: non-designed joints due to early set or production break during pouring.

The above influences can result in severe loss of concrete durability. To a large degree they may be avoided by preventive site practices employed during placing, compaction and curing.

Structural movement

Cracking and spalling due to movement with the structure in service can result in severe breakdown of durability at the surface or throughout the section of members. Movement can take the form of shearing, flexure, displacement or settlement.

Joint deterioration

The deterioration of joints in a containment structure whether through expansion, contraction or construction, can result in severe localised durability problems. It is known that joint sealants can have a very limited design life particularly in an aggressive environment and whilst these are generally not the primary source of waterproofing, this breakdown can result in severe ingress and egress through a structure. This is particularly serious when ingress results in pollution of a potable water supply and egress results in external pollution from a sewage or a chemical process or loss of stored potable water.

Freeze-thaw action

This is usually a secondary decay process, the loss of durability being initiated by chemical or physical influences. The saturated surface of concrete is subject to expansion and disintegration due to the repeated freeze-thaw cycle.

Mechanical damage

There are many instances of mechanical damage to structures such as erosion, abrasion, fire damage and explosive damage. Where these are predictable, consideration should be given to enhanced durability provision by the use of specialised concrete mixes such as microsilica concrete or densified concrete.

METHODS OF MAINTENANCE AND REFURBISHMENT

There are many good proprietary concrete repair materials and coatings available to the refurbishment industry. Decayed and carbonated areas of reinforced concrete can be removed by mechanical means, the rebar treated against corrosion and the broken-out areas replaced with polymer modified cementitious repair mortar. The entire area can then be treated with a pore-filling render and finally coated in a cementitious anti-carbonation coating to give future uniform corrosion potential.

There are various advanced electrochemical correction and protection processes currently being utilised such as "desalination" which is the removal of free chloride ions from chloride infested concrete. "Realkalisation" is a process whereby the alkaline protective zone to rebar is re-established in the body of the concrete. Finally "cathodic protection" involving the use of sacrificial anodes, is well established for bridge and harbour works maintenance. The above methods go a long way to correcting and giving future protection against durability breakdown in concrete due to chemical and physical influences detailed in this Paper.

Treatment of joints and cracking in structures is a less well-established science. Designed joints need to be waterproof and durable whilst still accommodating future movement. Cracks are in a sense non-designed movement joints. Severe structural decay can occur at wall and floor joints in articulated liquid retaining structures and at the wall to roof joint bearing surface. In addition, water ingressing through joints and cracking in a water retaining structure can cause pollution to stored potable water supplies whilst deteriorating surfaces. The elastomeric sealants within joints can be replaced, or as an alternative, overcoated with an elastomeric bandage adhered to the adjacent surfaces using a compatible epoxy resin. In either case, the complete and continuous tanking of joints below the water level is necessary.

Cracks which have ceased to exhibit further movement, known as 'dead' cracks, can be injected with epoxy resins or polyurethane resins. On the other hand, cracks which continue to move and are 'live' as shown by monitoring exercises, can be treated as joints and as such a sealant may be inserted or the crack trend may be overbandaged as described above.

The treatment of internal surfaces of liquid retaining structures can take many forms depending on the degree and nature of the degradation. Screeding and rendering systems include elastomeric cementitious coatings, sprayed concrete and hand-applied polymer-modified coatings. In addition, polymeric materials such as polyurethanes may be used and these may be hand-applied or sprayed. When the liquid retained is drinking water; the coating materials must be suitable for contact with potable water. In England and Wales the controlling authority with respect to these materials is the Drinking Water Inspectorate (DWI).

The waterproofing of external surfaces of liquid retaining structures and in particular the external roof surface; is the subject of the two CIRIA reports referred to in the References to this Paper. The membranes employed fall into two categories; namely loose-laid and liquid-applied. The waterproofing method employing loose-laid membrane such as polypropylenes, is generally used where the upper surface of the roof structure is so badly degraded that the application of a bonded system is inappropriate or there is a high degree of articulation in the form of joints as with a pre-cast roof These roof structures are covered in gravel or topsoil. The other category is the liquid applied system where the material type is generally polymeric and elastomeric such as polyurethanes or polyureas. This material is hand or spray applied to the roof surface which is generally left exposed. Prior treatment of any movement joints is required as described above.

MAINTENANCE AND REFURBISHMENT STRATEGY

Within the Water Projects Group of MWH at Warrington, England there are specialist civil engineers who have developed an expertise in the repair and refurbishment of structures and pipelines constructed in a variety of materials such as cast iron, steel, concrete, masonry and brickwork.

The structures range from aqueduct bridges and water retaining structures to pipelines, some of which are over one hundred years old. During the design life of these structures various forces of mechanical and chemical degradation act on them resulting in all forms of corrosion ultimately leading to structural failure and in the case of water retaining structures; leakage. The corrosion exhibits itself in many forms such as decayed concrete with rusting reinforcement, breakdown of paint systems, rusting of steel bridge members and leaking joints within structures and pipelines.

In the initial stages of a project involving refurbishment, MWH specialist civil engineers carry out a detailed inspection of the structure under examination and set up a testing regime to determine the root cause of degradation and then they carry out structural checks to determine how safe the structure is. There follows an intense period of investigation into which materials and methodologies are suitable for use in the repair and refurbishment process. In order to meet the refurbishment needs, cost effective, high performance and state of the art repair materials are chosen. These materials range from loose and liquid applied waterproofing membranes, concrete repair compounds, spray-applied concrete, elastomeric joint sealants, durable paint systems, carbon fibre reinforcement and leak-stopping compounds.

The water company is presented with a series of costed options for refurbishment with clear recommendations on which decision making can be based.

As a means of expediting the construction period, MWH civil engineers have an established cost-saving form of contract known as the "Framework Agreement" with selected specialist contractors in the field of refurbishment. Model documentation, consisting of a detailed specification, drawings and a schedule of agreed prices accurately determine the work content and contract value, prior to construction. The result is fast-tracking of the design and procurement process and minimum site supervision from which considerable savings accrue.

The "Framework Agreement" is based on the following criteria.

- The work is definable and low risk.

- Similar scopes of work apply to each water asset.

- There is continuity of work.

- It is important to capture and build on expertise gained.

- A short period from receipt of the water company's commission to contract award is required.

- Individual design and procurement costs must be minimised.

- The water company's operating considerations requires contractor flexibility and commitment.

Essentially a Framework Agreement is an agreement to enter into a future contract based on pre-agreed terms, conditions and rates. Each contract awarded under a Framework Agreement is a stand-alone contract with all the necessary components such as conditions of contract, pricing regime, scope of work and working drawings.

The Water Projects Group, who have carried out a number of successful refurbishment contracts on water retaining structures and aqueduct bridges, put together an initial scope of work and specification for a number of repeatable and standard refurbishment work items.

Experience with the use of the Framework Agreement is well proven and the following benefits have already been realised such as:

- Shorter design and procurement cycle resulting in shorter period from receipt of the water company's commission to contract award.

- Corresponding reduction in design manhours in the region of 60%.

- Improved feedback from contractors with respect to "state of the art" materials and better methods of working; both resulting in lower costs.

- Identification of efficiencies allowing Framework Rates to be reduced for future work.

- Standardisation of refurbishment methodology for each type of structure.

CONCLUSIONS

The combination of engineering expertise during investigation and design together with good procurement practice and efficient site management under the Framework Agreement is reaping rewards and design and construction costs are lower than with more traditional forms of contract.

During the last, five-year duration Asset Management Plan with United Utilities PLC, more than twenty vital aqueduct bridges carrying water supplies from upland sources to major conurbations have been repaired and the associated pipelines refurbished and sealed against leakage; which is currently an issue of great public concern.

Many reinforced concrete, masonry and brickwork water retaining structures have been sealed against the infiltration of pollutant water through their roofs and outward leakage from their walls and floors. In addition through refurbishment, structures have had durability and strength properties returned to them resulting in the water company being guaranteed further design life for water assets generally in excess of 25 years.

REFERENCES

1. ALLEN, R T, EDWARDS, S C, Repair of Concrete Structures, Third Edition, 1997, Chapters 3 & 9.

2. CIRIA, Waterproofing and Repairing Underground Reservoir Roofs, Technical Note 145, 1991.

3. CIRIA, Underground Services Reservoirs, Waterproofing and Repair Manual, Report 138, 1995.

A TECHNOLOGICAL MODEL FOR PREDICTING REBAR CORROSION PRODUCED BY COVERCRETE CARBONATION

A Giovambattista

L Eperjesi

L Ferreyra Hirschi

National University of La Plata

Argentina

ABSTRACT. In this paper a simple technological model is developed for the service life prediction of reinforced concrete structures built in the local environment with rural and urban characteristics. The model contemplates the steel corrosion produced by covercrete carbonation. It was developed starting from the report of structures in service. The obtained prediction model correlates the parameters that characterise the covercrete with useful service life. The structures evaluated (bridges, buildings and ducts) are placed in the Province of Buenos Aires, Argentina, whose environment is representative of an area of 570000 km^2. The structures have an age that varies between 15 and 60 years. Many of these have rebar corrosion induced by covercrete carbonation. The behaviour of structures in service is analysed in relation to different parameters linked with the quality of covercrete, such as effective porosity, density, coefficient of capillary absorption and carbonation depth. Simultaneously, concrete mixes were prepared in the laboratory and the covercrete parameters were determined. These mixes cover a wide range of water/cement ratios and compression strength, including those corresponding to the structures in service evaluated. These values were applied to the prediction model and the results obtained are compared with the effective performance of the evaluated structures. It is observed that the detected parameters match the behaviour in service and the history of the structure. Therefore, they are useful for durability design.

Keywords: Durability design, Service life prediction, Steel corrosion, Carbonation, Quality of the covercrete, Capillary absorption, Effective porosity, Density.

Professor A Giovambattista, is a full Professor and Researcher, Faculty of Engineering, UNLP. His research focuses on concrete durability, repair and maintenance.

Dr L Eperjesi, is a Lecturer and Researcher, Faculty of Engineering, UNLP. Her research focuses on pathology and durability design of concrete structures.

Mr E Ferreyra Hirschi, is an Assistant Lecturer, Faculty of Engineering, UNLP. His research focuses on pathology and durability design of concrete structures.

INTRODUCTION

General Concepts

From records of performance in service, it can be stated that carbonation depth in the time limit of useful service life design ($d_{c,sl}$) is:

$$d_{c,sl} = (2.g_1.g_2.g_3.\Delta c.(D_{nom}/a))^{0.5} (t_0/t)^n t^{0.5} \qquad (1)$$

where

g_1: parameter for microclimatic conditions, describing the mean moisture content of concrete.
g_2: parameter to describe curing conditions.
g_3: parameter to describe the effect of water separation (local w/c ratio).
n: parameter for microclimatic conditions, describing wetting and drying.
t_0: reference period, $t^{0.5}$-law valid, years.
t: time, years.
a: the amount of CO_2 for complete carbonation, Kg/m^3.
D_{nom}: the diffusion coefficient of dry concrete for carbon dioxide in a defined environment (20°C, 65% relative humidity), mm^2/year.
Δc: the concentration difference of carbon dioxide at the carbonation front and in the air, which usually means the carbon dioxide content of the surrounding air, Kg/m^3.

Equation (1) has a wide field of application that covers different microclimates and types of concrete. It also allows viewing and quantifying all variables involved in the process [1]. However, a simple technological model can be developed from this Equation to design structures placed in areas where climatic conditions are defined.

For a sufficiently large t_0, equal to or greater than 15 years, the parameters of Equation (1) are implied in the porous structure that conditioned the ingress of water and CO_2.

By measuring carbonation depth d_c in time, it can be stated that

$$d_c = C \, t^{0.5} \qquad (2)$$

Where C is a carbonation constant dependant on covercrete characteristics. In this work covercrete is characterised by its specific coefficient of capillary absorption S. This is to say that C=f(S), and then it follows that

$$d_c = f(S) \, t^{0.5} \qquad (3)$$

Coefficient of Capillary Absorption

Capillary absorption first appeared in the literature on building materials in relation to soil evaluation. Then, the theory of unsaturated flow (where capillary absorption is a well-defined amount) started to be applied to the movement of water in porous solids. However, it was not until the 80's that studies started to be applied to cementitious materials.

The work by Ho and Lewis [2] showed that capillary absorption varies with concrete and mortar composition.

Currently, there is special interest in characterising concrete in terms of its durability. In deterioration processes, water acts as a vehicle for the transport of aggressive substances. At first, we could assume that capillary absorption reflects the ability of the material to absorb and carry water by capillarity, representing the phenomenon as follows:

$$AC = k\, t^{\,0.5} + AC_o \qquad (4)$$

Where **AC** is the apparent capillary absorption (called capillary absorption in the literature) and represents the volume of water absorbed per unit area; **k** is the coefficient of apparent capillary absorption (sorptivity); **t** is the time for which capillary absorption is measured and AC_o is the initial apparent capillary absorption.

In available previous records capillary absorption is generally assessed by tests using a short period of time (between 1 and 4 hours). However, according to Martys [3] and our own experience, a longer time is required for the test to achieve results that will allow modelling capillary transport with a view to predicting performance in service.

Covercrete is better characterised by relating k to porosity **p**. A parameter then results, which we have called effective capillary absorption coefficient **S** (capillary sorption according to DURAR [4]), defined as S=k/p. Physically, S stands for capillary rise height for unit time in a model of parallel capillary tubes.

Taking into account the above considerations, in this work we have introduced some changes in the test methodology and in the interpretation of the capillary absorption phenomenon. Such methodology was applied to the study of different types of structures and related to their field performance.

EVALUATION OF CONCRETE STRUCTURES IN SERVICE

AC was evaluated on concrete cores extracted from the structures being studied. Test conditions are the same as those reported in former research papers [5], [6], except that in this case the duration of the test was extended up to a maximum of 72 hours. AC determinations were supplemented by measurements of covercrete thickness; absorption and density, according to ASTM C 642; carbonation depth by phenolphthalein spraying. From this latter value C was calculated using Equation (2). The rebar condition was also evaluated for quality.

Studied concretes correspond to structures in service with pathological signs of different intensity. The composition of mixes is not known as the ages of the structures range between 15 and 60 years. Analysed structures are located in the Province of Buenos Aires and placed in moderately aggressive environments. According to Köppen's classification, the climate can be described as mild and humid, rainy all year through, with a mean annual rainfall of the order of 1000 mm, mean annual ambient relative humidity of 75% and mean annual temperature below 18°C. Structures have been identified as follows:

I: facilities in a factory using non-contaminating processes.
II: garage area in a block of flats.
III: underground aqueduct.
IV: bridge on a stream on a provincial route.
V: cantilevered area in a disco.
VI: unfinished building structure.

TEST RESULTS AND CALIBRATION OF THE PREDICTION MODEL

Table 1 shows, for each structure analysed, their age at the time measurements were made, carbonation constant $C(mm/year^{0.5})$, covercrete thickness $CT(mm)$ and the values obtained for $k(mm/h^{0.3})$, $p(\%)$ and $S(mm/h^{0.3})$. The state of rebars (active or passive) related to the corrosion process, which was evaluated by visual inspection, is also included.

Table 1   Summary results of concrete physical properties

STRUCTURE	AGE (YEARS)	K (mm/h^n)	P (%)	S (mm/h^n)	CT (mm)	C $(mm/year^{0.5})$	REBAR CORROSION
		1.17	7.8	15.0	12	4.7	Yes
		0.29	8.0	3.7	>110	0.8	*
A	60	0.26	7.2	3.7	>100	1.0	*
		0.12	7.0	1.8	>120	1.8	*
		0.17	7.8	2.1	>55	0.7	*
		1.15	16.2	7.1	2	3.4	Yes
B	15	0.96	16.3	5.9	8	2.3	Yes
		0.58	14.9	3.9	10	1.8	Yes
C	43	0.68	19.3	3.5	8	0.3	No
		0.19	17.5	1.1	9	0.3	No
		2.85	13.1	21.8	---	10.1	**
		2.47	10.8	23.0	16	7.6	Yes
		1.71	9.3	18.4	22	7.8	Yes
D	60	2.65	10.6	24.9	11	5.9	Yes
		1.51	8.3	18.2	17	6.1	Yes
		1.87	11.1	16.9	---	6.3	**
		3.04	11.1	27.3	---	6.6	**
		0.83	8.9	9.3	---	6.2	**
		0.78	9.4	8.3	176	7.4	No
E	53	0.45	8.7	5.2	143	3.7	No
		1.51	10.3	14.6	170	7.7	No
		1.62	13.0	12.5	10 to 20	6.7	Yes
		4.13	14.6	28.2	10 to 20	11.2	Yes
		2.30	11.3	20.4	10 to 20	11.2	Yes
		1.63	11.7	13.9	10 to 20	6.7	Yes
F	20	2.32	15.0	15.5	10 to 20	6.7	Yes
		4.73	16.0	29.5	10 to 20	8.9	Yes
		1.47	11.8	12.5	10 to 20	5.6	Yes
		0.78	14.1	5.6	10 to 20	2.5	Yes
		1.95	16.9	11.6	10 to 20	2.0	Yes
		1.15	13.6	8.4	10 to 20	3.4	Yes

* Rebar was not reached
** Normal concrete

The best fit for Equation (4) was obtained by considering a power of time t whose exponent varies from 0.25 to 0.50, depending on the material characteristics. This was also noted by N. Martys [3]. The values of k included in Table 1 correspond to a correlation curve of the power law family with a variable exponent.

Figure 1 represents the values of C and S calculated from experimental measurements. Such parameters are correlated, according to Equation (5), with r = 0.88.

$$C = 0.375 \, S \qquad (5)$$

The reliability limits for an 80 % probability (C = 0.375 S ± 2) is also shown on this graph.

From a mathematical point of view Equation (5) has an acceptable correlation, as it was derived from determinations on concrete structures in service. However, for design and specification purposes, it is advisable to use the upper curve of the reliability limit that corresponds, in probabilistic terms, to 20 % of faulty elements. The curve then needs to be adjusted with more data.

Equations (2), (3) and (5) are suitable tools for durability design. Having established service life as a boundary condition, covercrete thickness depends on concrete quality, which is characterised by S. Thus, it follows that

$$CT \geq dc = (0.375 \, S \pm 2) \, t^{0.5} \qquad (6)$$

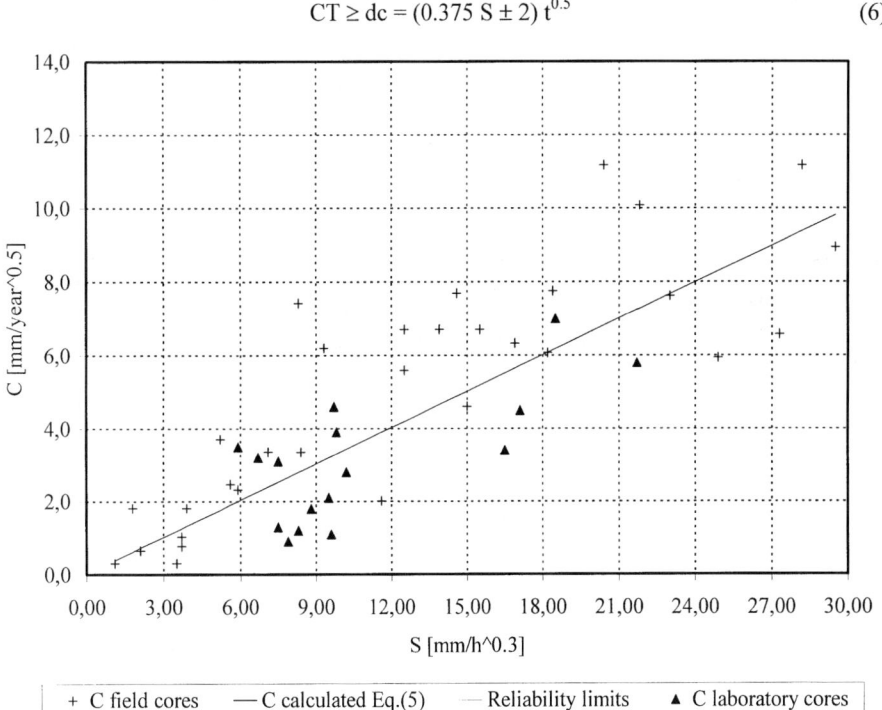

Figure 1   Values of C and S from experimental measurements

We would also like to comment on reference values indicated by DURAR [4]. For moderately aggressive environments, this manual recommends the use of concretes with $S \leq 6$ mm/h$^{0.5}$ together with 30-mm covercrete thickness, without referring to the resulting useful service life. By applying a power law fit in our work we have noted that the stated threshold valued of S is equal to 10 mm/h$^{0.3}$, a value that applied to Equation (6) results in 50 years of service life.

CONCRETE MIXES PREPARED IN THE LABORATORY. ANALYSIS OF THEIR COMPLIANCE WITH RESPECT TO THE PREDICTION MODEL

Concretes with different water/cement ratios (0.37, 0.50 and 0.66) and different types of cement (ENV 197-1 CEM I 42.5, II/B-P 42.5 and II/B-L 42.5) were prepared in the laboratory. After 7 days of standard curing, concretes were stored up to test ages, subject to variations in temperature and humidity typical of our environment. Concrete characteristics, and S and C values are included in Table 2.

Table 2   Characteristics of laboratory prepared concretes
Values obtained for S and C

TYPE OF CEMENT	W/c RATIO	S (mm/h^n)	C (mm/year$^{0.5}$)
CEM I 42.5	0.37	13.8 10.1	0 0.6
	0.50	7.5 6.7	3.1 3.2
	0.66	9.5 8.3	2.1 1.2
CEM II/B-P 42.5	0.37	19.1 18.4	2.8 2.5
	0.50	16.5 17.1	3.4 4.5
	0.66	18.5 21.7	7.0 5.8
CEM II/B-L 42.5	0.37	10.2 10.2 13.3 11.9	0 0 0 0.5
	0.50	8.8 8.2 5.9 7.9	1.8 2.4 3.5 0.9
	0.66	9.8 9.7 7.5 9.6	3.9 4.6 1.3 1.1

If we admit that the measurement of depth of carbonation has an error of ±1 mm, the values of S and C, measured on laboratory concretes with a w/c ratio of 0.50 and 0.66, are within the 80% reliability limits of Equation (5).

From a new adjustment with all the samples, the following Equation results:

$$C = 0.345 \, S \qquad (7)$$

Concretes with a w/c ratio of 0.37 do not fit the expression proposed by Equation (5). Changes in the porous structure are likely to modify the process of CO_2 diffusion trough concrete. Therefore, the method for evaluating porosity should be adjusted to allow extending the methodological set out of this work to concretes with low w/c ratios, similar to those used in high performance concretes.

CONCLUSIONS

This work shows the feasibility of making a simple technological model for durability design of reinforced concrete structures subjected to steel corrosion due to concrete carbonation.

The model allows linking service life with quality and thickness of rebar covercrete. Concrete quality is characterised by its coefficient of effective capillary absorption and it applies to certain environments.

This model was adapted for a region in our country that has defined climatic characteristics. It was calibrated with samples extracted from structures, which had been exposed for 15-60 years, built with conventional concrete. Equation (6) results from such calibration.

This Equation should be used as a guide until calibration can be improved with data from a larger number of structures in service.

ACKNOWLEDGEMENTS

The authors are grateful to Valeria Cifre Carrillo for helping in the experimental determinations.

REFERENCES

1. CEB Bulletin 238, New approach to durability design, An example for carbonation induced corrosion, 1997.

2. HO, D., Influence of slag cement on the water sorptivity of concrete, Fly ash, silica fume, slag and natural pozzolans in concrete, American Concrete Institute, SP-91, 1986, p 1463-1473.

3. MARTYS, N., FERRARIS, C., Capillary transport in mortars and concrete, Cement and Concrete Research, 1997, Vol. 27, No 5, p 747-760.

4. CYTED, DURAR, Manual de inspección, evaluación y diagnóstico de corrosión en estructuras de hormigón armado, 1997.

5. EPERJESI, L., FERREYRA HIRSCHI, E., SARALEGUI, G., GIOVAMBATTISTA, A., Influencia de las adiciones activas en la calidad del hormigón superficial, Primer Congreso Internacional de Tecnología del Hormigón, AATH, 1998, Vol 1, p 243-252.

6. HALL, C., Water sorptivity of mortars and concretes: a review, Magazine of Concrete Research, 1989, Vol 41, No 147, p 51-61.

INFLUENCE OF FIRE ON THE MECHANICAL BEHAVIOUR OF CONCRETE SPECIMENS

S Kumar

Harcourt Butler Technological Institute

India

ABSTRACT. In building design it is useful for structural designer to know the effects of fire on strength of various building materials especially on concrete and mortar and structural elements. In the present investigation, an attempt is made to experimentally verify the effects of fire on strength of concrete within a specified duration. The variables of study were mix proportion, water-cement ratio and flame exposure duration. The mix proportions of concrete considered in this study were 1:1:2, 1:1.5:3 and 1:2:4 by weight of ingredients. Three different water-cement ratios were taken for all the mixes. The experimental study reported here was planned to investigate the mechanical behaviour of concrete specimens after exposure to flames for different periods of time. The concrete specimens of M-25 and M-30 grades were subjected to elevated temperatures ranging from 200^0 C to 1000^0 C. It was observed that the mix proportion, water-cement ratio and flame exposure duration have significant influence on the concrete strength under fire exposure.

Keywords: Accidental damage, Buildings, Degradation of concrete, Durability, Fire damage, Fire proofing, Fire resistant structures, Fire safety, Structural fire engineering.

Dr Sunil Kumar, is Assistant Professor in Civil Engineering at Harcourt Butler Technological Institute, Kanpur, India. He received his B.Sc. Engineering in Civil Engineering from Aligarh Muslim University, Aligarh, M. Tech. From Kurukshetra University, Kurukshetra and Ph. D. from Kanpur University, Kanpur. His main research interests include the durability of concrete structures in aggressive environments, structural fire engineering, probabilistic design loads in buildings and waste management. Dr. Kumar has published research papers widely in National and International Journals and Seminars. He has several years of experience in designing a number of concrete structures including tall buildings, water retaining structures, and water and sewage treatment plants in India.

INTRODUCTION

During the lifetime of a structure the possibility of accidental loading is always present. Fire is one of the most destructive accidental loads that a structure can be subjected to in its lifetime. In a typical fire the temperature reaches 500^0 C in about 10 minutes and 950^0 C in one-hour [1]. The role of the structural engineer regarding design for fire safety in buildings is concerned mainly with protection of structure. The fire effect on the structural components of a building depends on the materials of construction and also on the duration of fire and the temperature attained. Concrete is a non-combustible material and generally resists fire fairly well. But, like any other non-combustible material, concrete can be damaged by fire if it is exposed to enough heat for a sufficient time. The need to understand its behaviour in fire has resulted in considerable research in to the effect of elevated temperatures on the properties of reinforced concrete [2-7].

The structural properties of concrete are modified by thermal exposure. The fire damage that causes utmost concern is the reduction in compressive strength of concrete followed by internal and external cracking. The effect of increase in temperature on the strength of concrete is small and somewhat irregular below 250^0 C [8], but above 300^0 C a definite loss of strength takes place. Harda [9] has shown that concrete, which has been heated at a temperature below 500^0 C, will rehydrate when cooled down and gradually regain most of its strength. Concrete is believed to lose 25 % of its unfired compressive strength when heated to 300^0 C, and 75 % at 600^0 C [10]. The loss in strength at higher temperature is greater in saturated than in dry concrete, and it is the moisture content of the concrete that is the most important factor determining its structural behaviour at higher temperatures [11].

Siliceous aggregate concrete is known to exhibit a change in colour when subjected to elevated temperature; it becomes pink at 300^0 C and whitish grey at 600^0 C [10]. In concrete, the aggregates undergo a progressive expansion on heating while the hydrated product of the set cements, beyond the point of maximum expansion shrinks. These two opposing actions differ considerably in their behaviour under exposure to flaming.

Concrete is subjected to a form of damage due to excessive heat, which is known as spalling. The build up of steam pressure in the concrete voids during fires results in explosive spalling, which causes significant pieces of concrete to break from the concrete surfaces [1]. This phenomenon usually occurs within the first 30 minutes of exposure to heat. The gradual separation of pieces – sloughing off – subsequently takes place because of the formation of continuous fracture planes. When fire causes spalling, the reinforcing steel may be exposed and the heated steel will lose its strength.

In literature most of the experimental studies on fire are based on ASTM standard fire tests. The experiments were conducted following a standard time-temperature curve. There is a substantial evidence to show that the laboratory fire test may be inappropriate indicator of how a structural element behaves during an actual fire [12-19]. The most important difference between the laboratory tests for fire resistance and an actual fire concerns the nature of the heating regime in the two cases. There are some situations in which concrete elements are subjected to a constant elevated temperature for a very long duration, e.g., in nuclear reactors; concrete walls withstand high temperatures for a very long period. The experimental study reported here was planned for such a situation. The mechanical behaviour of concrete specimens after exposure to flames for a particular elevated temperature for a constant period of time is reported in this paper.

EXPERIMENTAL PROGRAMME

Materials and Mix Proportion

To study the effects of water–cement ratio and mix proportion on fire resistance of concrete, three nominal mixes of concrete designated as mix A, mix B and mix C were adopted in this investigation. The details of the mixes are given in Table 1. Coarse aggregate of maximum size 20 mm and fineness modulus 7.12, river sand of fineness modulus 2.36 as fine aggregate and ordinary Portland cement were used in making concrete with mixes A, B and C. The 28 days compressive strength of different mixes are given in Table 2. To study the effect of elevated temperatures ranging from 200^0 C to 1000^0 C, cement concrete specimens were made with design mixes of M-25 and M-30 grades having characteristic compressive strength of 25 and 30 N/mm^2 respectively.

Preparation of Specimens

Concrete cubes of size 150 x 150 x 150 mm were cast and demoulded after 24 hours. Then these cubes were cured in water for 28 days. After 28 days of curing, the specimens were allowed to dry for 7 days in atmosphere before conducting fire test.

Testing of Specimens

Since, in the present study a criteria is to be established in selecting mix proportion and water-cement ratio for concrete exposed to fire, mixes A, B and C were subjected to a constant temperature of 700^0 C for a specified exposure durations of 10 minutes, 30 minutes and 1 hour. All the cubes of concrete were placed in the furnace and then they were slowly heated to the required temperature, which was maintained for the required exposure period.

The concrete specimens of M-25 and M-30 grades were exposed to constant temperatures of 200^0 C, 400^0 C, 600^0 C, 800^0 C and 1000^0 C for exposure durations of 1 hour, 3 hours and 5 hours. Initial weights of these cubes were noted. The cubes after fire test were allowed to cool in the atmosphere. They were weighed again for loss in weight. The cubes after exposure to flames were tested as per the standard procedure to get the compressive strength.

EXPERIMENTAL RESULTS AND DISCUSSION

Effect of W/C Ratio and Mix Proportion on Fire Resistance of Concrete Specimens

Figures 1-3 show the effect of water-cement ratio on the fire resistance of concrete specimens. The concrete specimens were exposed to 700^0 C for flame duration of 10, 30 and 60 minutes. It was observed that the estimated compressive strength decreases as the fire exposure time increased. It is also noted that there is an increase in loss of compressive strength of concrete with the increase in water-cement ratio. The percentage loss in compressive strength is calculated with reference to the strength of concrete in unfired concrete specimens at that age. It is seen from the Figures 1-3 that for the same water-cement ratio, percentage reduction in compressive strength is lower with mix A as compared to mixes B and C. Hence from Figures 1-3 it can be interpreted that leaner mixes appear to suffer a relatively lower loss of strength than richer ones.

Table 1   Details of concrete mixes

MIX DESIGNATION	MIX RATIO (BY WEIGHT)			WATER-CEMENT RATIO
	Cement : Fine aggregate : Coarse aggregate			
A	1	2	4	0.5, 0.6, 0.7
B	1	1.5	3	0.5, 0.6, 0.7
C	1	1	2	0.4, 0.45, 0.5

Table 2   Compressive strength of controlled concrete specimens

MIX DESIGNATION	WATER-CEMENT RATIO	COMPRESSIVE STRENGTH (N/mm^2)
A	0.5	27.8
	0.6	24.5
	0.7	18.8
B	0.5	29.0
	0.6	25.1
	0.7	20.5
C	0.4	34.5
	0.45	33.0
	0.5	29.5

Effect of Fire Exposure Temperature, Duration and Mix Proportion on Fire Resistance of Concrete Specimens

Figures 4 and 5 show the percentage loss in compressive strength of M-25 and M-30 grades of concrete exposed to elevated temperatures ranging from 200^0 C to 1000^0 C for three time intervals. It was observed that rich concrete mix has shown relatively higher percentage of strength loss as compared to leaner concrete mix. Residual strength of M-25 mix was 65% whereas it was 56% for M-30 concrete after these were subjected to elevated temperatures of 600^0 C for 5 hours. Difference in the observed percentage loss in compressive strength for two types of concrete subjected to elevated temperatures of 200^0 C to 400^0 C was only up to 5%. However, for temperature more than 600^0 C and above this difference was substantial. The reason for greater loss of strength in high strength concrete may be due to more internal stresses set up resulting in more reduction in concrete strength.

Effect of Fire Exposure on Weight Loss of Concrete Specimens

Concrete specimens were weighed before and after removal from furnace. Figures 6 and 7 show the percentage loss in weight of concrete specimens of grade M-25 and M-30 exposed to temperature ranges from 200^0 C to 1000^0 C. As the temperature is raised loss in weight of the concrete specimens increases in both mixes. The loss in weight was observed gradual up to a temperature of 600^0 C. But at a temperature of 600^0 C and beyond there was a significant drop in weight of concrete specimens in both the cases.

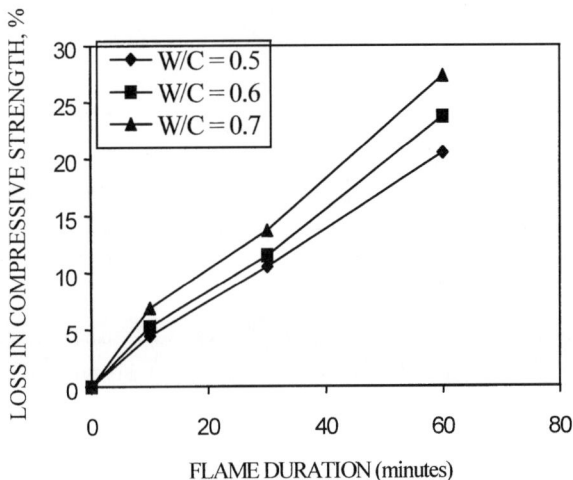

Figure 1   Effect of water-cement ratio on fire resistance of concrete (mix A)

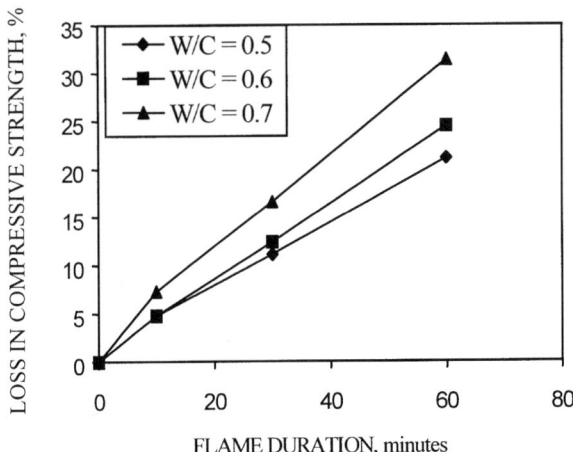

Figure 2   Effect of water-cement ratio on fire resistance of concrete (mix B)

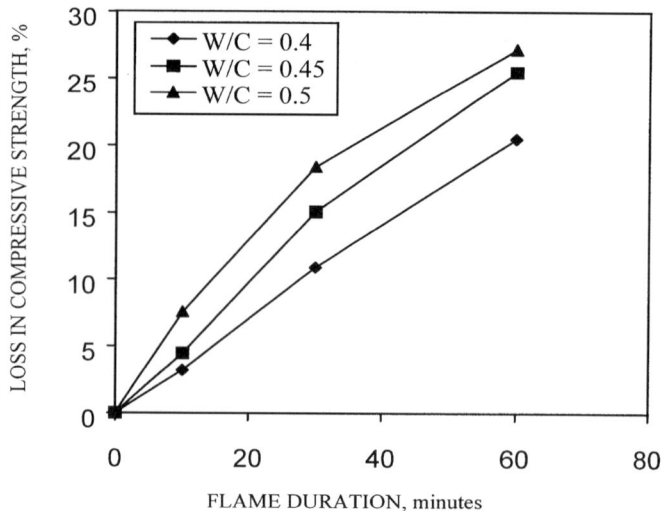

Figure 3   Effect of water-cement ratio on fire resistance of concrete (mix C)

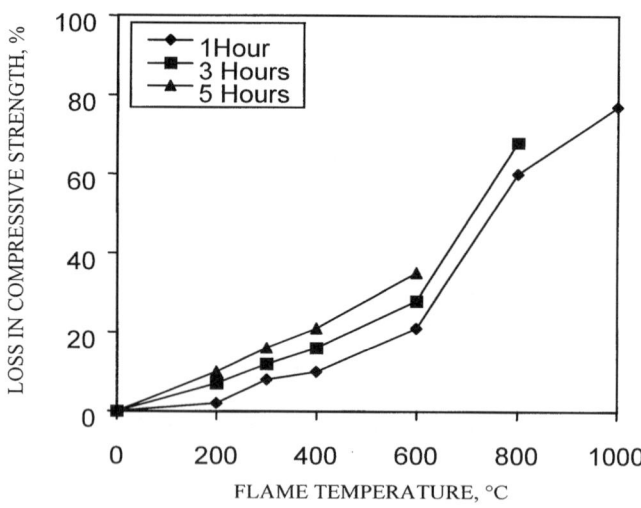

Figure 4   Effect of flame temperature on fire resistance of concrete (mix M-25)

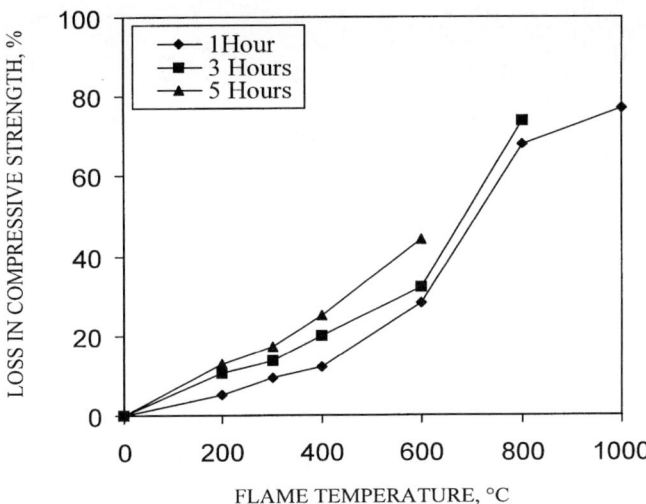

Figure 5   Effect of flame temperature on fire resistance of concrete (mix M-30)

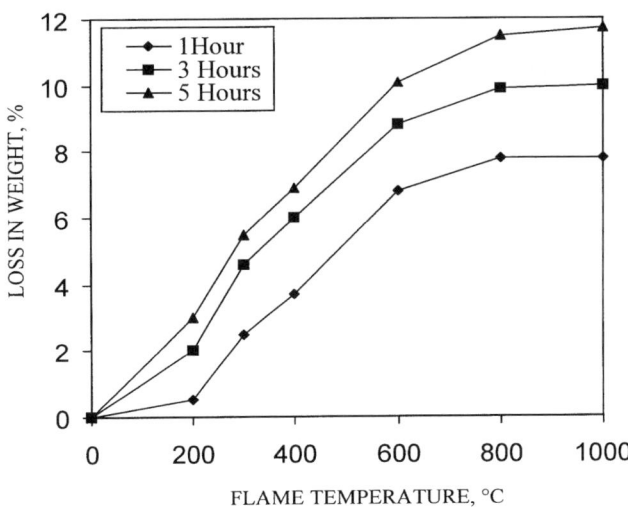

Figure 6   Effect of flame temperature on weight of concrete (mix M-25)

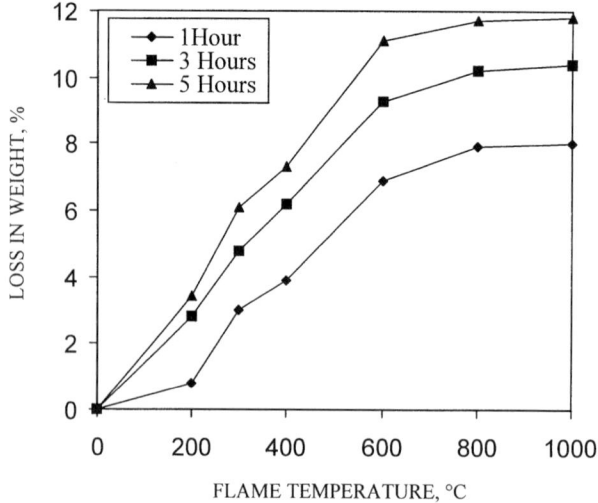

Figure 7 Effect of flame temperature on weight of concrete (mix M-30)

CONCLUSIONS

Based upon the results of the experimental study reported in this paper, the following conclusions are drawn.

1. The compressive strength of concrete decreases as the fire exposure time increases.

2. Cement concrete with the leaner mix indicated more residual strength as compared to richer mixes giving an indication that leaner mixes are relatively better fire resistant when subjected to elevated temperatures.

3. There was a gradual loss in weight of concrete specimens till a temperature of 600^0 C and thereafter a significant drop was observed in both the grades of concrete.

REFERENCES

1. NASSIF, A.Y., ET AL., A new quantitative method of assessing fire damage to concrete structures, Magazine of Concrete Research, 47 (172), 1995, pp 271-278.

2. INSTITUTION OF STRUCTURAL ENGINEERS, LONDON. Fire resistance of concrete structures, 1975.

3. INSTITUTION OF STRUCTURAL ENGINEERS, LONDON. Design and detailing of concrete structures for fire resistance, 1978.

4. MALHOTRA, H.L., Design of fire resisting structures, Surrey Univ. Press, London, 1982.

5.   DIAS, W.P.S., Some properties of hardened cement paste and reinforcing bars upon cooling from elevated temperatures, Fire and Materials, 16, 1992, pp 29-35.

6.   MOETAZ, M., EL-HAWARY., ET AL., Effects of fire on flexural behaviour of r. c. beams, Construction and Building Materials, 10 (2), 1996, pp 147-150.

7.   ROBERTO, F., GAMBAROVA, P.G., Effect of high temperature on the residual compressive strength of high strength siliceous concretes, A.C.I. Materials Journal, 85, 1998, pp 395-405.

8.   SAEMANN, J.C., WASHA, G.W., Variation of mortar and concrete with temperature, A.C.I. Materials Journal, 54, 1957, pp 385-395.

9.   HARDA, T., Research on fire proofing of concrete and reinforced concrete construction, Report, Tokyo University of Technology, Japan, 1961.

10.   THE CONCRETE SOCIETY, LONDON. Assessment and repair of fire-damaged concrete structures, Technical Report 33, 1990.

11.   LANKARD, D.R., ET AL., Effects of moisture content on the structural properties of Portland cement concrete exposed to temperatures up to 500^0 F, Temperature and Concrete, A.C.I. Sp. Publn. No. 25, 1971, pp 59-102.

12.   CONNOR, D.J.O., Structural engineering design for fire safety in building, The Structural Engineer, 73 (4), 1995, pp 53-58.

13.   GUSTAFERRO, A.H., LIN, T.D., Rational design of reinforced concrete members for fire resistance, Fire Safety Journal, 11, 1986, pp 85-98.

14.   HARMATHY, T.Z., Fire resistance versus flame spread resistance, Fire Technology, 12 (4), 1976, pp 290-302.

15.   LIE, T.T., Characteristic temperature curves for various fire severities, Fire Technology, 10 (4), 1974, pp 315-326.

16.   LIE, T.T., STANZAK, W.W., Structural steel and fire-more realistic analysis, Presented at the A.S.C.E. National Structural Engineering Meeting held in Cincinnati, Ohio in April 1974 (Pre-print 2256).

17.   MAGNUSSON, S.E., THELANDERSSON, S., Temperature-time curves for the complete process of fire development, Acta Polytechnica Scandinavica, Civil Engineering and Building Construction Series No. 65, Stockholm, Sweden, 1970.

18.   PETTERSSON, O., Theoretical design of fire exposed structures, Division of Structural Mechanics and Concrete Construction, Bulletin 51, Lund Institute of Technology, Lund, Sweden, 1976.

19.   ELLINGWOOD, B., SHAVER, J.R., Effects of fire on reinforced concrete members, Journal of Structural Division, A.S.C.E., 106 (11), 1980, pp 2151-2166.

MECHANISM OF DAMAGE FOR THE ALKALI-SILICA REACTION: RELATIONSHIPS BETWEEN SWELLING AND REACTION DEGREE

M J Riche   **M E Garcia-Diaz**

M D Bulteel   **M J M Siwak**

Ecole des Mines de Douai

C Vernet

Lafarge Company

France

ABSTRACT. In this study, a damage mechanism is proposed for the alkali-silica reaction. In a first time, a new chemical method have been developed for quantitative measurement of reaction degree in a "reactor model" without cement paste, and in a second time this method have been adapted on mortar bar. Two reaction steps are taken into account in the mechanism: formation of Q_3 sites made by breaking up siloxane bonds, and the dissolution of these Q_3 sites. It shows that the formation of Q_3 sites prevails over dissolution during the swelling step. A relationship is established between the reaction degree characterised by the quantity of Q_3 sites in the aggregate and its physical properties characterised by absolute density and specific pore volume. During the swelling step, the density is near-constant while the specific pore volume increases. The physical properties of the aggregate are used to value its expansion, and relationships are made between this expansion and the swelling of mortar bar. It enables to measure a restoration factor of the chemical swelling, which for a stiff and low porous matrix (ratio Water/Cement = 0.35) is high (about 3). These results show that it seems to be the siloxane bonds breaking up by ions OH⁻ of the pore's solution which causes a structural swelling of the siliceous aggregates. From this works, this methodology will be apply on concrete to perfect a diagnosis method.

Keywords: Alkali-silica reaction, Mortar, Reaction degrees, Silanol, Aggregate porosity, Swelling mechanism.

M J Riche is a PhD student in Civil engineering Department of the Ecole des Mines de Douai.

M E Garcia-Diaz is an Associate Professor in Civil Engineering Department of the Ecole des Mines de Douai. His research field concern chemical reaction in cement based materials.

M D Bulteel is a Research Engineer in Civil Engineering Ddepartment of the Ecole des Mines de Douai. His research field concern chemical reaction in cement based materials.

M J M Siwak is responsible for the Civil Engineering Department of the Ecole des Mines de Douai.

M C Vernet is a Research Senior Manager in Lafarge Company.

INTRODUCTION

The chemical reaction between certain forms of silica present in aggregate and the cement paste better known as the Alkali-Silica Reaction (ASR) entails damage in concrete structures. The ASR was largely studied, the mechanism was described using different models [1-4] and can be written following two main steps [5]:

- Formation of Q_3 sites (step 1), due to a first siloxane bonds breaking up by hydroxide ions attack :

$$2SiO_2 + OH^- \rightarrow SiO_{5/2}^- + SiO_{5/2}H \tag{1}$$

Here, on a structural point of view, SiO_2 represents Q_4 silicon tetrahedron sharing 4 oxygens with 4 neighbours, and using a simplified notation, $SiO_{5/2}^-$ represents the Q_3 sites negatively charged in a basic solution.

The neutralisation of these Q_3 sites follows the equilibrium:

$$SiO_{5/2}H + OH^- \Leftrightarrow SiO_{5/2}^- + H_2O \tag{2}$$

In contact with an alkaline solution the preponderant form is $SiO_{5/2}^-$. The Q_3 sites negatively charged are neutralised by Na^+, K^+ and Ca^{2+} cations.

- Dissolution of silica (step 2), due to continued hydroxide ions attack on the Q_3 sites to form silica ions $H_2SiO_4^{2-}$, $H_3SiO_4^-$ and small polymers.

$$SiO_{5/2}^- + OH^- + \frac{1}{2}H_2O \rightarrow H_2SiO_4^{2-} \tag{3}$$

Afterwards, precipitation of silica ions by the cations of the pore solution of concrete is liable to C-S-H and/or C-K-S-H phases formation.

Bulteel et al [5] have developed a chemical method to measure the reaction degree in a concrete sub-system involving the main ASR reagents: ground aggregate, $Ca(OH)_2$ and KOH. This method has enabled us to quantify two reaction degrees specific to ASR based on equation [2] et [3] :

- Q_3 sites content :

$$n* = \frac{\text{moles of } Q_3 \text{ sites}}{\text{moles of residual silica}} \tag{4}$$

- Dissolution degree :

$$\alpha = \frac{\text{moles of dissolved silica}}{\text{moles of initial silica}} \tag{5}$$

From these reactions, different theories have been proposed to account for the swelling mechanism induces by the ASR: the theory of imbibition pressure or osmotic pressure [6], the theory of ions diffusion [7], the theory of crystallization pressure [8], the theory of the gel dispersion [9], the theory of the electrical double-layer repulsion [10] and the theory of the swelling of porous body [11]. But none of them have really gained credence and could explain all the mechanisms and the experimental results.

The aim of this paper is to propose a damage mechanism for the Alkali-Silica Reaction based on the measurement of the chemical reaction degree and the physical characteristics of the aggregate. In a first time, we must adapt the chemical method developed by Bulteel et al [5] to measure the reaction degrees in micro-mortar. In a second time, we will evaluate the consequences of the ASR in terms of absolute density and specific pore volume of the aggregate. Finally, we will propose a model for the swelling based on the relationships between reaction degrees and the aggregate physical properties evolution.

CHARACTERISATION OF THE AGGREGATE

The material used is a flint aggregate from north of France. A complete characterisation has been given by Bulteel [12]. This aggregate is ground to obtain a [0.16-0.63 mm] size distribution. The X-ray fluorescence examination gives a composition close to 99% SiO_2 and this silica is constituted of Q_4 tetrahedra (SiO_2) and Q_3 tetrahedra ($SiO_{5/2}H$). The Q_3 mole fraction measured by thermogravimetry is close to 0.07.

The flint aggregate has a specific area close to $1 m^2/g$ and a specific pore volume of 0.0025 cm^3/g. Both are measured on a ASAP (Accelerated Surface Area and Porosimetry) analyser based on the physical adsorption of Nitrogen on a solid material. Thanks to the adsorption curve, we could value the specific area by the BET method (Brunauer, Emmet, Teller). The specific pore volume is obtained by the BJH method (Barrett, Joyner, Halenda) from the adsorption and desorption curves. The absolute density of the flint aggregate measured by Helium pycnometer analysis is about 2.595 g/cm^3.

DESCRIPTION OF MICRO-MORTAR BAR METHOD

The Micro-mortar bar method is based on a part of Microbar test: AFNOR P18-588 standard [13]. The formulation of the mortar is done with this characteristics: ratio Water/Cement = 0.35, ratio Cement/Aggregate = 2. To grow the reaction degree, the mixing water is replaced by KOH solution of 1.7 Mol/l concentration. After 24 hours hardening times, mortar bars are measured to determine their initial length and volume by hydrostatic thrust, before putting them into steam treatment during 4 hours at 100° C. After this treatment, Mortar bars are introduced in a closed stainless steel container on a stand with 20 ml of water. The container is then autoclaved during a given time at 80° C to accelerate ASR under controlled temperature. After each given time, mortar bars are placed at 20° C and 100% Relative Humidity to cool down during 4 hours before measuring their length and volume. In the same time, we measured the reaction degrees.

To quantify the reaction degrees, the aggregate must be extract from the micro-mortar by dissolution of the cement paste and reaction products of the ASR (C-S-H or C-K-S-H). For this, we used a selective treatment in 2 steps: a first acid treatment to remove cement paste and reaction products, in a second time a complexant solution treatment to remove still impurities (Fe_2O_3, MgO,). On this extracted aggregate, we can apply the chemical method developed by Bulteel et al. [5] which is an acid treatment to protonated the Q_3 sites in $SiO_{5/2}H$ forms. The efficiency of this treatment is controlled by X-ray fluorescence: the SiO_2 content of the remaining silica must be at least 99%.

The measurement of the Q_3 content is obtain by a thermal treatment of the residual silica at 1000° C, the silanols groups are condensed to give back silica Q_4 and release water following :

$$2SiO_{5/2}H \xrightarrow{1000°C} 2SiO_2 + H_2O \qquad (6)$$

The release water measurement by thermogravimetry allows the calculation of the quantity of Q_3 tetrahedra in the aggregate sample (n*[equation 4]).

The weight of the residual silica at given times allows to determine by difference with the residual silica after 24 hours hardening, the quantity of dissolved silica (α[equation 5]).

In the same time, the physical properties of the residual silica are measured before the thermal treatment by BET and BJH analysis for specific area and specific pore volume, and by helium pycnometer analysis for absolute density.

RESULTS AND DISCUSSION

Expansion of mortar bars

Expansion determined by lengthways measurement and hydrostatic thrust are given in Figure 1:

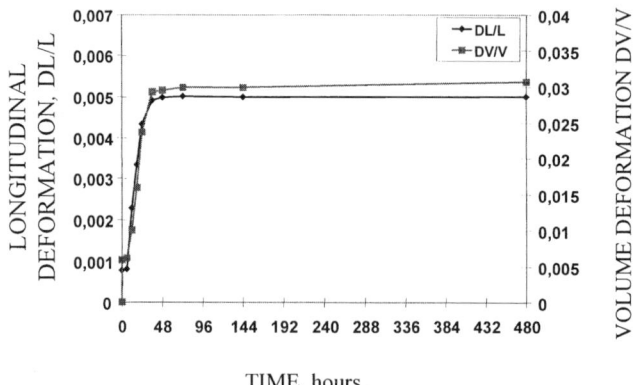

Figure 1   Expansion as function of time

The swelling step corresponds to the first 36 hours of reaction. After this, there is no more external expansion measured on mortar bars. The obtained swelling is 4 to 5 times higher than reactivity limits given by LCPC recommendations [14]. The measurement done after the steam treatment of 4 hours at 100° C explain the first expansion at t_0 before the beginning of the autoclaving step.

The reaction degrees

- Evolution of α :

The dissolution degree α according to time is given in Figure 2:

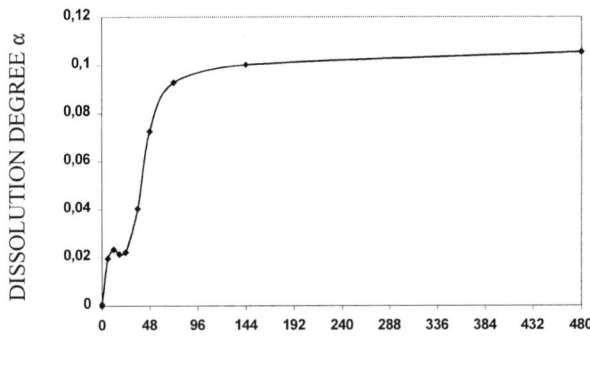

TIME, hours

Figure 2 α as function of time

During the steam treatment, we didn't observe dissolution whereas the swelling began. Then, the dissolution degree reach about 2% and still near-constant during the major part of the swelling step before increases to an asymptotic value about 10%.

Evolution of n* :

The evolution of n* according to time is given in Figure 3:

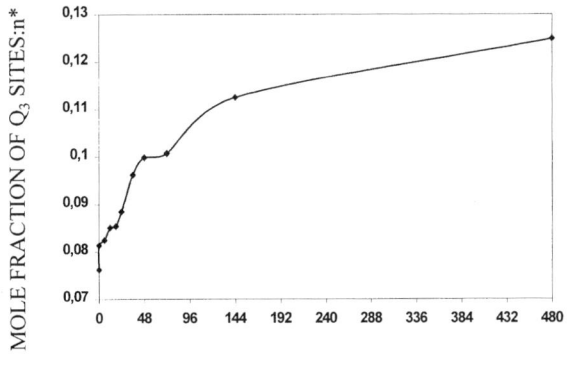

TIME, hours

Figure 3   n* as function of time

Contrary to the dissolution, the mole fraction of Q_3 sites in the aggregate increases from the beginning of the steam treatment and continues to increase during all the experiment. The swelling step corresponds to the increasing of n* from 0.076 to 0.096.

The evolution of reaction degrees shows that the formation of Q_3 sites prevails over dissolution during the swelling step. So, we will study the relationships between physical properties of this altered aggregate and its Q_3 sites content.

Relationships between reaction degree, physical properties of the aggregate and mortar swelling :

- Evolution of pore volume :

The specific pore volume V_{pore}(n*) of the aggregate measured by BJH analysis according to its mole fraction of Q_3 sites (n*) is given in Figure 4:

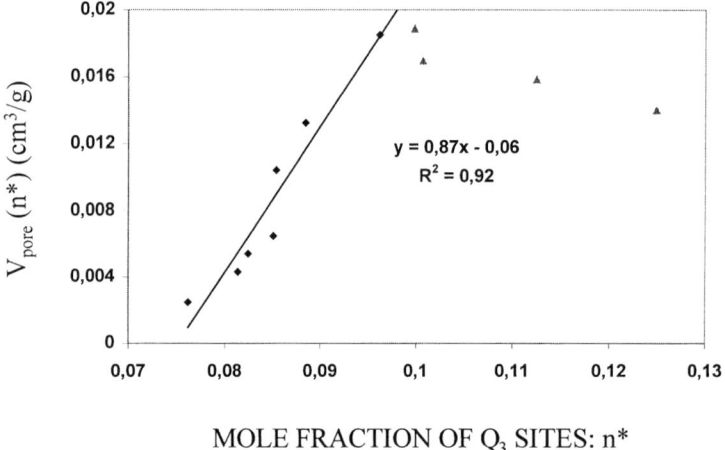

MOLE FRACTION OF Q_3 SITES: n*

Figure 4   Specific pore volume as function of n*

During the swelling step, we obtain a linear correlation between the increase in pore volume and the increase in Q_3 sites. Afterwards, when the swelling reach its asymptotic value, the specific pore volume decreases slowly.

- Evolution of absolute density :

The absolute density d_{abs}(n*) of the aggregate measured by Helium pycnometer according to its mole fraction of Q_3 sites is given in Figure 5:

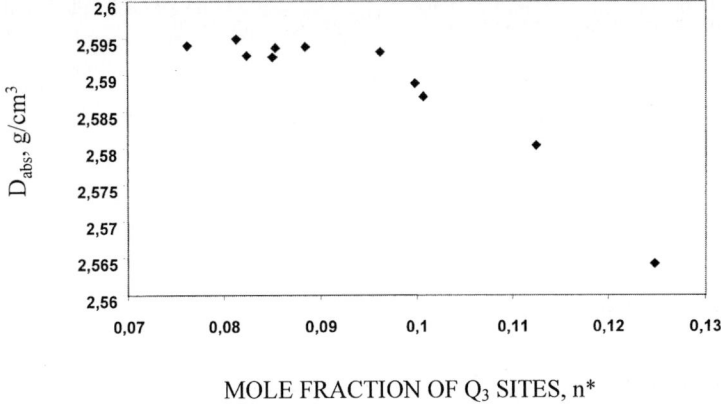

MOLE FRACTION OF Q$_3$ SITES, n*

Figure 5   Absolute density as function of n*

The absolute density of the aggregate is near-constant during the swelling step and finish to decrease for a Q$_3$ contents of 0.096.

- Evolution of mortar swelling :

The volume deformation of mortar bars according to the mole fraction of Q$_3$ sites is given in Figure 6:

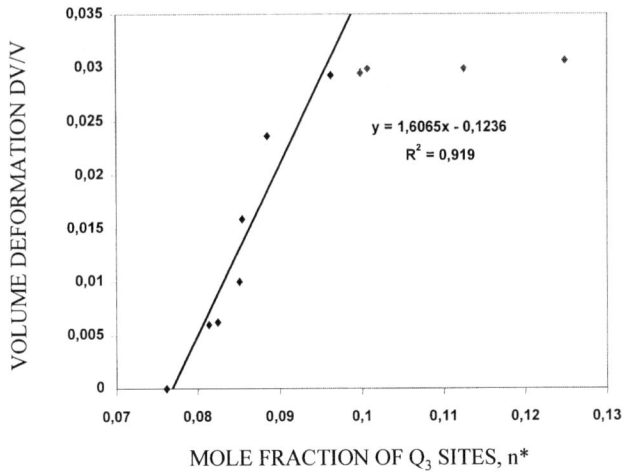

MOLE FRACTION OF Q$_3$ SITES, n*

Figure 6   Expansion measured as function of mole fraction of Q$_3$ sites

During the swelling step, we observe a correlation between the increase in Q_3 sites of the aggregate and the swelling of mortar bars. The increase of the specific pore volume is correlate to the Q_3 sites formation in the aggregate. So, the siloxane bonds breaking up induce a structural swelling in the aggregate which causes the mortar swelling. To validate this hypothesis we used the absolute density and the specific pore volume to value the expansion of the aggregate and we correlate it to the swelling of mortar bar.

Relationships between the expansion of the aggregate and the expansion of mortar bars:

-       Determination of the expansion of the aggregate : swelling model :

This model is based on the valuation of the apparent volume of the aggregate $V_g(n^*)$ given by the addition of its absolute volume (characterised by its absolute density $d_{abs}(n^*)$) and its specific pore volume $V_{pore}(n^*)$ characterised by BJH analysis.

For 1g of aggregate, its apparent volume as function of its mole fraction of Q_3 sites is following:

$$V_g(n^*) = \frac{1}{d_{abs.}(n^*)} + V_{pore}(n^*) \tag{5}$$

From this relation, we could value the expansion of the aggregate as function of its mole fraction of Q_3 sites:

$$\frac{DV_g}{V_g}(n^*) = \frac{\frac{1}{d_{abs.}(n^*)} + V_{pore}(n^*)}{\frac{1}{d_{abs.}(n_0^*)} + V_{pore}(n_0^*)} - 1 \tag{6}$$

From the expansion of the aggregate, the swelling of mortar bars could be value providing that the volumic fraction of aggregate in mortar bar is determined :

$$\left[\frac{DV}{V}\right]_{mortar\ bar} = \left[\frac{DV}{V}\right]_{aggregate} \times \%_{volumic} Aggregate \tag{7}$$

The expansion given by this model is liken to the expansion measured on mortar bar :

The correlation between the measured expansion and the calculated expansion shows that the swelling of the granular skeleton is liable for the mortar swelling. As the aggregate swelling is due to the siloxane bonds breaking up which induce an increase of Q_3 sites, we could contend that the ASR swelling has a structural part.

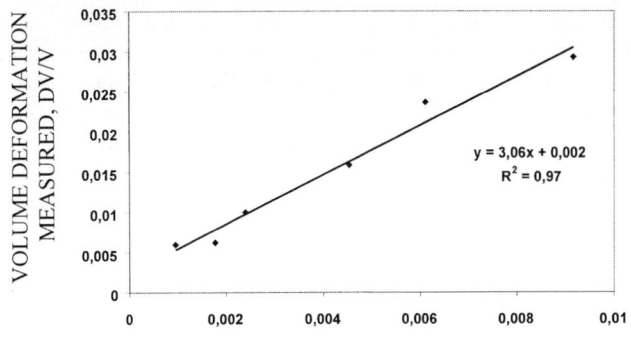

VOLUME DEFORMATION CALCULATED, DV/V$_{mortar\ bar}$

Figure 7   Expansion measured as function of expansion calculated

CONCLUSIONS

The aim of this work was to propose a swelling mechanism for the Alkali-Silica Reaction. We used a new chemical method for quantitative measurement of reaction degree in mortars. This method enables us to show that the siloxane bond breaking up to form Q_3 sites prevails over dissolution during the swelling step. The formation of Q_3 sites entails an increase of the specific pore volume and a few increase of the absolute volume of the aggregate. From the variation of the physical properties of the aggregate, we have developed a specific swelling model. The application of this model on mortar bars enables us to establish a correlation between the mortar swelling and the granular skeleton swelling. This results shows that it is the siloxane bond breaking up which is liable of the swelling of the aggregate, so we could affirm that the swelling mechanism has a structural part. The slope of the correlation between mortar and granular skeleton expansions gives the restore factor of the chemical swelling to the mortars. This restore factor is high (about 3) and express the stiffness of the low porous matrix which amplify the swelling. Now, we will try to adapt this methodology on concrete to improve diagnosis methods on concrete structures.

REFERENCES

1. DENT GLASSER, L S, AND KATAOKA, N, 1981. The chemistry of Alkali-Aggregate Reaction. *Proceedings of the 5th International Conference on Alkali-Aggregate Reaction*, Cape Town, 1981, Paper S252/23.

2. POOLE, A B, 1992. Alkali-Silice Reactivity mechanisms of gel formation and expansion. *Proceedings of the 9th International Conference on Alkali-Aggregate Reaction*, London (England), 1992, Vol 1, pp 782-789.

3. WANG, H, and GILLOT, J E, 1991. Mechanism of Alkali-Silica Reaction and significance of calcium hydroxide. *Cement and Concrete Research*, Vol 21, 1991, pp 647-654.

4. DRON, R, 1990. Thermodynamique de la réaction alcali-silice, *Bulletin de liaison des Laboratoires des Ponts et Chaussées*, No 166, 1990, pp 55-59.

5. BULTEEL, GARCIA-DIAZ, SIWAK, VERNET, ZANNI, 2000. Alkali-Aggregate Reaction : A method to quantify the reaction degree, 11th International Conference on Alkali-Aggregate Reaction in Concrete, Canada, Québec, pp 11-20.

6. DENT GLASSER L S, Osmotic pressure and the swelling of gels, Cement and Concrete Research, 1979, 9, pp 515-517.

7. CHATTERJI S, Mechanisms of alkali-silica reaction and expansion, 8th International Conference on Alkali-aggregate Reaction in Concrete, Kyoto, 1989b, London and New York, Elsevier Alied Science, pp 101-105.

8. DRON R, BRIVOT F, CHAUSSADENT T, Mécanisme de la réaction alcali-silice, Bulletin de liaison des Ponts et Chaussées, mars-avril 1998, 214, réf. 4175, pp 61-68.

9. JONES T N, A new interpretation of alkali-silica reaction and expansion mechanisms in concrete, Chemistry and Industry, 1988, pp 40-44.

10. PREZZI M, MONTEIRO J M, SPOSITO G, The alkali-silica reaction, Part 1 :Use of the double-layer theory to explain the behaviour of reaction-products gels, ACI Materials Journal, Technical Paper, jan-febr. 1997, 94.M2.

11. COUTY R, MARTIN S, DESSINGE M, Etude du gonflement des corps poreux en fonction des solutions diffusantes, rapport de recherche ESPCI, 1998.

12. BULTEEL, D, 2000, Quantification de la réaction alcali-silice : application à un silex du nord de la France, Thèse de doctorat : Université de Lille 1.

13. AFNOR N F P 18-588, 1991, Stabilité dimensionnelle en milieu alcalin (essai accéléré sur mortier MICROBAR) - Granulats.

14. LCPC, 1994, Recommandations pour la prévention des désordres dus à l'alcali-réaction, Laboratoire Central des Ponts et Chaussées.

RECONSTRUCTION OF RUNWAY 9R-27L AT HARTSFIELD ATLANTA INTERNATIONAL AIRPORT, THE 33 DAY WONDER – A CASE HISTORY

D Molloy
Hartsfield Atlanta
International Airport

S R Kuchikulla
R & D Testing &
Drilling Inc

F O Hayes
Aviation Consulting
Engineers Inc

A Saxena
Trinidad Engineering & Design Inc

D S Saxena
ASC Geosciences Inc

United States of America

ABSTRACT. The William B Hartsfield Atlanta International Airport (ATL) is the world's busiest airport handling some 80 million passengers annually and serves major international and domestic markets around the globe. The airfield itself consists of roughly 4,200,000 sq m (5,000,000 sq yds) of pavement and includes 4 parallel runways measuring from 2,740 to 3,620 m (9,000 to 11,889 ft) long. Runway 9R-27L was initially constructed over 27 years ago and has been in service well beyond its design life. Due to pavement deterioration accelerated by alkali-silica reactivity (ASR) and other distress, the owner decided to remove Runway 9R-27L in late 1999 under a fast-track 36 day schedule. A joint venture of major contractors bid $52 million to complete the project under this fast track approach. A conventional approach was estimated to cost between $15 and $20 million without factoring the impact of a closed runway at some $475,000 per day over a normal 6 month construction timeframe.

Keywords: Aviation, Runways, PCI, ASR, Quality cont rol testing, Flexural strength.

Mr Subash Reddy Kuchikulla is Quality Assurance Engineer for ATL Geotechnical and Materials Programs at R & D Testing & Drilling, Inc, 2366 Sylvan Road SW, Atlanta, Georgia, 30344, USA.

Mr Daniel Molloy, P E, is Assistant Aviation General Manager of Facilities at Hartsfield Atlanta International Airport, Department of Aviation, P O Box 20509, Atlanta, Georgia, 30320, USA.

Mr Frank O Hayes is Aviation Pavement Specialist with Aviation Consulting Engineers, Inc, 1400 Aviation Boulevard, Atlanta, Georgia, 30340, USA.

Mr Anupam Saxena, P E, M ASCE, is Technical Services Manager of Trinidad Engineering & Design, Inc, 2260 Godby Road, Atlanta, Georgia, 30349, USA.

Mr D S Saxena, P E, F ASCE, is Senior Principal and President of ASC geosciences, inc, 3055 Drane Field Road, Lakeland, Florida, 33813, USA.

BACKGROUND/FUTURE OF ATL

The William B Hartsfield Atlanta International Airport (ATL) is the world's busiest airport handling over 80 million passengers annually and serves major international and domestic markets around the globe. ATL, owned by the City of Atlanta, is located on 1,518 ha (3,750 ac) located some 16 km (10 mi) south of downtown Atlanta. ATL is home to the largest airline hubbing operations in the world, Delta Air Lines, and its Central Passenger Terminal Complex (CPTC) is situated on 53 ha (130 ac) with approximately 529,500 sq m (5.7 million sq ft) of facilities including 6 concourses with 168 gates served by a central, underground airport people mover system and pedestrian mall. Currently, some 34 passenger airlines and 23 all-cargo airlines serve HAIA. The airfield itself consists of roughly 4,200,000 sq m (5,000,000 sq yds) of pavement and includes 4 parallel runways measuring from 2,740 to 3,620 m (9,000 to 11,889 ft) long. These east/west runways handle over 2,000 daily takeoffs and landings. ATL has begun an intensive Capital Improvements Program totaling some $ 5.4 billion over the next 10 years. This CIP includes 4 major components: 1) a fifth runway; 2) an International Terminal; 3) a Consolidated Rental Car Facility; and, 4) a South Terminal.

ATL AIRFIELDS

History of Runways

In anticipation of considerable growth during the 1970s and 1980s, Runway 8R-26L was reconstructed in 1969 utilizing a 24/7 schedule over 40 days and nights. Construction was expedited in this fashion to reopen the runway as soon as possible. In 1972, 9R-27L was constructed, and in 1974 9L-27R was reconstructed. Runway 8L-26R was constructed in 1984, with east and west extensions added to 9L-27R.

Throughout the 1980s and well into the 1990s, several pavement evaluation studies were conducted to look more closely at the performance of ATL airfield pavements and address concerns about their structural condition, pavement surface distress, concrete characteristic strengths, and age.

Pavement Evaluation Programs

1997 Pavement Evaluation

An exhaustive pavement evaluation program was performed in 1997 due to concerns about the relative age of some of the runways. The 1997 study employed several investigative techniques, including:

- Falling Weight Deflectometer (FWD)
- Visual Survey
- Runway Ride Quality Measurement
- Field Delamination Testing including Impact Echo Testing and Ground Penetrating Radar
- Field Drilling and Sampling
- Laboratory Testing including Petrographic Examination

Results of the 1997 study indicated the Pavement Condition Index (PCI) of the runways ranged from 53 to 97 with most values in excess of 70. The predominate distress observed was widespread map cracking in each of the runways. In Runway 9R-27L in particular, the joint distress was manifested into localized spall features, which were increasing at an accelerating rate, as illustrated in Figure 1. It was noted that while the overall condition of Runway 8R-26L would appear lower than that of Runway 9R-27L, the results of PCI data indicate that 9R-27L was deteriorating at a faster rate than Runway 8R-26L, particularly with respect to the pavement keel or center section. Table 1 summarizes 1997 PCI Data for all ATL runways.

Table 1   Pavement Condition Index for ATL Runways, 1997

RUNWAY	PCI RATING	RANGE OF PCI VALUES	PROJECTED YEAR IN WHICH PCI BECOMES CRITICAL (= 50)
9R-27L	Fair to very good	42 to 80	1999
8R-26L	fair to good	42 to 79	2001
9L-27R	Good to excellent	57 to 99	2003
8L-26R	Very good to excellent	59 to 100	2010-2015

Pavement Condition Scale	
Excellent	85-100
Very Good	70-84
Good	55-69
Fair	40-54
Poor	25-39
Very Poor	10-24
Failed	0-9

The 1997 pavement evaluation study concluded that a pavement exhibiting a PCI value below 70 would require significant maintenance and one below 50 would require major rehabilitation or replacement.

1999 PCI Survey

Amongst growing concerns about the condition of Runway 9R-27L, a pavement evaluation study was again performed utilizing a heavy weight deflectometer (HWD) in lieu of the FWD. Results of the 1999 study indicated PCI values ranging from 25 to 79 with the center or keel portions of the runway again exhibiting significantly lower values. Figure 2 illustrates chronological trends of PCI for center pavement sections of the ATL runways.

Figure 1   Typical joint distress

Table 2   Summary of 1999 PCI Data for 9R-27L

SECTION	SAMPLE UNIT	LEFT (L) AB-LINE	CENTER(C) CD-LINE	RIGHT (R) EF-LINE
01	01	72	72	70
left section	02	76	67	62
	03	74	57	61
	04	79	46	58
	Average	*75*	*61*	*63*
02	05	79	32	74
keel section	06	76	45	74
	07	76	50	48
	Average	*77*	*42*	*65*
03	08	64	53	63
right section	09	25	44	65
	10	53	67	63
	11	45	70	61
	12	59	71	63
	Average	*49*	*61*	*63*

Deterioration from ASR

During both the 1997 Pavement Evaluation and 1999 PCI Survey, evidence of alkali-silica reactivity (ASR) was observed in concrete cores from all 4 runways, as illustrated in Figure 3.

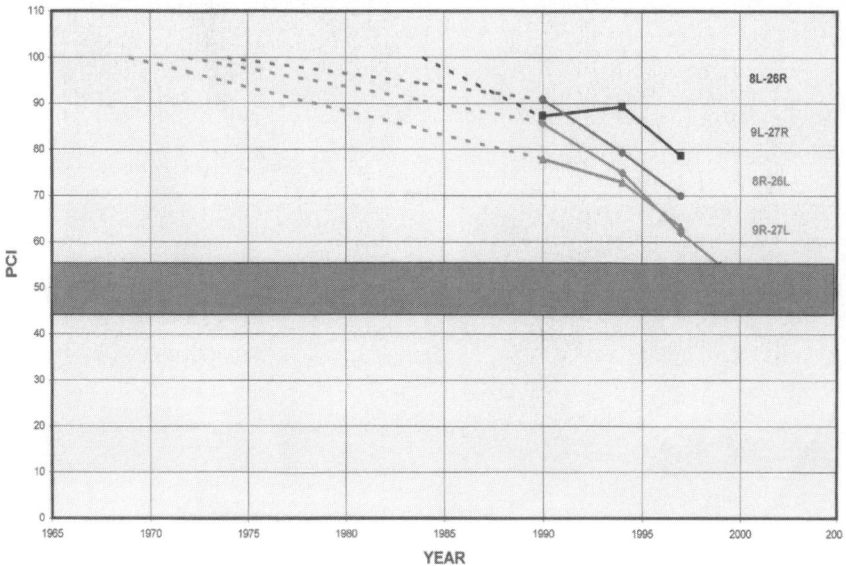

FIGURE 10 - CENTER SECTIONS - PCI

Figure 2   Trend comparisons for PCI values of all runways at ATL

The reactions appeared most severe in Runways 9R-27L and 8R-26L. Voids in the concrete matrix in these runways appeared nearly filled with the gel product of ASR. Furthermore, horizontal cracks within the top 15 cm (6 in) were observed in cores from Runway 9R-27L. The spalling and surface cracks observed in Runway 9R-27L were load-associated deterioration from ASR. It was further noted that the ASR appeared worse in concrete pavement using Portland cement and less severe in concrete using slag cement. Together, these observations in Runway 9R-27L indicated ASR had in fact contributed to the deterioration of the concrete pavement to a point where replacement of the runway was considered necessary.

RECONSTRUCTION OF RUNWAY 9R-27L

Overall Construction Approach

While replacement of Runway 9R-27L was conceptualized in early 1998, a typical reconstruction schedule of some 6 to 7 months was not acceptable because the runway handled one-fourth of the daily flight operations and was also only one of two CATIIIB certified runways that provided pilots with the best landing capability in severe weather. The project designer, Aviation Consulting Engineers, Inc (ACE) in association with the Department of Aviation is credited with the planning, design, and construction management of the Runway 9R-27L Reconstruction Program. An unprecedented planning effort began over 18 months in advance of construction and included close coordination and consultation with the City of Atlanta, major airlines, and FAA.

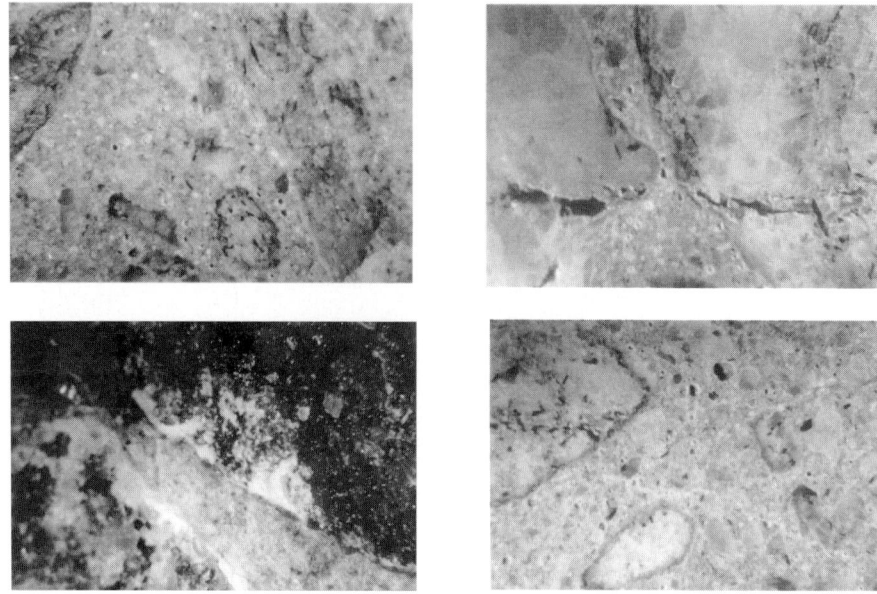

Figure 3   Indications of ASR Deterioration

A joint venture of major contractors included APAC Ballenger, C W Matthews Contracting, Swing Construction, and Mitchell Construction. Together, this team was selected to remove and replace the 2,740-m by 46-m (9,000-ft by 150-ft) wide Runway 9R-27L, over 167,000 sq m (200,000 sq yd) of 400- to 560-mm (16- to 22-in.) thick concrete pavement in only 36 days.

During this period, normal operations would be transferred to Taxiway R, just south of Runway 9R-27L, that was readied to serve as a temporary runway during this period. This fast-track alternative approach was bid at $52 million.   A conventional approach was estimated to cost between $15 to $20 million, without factoring the impact of a closed runway at some $475,000 per day over a normal 6 month construction time frame.   So actually, the fast-track bid was significantly less expensive to the owner. Because time was so critical, penalties for not meeting the construction schedule ran as high as $200 per minute.

Removal of Existing Pavements

A critical phase of the construction project was the removal of some 21,150 concrete panels, each measuring 2.3 m by 3.7 m (7.5 ft by 12 ft) by 406 mm to 559 mm (16 to 22 in) thick. A specialty contractor, Penhall Company, was chosen for this task based on their experience in California with removal of earthquake-damaged bridges and interstates following major California earthquakes. Penhall mobilized multiple teams of operators and equipment from across the country to complete the task. Once the existing concrete slabs were cut, unique pavement removal buckets were utilized to carry away a maximum of three 9,150-kg (20,150-lb) panels onto each flat bed truck for transport to a 5.7-ha (14-ac) stockpile yard accommodating some 1,900,000 kN (213,000 tons) of panels stacked on top of one another, sometimes as high as 13 panels (refer to Figure 4).

Figure 4   Removal of Existing Pavements

Concrete Mix Design and Construction Operations

A total of 13 trial concrete mix design were formulated. Flexural strength data from this mix design phase indicated that a target 7-day strength of 4,130 kPa (600 psi) as well as a target 28-day strength of 4,480 kPa (650 psi) could easily be achieved. The newly-constructed runway was designed to have an average thickness of 46 cm (18 in) and a design life cycle of well over 20 years. The resource needs for this project were significant, including: i) 1,200+ workers working 24 hours per day, 7 days per week at the height of construction; ii) 200 to 300 truck loads of materials per day; and, iii) three on-site concrete batch plants (refer to Figure 5), plus an existing off-site commercial batch plant as backup, each capable of producing as much as 7.7 cu m (10 cu yds) per min.

Phase I operations for the project focused on a temporary modification of an adjacent taxiway which would serve as a replacement runway during construction. This phase lasted 70 days and included procurement and mobilization. Phase II comprised the actual runway reconstruction for a planned 36 day timeframe; this phase was actually completed in a record 33 days. Phase III was scheduled for 30 days to convert the temporary runway back to a taxiway, to remove runway lighting and install taxiway lighting, convert runway striping to taxiway markings, and remove temporary exits.

Figure 5   Batching and Placement of Concrete

QUALITY CONTROL TESTING

Quality Assurance testing for the owner and Quality Control testing for the contractor were performed during several key phases of construction, including subgrade preparation, installation of a soil-cement base, bituminous pavement placement and, of course, concrete pavement placement.   In accordance with FAA P-501 specifications, testing during construction included casting beams for flexural strength. Between QA and QC functions, over 2,000 beams were cast. Target design flexural strengths were 4,130 kPa (600 psi) at 7 days and 4,480 kPa (650 psi) at 28 days when tested in accordance with ASTM C 78. Tables 3 and 4 summarize QC strength data.

Table 3   Summary of Quality Control Flexural Strength Data

TEST AT DAYS	N, SAMPLE SIZE	RANGE kPa/psi	AVERAGE kPa/psi	S D kPa/psi
1	166	1,930 – 4,892	3,197	544
		280 – 710	464	79
3	166	2,825 – 5,994	611	599
		410 – 870	4,210	87
7	318	3,514 – 6,201	665	482
		510 – 900	4,582	70
28	349	4,823 – 7,648	811	530
		700 – 1,110	5,588	77

Table 4   Summary of Quality Control Compressive Strength Data

N, SAMPLE SIZE	RANGE kPa/psi	AVERAGE kPa/psi	S D kPa/psi
56	25,011 – 41,340	31,350	3,569
	3,630 – 6,000	4,550	518

Results of the strength testing for flexural strength on beams indicated that the ultimate 28-day target strength of 4,480 kPa (650 psi) was generally achieved between 3 days and 7 days.

ACKNOWLEDGEMENTS

The authors acknowledge the cooperation of the entire design and construction team which was essential to the success of the 9R-27L Reconstruction project. Planning, design, and construction management were performed by Aviation Consulting Engineers, Inc (ACE) in association with the Department of Aviation on behalf of the owner, City of Atlanta. A plaque was placed on the completed runway project to recognize and honor Mr Frank Hayes with ACE for over 25 years of engineering service at ATL. The construction team was led by a joint venture of 4 firms, including APAC Ballenger, C W Matthews Contracting, Swing Construction, and Mitchell Construction, all from Georgia, USA.

DURABILITY OF CONCRETE FOR TRANSPORT CONSTRUCTION

L A Fedner   S N Efimov   A B Samohvalov

Scientific and Technical Research Laboratory

J V Nikiforov

Moscow Automobile and Highway Construction

Russia

ABSTRACT. The paper discusses requirements to the concrete for transportation construction. The main attention is paid to freeze/thaw resistance. The methods of augmentation of concrete freeze/thaw resistance and related requirements are described. Chemical and mineral composition and fineness of cement play the important role in the provision of the concrete durability. Mineral admixtures, except blast-furnace granulated slag under certain conditions, and high containment of C_3A mineral have a negative influence on freeze/thaw resistance. The mix proportioning and characteristics of durable concretes for transportation construction are shown. The results given here and practice have ascertained that by the application of complex super-plasticizing and air entraining admixtures and cements with normative mineral composition it is possible to obtain high strength and frost resistant concretes.

Keywords: Transportation construction, Durability, Freeze/thaw resistance, Chemical and mineral composition of cement, Additives, Engineering properties.

Professor L A Fedner, is Scientific Director of Scientific and Technical Research Laboratory "CEMENT" of Moscow Automobile and Highway Construction Institute (Technical University). He specialises in the problems of interrelation between the properties of cement and concrete and of the durability of concrete.

Dr J V Nikiforov, is Chief Editor of "Cement and Its Applications" magazine and also Research Fellow of Moscow Automobile and Highway Construction Institute (Technical University). He specialises in technology of binder materials and products based on them and also works in the field of the usage of by-products in cement industry. He is the author of more than 200 scientific works and the participator of many International Congresses and Conferences.

Dr S N Efimov, is Research Fellow in Scientific and Technical Research Laboratory "CEMENT". His interests include the problems of concrete technology, especially the problems connected with the usage of concrete in transportation constriction and the effective application of chemical admixtures.

Dr A B Samohvalov, is Research Fellow in Scientific and Technical Research Laboratory "CEMENT", interested in technology, strength and freeze/thaw resistance of concrete.

INTRODUCTION

The advantage of concrete is its resistance to changes of physico-mechanical properties, among them deformation over a wide range of environmental and loading conditions. In addition to that, compressive and bending strength of concrete can increase during all period of service, which is very important considering constantly increasing loading. As a rule, the possible destruction of concrete could take place as a result of aggressive action of outward medium. On the whole, the durability of concrete for transportation construction is determined by its construction and service properties, its texture and structure, which should fit the service conditions.

REQUIREMENTS OF CONCRETE FOR TRANSPORTATION STRUCTURES AND FACTORS INFLUENCING ITS DURABILITY

The basic requirements to the concrete for transportation structures (e.g. road and airport pavements, bridges, viaducts etc.) are strength, strain capabilities and durability in severe working conditions. Concrete should remain durable in spite of alternating cycles of moistening and drying, freezing and thawing, corrosion due to anti-icing salts.

National codes and standards in Russia have following requirements to concrete for transportation construction:

— W/C ratio in concrete mixtures for single-layer or upper layer of two-layer road and airdrome pavements should not exceed 0.5; for steel-concrete bridge structures with frost-resistance index F300 (300 cycles of freezing and thawing in salted water in testing conditions) – not more than 0.45; in practice for concreting of such structures W/C usually ranges from 0.35 to 0.40, which is connected with the usage of plasticizing and complex chemical admixtures.

— The volume of entrained air in concrete mixtures should be in the following range: 5 to 7% for the concrete of road and airdrome pavements and 2 to 5 % for the concrete of bridge structures.

— The minimum design classes of concrete, in dependence on its assignation, type and service conditions of the structure, for road and airdrome pavements are: for the bending strength – $B_{tb}2.4$ to $B_{tb}4.0$; for compression strength – B15 to B30; for bridge structures: for compression strength – B 25 to B 45 (class indicates guaranteed strength in N/mm^2).

— Design frost resistance indexes of concretes, depending on climate, assignation of concrete, disposition and type of the structure, for road and airdrome pavements are – F50 to F300; for bridges – F100 to F500.

The most effective method of enhancing frost resistance of concrete, especially in the case of joint influence of cold and deicing agents, is to create in the structure of the cement stone the system of air conventionally closed pores with the size of 50×10^{-3} to 250×10^{-3} mm removed to 200×10^{-3} to 250×10^{-3} mm one from other – "distance factor" [1, 2]. The enhancement of frost resistance is also promoted by the limitation of the water-cement ratio (W/C) in order to decrease the capillary porosity of concrete and compact its structural component – cement stone.

The important requirement for concrete in road and airdrome pavements is resistance to severe reiterating cycles of transportation and temperature stresses causing the fatigue of concrete. In such case durability of concrete is characterized by its fatigue bending strength. The bending strength is one of the basic design characteristics for concrete pavements. The air entraining admixtures slightly reduced the strength of concrete. According to [3] the bending strength of concrete is:

$$R_{tb}=0{,}39R_{cb}(C/W-0{,}1)\times(1-0{,}025V_a) ,$$

Where: R_{cb} = the bending strength of cement,
              C/W = cement-water ratio,
              V_a = volume of the entrained air, %.

Our data indicate that the increase in the entrained air volume by 1 % causes the decrease in concrete compression strength by 6 to 8 % and bending strength up to 3 to 5 %.

The increase of entrained air volume and decrease of water containment are achieved by doping of the chemical admixtures in the water. The prevailing admixtures used to provide required air entrainment in Russia are agents, based on processed by-products of wood-working industry; super-plasticizers of naphthalene-formaldehyde type have recommended themselves as good water-reduction admixtures.

Road and airdrome pavements, bridge slabs made from cast-in-situ concrete have large surface. In connection with this such technological operation as fresh concrete treatment aimed to prevent water evaporation and provide fully hydration becomes particularly important. The treatment is managed by different film-formation agents and other means.

The choice of cement for the concrete for transportation construction deserves the special attention. Effectiveness of the usage of cements in concretes depends on many factors and consists in not only that or other dosage of cement in concrete, but in providing stable work of cement stone in different working conditions, namely its durability. The chemical composition of cement has grate importance for the freeze-thaw resistance of concrete. It is known that all mineral additives in more or less degree decrease the freeze-thaw resistance.

Nevertheless, according to our data, the frost resistance of steamed concrete increases if, other things being equal, it is made on fine cement with slag additive. In this case the considerable improvement of pore structure of cement stone is achieved, the basicity of hydrosilicate phases decreases, which contributes to the rise of durability.

For example, steamed concretes on fine slag Portland cement with blast-furnace slag containment of 60-70 % and fineness of 450-500 m^2/kg (according to the adopted in Russia method of fineness evaluation by the air permeability of cement powder, which indicates less values than Blein method) proved to be highly frost resistant (more than 700 cycles) in natural conditions of the White Sea. Laboratory researches and the practice of transportation construction from cast-in-situ concrete proved the possibility of effective usage of cements with up to 15 % additive of blast-furnace slag. The possibility of the usage of fine multi-component cements with additives of sand, carbonate rock and some other in transportation construction needs further investigation.

The influence of the three-calcium aluminate to the frost and frost-salt resistance of concrete is widely discussed in literature. According to [4] for the concretes without air entrainment admixtures there is no statistically trustworthy interrelation between the mass loss during freeze-thaw resistance test and the containment of C_3A. Yet the majority of researchers (Shestopiorov S.V., Satalkin A.V., Ivanov F.M., Gorchakov G.I., Kuntcevitch O.V., Powers T., Valenta M. and others) have advanced the synonymous opinion about negative influence of C_3A on the synthesis of frost resistant structure of cement stone.

According to active Russian standards for road and airdrome pavements and bridge structures it is necessary to use the cement made from the clinker of regulated composition with not more than 8 % C_3A mass containment. As addition it is possible to use only blast-furnace slag in not more than 15 % quantity, and the specific surface of cement should be not less than 280 m^2/kg (according to the method adopted in Russia). The cement setting should begin not earlier than 2 hours.

In addition to that cements for transportation construction should meet some other requirements. It is necessary to regulate the bending strength class (index) of road cement and also take into consideration the kinetic of strength growth along with strength in the age of 1 day. It is connected with the fact that the temperature-shrinkage joints cutting can be fulfilled only after the compressive strength of the concrete would achieve 10 N/mm^2. The requirements to frost resistance for structures with large surface are rather severe, so segregation (bleeding) of the fresh mix, which is peculiar to them, should be controlled by putting proper requirements to the properties of cement. Also it is expedient to put some additional requirements to C_3A content in accordance with the service conditions of concrete: for example, the upper level of C_3A content in clinker for cement used for surface layers of road or airdrome pavements should not exceed 7 %, but for basements it is possible to admit 8 %.

EXPERIMENTAL DETAILS

The solution to problems of the effective usage of different cements for the creation of durable concretes lays in systematic study of construction and service properties of the latter in dependence on the properties of the former. The typical characteristics of cements used in our study are given in Table 1. All of them are from Russian manufacturers and destined for transportation construction.

Table 1   Properties of cements

CEMENT	COMPOSITION, % by mass				WATER DEMAND %	*SPECIFIC SURFACE, m^2/kg	SETTING TIME, hours-minutes	
	C_3S	C_2S	C_3A	C_4AF			Initial	Final
1	64.8	14.0	6.0	14.1	23.5	390	2-50	3-55
2	59.7	14.2	7.8	13.0	25.0	350	2-30	3-30
3	66.1	9.5	3.4	17.5	24.75	320	2-40	3-45
4	55.0	18.0	7.0	12.0	25.5	360	2-40	3-35
5	51.0	27.0	5.5	13.4	24.25	380	2-45	4-35

* according to Russia testing procedure (ГОСТ 310.2-76)

The materials, used for concretes beside cement, were:

– aggregates: natural medium grade sand with 2,3 size modulus and 5-20 mm size granite crushed stone;

– chemical admixtures: naphthalene-formaldehyde super-plasticizer and air entraining agents, widely used for transportation construction.

A typical set of mix proportions, fresh properties of mixes, kinetic of the strength growth for different conditions (normal or steaming by the 3+3+8+4 hours regime with the temperature of isothermic heating 60°C) are given in Table 2.

Table 2   Mix proportions, properties of fresh concrete and development of strength

MIX N°			1	2	3	4	5	6	7	8	9	10
CEMENT (See Table 1)			1	1	2	2	3	3	4	4	5	5
MIX PROPORTIONS, kg/m^3	PC		350	415	330	365	415	370	335	365	325	460
	Sand		765	705	800	800	695	730	705	780	760	645
	Coarse aggregate		1105	1085	1110	1120	1090	1085	1170	1110	1065	1090
	Water		170	160	145	155	160	170	160	165	150	165
	W/C		0.49	0.39	0.44	0.42	0.39	0.46	0.48	0.45	0.46	0.36
ADMIXTURE, % of cement by mass	Super-plasticizer		0.6	0.8	0.6	0.6	0.8	0.8	0.8	0.6	0.8	0.6
	Air-entraining agent		0.03	0.03	0.03	0.02	0.02	0.02	0.02	0.03	0.02	0.02
SLUMP, mm,	5 minutes		110	70	45	60	50	45	50	70	50	40
	30 minutes		40	60	15	15	40	25	20	25	30	30
	60 minutes		30	55	5	5	20	15	10	20	10	15
ENTRAINED AIR, % of the volume of the mix	5 minutes		7.5	8.0	4.6	3.7	7.6	6.2	6.3	5.0	6.3	6.2
	30 minutes		6.0	6.5	4.5	3.0	6.2	5.9	5.7	3.9	6.0	5.9
	60 minutes		5.3	5.6	4.3	2.4	5.8	5.7	4.9	3.2	4.9	4.8
COMPRESSIVE STRENGTH, N/mm^2, days	Normal	3	23.0	35.8	29.3	40.9	31.0	20.0	29.8	33.5	20.0	27.4
		7	35.1	44.5	43.9	52.6	45.1	33.4	39.8	49.0	32.0	54.0
		28	44.0	53.0	56.7	68.0	54.8	42.8	56.1	61.6	42.6	62.7
	Steam	1	30.5	37.5	33.5	39.8	32.7	24.9	34.1	42.0	21.2	40.5
		28	42.5	52.5	51.0	60.0	46.9	37.6	49.5	58.1	35.6	56.3
BENDING STRENGTH, N/mm^2, days	Normal	7	3.3	3.8	3.9	4.8	4.0	3.1	5.1	5.5	3.5	4.3
		28	4.6	5.5	5.9	6.7	5.6	4.3	7.0	7.8	4.3	6.6
	Steam	28	4.4	5.4	5.4	6.1	4.8	4.1	6.6	7.2	3.9	6.1

During the placing of the concrete by slipping forming devices for road and airdrome construction the workability of mixture is fixed within the limits of 20-40 mm slump on the site; for the bridge construction the slump usually is 80-120 mm. As the time of transportation of the concrete mixture to the site in airdrome construction don't exceed, as a rule, 30-40 minutes, then for road and bridge construction this time is 1-1,5 hour and more. This is connected with changes of rheological properties of the mixtures and the decrease of entrained air volume.

DISCUSSION AND CONCLUSIONS

The data given in Table 2 show that it is possible to obtain the concrete mixtures with proper properties on all studied cements.

The strength kinetic of the concretes used in study is:

- For the normal conditions the compression strength in the age of 3 days equals 44 to 68 % of the strength in the age of 28 days;
- For the normal conditions in the age of 7 days the compression strength equals 71 to 86 % and the bending strength – 65 to 81 % of the corresponding strength in the age of 28 days;
- The compression strength on the 1 day after steaming equals 60 to 72 % of the strength in the age of 28 days.

The compression strength to bending strength ratio, which is indirect characteristic of the crack resistance, equals: for normal conditions – 8 to 10; after steaming – 7.5 to 10.

In order to obtain high quality concrete for transportation construction it is necessary to optimize the composition of concrete and the dosages of admixtures for every particular cement taking into consideration the designation and service conditions of the structure. So, in examples № 1 and 2 from Table 2 the 0.03% dosage of air entraining agent proved to be sufficient, but in example № 8 – insufficient to provide air entrainment, required for upper layers of road and airdrome pavements.

The usage of super-plasticizer of naphthalene-formaldehyde type in the dosage of 0.8 % of cement mass slows down the setting of concrete mix, especially in the first 30 minutes after mixing. The entrained air volume in 30 and 60 minutes after mixing decreases by 13 and 24 % from the first value.

The usage of studied cements makes possible to obtain concretes for transportation construction with compression strength classes B30 to B45 and bending strength classes – B_{tb} 4,0 to B_{tb} 6,0 (mixes 2, 3, 4, 5, 7, 8 from Table 2).

The porosity characteristics of concretes (mixes 1, 2, 5, 6 from Table 2) are following: total porosity lays in the limits 13.8-18 %, open capillary porosity – 8.7-12.5 %, conventionally closed porosity – 4.2-6.0 %.

The samples-cubes of concretes were tested on frost resistance. The conditions of testing: freezing to the temperature of -50±5°C and thawing on the temperature of +18±2°C in the 5% water solution of NaCl.

All series of samples endured more than 20 cycles without strength lessening or mass loss, which matches to frost resistance index F200 and more (according to the basic testing procedure).

The water-tightness indexes of concretes are greater than W6 (0,6 N/mm^2).

The values of elasticity modulus for 0.3 of compression strength stress level are 2.98×10^4 to 4.69×10^4 N/mm^2; the shrinkage deformations in the age of 180 days are 0.35 to 0.45 mm/m; creep characteristic - $\varphi = 1.2$ to 1.8.

So, the usage of cements with regulated chemical and mineral composure along with complex admixture (super-plasticizer and air entraining agent) and observance to optimal hardening conditions, including the steaming regimes, give opportunity to obtain concretes, which meet the requirements laid on the concretes to road, airdrome, bridge and other branches of transportation construction. Directed modification of the structure permits to create concretes with high strength and durability, which are commonly called "high quality concretes".

REFERENCES

1. MIELENZ R, WOLKODOFF V, BACKSTROM J, BURROWS R. Origin, evolution and effects of the air void system in concrete, Journal of Concrete Institute, 1958, Vol. 30, № 1, pp 18-22.

2. BLUMEL O, SCHPRINGENSCHMID P. Grundladen und Practis der Herstellung und Uberwachung von Luftporenbeton, Strasse und Tiefban, 1970, Vol. 24, № 2, pp 2-6.

3. SHEININ, A.M., Concrete for road and airport pavements, "Transport", Moscow, 1991, 150 pp. [in Russian].

4. STARK, J., Interrelation between the hydration of cement and durability of concrete, Cement, 1999, Special Issue "I (9) International congress on the chemistry and technology of cement - Moscow", pp. 39-45 [in Russian].

ULTRA THIN WHITETOPPINGS:
A 2-D FINITE ELEMENT APPROACH

W De Corte

D De Leersnyder

E De Winne

Ghent University

Belgium

ABSTRACT. Ultra Thin Whitetopping (UTW) is a relatively new technique for resurfacing deteriorated asphalt pavements. Very thin concrete slabs are placed on old asphalt layers to form bonded composite pavements. The very small thickness of the concrete slab is justified by the use of close joint spacings, bond between the concrete and the existing pavement and the use of high quality concrete. The technique has many advantages such as absence of rutting, quick opening to traffic, a competitive cost, etc.. On the other hand UTW is a relatively new technique, certainly in Belgium and Europe and many features are as yet not known. Although UTW is used in practice, a definitive design method is still to be developed and agreed upon. The paper describes the results of a two dimensional finite element calculation for this type of pavement. The results confirm that the bonding effect between the two top layers, the slab length and the relative and absolute thickness of both layers. Concluding it can be stated that, although not every aspect is clear, UTW has a promising future. Maybe the results shown in this paper may contribute to the further development of this technique.

Keywords: Ultra thin whitetopping, Inlay technique, Finite element calculation.

W M L De Corte is civil engineer and working as a researcher at Ghent University, Department of Roads and Bridges. His research focuses mainly on bridge deck technology and field measurements on large infrastructure systems.

D De Leersnyder is civil engineer and has performed research in the field of Ultra Thin Whitetopping as part of his Masters Thesis.

E De Winne is civil engineer and Professor at Ghent University, as well as General Inspector at the Flemish Road Administration. His research focuses mainly on pavement structures and mobility related problems.

INTRODUCTION

For many years, concrete overlay on asphalt pavement is known as whitetopping. During the past ten years thinner overlays are being studied. These have been commonly referred to as ultra-thin whitetoppings (UTW). This composite pavement provides a durable, rut-resistant surface for municipal streets, urban intersections, country roads and similar applications. The technique has some special characteristics as there are :

• Thicknesses for the UTW range from 50 to 100 mm (2 to 4 in.).

• The design is based on a bond between the underlying asphalt layer and the new concrete layer. A milled asphalt surface is necessary.

• Apart from the small thickness, UTW uses very small pavement joint spacing. In this way, it tries to reduce bending stresses.

Based on examples in the USA and the studies in laboratory conditions, this technique has been developed and studied. In this paper, study is performed into the composite behaviour of this concrete overlay. The performance consequences of the characteristics given in the previous paragraph are investigated. In the following, a short description is given of the theory on Ultra-Thin Whitetopping.

The consequence of bonding the 2 layers is that a composite pavement is created. Because asphalt and concrete work together, a slab is composed by a part of concrete and a sublayer asphalt. Therefore, the neutral axis shifts downwards and reduces the tensile stresses, in the middle of a concrete slab. Lowering the neutral axis can have just the opposite effect on a corner of a slab. When a wheel load is put on the end of a slab, tensile stress can appear on the surface. A lower neutral axis creates a larger tensile stress in the concrete. This is shown in Figure 1. This might be the reason for corner cracking which is often a problem with UTW.[1,2]

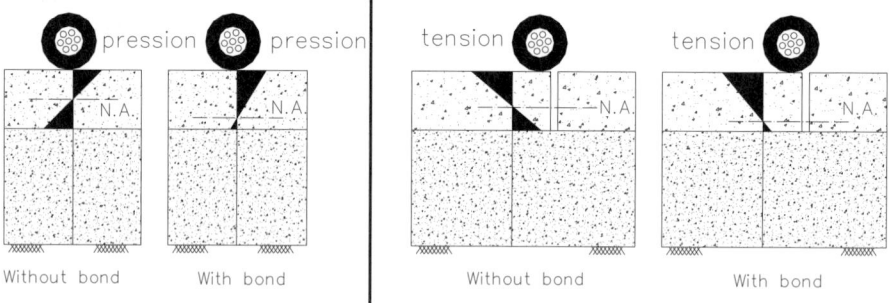

Figure 1   Results of neutral axis shift.

The bond between concrete and asphalt is influenced by the treatment of the asphalt surface, the composition of the concrete, weather conditions. By using short joint spacings, relatively small slabs are created (0.6m to 1.5m). A wheel load put on this slab is dispersed on the underground rather by compression than bending.

The joint spacing reduces the moment arm of the applied load and reduces the stresses due to bending. Curling of the slab by temperature- or moist gradient is not that explicit. The road sections built in America indicate that a joint spacing of 12-15 times the concrete slab thickness is recommended in both directions. Figure 2 describes the principle.

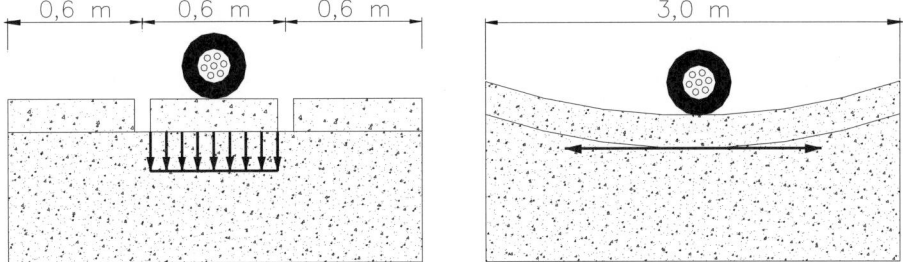

Figure 2   Short joint spacing : Stress reducing principle

Apart from these theoretical considerations, other considerations have to be borne in mind.

- It is believed that for an effective composite section the underlying asphalt layer has to be at least 75 mm. The thickness of this layer has an important influence on the lowering of the neutral axis in the UTW-section.

- Another important issue of UTW is the point in time of sawing the concrete slab. The joints should be sawed as early as possible to avoid cracking of the slap by shrinking under hard weather conditions. This has to be completed without damaging the borders of the joints. The sawing depth is ¼ a 1/3 of the concrete slab thickness.

- High-strength-concrete is frequently used for UTW-applications. It offers the possibility to repair an asphalt surface in one weekend and cause, in that way, less bother to the traffic. Fibers can improve the toughness of the concrete and give a higher bearing capacity to the UTW.

2-D FINITE ELEMENT CALCULATIONS

Finite Element Model

In order to provide some understanding of the various empirical formulae concerning Ultra Thin Whitetoppings, a Two Dimension Finite Element Model was developed consisting of plane strain membrane elements which represent the following layers : soil, sub-subgrade, subgrade, asphalt layer, concrete layer. The following characteristics from Table 1 have been used. All calculations have been done through the linear elastic method for simplification purposes and adopting a single asphalt modulus since heavy temperature change are not expected due to the thermal inertion of the concrete top layer. A factor a major importance in the design of Ultra Thin Whitetoppings is the effect of the bond between the asphalt and concrete layers. Using various techniques the contractor will try to obtain a full bond on which the designer will count and which is the base of U.T.W.

Table 1   Material Characteristics

MATERIAL	YOUNG'S MODULUS (Mpa)	POISSON'S MODULUS	THICKNESS (mm)
Concrete	35000	0.15	Var
Asphalt	12500	0.30	Var
Subgrade	500	0.5	300
Sub-subgrade	250	0.5	300
Soil	50	0.5	1200

This type of connection is generated in the model through a system of vertical and quasi horizontal spring elements. The stiffness of the horizontal springs reflects the bond between concrete and asphalt layer. A low spring stiffness equals almost no bond and a very high stiffness equals full bonding. Although both calculations have been done only the bonded results are relevant for UTW technique and are presented in this paper. Adjacent concrete slabs between saw-cuts are not attached in a vertical neither horizontal way. Since the saw cut is at least 3 mm wide, it is found that all horizontal deformations are smaller than this distance. Therefore no transfer of forces is taken into account. However, only accurate field test may give information regarding the true mechanism in these saw-cuts. Results which are however unavailable to the writer of this paper.

Regarding the sublayers two extreme circumstances are taken into account (Figure. 3 and Figure 4). The first represents a concrete top placed on a base with very little side support, representing a normal pavement with however poorly placed substructure. The second represents a concrete top very well enclosed between sublayers, stretching wide in both directions, representing a cross-roads type of U.T.W. rehabilitation. All other configurations will find themselves between these systems. In each case a 4 m wide lane is considered consisting of 4 concrete slabs of 1 m wide. The applied force consists of a 250 mm wide constant stress of 0.6 N/mm², representing a single axis. Since a linear elastic calculation was done the results are easily transformable to other wheel configurations and axle loads.

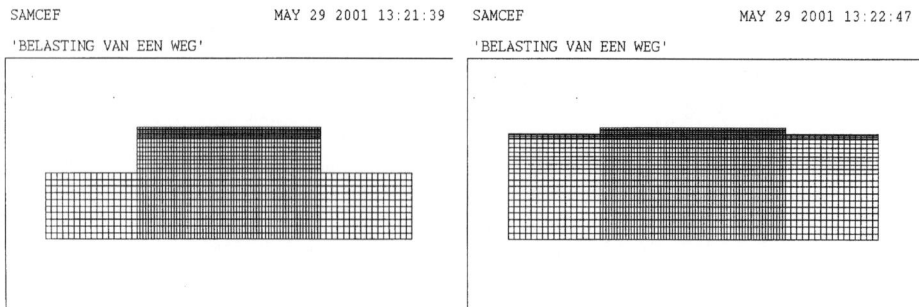

Figure 3   Configuration 1   Figure 4   Configuration 2

RESULTS

Firstly for both systems a calculation is done on a 100 mm concrete, 100 mm asphalt system. This represents a typical U.T.W. situation. On this configurations, the applied force is placed on 15 different locations : for each of the four slabs : left – center – right (12) and 3 times with the center of the force on the saw cut. In the resulting stresses focus is set on the maximum horizontal tensile stresses (or in absence of tensile stresses: compressive stresses) at the top of the four concrete slabs. Indeed it is found that the maximum tensile stress may occur at the top of a slab which is not subjected to direct pressure (Figure . 5). In addition the maximum horizontal tensile stress is retrieved on the bottom of the asphalt layer.

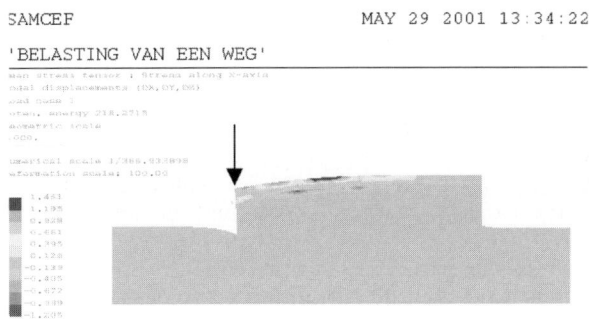

Figure 5   Horizontal Stresses due to edge loading

The results are given in figure 6 for configuration 1 and in Figure 7 for configuration 2. On the horizontal axis the location of the center of the applied force is given. On the vertical axis the maximum horizontal tensile stress or in absence of a tensile stress the maximum horizontal compressive stress in each concrete slab is given. Please note that the location of these maximum stresses is all but constant and may vary through the extents of the slabs and the asphalt layer.

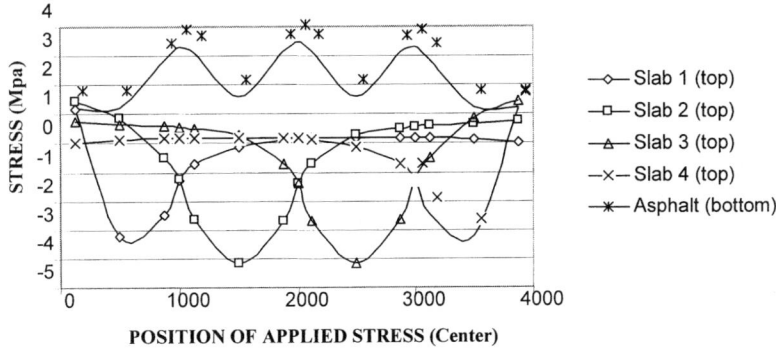

Figure 6   Horizontal Stresses Configuration 1

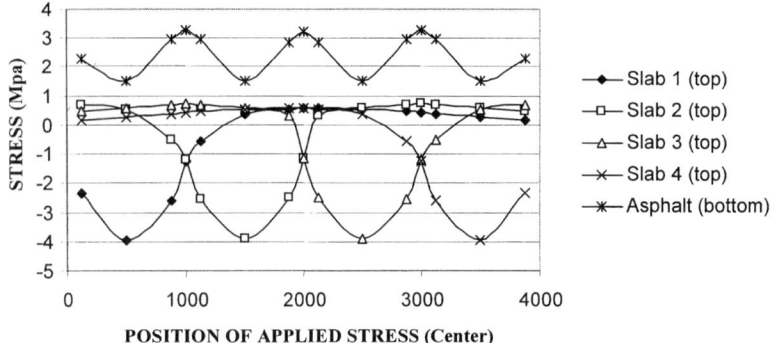

Figure 7   Horizontal Stresses Configuration 2

DISCUSSION

Maximum asphalt stresses

Maximum tensile stresses in the asphalt occur when the stress is applied directly above the saw cut, this due to the large decrease of stiffness at that location. There is no effect of the circumstances in transverse direction. Configuration 1 and 2 produce equal values. These results are comparable to Ref. 3.

Maximum concrete stresses

Comparing the results for configuration 1 and 2 it is clear that there is a strong influence of the asphalt and substructure, directly adjacent to the free edges of the concrete. When a spring model is used to simulate substructure and soil, it is not completely clear which of the 2 configurations mentioned is represented.

In general the largest horizontal tensile stresses are found on slab 2 when the load is on the edge of slab 1. This result indicates that although it is believed that UTW acts mainly in vertical stress transfer directly into the asphalt, due to the deformation of asphalt and substructure, the concrete needs following this deformation, causing the largest tensile stresses in the 2nd slab where the deformation curvature is the largest. For configuration 2 there is no side effect visible and the stress curves for the 4 slabs are equal. In this case the largest horizontal tensile stresses occur as well in the not directly stressed slabs, the maxima correspond to the stresses in slab 3 for configuration 1. The results are comparable to Ref. 3, taking into account that this is a 2-D method, and for both calculations, the exact location of the maximum tensile stresses are not given.

PARAMETRIC STUDY

A parametric study has been performed using the model described in paragraph 2.1. Both asphalt and concrete thickness have been varied between 75 and 125 mm. Taking into account the recommendation given in paragraph 2.3, focus is set on :

- The maximum horizontal tensile stress at the bottom of the asphalt layer, at the saw cut location with a load exactly above (configuration 1 or 2). Results given in Figure 8.

- The maximum horizontal tensile stress at the top of the concrete layer in slab 2, with the load at the free edge of slab 1 (configuration 1). Results given in Figure 9.

- The maximum horizontal tensile stress at the top of the concrete layer in slab 3, with the load at the free edge of slab 1, (configuration 1) corresponding to the maximum horizontal tensile stress at the top of the concrete layer in configuration 2. Results given in Figure10.

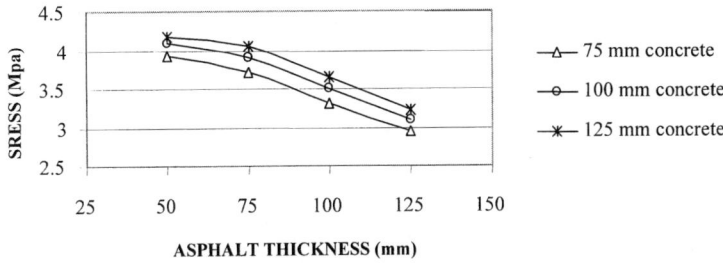

Figure 8   Maximum Horizontal Asphalt Stress (bottom of layer)

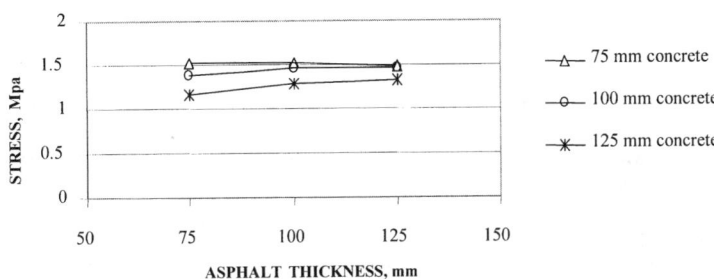

Figure 9 Maximum Horizontal Concrete Stress on slab 2 (top of layer)

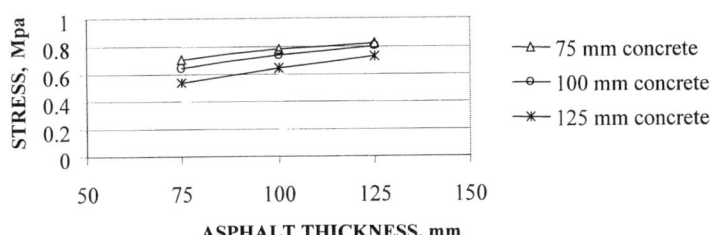

Figure 10   Maximum Horizontal Concrete Stress on slab 3 (top of layer)

DISCUSSION

Asphalt stresses

The maximum asphalt stresses will reduce considerably with an increase in asphalt thickness. However in contradiction to what could be believed, it is found that the maximum stresses will increase when applying larger concrete thickness. These results are not that contradictory since the saw-cut will reduce the stiffness only locally, whereas the stresses result for a total stiffness system. Therefore increasing the concrete thickness will increase global stiffness and therefore the bending moments, whereas the resulting stresses are calculated through local stiffness.

Concrete stresses

The maximum concrete stresses will reduce considerably with an increase in asphalt thickness. Again however in contradiction to what could be believed, it is found that the maximum stresses will increase when applying larger asphalt thickness. The effect however is not very large and may even change of sign when applying concrete thickness smaller then 75 mm.

Concluding it is found that no optimum situation is immediately clear since when increasing concrete thickness, the asphalt stresses increase and when increasing the asphalt thickness, the concrete stresses increase. A fatigue and tensile strength calculation may be necessary.

However it must be clear that these calculations are not meant as design rules and need confirmation by test measurements. They have as only purpose the further understanding of the complicated behaviour of the Ultra Thin Whitetopping rehabilition system.

CONCLUSION

Ultra Thin Whitetopping is a promising new method for rehabilitation of asphalt surfaces. The design and working principles are however complicated and not completely clear at this time. A 2-D finite element calculation has been performed on this type of pavement using simple plane strain linear elastic elements. The results however indicate that increase the stiffness of one layer will certainly reduce stresses in that layer but may create considerably higher stresses in the other layer. The results are based on various assumptions and need confirmation through wide scale or prototype testing in order to confirm these assumptions. In spite of the lack of confirmation, hopefully these results may contribute to the further understanding of the promising technique of Ultra Thin Whitetopping.

REFERENCES

1.   COLE, L.W.,MACK, J.W. PACKARD R.G. Whitetopping and Ultra Thin Whitetopping – The US Experience, ACPA, A Reference Manual, 1999.

2.   CABLE J. K., Iowa Ultra Thin Whitetopping Research, A Performance Update, Washington Research Board, 1998.

3.   WU C. L., TARR S., REFAI T. Model Development and Interim Design Procedure Guidelines for Ultra-thin Whitetopping Pavements, ACPA, A Reference Manual, 1999.

THAUMASITE IN CONCRETE STRUCTURES: SOME UK CASE STUDIES

D E Wimpenny

D Slater

Halcrow Group

United Kingdom

ABSTRACT. Thaumasite is a calcium sulphate carbonate silicate hydrate formed from the reaction of sulfates with silicate and carbonate in concrete exposed to cold wet conditions. The reaction can lead to softening, expansion and cracking of concrete. Unlike conventional sulfate attack, in which the calcium hydroxide and calcium aluminate hydrates react with sulfates to form gypsum and ettringite, in the case of thaumasite formation the calcium silicate hydrates in the cement paste can also be attacked. As a consequence, even concrete containing sulphate-resisting Portland cement and designed in accordance with the recommendations of BRE Digest 363 may be affected.

Following the identification of the thaumasite sulfate attack (TSA) to a bridge foundation in Gloucestershire in February 1998, over forty other sites have been investigated by Halcrow to confirm the extent of this form of sulfate attack. These investigations have found thaumasite formation in a variety of member types, concrete grades and exposure conditions. In some cases the sulfate reaction products were found to be almost pure thaumasite, whilst in other cases it was mixed with gypsum and ettringite. The pattern of deterioration has also varied widely from thaumasite formation without disruption to severe TSA related deterioration requiring repair and replacement of members.

This paper discusses four case studies in England and Wales, encompassing buried and non-buried members, mixes with Portland cement and sulfate-resisting Portland cement and disturbed and undisturbed ground. The paper discusses some key factors and features associated with the deterioration observed in each case in order to provide practical guidance to engineers and infrastructure owners and managers.

Keywords: Sulfate attack, Thaumasite, Bridge foundations, Diagnosis, Repair.

Mr D E Wimpenny, is a Senior Materials Engineer in the Materials Technology Unit, Halcrow Group. His interests include the specification and testing of concreting materials, the diagnosis of construction material failures and the design of remedial works.

Mr D Slater, is Technical Director, Materials for the Halcrow Group. His interests include the design of concrete structures for durability. He is a member of the Thaumasite Expert Group, Concrete Society Design Group and BSI Working Group B/517/1/1 on concrete durability.

INTRODUCTION

In February 1998 during bridge strengthening works, attack was discovered to the foundations of a 30-year old overbridge to the M5 in Gloucestershire. The thaumasite form of sulfate attack was suspected in view of the high quality of the concrete and the Building Research Establishment confirmed the diagnosis. Prior to this discovery the known incidence of thaumasite sulfate attack (TSA) in the UK had been limited to a small number of cases in non-structural concrete exposed to cold wet conditions. However the potential for such attack had been established in the laboratory and the UK guide on sulfate attack, BRE Digest 363, published in 1996 briefly referred to the phenomenon [1].

TSA is distinct from conventional sulfate attack in that the calcium silicate hydrates react with sulfates to form calcium sulfate carbonate silicate hydrate instead the conventional reaction with calcium hydroxide and calcium aluminate hydrates form gypsum and ettringite. This difference in the form of attack means that even concrete containing sulfate-resisting Portland cement and designed in accordance with BRE Digest 363 may be affected.

In response to the discovery of TSA the Highways Agency commissioned an investigation of twenty-eight structures in Gloucestershire and the Minister for Construction set up an Expert Group to develop guidance and advice for new and existing structures. The Expert Group, using information from the Highways Agency studies and other sources, published their findings in January 1999 [2] and is issuing a yearly review of the situation [3].

To date over forty sites In England and Wales, encompassing a variety of member and concrete types and exposure conditions, have been investigated by Halcrow to determine the extent of this form of sulfate attack. Over half of these sites were found to have TSA and this paper discusses four of these cases:

1)   Buried in situ reinforced concrete cast in shutters
2)   Buried in situ reinforced concrete cast against undisturbed ground
3)   Buried in situ reinforced concrete cast against disturbed ground
4)   Non-buried in situ reinforced concrete exposed to sulfate bearing water.

CASE STUDIES

Case Study 1 – Buried in situ reinforced concrete cast in shutters

Description

This case study comes from the M5 overbridge which initiated the original concern over thaumasite in the UK. The bridge has three slender columns at each pier, 750mm x 450mm in cross-section and 13m long resting on a spread footing approximately 5.5m below ground level. The columns were formed from in situ reinforced concrete cast in shutters within a cutting through undisturbed Lower Lias clay. Excavated Lower Lias clay was stockpiled and later used to backfill around the columns and form a 2m embankment either side of carriageway. Oxidation of the pyrite in this reworked clay is thought to have led to high sulfate levels being developed in the ground.

A condition survey was carried out in which softened areas were mapped and the depth of expansion and softening were measured to each face of the columns. Concrete cores and dust samples and fragments were subject to petrographic examination, Scanning Electron Microscope (SEM) microprobe analysis, compressive strength testing and chloride and sulfate determination.

Main Findings

The attack was principally evident due to softening of the surface of the columns exposed during strengthening work. The distribution of the softening varied, with no attack within 1m of ground level, partial attack in the form of patches or bands of softening up to 500mm in size approximately 2m below ground level and attack across the full area of all the faces below 4m depth. The reaction products, which were initially very soft, become hard and friable after being exposed for several days. Rust staining was present within some of the attacked areas and breakouts revealed deep pitting corrosion to the reinforcement, representing 30% loss of section, and black and green reaction products which turned brown on exposure characteristic of anaerobic chloride-induced corrosion.

The attacked concrete was observed to have a soft, white pasty layer, typically 5mm in thickness, but no visible cracking. Below this layer, the concrete was found to have cracks running sub-parallel to the surface and the coarse aggregate particles were surrounded by white 'halos' of reaction products. The frequency and size of the halos reduced with depth below the exposed surface such that typically at 20mm depth only occasional white specks of reaction products were present in pores without any physical disruption to the cement matrix; this is described as Thaumasite Formation (TF) rather than TSA (see Figure 1).

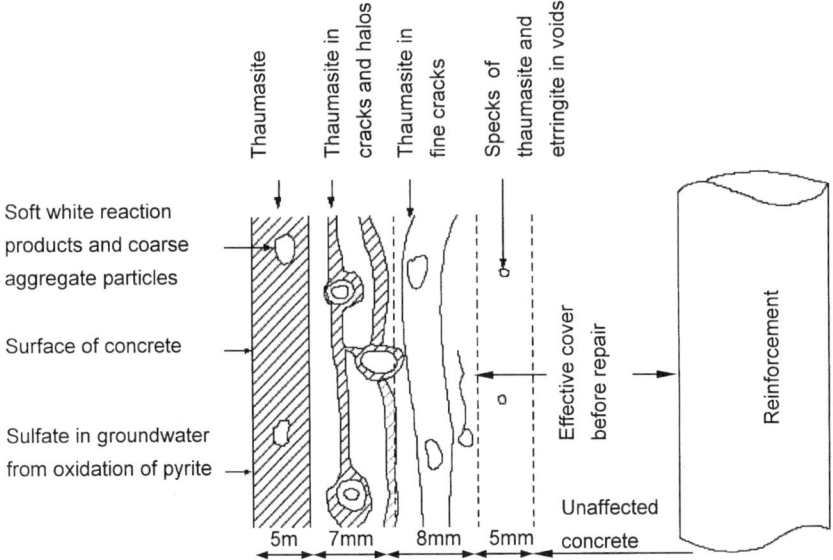

Figure 1   Schematic representation of TSA affected concrete

Petrographic examination indicated relatively pure thaumasite with little intermixing with other sulfate reaction products such as ettringite. The SEM microprobe analysis confirmed this finding with the thaumasite being associated with SiO_2/Al_2O_3 ratios of greater than 9, compared with a ratio of less than 0.4 for ettringite. Coarsely crystalline carbonate was present within the concrete and at its surface. Some carbonation of the surface of the concrete would be expected between casting and backfilling. However, the observed carbonation is different than normal atmospheric carbonation and suggests that carbonates may be penetrating from the groundwater consistent with the postulated mechanism of alkali-carbonation [4].

Petrographic examination, including point counting, indicates that the concrete contains 300-400kg/m^3 of Portland cement and has a water-cement ratio of 0.48-0.57. This would meet Class 1 or 2 sulfate conditions to BRE Digest 363 [1]. The major rock type in the coarse aggregate is dolomitic limestone. The fine aggregate is quartz/metaquartzite with 25% limestone fines. In situ cube strength values of 60.5-83.0N/mm^2 and estimated proportions are consistent with a grade C40 concrete.

Sulfate and chloride determinations from 10mm depth increments of a 100mm diameter core are shown as profiles in Figure 2. The depth of TSA given by the depth of cracks infilled with thaumasite from the petrographic examination is also indicated. It can be observed that the depth of TSA approximately corresponds to point at which sulfate values exceed approximately 5% SO_3 by mass of cement, equivalent to 6% SO_4 by mass of cement. The chloride profile shows a peak of over 0.8% chloride ion by mass of cement 5-10mm deeper than the depth of TSA. This together with the evidence of corrosion to the reinforcement suggests that chloride induced corrosion may be an underlying problem in areas where TSA is present.

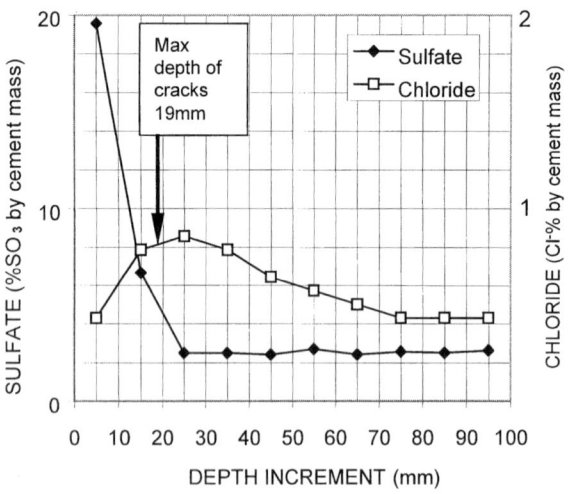

Figure 2 Sulfate and chloride profiles in TSA affected concrete

Comparison of the distribution of attack with the results of piezometric monitoring over an 18-month period indicate there is no TSA above the maximum ground water level and full attack to concrete below the minimum water level. Partial attack at one pier coincides with the location of carrier drains through the backfill.

The distribution of softening values to the columns indicates there is reduced softening furthest from the oncoming traffic and to faces furthest away from the carriageway. The above indicates that the pattern of attack is closely associated with the presence of water, either as natural ground water or derived from other sources such as run-off and traffic spray from the carriageway.

The typical pattern of softening and expansion to the columns is indicated in Figure 3. Areas of expansion and softening are closely associated. The maximum net loss of cross-section for this site was 33mm, which is less than the value associated with a risk of buckling failure to the columns. However, in view of the possibility of further deterioration during the remaining service life it was decided to replace the affected lower portion of the columns and recast a new spread footing above the old base [5].

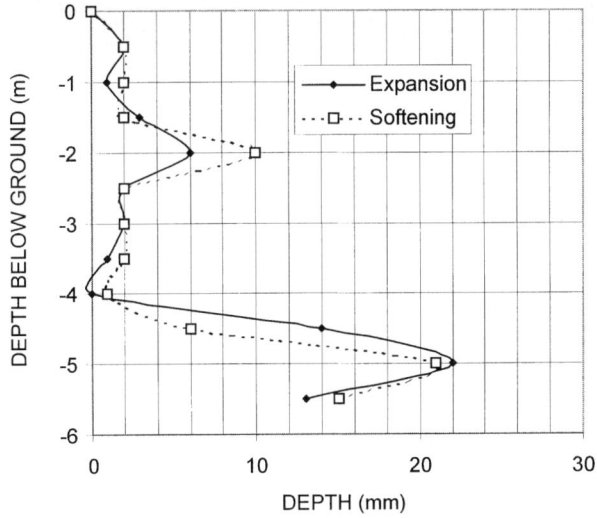

Figure 3   Typical pattern of softening and expansion to columns

Case Study 2 - Buried in situ reinforced concrete cast against undisturbed ground

Description

The structure is an M5 overbridge (approximately 30km south of the site in Case Study 1) having three 760mmx450mm columns approximately 10m long, at each pier resting on a spread footing approximately 3m below ground level. The 900mm thick spread footing is founded on undisturbed Lower Lias Clay and reworked Lower Lias Clay was used to backfill around the columns and to form a 1m high embankment at the southbound hardshoulder.

The structure was constructed in 1970 and was being excavated in 1998 for strengthening work when TSA was identified as a risk to highway structures in Gloucestershire.

A condition survey was undertaken at the central reserve and hard shoulders. Core samples were taken to the columns and through the spread footing to determine the condition of underside of the base in contact with undisturbed Lower Lias Clay.

Main Findings

The survey found severe attack to the columns and top surface of the spread footing with greater than 25% of surface areas softened to over 25mm depth in places. The deterioration to the upper surface of the base had resulted in the reinforcement being exposed over a length of approximately 1m. Black corrosion products and deep pitting to the reinforcement, representing 100% loss of section, were evident in one location. The affected portions of the columns were replaced and a new spread footing was cast on top of the existing base.

At some highway structures in Gloucestershire, the blinding concrete was found to contain a different cement type, ie sulfate-resisting Portland cement, than the adjacent structural element. However, at this site there were no such differences in the composition of the concrete at the upper and lower surfaces of the spread footing. Petrographic examination, indicates that the concrete contains 270-360kg/m^3 of Portland cement and has a water-cement ratio of 0.56-0.60. This would meet Class 1 sulfate conditions to BRE Digest 363 [1]. The major rock type in the coarse aggregate is limestone (dolomitic, bioclastic and oolitic). The fine aggregate is quartz/metaquartzite with 20% limestone fines, recrystallised sandstone and chert. The mean estimated in situ cube strength values of 60.5-71.5N/mm^2 and proportions are consistent with a grade C35 concrete.

Petrographic examination of cores taken through the base found cracks infilled with thaumasite at a maximum depth of 16mm and sulfate levels greater than 5% SO_3 by mass of cement to 20mm depth. In contrast, the base of the spread footing was found to have no evidence of TSA despite sulfate levels in excess of 5% SO_3 by mass of cement to 50mm depth. This finding is consistent with cores taken through the spread footing to the bridge in Case Study 1; seven of which indicated no TSA and four of which show thaumasite formation in voids without damage to the concrete.

The absence of TSA to the underside of the base cast against the undisturbed Lower Lias clay suggests that even where sulfates are present at a site and the concrete has not been designed to resist sulfate attack, the relative accessibility of concrete surfaces to sulfates and to water are key factors in determining if TSA occurs.

Case Study 3 - Buried in situ reinforced concrete cast against disturbed ground

Description

The structure is an overbridge at the interchange between the M5 and A40 trunk road constructed in 1968-69 in a 7m deep cutting. Strengthening works on the bridge, involving the construction of new inclined steel columns to form 'V' piers, were undertaken in 1998-1999. TSA was identified to the pile cap supporting the existing pier and an investigation was undertaken to establish the condition of other elements of the structure, including the abutments.

The abutments to the bridges consist of bank seats founded above the original ground level on an embankment formed from reworked Lower Lias clay and supported on 900mm diameter bored in situ concrete piles. Trial pits were constructed adjacent to the abutments in order to expose the top 3m of the pile. The half-joints over the abutments were observed to be leaking and within 24 hours of excavation one of the trial pits filled to approximately 2m depth with water.

Core samples were taken for petrographic examination, chemical analysis and compressive strength testing.

Main Findings

The concrete surface appeared free of deterioration except for a 55mm deep gouge, coinciding with the position of a granular layer within the backfill. A hammer soundness survey identified areas of softening to the concrete and removal of the apparently attacked surface revealed a thin layer of white reaction products and halos around the exposed aggregate particles.

The depth of maximum softening measured by drilling was 20mm. Black and green corrosion products were visible on the surface of the pile and a covermeter survey found zero depth of cover at these locations suggesting displacement of the reinforcement cage during construction followed by anaerobic chloride-induced corrosion.

The interface between the undisturbed and disturbed Lower Lias clay was evident at approximately 2m depth below the soffit of the pile cap within one of the excavations. The concrete in the disturbed ground was found to have softening to over 15% of the exposed surface whereas there was no evidence of softening or attack to the concrete cast against the undisturbed ground below (Figure 4).

The petrographic examination indicates the concrete to contain approximately 290-420kg/m^3 of sulfate-resisting Portland cement and have a water/cement ratio of 0.48-0.55. This would meet Class 2-3 sulfate conditions to BRE Digest 363 [1]. The coarse aggregate is 100% limestone (dolomitic, oolitic and recrystalised limestone). The fine aggregate contains 5-80% limestone in combination with quartz/metaquartzite. The in situ cube strength values of 33.5-41.5N/mm^2 and estimated proportions are consistent with a grade C35 or C40 concrete.

The petrographic examination of the cores found cracks up to 24mm deep infilled with thaumasite. The sulfate levels in excess of 5% SO_3 by mass are present at 10-40mm depth. These results confirm the presence of TSA where relatively sulfate resistant concrete has been cast against disturbed Lower Lias clay. Prior to the investigation the higher elevation of the abutment compared to the piers within the cutting was expected to lessen the risk of attack due to a reduction in the available groundwater.

However, in this case the leaking half-joints appears to represent an important source of water for the reaction.   No remedial work was carried out to the abutments due to the restricted programme for the strengthening work but the abutments were identified for future monitoring and maintenance.

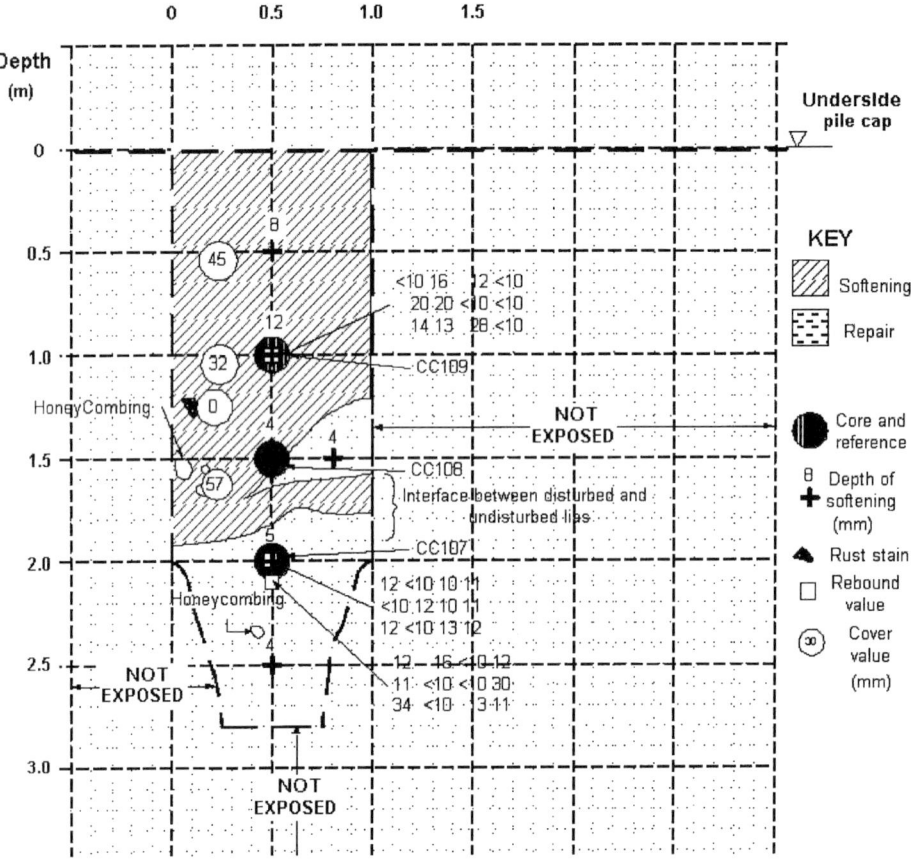

Figure 4   Defect mapping for bored in situ concrete piles

Case Study 4 - Non-buried in situ reinforced concrete exposed to sulfate bearing water

Description

The structure is a drainage adit constructed in 1985/1986 as part of works to stabilise a landslip affecting a trunk road. It comprises a main adit and five branches totalling approximately 625 linear metres of hard rock tunnel, 3m wide by 2.5m high. The adit has a concrete lining to the first 70m.

The remaining 550m is supported on RSJ supports at 900mm centres connected by tubular struts and has a reinforced concrete invert with 200mm high integral kerbs (Figure 5). The structure is inspected annually and deterioration was first noted approximately 10 years after construction.

Surveys of the structure were undertaken in the winter of 2000/2001. The deterioration of the kerbs was found to be in the form of white deposits on the concrete surface, halos around the aggregate particles and total loss of cementitious strength. In the worst affected areas the surface had fragmented to approximately 10mm depth with a total loss of the arris to the kerb and expansion of over 50mm to the top surface. These areas coincide with a change in the ground from predominantly sandstone to boulders, clay and mudstone. The local Rhaetic mudstsones of the Westbury Formation are identified as sulfide bearing in the TEG Report [2] and there a close association was observed between of areas of deterioration with brown, black and occasionally yellow coloured staining from seepage from the bedding planes and fractures in the mudstone. Analysis of water draining from a well-point in the adit showed it to have a sulfate content of 13mg/l and pH of 8.0 indicative of Class 1 sulfate conditions to BRE Digest 363, 1996 [1]. This is consistent with an absence of attack to the invert and confirms the likely source of sulfates is gradual seepage rather than the bulk of the water draining from the adit.

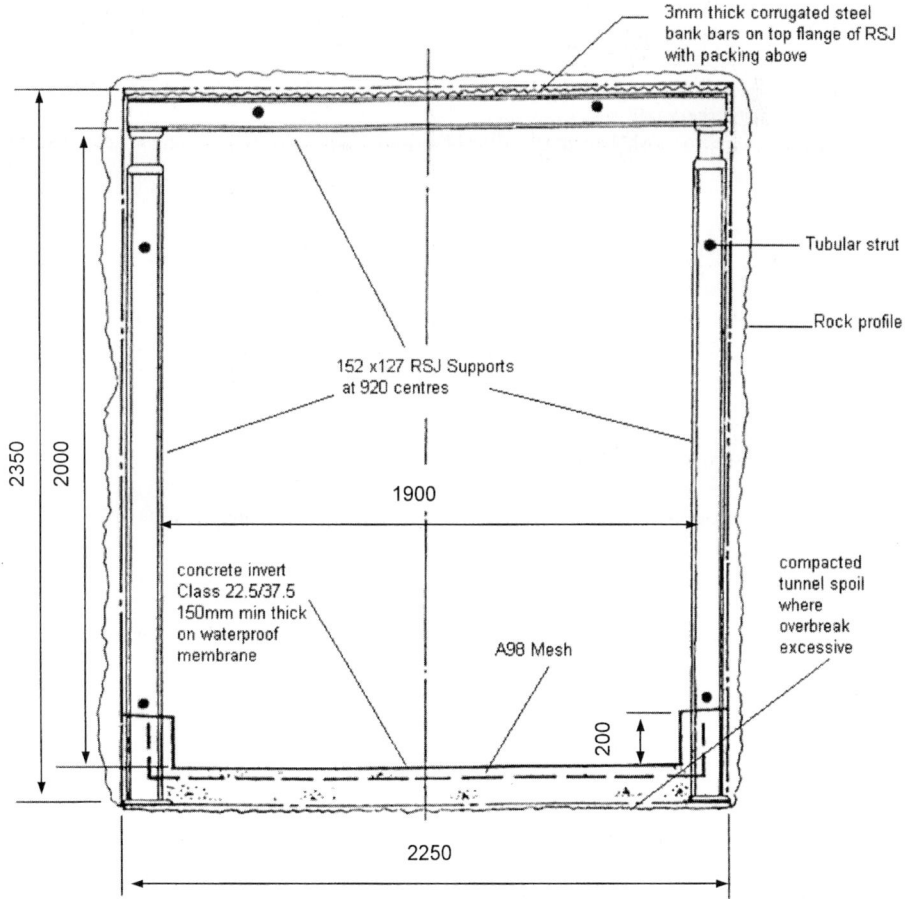

Figure 5   Cross-section through adit

Following a condition survey, in which the extent of the disintegration was mapped, holes were drilled at the interface between the rear of the kerb and the rock. A boroscope examination found no evidence of sulfate attack at the interface. Cores and dust samples and fragments of deteriorated concrete were taken for petrographic and SEM microprobe examination and sulfate determination.

Main Findings

Petrographic examination of the fragments indicated expansive growth of gypsum and to a lesser extent thaumasite and ettringite in the cement paste. The cores were found to have abundant cracking around aggregate particles and parallel to the external surface infilled with a material resembling thaumasite. SEM Microprobe analysis of these reaction products indicates normalised sulfate contents (SO_3) of over 20% and a high SiO_2/Al_2O_3 ratio characteristic of thaumasite.

Petrographic examination and point counting indicates the concrete contains approximately 450kg/m^3 of sulfate-resisting Portland cement and has a water-cement ratio of 0.49-0.55. This would meet Class 2 sulfate conditions to BRE Digest 363 [1]. The fine aggregate and coarse aggregate are both recrystallised limestone. Coarse and granular carbonation are present at the ends of samples, in some cases as multiple vaterite layers, suggesting an external source of carbonate is present.

The transition is very sharp between the unaffected and TSA affected concrete and the sulfate levels fall rapidly below the surface such that removal of 10-20mm of material beyond the deteriorated zone would remove all material with a sulfate content (SO_3) in excess of 5% by mass of cement. Where water seeping from the rock has dripped onto the upper surface of the kerbs, mounds 10-50mm high are present of expanded and disintegrated concrete adjacent to unaffected concrete. This intense local disintegration suggests that an increasing cycle of chemical attack followed by physical disruption has occurred in these locations.

It was recommended that the kerb was reinstated using a non-Portland based cementitious repair mortar protected by a coating to overlap the adjacent unaffected concrete. The unexposed faces of the kerb in contact with the rock were not considered to be at high risk of TSA.

CONCLUSIONS

The four case studies suggest that high levels of sulfate and the availability of water predominate over the factors such as cement type and concrete quality in determining the extent of TSA. Concrete designed for Class 2-3 sulfate conditions to BRE Digest 363 were attacked whilst concrete meeting only Class 1 was unaffected.

High carbonate content aggregates were used in the concrete for all case studies. However, the presence of coarse carbonation at surfaces suggest an external source of carbonates may sometimes be present. Concrete surfaces cast against undisturbed ground had no evidence of TSA whereas surfaces of buried in reworked Lower Lias Clay or having contact with sulfate bearing water seeping from Rhaetic Mudstone suffered substantial attack requiring repair or replacement of members.

Petrography in conjunction with SEM Microprobe analysis confirmed TSA in three of the case studies. The reaction products varied from almost pure thaumasite to mixtures with other sulfate reaction products. The TSA affected concrete was sharply defined by the evidence of damage in terms of cracks infilled with thaumasite and excessive sulfate levels.

The presence of high chloride levels in the concrete and staining and pitting to the reinforcement suggests chloride induced corrosion is a potential underlying problem in highway structures affected by TSA.

REFERENCES

1. BUILDING RESEARCH ESTABLISHMENT, Digest 363, Sulfate and acid resistance of concrete in the ground, January 1996, p 12.

2. THAUMASITE EXPERT GROUP, The thaumasite form of sulfate attack: Risks, diagnosis, remedial works and guidance on new construction, DETR, January 1999, p 180.

3. THAUMASITE EXPERT GROUP, One-year review, Concrete, Vol 34, No 6, June 2000, p 51-53.

4. GAZE, M E, CRAMMOND, N J, The formation of thaumasite in a cement:lime: sand mortar exposed to cold magnesium and potassium sulfate solutions, BRE Report No 80002, November 1999.

5. WALLACE, J, Strengthening thaumasite-affected concrete bridges, Concrete, Vol 33, No 8, September 1999, p 28-29.

ANALYSIS OF A LARGE DATABASE OF REINFORCED CONCRETE BRIDGE INSPECTION RECORDS

C McParland J F Lyness

A Nadjai

University of Ulster

A Abu-Tair

Al-Quds University

Palestine

ABSTRACT. This paper describes the use of a large database of reinforced concrete bridge inspection records. To determine what data is useful an extensive analysis is undertaken. The analysis shows that the age and associated exposure conditions produce fundamentally deleterious effects on reinforced concrete structures. It shows how the structure type and member type influence the rate of degradation. Subsequently, a spreadsheet template was developed to present the analysis results. The service life of the bridges can be classified according to five defect phases, namely defect free, minor cracking, major cracking, spalling and failure. The likelihood of a structure or member having deteriorated to one of these defect phases is presented in the spreadsheet.

Keywords: Database, Defect history, Concrete bridges, Service life, Modelling.

Mr C McParland, is a Civil Engineer and PhD Research Student at the University of Ulster, Jordanstown. His research focuses on the degradation of reinforced concrete bridges.

Dr J F Lyness, is a Chartered Civil Engineer and Reader in the Faculty of Engineering, School of the Built Environment, at the University of Ulster, Jordanstown. His research focuses on the use of numerical methods in analysis and design.

Dr A Abu-Tair, is a Civil Engineer and Lecturer in structural engineering at Al-Quds University, Palestine. His research focuses on reinforced concrete performance.

Dr A Nadjai, is a Structural Engineer and Lecturer in the Faculty of Engineering, School of the Built Environment, at the University of Ulster, Jordanstown. His research focuses on structural design.

INTRODUCTION

Over past decades there has been considerable interest in the durability and maintenance of concrete bridges. In the 1970's there was a growing awareness, in many countries, of the need to preserve their stocks of bridges, as they are key elements in their highway systems.

Concrete bridges are required to maintain their serviceability over the specified design life, usually 100 years or more. Durability aspects of serviceability are illustrated in the design codes by deterministic rules for concrete cover and crack widths. An optimal maintenance strategy is necessary if bridges are to meet their service life requirements. Before deciding on suitable repair methods, it is necessary to determine the cause and the degradation mechanism. Only then can corrective maintenance costs be realistically estimated.

Most reinforced concrete structures perform satisfactorily during their service lives but there is a significant proportion that suffer major durability problems. The main cause of deterioration is corrosion of reinforcing steel, which results in the accumulation of corrosion products at the steel/concrete interface.

The high pH of hydrated Portland cement (between 12 and 13) provides passivity to reinforcement against corrosion and can ensure high resistance against this mechanism of deterioration. However, the processes of carbonation and chloride penetration can depassivate the reinforcement and result in corrosion.

The inherent risk of deterioration of reinforced concrete structures due to corrosion has highlighted the importance of developing service life prediction models so that optimal strategies for their maintenance and repair can be developed [1]. Approaches have recently been developed to estimate the service life of concrete structures. To date, they are based on the concept of delay time, the Factor Method, or a stochastic/ probabilistic approach. These models rely on sound engineering judgement or statistical information in order to determine the structure's deterioration rate. A key factor missing is that no 'actual data', collated over the past number of decades, has been used in their development.

This paper describes the contents of a large database of concrete bridge inspection records [2-5]. The defect histories of 725 concrete members from 439 bridges in the London area were obtained. Of the information recorded and stored in the database, perhaps the most relevant data was the component age and exposure conditions. As a structure ages it will deteriorate, and the rate of degradation will be accentuated under ever increasing aggressive environments. Through its service life the various components of a structure will undergo different stages of deterioration, namely defect free, minor cracking, major cracking, spalling and eventually failure.

It is evident that if enough data were available it would be possible to establish the age at which a typical structural component will have deteriorated to a specific phase in its service life. Consequently, this paper reports the extensive analysis undertaken on the database. Details of the development of a simple spreadsheet are provided. This spreadsheet was developed to present the findings from the analysis of faults. It is used to display the crack/fault most likely to develop in a component given its age and related exposure condition. Subsequently, a critical appraisal of the spreadsheet is offered in order to evaluate its suitability in the field of service life prediction.

BRIDGE DEFECT DATABASE

The database contains the defect histories of over 430 concrete bridges situated in the London area. The life span of many of these bridges goes back to the early 1930's with the associated inspection records covering most of this period. Indeed, a number of these bridges were constructed as long ago as 1880.

The data collected from the inspection records were organised into three Tables. Tables 1 and 2, below, show a typical layout of the data contained in the database. Research undertaken by Rigden et al. [5] presents further details and associated meanings of the information contained in this database.

Table 1   Typical layout of structure table

CODE	REFERENCE	TYPE	YEAR BUILT	CONSTRUCTION	SPANS	SPANNING
TC001	CW2	5	1937	0	0	0
TC002	CW2	5	1937	0	0	0
TC003	CW2	5	1937	0	0	0
TC004	CW2	5	1914	0	0	0
TD001	D12	2	1926	0	0	0

Table 2   Typical layout of defect table

CODE	INYR	FAULT	SIZE	CAUSE	MEMBER	EXP	URG	COMM
D037	1935	1	0.75	1	2	2	1	SEP
TC001	1966	7	4500	1	1	2	2	30D
TC001	1970	11	0	1	1	2	1	SEP
TC002	1970	8	0	1	1	3	1	30D
TC002	1976	9	0	1	1	3	1	

The database package used was Microsoft Access. This has the advantage of being compatible with other spreadsheet software so that statistical information can be produced from any analysis undertaken on the database. The Tables were viewed to determine which fields could be queried in order to extract useful and meaningful information.

By combining certain fields from each Table a further Table was produced, Table 3.

Table 3   Typical layout of query table

CODE	YEAR BUILT	INYR	TYPE	MEMBER	EXP	URG	FAULT	AGE (YEARS)
TC001	1937	1966	5	1	2	2	7	29
TC001	1937	1970	5	1	2	1	11	33
TC002	1937	1970	5	1	3	1	8	33
TC002	1937	1976	5	1	3	1	9	39
TC002	1937	1982	5	1	3	3	9	45

A key to the headings used is listed below.

Database Query   Code; Year Built; Inspection Year; Type of Structure; Component
Type; Exposure Condition; Urgency; Fault/Action Code; Age (years).

This Table contains the relevant data used to carry out the analyses.

EXTENSIVE ANALYSIS OF THE DATABASE

In order to appreciate the amount and type of data contained within the database, a number of
graphs and charts were produced to portray this information. These are outlined below.

BRIDGES INSPECTED

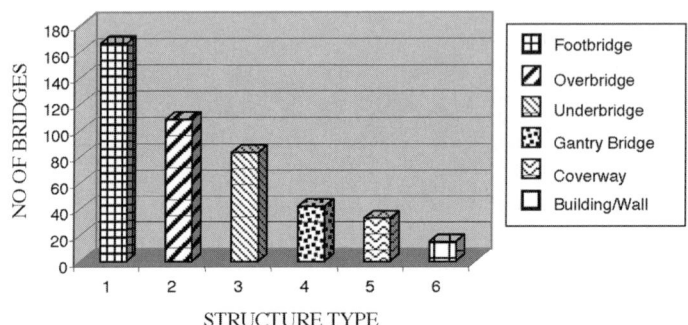

MEMBERS INSPECTED

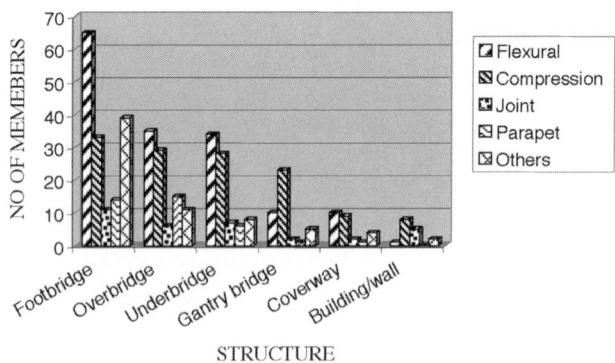

Figure 2   Members inspected

- Figure 1 illustrates the range of structures that have been inspected.
- Figure 2 highlights the type of member inspected on each structure and the frequency of that occurrence.

Perhaps the most significant data recorded during an inspection is the defect of the particular component under consideration.   The defect can be classed under different headings including defect free, minor cracking, major cracking, spalling and failure. Figure 3 displays the number of each type of defect recorded during the numerous inspections.   As this information is recorded for each component the degradation of a particular member can be charted throughout its service life. This information will be invaluable in the development of a model predicting the deterioration rate of various components.

DEFECTS RECORDED

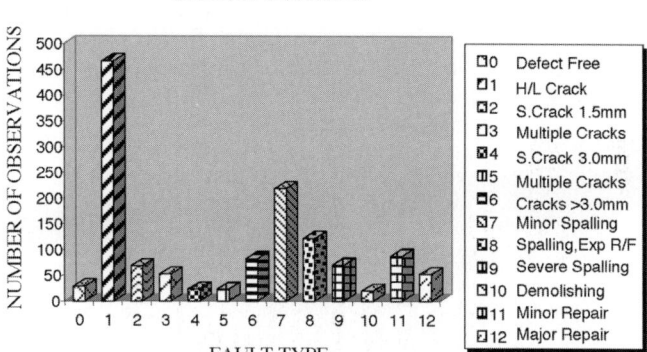

Figure 3   Defects recorded

Exposure Condition and Urgency

Adverse environmental conditions will have a deleterious effect on concrete so that the rate of degradation will increase as the conditions become more severe.   Figure 4 shows the percentage of components exposed to mild, moderate or severe conditions. A description of the exposure conditions is provided in BS8110 [6].

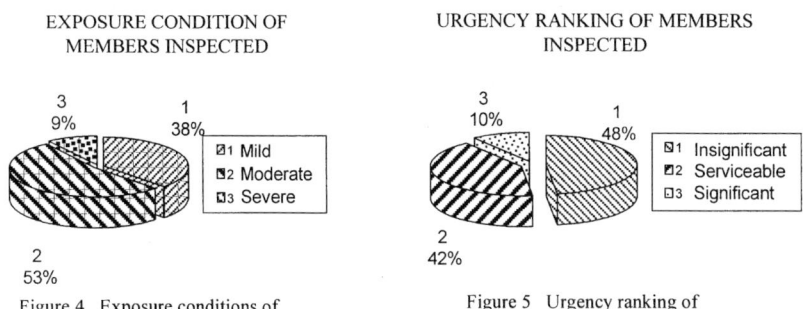

Figure 4   Exposure conditions of members inspected

Figure 5   Urgency ranking of members inspected

As a component degrades the urgency of repair will intensify. During an inspection each member under observation is recorded in terms of urgency ranking as significant, serviceable or very significant. Figure 5 illustrates the percentage of members that can be categorised using these rankings. At this stage two important factors have been ascertained through analysis of the database, that is the defect type and the associated exposure conditions.

Age of Component

The database was then interrogated further to extract more detailed information on each component. It was necessary to determine how the defect condition changed with age. By producing a bar chart it was clear that the defect type increased with age, i.e. components were deteriorating. This chart was also produced to show how exposure conditions 1 (mild), 2 (moderate) and 3 (severe) contributed to the degradation of the structure. Only records where the first occurrence of a fault was observed were used in this analysis. The chart for exposure to moderate conditions is shown in Figure 6. The urgency rating was also determined in comparison with the above exposure conditions. The chart displaying this data under moderate conditions is illustrated in Figure 7.

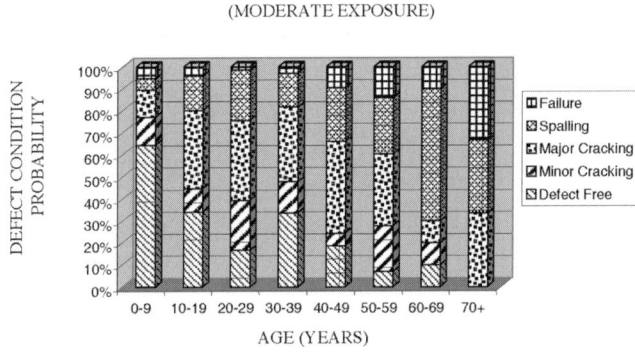

Figure 6   Defect conditions recorded with age (moderate exposure)

The same approach was applied for each individual structure and member. The information gives an appropriate relationship as to how different elements of a structure will degrade with age and the associated exposure. This was deemed as having particular relevance if a mathematical model was to be developed. The results provide a firm basis for this development

SPREADSHEET PROGRAM

Development

From the results obtained by the interrogation of the database it became apparent that there is a definite pattern in the way the defect information changes under certain conditions. It confirmed that, for a given structure or member, the age and exposure conditions were key factors in determining the deterioration rate of the element. Subsequently, a spreadsheet was developed to present the findings from the analysis.

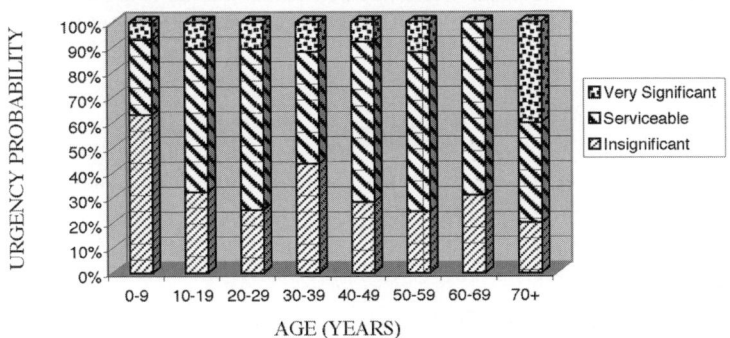

Figure 7   Urgency recorded with age (moderate exposure)

There were two forms produced, one for the overall structure and the other to be used for each individual member. Form 1 was designed to display defect information for the overall structure. There were three fields in which data input was required. The first field was concerned with the structural type. A list of the different structural types is given along with a prompt to enter the number corresponding to the preferred option. The second area of the form related to the age of the structure. A list of ranges of years was provided and, again, a prompt is given to enter the corresponding number that appeared adjacent to the appropriate age range. The final piece of information required related to the exposure conditions to which the structure was subjected. The three exposure conditions provided were mild, moderate and severe. There was very little information on structures subjected to very severe and extreme conditions, so these were not taken into account. It was required to enter the number corresponding to the desired exposure condition in the box provided. The input section of Form 1 is shown in Figure 8.

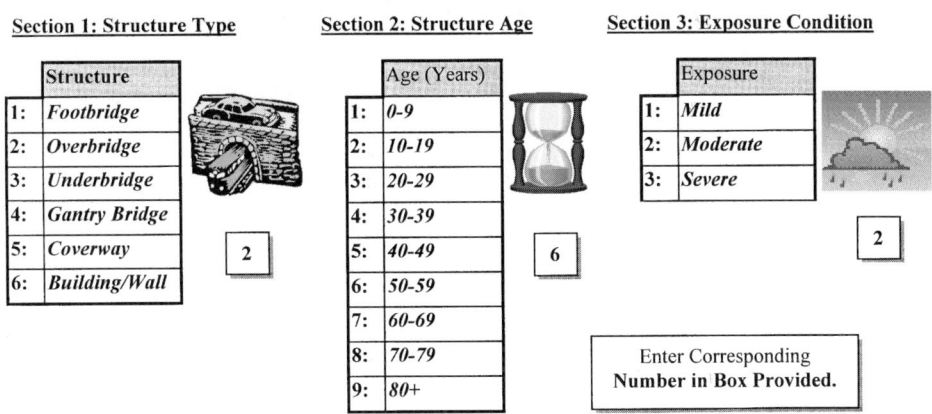

Figure 8   Input section of Form 1

The information displayed outlines the probability of each defect condition occurring, along with the urgency of this matter, given the specified data input. Where there is no information related to a certain combination, the message 'No data' appears. This is mainly relevant to gantries, coverways and buildings. There was not sufficient data available to provide any realistic information on these structural types. The output section of Form 1 is illustrated in Figure 9.

Output

Overbridge Aged Between 50-59 Years, Subject to Moderate Exposure.

The Following Output is Probable:

Crack/Fault	Urgency Ranking

Phase	Probability
Defect Free	1.82%
Minor Cracking	7.27%
Major Cracking	14.55%
Spalling	23.64%
Failure	52.73%

Urgency	Probability
Insignificant	8.16%
Serviceable	36.73%
Very Significant	55.10%

For Information on Individual Members Refer to Form 2

Figure 9   Output section of Form 1

Form 2, used for individual members, is similar to Form 1, the principal difference is in the first field where a prompt is displayed for data relating to the member type instead of the structure type. The values in the second and third part of this form are defaulted to be equal to those entered in Form 1. This was done for the majority of the inspections covered, because the age and exposure condition of the member will be the same as that for the overall structure. If this is not the case the data can be altered as before. Similar information to that in Form 1 will be displayed.

Appraisal of Spreadsheet

The development of a spreadsheet program offers a number of benefits for the presentation of such an extensive analysis:

- The spreadsheet has been developed in such a way that once the input section has been completed, the relevant output is displayed immediately. This avoids the need to search the database.

- It is user-friendly and it does not require an expert. Someone with limited knowledge in bridge management could use this approach to ascertain the defect likely for a given structure and/or member with minimal instruction.

For the structure and individual members, the following advantages are apparent:

- There is a substantial amount of data available to provide a high degree of confidence that the spreadsheet output accurately reflects the defect conditions that will develop in the total population of bridges.

- As more data is made available, the spreadsheet will provide a more accurate probability of the defect condition that will occur for the given structure or member type.

Following the development of a spreadsheet it is apparent that there are limitations to the use of this program, such as:

- For some structures there is limited data available to calculate the probabilities displayed in the output forms. Thus, the output is more sensitive to outliers. This problem can only be minimised when more data is made available.

- The data needs to be analysed further to remove these outliers. Although they are true observations encountered during the inspections, their occurrence is rare and does not reflect the general trend.

- The information displayed on the form should only be used as a guide. It can not predict all possible outcomes, only the most common cases will be highlighted.

CONCLUSIONS

This paper highlights the benefits of recording and modelling the deterioration process of stocks of concrete structures. The age and exposure conditions of concrete structures are important as part of the modelling process. They can be used to provide a classification of the deterioration of a structural element along with the associated urgency of maintenance required. The condition of each element can be described as 'defect free', showing signs of 'minor cracking', 'major cracking', 'spalling' or 'failure'. The urgency of repair is classed as insignificant, serviceable or very significant.

The spreadsheet program developed provides a useful tool for the engineer whenever an inspection is to be carried out. The information shows the probability of each defect condition occurring before inspection. It will also help the engineer to determine the most appropriate maintenance and repair techniques to use, if any. However, it does have limitations.

The defect classification is only as accurate as the data provided by the records. Although it may not be the most accurate model currently available, it does provide realistic results. As more records are made available then the accuracy of the information for the structure and / or component will increase and have more significance. The information obtained from the spreadsheet will be used in the development of a deterioration model.

Overall, the spreadsheet was developed to present the findings from the analysis of the bridge inspection database. The information obtained will be used to establish a more complete deterioration model that can readily simulate the degradation of the structures or members described in the database.

REFERENCES

1. MANGAT, P S, Characterisation of chloride induced corrosion for service life prediction of reinforced concrete, Bulletin of Electrochemistry 1995, Vol 11, Part 12, p 556-564.

2. CHRISTER, A H, et al, Modelling the deterioration and maintenance of concrete structures, Proceedings of the International Conference of Operation Research, Lisbon, Portugal, July, 1993.

3. REDMOND, D F, et al, O.R. modelling of the deterioration and maintenance of concrete structures, European Journal of Operational Research, 1997, Vol 99, No 3, p 619-631.

4. RIGDEN, S R, et al, Predicting future performance of concrete bridges using long term inspection records, Proceedings of the 5th International Conference on Structural Faults & Repair, Edinburgh, Scotland, edited by Forde, M C, 1993, Vol 1, p 43-46.

5. RIGDEN, S R, et al, Service life prediction of concrete bridges, Concrete Repair, Rehabilitation and Protection, edited by Dhir, R K & Jones, M R, E & FN Spon, London, 1996, p 705-714.

6. BS 8110 Structural use of concrete: Part 1, Code of practice for design and construction, 1985.

PERFORMANCE EVALUATION OF BRIDGE BALUSTRADES IN SOUTH AFRICA

C E Ackerman
Pretoria Technikon
South Africa

ABSTRACT. This paper describes the current practice pertaining to the use of bridge balustrades and simulated vandalism testing of standard South African National Roads Agency (SANRA) precast pre-tensioned concrete bridge balustrades, in order to create feasible guidelines from which existing design and maintenance methods for bridge balustrades can be improved. This investigation was prompted by the occurrence of numerous cases of vandalism to in-service bridge balustrades and their subsequent structural failure coupled with high repair costs.

In order to shed some light on the performance of in-service, precast pre-tensioned concrete balustrades a simulated vandalism scenario was developed for testing. The tested balustrades failed at 55,88 % of the maximum load of 61,88 kg exerted by an average adult person. Together with the failure of the loaded post, two types of failure patterns occurred. Either the bonding between the posts and handrail, or the top end of the posts of a complete unit, failed. These failure patterns provide insight into and practical understanding of the in-service performance of bridge balustrades exposed to vandalism.

Based on the current practice pertaining to the use of bridge balustrade and the failure patterns obtained from the test results, the critical governing factors in the design and maintenance of bridge balustrades were identified. Hence, a pro-active design approach can be adopted to control the location of structural failure and confine vandal damage to local element failure. Recommendations are made on how to minimize the maintenance required in terms of vandalized balustrades.

Keywords: Bridge, Balustrade, Vandalism, Performance, Evaluation.

C E Ackerman obtained a Masters Diploma in Technology from the Pretoria Technikon in 1996 and the topic for his research was the "Finite Element Analysis and Experimental Validation of the Dynamic Behavior of a Bridge Structure". Currently he is a Senior Lecturer at the Department of Civil Engineering, Pretoria Technikon and is lecturing primarily in structural analysis and design. He is a candidate for Doctor's degree in Technology and the research topic is: "Structural Deterioration Prediction in a Bridge Management System Domain".

INTRODUCTION

A large number of bridges on the South Africa road network were built in the 1960's and 70's. During this period, the emphasis in bridge balustrade design was on structural strength and therefore numerous pedestrian and road bridges were equipped with balustrades manufactured from profiled steel sections. These were usually bolted to the top or sides of the kerbs and copings. However the exposed steel surfaces were affected by adverse environmental conditions that resulted in corrosion and rust stains, and remedial measures were required. Repeated painting and structural maintenance of these steel balustrades to maintain them to an acceptable standard proved to be laborious and costly. Furthermore, galvanized coatings on profiled steel sections deteriorate at an accelerated rate and their life is reduced in humid conditions and in salt-laden marine atmospheres (SABS 763 [1]) such as the east and south coast of South Africa. Moreover assessing the condition of concealed surfaces exposed to corrosive environmental conditions remains problematic.

A different design approach that emphasizes low maintenance but still maintains structural strength was then adopted for bridge balustrades. Aluminium balustrades proved to be maintenance-free and subsequently were installed on most of the pedestrian walkways on bridges in South Africa. The social retrogression of a portion of the South African population has lead *inter alia* to the escalation of vandalism to bridge balustrades. Components of aluminium balustrades are removed by means of cutting, and in the case of profiled steel balustrades by unbolting and then sold for its scrap metal value.

To address the problems experienced with these balustrades, and to curb other anti-social acts, a number of road authorities introduced the use of steel mesh enclosures over pedestrian bridges instead of balustrades. These enclosures have not gained universal acceptance, mainly due to financial considerations. During the late 1980's a precast, pre-tensioned balustrade was developed for the rehabilitation of bridge balustrades on the N2 Durban Outer Ring Road and the N3 from Durban to Mariannhill. The balustrades were developed to be unattractive to vandals, and cost-effective in terms of initial cost and ease of repair and maintenance (*SA Construction World*, April [2]). The balustrade was incorporated into a departmental standard plan by the Department of Transport and has been used since to install balustrades on new bridges and rehabilitate existing damaged balustrades.

In most European countries and North American states, aluminium, steel and concrete are extensively used for bridge balustrades and parapets [3]. Interestingly the Load and Resistance Factor Design (LRFD) bridge design specification of the American Association of State Highway Transportation Officials (AASHTO) [4] makes no specific reference to or provision for the prevention or minimizing of vandalism to bridge balustrades in the design requirements.

PROBLEMS RELATING TO BRIDGE BALUSTRADE
DESIGN AND MAINTENANCE

In order to determine the extent of vandalism to bridge balustrades and the various remedial actions currently in use, an inspection was carried out on a number of bridges on major national routes. The inspection was necessitated by the fact that certain bridge management systems available to the various authorities responsible for bridges, which could have assisted in developing a database of damaged bridge balustrades, were not yet in place. In addition, the cycle of inspections on which bridge management systems are based may extend up to 5 years for any one bridge, particularly for relatively new bridges.

Aluminium Balustrades

The inspections revealed that a large number of bridges equipped with aluminium balustrades showed some degree of damage. Temporary repairs effected to these damaged balustrades are usually of a very low standard. These repairs sometimes consist of as little as plastic coloured barrier tape, wire mesh or guardrails (Figure 1) mainly due to the low cost involved and the need for the urgent protection of pedestrians.

Figure 1   Guardrails used to repair balustrades of Bridge B488 over the N3

Other solutions for the temporary repair to damaged balustrades often consist of the use of any available material such as steel posts, wire mesh and razor wire secured to the bridge by various means. Core-drilling as a means of securing replacement members is expensive and the temporary installation of elements by means of bolting or wiring onto the remaining balustrade members is not an optimum solution. In the design of the older type of balustrades, no provision was made for cost-effective repair in the event of vandalism or other damage to the balustrades. Although these repairs are intended as temporary measures, they often become a permanent solution due to budgetary constraints and a lack of feasible vandal-resistant replacement alternatives.

The realization that there are no suitable alternatives for the repair of damaged aluminium balustrades, resulted in pro-active measures being developed to protect aluminium balustrades that had not yet been damaged. In some instances balustrades were even painted to disguise the aluminium. In other instances PVC-coated wire mesh was tied to the aluminium balustrades on some pedestrian bridges in the Cape Peninsula. However, it served little purpose since the mesh is easily removed, leaving the balustrades exposed to vandalism.

The *in situ* casting of a horizontal concrete beam at approximately half the post height, or a reinforced mortar fill between the flange and web of the H-section post, proved to be the most practical solution to protect undamaged aluminium balustrades.

Precast, Pre-tensioned Bridge Balustrades

Although the precast, pre-tensioned concrete balustrades were developed to be vandal proof, the in-service performance of these balustrades over the past 10 years has proved that it is still vulnerable to vandalism (Figure 2). Note that some of the posts in Figure 2 failed at mid-span indicating that the handrail was intact at the time of the act of vandalism.

Figure 2   Damage to balustrades of Bridge B1831 over the N4

In some geographical areas it appears that stolen handrails are used as lintels because of the dimensional similarity between the handrails and lintels. The handrail is normally not damaged during the act of vandalism due to its high concrete strength of 60 MPa and the presence of the prestressing strands. However once the handrail has been removed, the individual cantilevered posts are easily broken. The total failure of the prestressed units poses a serious risk to pedestrians.

The methods necessary to replace vandalized or damaged prestressed balustrade members of the design currently in use are complicated and expensive. It involves the removal of the remainder of the damaged posts, modifications and repairs to the coping or kerb and subsequent replacement of the posts. If the handrails are still in position, it has to be removed before the posts can be repaired. The required 300mm embedment depth of the post into the kerb as specified by the Code of Practice for the Design of Highway Bridges and Culverts in South Africa TMH 7 Part 3 clause 4.8.4 [5] contributes to the high costs of the repairs. As a result Van Niekerk, Kleyn & Edwards [6] prepared a report for the South African Roads Board to investigate a possible reduction of the embedment depth of the posts from 300mm to 150mm. The report concluded that the embedment depth of the post can not be reduced.

Summary

The in-service performance of profiled steel, aluminium and precast, pre-tensioned concrete balustrades showed that structural strength, serviceability, stability and durability requirements are not sufficient to protect pedestrians. The repair solutions for damaged balustrades are often temporary, vandal-prone and costly.

SIMULATED VANDALISM TESTING OF PRECAST, PRE-TENSIONED BRIDGE BALUSTRADES

In order to establish the most vulnerable elements of the precast prestressed balustrades, it is necessary to perform tests on in-service balustrades to determine actual failure patterns. These tests are performed to elucidate the performance of the balustrade when subjected to unusual loading conditions. The failure patterns obtained from the tests are then analyzed and the critical governing factors in the balustrade design and maintenance identified.

To perform a simulated vandalism test, it is necessary to define and quantify whatever acts of vandalism may be committed to damage a bridge balustrade. However it is almost impossible to define all the scenarios of vandalism that could possibly occur. The physical ability of human beings to destroy, combined with the tools available, is unlimited. Quantifying such scenarios seem even more difficult if one takes into account the variety in the determinant factors governing the magnitude of vandalism. Therefore not all the possible acts of vandalism that may be committed are described in this paper.   Moreover the performance of the bridge balustrade as it is subjected to a random act of vandalism outweighs in importance the definition of the specific act of vandalism that may be committed.

The Selected Simulated Vandalism Test

In order to select a suitable simulation of vandalism for testing, an inspection of existing vandalized bridge balustrades was performed. This approach seemed to be more practical than an investigation of all the possible acts of vandalism that could be committed. The balustrades of Bridge B1831 over the N4 national route near Pretoria, which was chosen for the test site, were extensively vandalized during 1998. An examination of the damage to the bridge balustrades revealed two distinct failure patterns.

Firstly, the bonding between the posts and handrail failed, which indicates that a vertical and upward loading was applied to the handrail, probably by means of kicking or by impact blows. The balustrade unit consists of sections comprising 10 posts epoxied to each handrail section. At the damaged parts of the balustrade, the entire handrail over the ten-post section was removed. This suggests that the ten posts together with the handrail, act as a unit to the applied loading. Inspection of dislodged handrails found at the bridge site, showed that each rail had been subjected to only one major impact. The transfer of load to the epoxy resin bonding between the post and handrail - which is not that strong in tension, mainly due to the smooth surfaces of the contact areas - is assisted by the high strength of the prestressed concrete handrail.

In the second failure pattern observed, a single post of the ten-post unit failed at its mid-height, indicating that a horizontal loading was applied to the approximate mid-height of the post, probably by means of kicking or by impact blows. One damaged post, with handrail removed, showed rubber markings probably caused by motorcycling over the bridge. The presence of a number of ten post units where the handrail has been dislodged where only one post had suffered mid-span failure, could not be explained and therefore this failure pattern has been selected to be simulated under controlled conditions.

Bridge Balustrade Properties and Instrumentation

The decision to perform the test on location on an in-service bridge balustrade, instead of laboratory testing under controlled manufacturing and installation conditions, was two-fold. Firstly, neither the manufacturing process nor the installation procedure of the balustrades is suspected to be the reason why the balustrades fail. The specifications for the manufacture and installation of these balustrades are sufficient to control any manufacturing or installation irregularities. Moreover, possible irregularities form part of the in-service performance of the balustrades. Secondly, expensive specialized equipment is required to manufacture the precast, prestressed concrete posts and rails, and the concrete mix design is complex. This means that it can be assumed that quality control in the manufacturing process should be acceptable and very little can be gained by attempting to simulate the manufacture of the components.

The balustrades of the bridge B1831 over the N4 national route near Pretoria were selected for testing. This decision was made because the span configuration of the bridge meant that no traffic accommodation was required on the N4 under the bridge, and also because of the easy accessibility of the bridge site. The balustrades consist of 1800mm long units each, which comprise a 1790mm long, 60mm deep by 110mm wide 60MPa concrete handrail pre-tensioned with two 8mm prestressing strands, and 10 hexagonally shaped 60MPa concrete posts, 1100mm long by 55mm by 55mm, pre-tensioned with one 8mm prestressing strand. The handrail is bonded to the posts with filled epoxy resin and the posts are installed 300mm deep into an *in situ* concrete kerb at 180mm centers.

A schematic plan layout of the bridge instrumentation is shown in Figure 3. A 31,5mm diameter hexagonally shaped steel bar, 2 metres in length, fitted with a single gauge element at the point of contact with the post, was used to exert a horizontally applied loading to the balustrade. The 3-wire method was used to connect the single gauge element to a 14-channel amplifier. The voltage was analyzed and recorded through a Yokogawa 3655 4-channel Analyzer Amplifier at 5 millisecond intervals. Finally, the measurements were multiplied with a calibration factor determined beforehand in the laboratory.

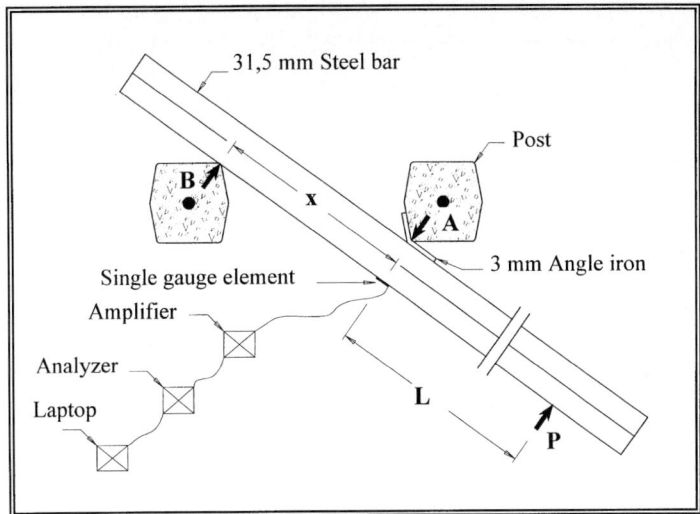

31,5 mm Steel bar

Post

B

x

A

3 mm Angle iron

Single gauge element

Amplifier

Analyzer

Laptop

L

P

Figure 3   Schematic plan layout of the bridge instrumentation

The Test Program and Discussion of Results

With the consent of the SANRA to perform in-service tests on their bridge, it was initially agreed that 24 posts would be sufficient for the tests. Tests were performed at two different post heights. Firstly, the steel bar was inserted between two adjacent posts just below the handrail. This position was chosen to verify the horizontal shear strength of the post and handrail bonding, which should be able to resist the applied loading, and to determine the maximum force that one average adult person can generate. The second series of tests was performed with the steel bar inserted between two adjacent posts at its mid-height. Three tests were performed with the steel bar inserted between two posts just below the handrail. The maximum loads measured at the strain gauge location A (Figure 3) for each of the three tests were 7,20kN, 7,10kN and 7,20kN. The generated load P (Figure 3) is calculated from the equation

$$P = \frac{x \times A \times 1000}{(L + x) \times 9,81}$$ (1)

where x is 155mm, 160mm and 157mm for test 1, 2 and 3 respectively and L is 1700mm for all three tests. According to Equation 1, the load P for the three tests yielded

$P_1 = 61{,}33\text{kg}, \ P_2 = 62{,}26\text{kg}, \ P_3 = 62{,}05\text{kg}$

It was not possible for an average adult person to damage the balustrades. Although the Code of Practice for the Design of Highway Bridges and Culverts in South Africa TMH 7 Part 2 clause 3.5.1.2 (c) [7] specifies a horizontal load of 27,52kg at mid-height for each post, the horizontal shear plane between the posts and handrail was able to resist an average maximum load of 61,88kg cantilevered by 1700mm.

For the second series of tests, it was initially anticipated that 12 tests (24 posts) with the steel bar at mid-height of the posts would be sufficient. However, only four tests were performed, due to the extensive damage caused by each test. For this test, the steel bar was inserted horizontally between two posts at mid–height and subjected to the loading. The loaded post failed during each test, and the maximum loads measured at the strain gauge location A (Figure 3) for the four tests were 3,790kN, 4,611kN, 4,187kN and 4,187kN. The load P (Figure 4) generated by the average person is calculated from Equation 1, where x is 155mm, 140mm, 155mm and 150mm for tests 1, 2, 3 and 4 respectively and L is 1700mm for all four tests. According to equation 1, the load P for the four tests yielded

$$P_1 = 32,282\text{kg}, \ P_2 = 35,763\text{kg}, \ P_3 = 35,663\text{kg}, \ P_4 = 34,606\text{kg}$$

The average load for the four tests is 34,579kg, which is 55,88% of the average maximum load of 61,88kg. Besides the failure of the post under loading, two separate failure mechanisms developed instantaneously just before the post failed.

Firstly, the bond between post and handrail for the complete 10-post unit failed during tests 1 and 4 (Figure 4). The bond probably failed due to the fact that the vertical distance between the handrail soffit and kerb for the posts bending under loading is shorter than the unloaded straight post length, which results in a vertical tension force at the post and handrail bonding. This tension force is transferred over the total unit length via the high-strength, prestressed handrail, and consequently hinges are formed at the bond under increased loading. Due to the hinge that develops at the bond between the loaded post and handrail, rotation of the post takes place. A considerable amount of additional moment is transferred to the point of load application, causing the failure of the post at mid-span.

Figure 4   Typical bond failure between posts and handrail (Tests 1 and 4)

The second and most significant failure mechanism developed at the top end of all 10 posts of the unit during tests 2 and 3 (Figure 5). Transmission length problems and eccentricity augmentation of the prestressing force in the post may be the cause of this failure pattern. The code specifies an embedment length of 300mm at the end of the post to compensate for the reduced force in the tendon brought about by insufficient strand anchorage at the post ends. The moments developed by the applied loading at the weak top end of the post is increased by additional moments generated by the eccentricity augmentation of the prestressing force in the post subjected to the applied load. The increased moments are transferred to the other posts of the unit through the high-strength, prestressed handrail, after which subsequent failure occurs in the transmission length zone. Due to the rotation of the post, a considerable number of additional moments are transferred to the point of load application, causing the failure of the post at mid-span.

Figure 5   Typical failure of top end of posts (Tests 2 and 3)

CONCLUSIONS AND RECOMMENDATIONS

The precast, prestressed concrete bridge balustrade performed well within the load specification of the code. However, the in-service performance of the balustrades showed that the code requirements are not sufficient to protect the balustrades against all types of vandalism. The simulated vandalism tests revealed an overall failure of a 10-post unit under one load application position. The presence of prestressing forces in the balustrades increased the magnitude of the applied loading, which resulted in complete failure of the prestressed posts. No provision can be made in the posts for the required transmission length of the prestressing force at the top ends, which should be at least 300mm.

The tests showed that the prestressed concrete balustrade is susceptible to serious damage as a result of vandalism and could pose a serious safety risk to pedestrians. It is the legal responsibility of the relevant road authorities to ensure the safety of road users and pedestrians. During 1994 a road user was seriously injured by a rock fall on Chapman's Peak Drive. The Cape High Court ruled that the Cape Metropolitan Council had been negligent in not placing special and effective signs to warn road users of the possible dangers of using the road and that the council failed to monitor and correct the impact of local weather conditions effectively. This ruling means that there is a legal obligation on the road authorities to ensure that items such as bridge balustrades are maintained to safe standards.

Since it is impractical to identify all the possible scenarios of vandalism and to quantify its results for possible inclusion in the code, the following guidelines are proposed to augment the current code requirements:

1. An appropriate specification for the design and manufacture of bridge balustrades should be compiled to address the problems identified during these tests.

2. Newly developed balustrades should be tested under specified loading conditions until failure occurs and the following criteria should be satisfied:

   The residual structure of the tested or damaged balustrade should remain sufficiently intact to protect pedestrians until rehabilitation or repairs can commence.

   It has to be accepted that balustrades may be damaged, therefore the design should lend itself to simple, fast and low cost repairs when such work is required.

3. Existing bridge management systems should be expanded to incorporate the following:

   An inventory module of pedestrian balustrades.

   A pedestrian balustrade evaluation form which focuses on signs of vandalism to balustrades for use by inspection officials.

   A regular inspection program should be developed and followed. Statistical prediction models can be used to optimize inspection programs.

REFERENCES

1. SOUTH AFRICAN BUREAU OF STANDARDS. Standard Specification for Hot-dip Zinc Coatings, SABS 763-1971, clause B-7, pp.17.

2. S A CONSTRUCTION WORLD. New designs adopted for bridge balustrades and balustrades. 1992. pp.33.

3. ARIZONA DEPARTMENT OF TRANSPORTATION. Combination Pedestrian-Traffic Bridge Railing. 1998. Drawing No SD1.04

4. AMERICAN ASSOCIATION OF STATE HIGHWAY TRANSPORTATION OFFICIALS. AASHTO LRFD Bridge Design Specification. 1998. 2nd ED.

5. CODE OF PRACTICE FOR THE DESIGN OF HIGHWAY BRIDGES AND CULVERTS IN SOUTH AFRICA. TMH 7 Part 3, clause 4.8.4. 1989. pp.108-109

6. VAN NIEKERK KLEYN & EDWARDS. Report on Performance Testing of Pre-tensioned Concrete Balustrade Posts for Reduced Embedment Length. Report No DR 91/26. 1991. Director-General: Transport.

7. CODE OF PRACTICE FOR THE DESIGN OF HIGHWAY BRIDGES AND CULVERTS IN SOUTH AFRICA. TMH 7 Part 2, clause 3.5.1.2. 1981. pp42-43

METHODOLOGY FOR CHARACTERIZATION OF ASR IN AGGREGATES IN CHIHUAHUA, MEXICO

C C Olague

G Wenglas

University Autonomous of Chihuahua

P Castro

CINVESTAV-IPN Merida

Mexico

ABSTRACT. This paper describes a study carried out to determine the potential reactivity of aggregates in the state of Chihuahua, México. The methodology development take into account: physiography, geology, climate, and hydrology, for chose sites for testing. 46 sites were analyzed: 22 gravel and 24 sand. According to the results of petrographic examination and X rays diffraction, 100% sites have mineral reactive like chalcedony, quartz crystalline, lithic rhyolites, andesites. The results of ASTMC 289 in sands indicates: 10 sites innocuous, 12 potentially reactive, and 2 reactive. In course aggregates indicates: 4 banks reactive, 11 potentially reactive and 7 innocuous. The ASTMC 1260 was used in 13 sands sites and 12 course aggregates, previously classified according to ASTM 289 like: 11 sands potentially reactive, 2 sands reactive, 10 gravels potentially reactive and 2 course aggregate reactive. The new classification results 10 sands sites reactive, 1 sand potentially reactive, 2 sands innocuous, 6 gravels sites reactive, 3 gravels sites innocuous and 3 gravels sites potentially reactive.

Keywords: Reactivity, Aggregates, Silica, Alkalis.

C Olague, is a Researcher Professor and Chairman of master programs in the Department of Civil Engineering at the University Autonomous of Chihuahua, Mexico. His research focuses mainly in concrete durability and repair and maintenance of concrete pavements.

P Castro, is a Researcher Professor in the CINVESTAV-IPN Mérida. His research focuses mainly in corrosion concrete.

G Wenglas, is a Researcher Professor in the Department of Civil Engineering at the University Autonomous of Chihuahua, México.

INTRODUCTION

The alkali silica reactive (ASR) is a two step physic-chemical reaction in portland cement concrete between alkalies and silica or silicates, in the presence of moisture [1]. The first step is a reaction between the alkalies and silica to form an hydrophilic expansive gel. In the second step, the gel absorbs moisture, swells, and creates an internal pressure that may crack the concrete [2].

The existence of ASR can be suspected through visual observation of crack patterns and past history of material sources. Confirmation of the presence of ASR, however, must be done in the laboratory using petrographic procedures. ASR can be avoided in new concrete with proper material selection [3,4]. This is done using the past history of materials or by testing new sources for ASR. New sources can be evaluated by quick chemical test (ASTM C 289), and rapid expansion mortar bars test (ASTMC 1260). When expansion exceeds 0.10% then remediation measures include selecting cement low alkali content, using non reactive aggregates, replacing a percentage of the cement with certain pozzolans, or using a chemical admixture such as a lithium compound.

The main purpose of this study is to present current knowledge and experience on ASR of aggregates fine and course in Chihuahua, México [5,6]. This study was based in four kind of test: petrographic examination (ASTMC 295), X rays diffraction, quick chemical test (ASTMC 289) and rapid expansion mortar bars (ASTMC 1260).

EXPERIMENTAL

In 1998, a program of studies on AAR was started by the University Autonomous of Chihuahua, México. These included investigations on the AAR potential of aggregates from a number of local quarries. The methodology used for selective sites tested is based according to severals factors like: physiography, geology, climate, and hydrology. This study considered two physiographic regions and nine provinces physiographic, that there are in the state of Chihuahua. According to the state geology, places where limestone and igneous stone are exploited were considered in this study. Chihuahua has 12 different climates, seven of them were considered.

In this study were considered 46 sites, 22 coarse aggregates and 24 fine aggregates were analyzed. The first essential step of the testing program is the petrographic examination (ASTMC 295), that consist in microscopic examination in a thin section complemented by X-Rays Diffraction. The X-rays diffraction test was used for detecting clays minerals and cryptocrystalline to microcrystalline quartz and poorly crystalline quartz. In some cases, a petrographer with strong AAR experience could recommend accepting an aggregate based solely on the knowledge of its petrographic characteristics, however, most of the time, additional testing is required to confirm preliminary indications from petrography. In a third step were analyzed the aggregates coarse and fine with quick chemical test (ASTMC 289). The quick chemical test evaluates aggregate reactivity by measuring the amount of dissolved silica and the reduction of alkalinity in the reaction alkali solution. In a four step fundamentally the aggregates that were considered potentially reactive according to ASTMC 289 were tested with rapid expansion mortar bars test (ASTMC 1260). Mortar bars, 25 by 25 by 285 mm in size, are prepared using specified aggregate size fractions, mixture proportions, and mixing procedures.

After 24 h moist curing in their moulds, the bars are placed in a sealed plastic container filled with water at 23° C, and the container stored in an oven at 80° C and measured hot periodically over a 2 week time period. The maximum 14 days expansion limit for non reactive aggregates ranges somewhere between 0.08% and 0.20%. In this study according to ASTMC 1260 it is considered 0.10% maximum expansion limit for non reactive aggregates.

The final expansion and expansion at intermediate points during the test was calculated as follows: calculate the difference between the zero comparatory reading of the specimen and the reading at each period to the nearest 0.001% of the effective gage length and record as the expansion of the specimen for that period, and it is reported average expansion of the three specimens of a given cement aggregate combination.

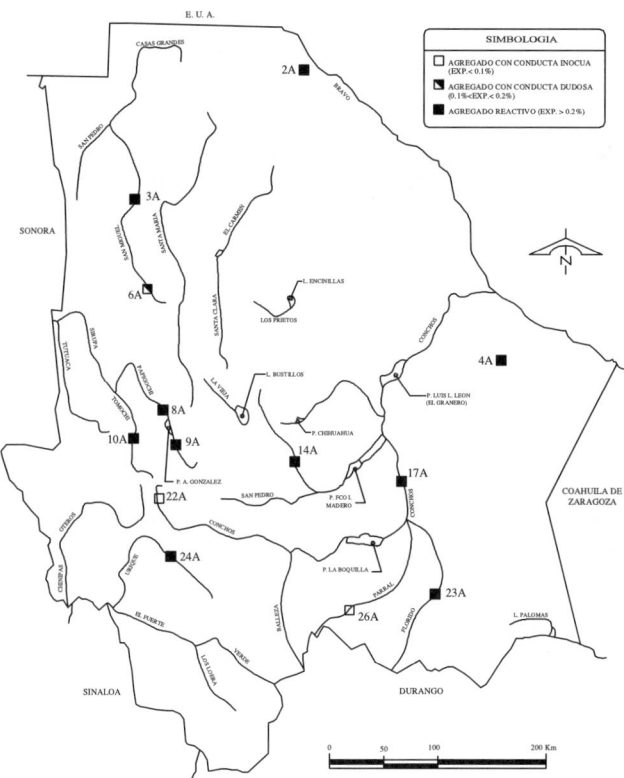

Figure 1   Maps of potentially reactive aggregates for sands according ASTMC 1260-94

The suggested limits of expansion in the ASTMC 1260 test are: <0.10 innocuous 0.10%-0.20% indeterminate, >0.20% reactive. It is not yet established if these limits can be applied universally (Mc Nally and Richardson 2000). Limits on the expansion of potentially reactive aggregates should be stated based on the particular experience, considering mineralogical composition of aggregates, amount of reaction products, and potential expansion of reaction products.

RESULTS AND DISCUSION

The Figures 1 and 2 present the selective sites for this study according to the methodology before described. Additionally is showed a classification of fine (see Figure 1) and course (see Figure 2) aggregates according to ASTMC 1260.

According to petrographic examination (ASTMC 295), the reactive minerals detected in coarse and fine aggregates were: quartz, andesites, chalcedony, lithic rhyolites, lithic andesites, cristoballite and opal [6]. These reactive minerals were detected in 100% of analyzed sites. The Table 1 shows the results of ASTMC 295 for each bank of gravel tested, complemented with X rays diffraction to detect if the quartz is crystalline or amorphous. The table 2 shows the results of ASTMC 295 and X rays diffraction for sand banks.

The quick chemical test (ASTMC 289) was used in order to have a preliminary characterization in terms of ASR for the state of Chihuahua Mexico. The results of reactivity for sand banks were: 2 reactive banks y 11 potentially reactive, this is showed in Table 2. In case of gravel banks the results were: 2 reactive and 10 potentially reactive, the Table 1 describes this [7, 8].

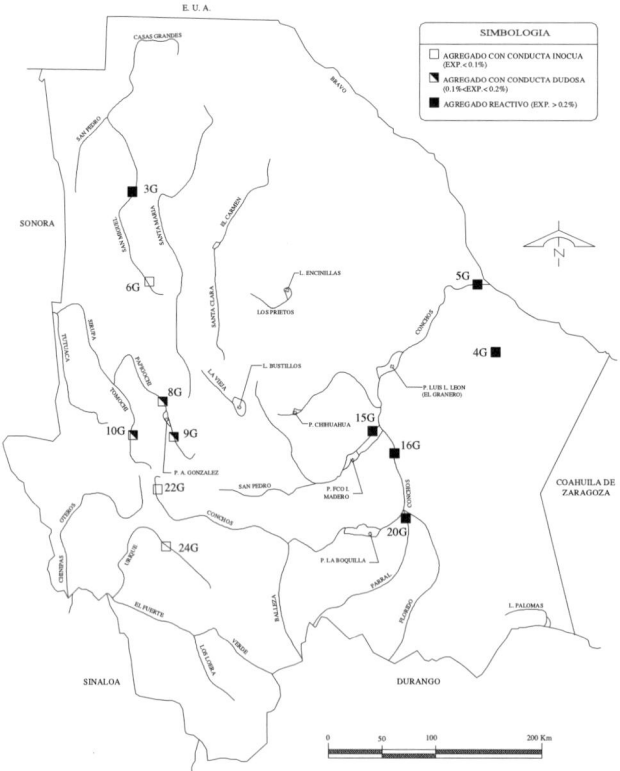

Figure 1   Maps of potentially reactive aggregates for gravels according ASTMC 1260-94

In order to have more realistic aggregates classification for ASR was proposed the rapid expansion mortar bar test (ASTMC 1260). The same sites of sands and gravel were tested with both test ASTMC 289 and ASTMC 1260. The Table 1 shows the results of sands: 10 reactive, 1 innocuous and 1 potentially reactive.

Table 1   Results of ASTMC 295, ASTM 289-94 and X-rays diffraction
for course aggregates

NUMBER OF BANKS	QUICK CHEMICAL TEST ASTMC (289)			X-RAYS DIFFRACTION	PETROGRAPHIC EXAMINATION (ASTMC 295)	
	Aggregates Classification	Sc	Rc	ASR	Reactive Minerals	%
3G	Potentially Reactive	719	226	quartz cryptocrystalline	andesites chalcedony quartz	1.1 3.4 11.3
4G	Potentially Reactive	448	278	quartz cryptocrystalline quartz	chalcedony quartz andesites	6.0 29.6 4.0
5G	Potentially Reactive	374	179	cryptocrystalline	chalcedony quartz	11.6 13.2
6G	Potentially Reactive	332	241	quartz cryptocrystalline	quartz lithic rhyolites andesites Chalcedony	10.7 6.0 4.5 1.0
8G	Potentially Reactive	478	228	quartz cryptocrystalline	Quartz lithic andesites lithic rhyolites Opal	10.8 1.6 1.6 1.0
9G	Potentially Reactive	495	128	quartz cryptocrystalline quartz	Quartz Andesites	18.8 1.4
10G	Potentially Reactive	368	182	cryptocrystalline	Quartz lithic rhyolites Chalcedony	12.8 12.9 0.6
15G	Potentially Reactive	531	304	no results	Cristobalite Quartz lithic rhyolites	1.6 7.8 6.2
16G	Reactive	509	110	quartz cryptocrystalline	Quartz	20.7
20G	Reactive	444	107	quartz cryptocrystalline	Quartz lithic rhyolites	16.6 8.4
22G	Potentially Reactive	319	247	quartz cryptocrystalline	Andesites Quartz	3.1 7.1
24G	Potentially Reactive	452	232	quartz cryptocrystalline	Andesites Quartz	3.7 16.1

The Table 1 shows the results of gravels: 6 reactive, 3 innocuous and 3 potentially reactive. The criteria used to determine deleterious potentially aggregates was: if the expansion is less than 0.1% the aggregates are considered innocuous, if the expansion is less than 0.2% are considered potentially reactive, and if is more than 0.2% are considered reactive.

Table 2　Results of ASTMC 295, ASTM 289-94 and X-Rays diffraction for fine aggregates

NUMBER	QUICK CHEMICAL TEST			PETROGRAPHIC EXAMINATION		X-RAYS DIFFRACTION
OF	ASTMC289			ASTMC 295		
BANKS	Aggregates Classification	Sc	Rc	Reactive Minerals	(%)	(ASR)
23-A	Potentially Reactive	511	199	andesites quartz Lithic rhyolites	5 10 35	Cryptocristalline quartz
8-A	Potentially Reactive	275	169	chalcedony quartz	20 20	Cryptocristalline quartz
10-A	Potentially Reactive	375	179	quartz Lithic rhyolites	20 30	Cryptocristalline quartz
26-A	Reactive	255	91	andesites quartz Lithic rhyolites	8 25 30	Cryptocristalline quartz
24-A	Potentially Reactive	448	143	andesites quartz Lithic rhyolites	7 50 20	Cryptocristalline quartz
4-A	Potentially Reactive	403	213	chalcedony quartz	10 35	Cryptocristalline quartz
17-A	Reactive	444	113	chalcedony quartz Lithic rhyolites	20 20 50	cryptocristalline quartz
3-A	Potentially Reactive	413	172	andesites chalcedony quartz	20 30 30	cryptocristalline quartz
9-A	Potentially Reactive	191	206	quartz Lithic rhyolites	15 30	cryptocristalline quartz
22-A	Potentially Reactive	354	268	quartz Lithic rhyolites	15 30	cryptocristalline quartz
6-A	Potentially Reactive	336	274	andesites quartz Lithic rhyolites	10 15 50	cryptocristalline quartz
2-A	Potentially Reactive	248	142	chalcedony quartz	10 50	cryptocristalline quartz
14-A	Potentially Reactive	623	194	quartz Lithic rhyolites	20 40	cryptocristalline quartz

Table 3   Results of expansions in sands at 16 days according ASTMC 1260-94

NUMBER OF BANKS	RAPID EXPANSION MORTAR BARS (ASTM C-1260-94)	
	Aggregates classification	Expansion
3G	reactive	0.215
4G	reactive	0.605
5G	reactive	0.499
6G	innocuous	0.079
8G	potentially reactive	0.195
9G	potentially reactive	0.102
10G	potentially reactive	0.152
15G	reactive	0.242
16G	reactive	0.265
20G	reactive	0.370
22G	innocuous	0.083
24G	innocuous	0.081

Table 4   Results of expansion in coarse aggregates at 16 days according ASTMC 1260-94

NUMBER OF BANKS	RAPID EXPANSION MORTAR BARS (ASTMC 1260)	
	Expansion	Aggregates classification
23-A	0.601	reactive
8-A	0.300	reactive
10-A	0.533	reactive
26-A	0.091	innocuous
24-A	0.461	reactive
4-A	0.511	reactive
17-A	0.755	reactive
3-A	0.477	reactive
9-A	0.424	reactive
22-A	0.029	innocuous
6-A	0.194	potentially reactive
2-A	1.028	reactive
14-A	0.498	reactive

CONCLUSIONS

The results of gravel banks shows that the banks 22G, 24G and 6G according with the results of ASTMC 1260 were considered innocuous but the results of ASTMC 295 and X-rays diffraction detected the presence of potential deleterious aggregates and the results of ASTMC 289 indicated that this aggregates must be considered potentially reactive. As a result of join test, the banks 22G, 24G and 6G must be considered potentially reactive and remediation measures must be taken. The same could be said for sand banks.

All testing methods are accelerated tests, it is clear that a combination of tests generally represents the best approach. The methods that have been proposed so far have their limitations. Some succeeds in identifying reactivity for certain aggregates whereas fails for others. Therefore it is difficult to ascertain an aggregate is absolutely non reactive using the currently available testing methods considered   individually. As a result of this study is recommended to realize several tests like: ASTMC 295, X rays diffraction, and ASTMC 1260 at least, complemented with inspection in situ of structures existents for detecting the potentially of ASR for new structures.

The ASR can take decades and produce extensive damage in structures. With the available knowledge base, there is no reason for potentially reactive aggregates to be used in concrete without an appropriate preventive action to minimize the risk of concrete distress due to ASR.

ACKNOWLEDGMENTS

The authors would like to acknowledge the support provided for the project by the National Council of Science and Technology of Mexico (contract 9704012). The authors are indebted with Emilio Caballero Morales and Jorge Luis Almaral for performing the test of this study and to UACH Grupo GCC, CEMEX and CINVESTAV IPN Unidad Merida.

REFERENCES

1. STANTON, T. (1940). Expansion of Concrete through Reaction Between Cement and Aggregate, proceedings, ASCE, V 66, Dec, pp 1781-1812.

2. STARK, D. (1994). Alkali Silica reactions in concrete, in: P.Klieger, J.F. Lamond (Eds), Significance of Tests and Properties of Concrete and Concrete-Making Materials, ASTM STP 169C, ASTM, Philadelphia, PA pp 365-371.

3. MIELENZ, R. (1958). Petrographic Examination of Concrete Aggregate to Determine Potential Alkali-Reactivity, Research Paper No 18-C, Highway (Transportation) Research Board, pp 29-38.

4. MIELENZ, R. (1978). Petrographic Examination (Concrete Aggregates), Significance of Tests and Properties of Concrete Making Materials, STP- 169B, ASTM, Philadelphia, Chapter 33, pp 536-572.

5. CABALLERO, E. (1999). Evaluations of materials for rule potentially Reactive Aggregate in Concrete Pavements. M.Eng. Thesis, Universidad Autónoma de Chihuahua, México 1999 pp 90.

6. ALMARAL, J. (1999). Physical and Chemical Characterization of aggregates for Concrete Pavements. MEng Thesis, Universidad Autónoma de Chihuahua, México pp 138.

7. OLAGUE, C., Castro P., López W. (2001). Alkali silica reactive of aggregates in the state of Chihuahua, México. International Congress of IRF, París France.

8. OLAGUE, C., LÓPEZ W., CASTRO P. (1999). Caracterización de la reactividad potencial de agregados para uso en pavimentos rígidos de la ciudad de Chihuahua en México. V CONPAT, Montevideo Uruguay.

CHLORIDE PENETRATION INTO SWEDISH ROAD BRIDGES EXPOSED TO SPLASH FROM DE-ICING SALTS

A Lindvall

Chalmers University of Techology

Sweden

ABSTRACT. This paper describes a study of chloride penetration into seven reinforced concrete road bridges in and around Göteborg on the west coast of Sweden. The bridges represent typical Swedish reinforced concrete motorway bridges with an age between 25-35 years and a w/b between 0.45-0.50. The chloride ingress is determined as chloride penetration profiles, where the quotient between chloride and calcium content at different depths from the surface is shown. The results show on large variations in chloride penetration both between different bridges but also on one single bridge. The variations in chloride ingress can be explained mainly by variations in the environmental actions but also variations in the material properties and execution during construction.

Keywords: Chloride penetration, Environmental actions, De-icing salt, Motorway bridges, Reinforced concrete.

Mr A Lindvall, is a Doctoral-Student at the Department of Building Materials at Chalmers University of Technology in Göteborg. His research focuses mainly on the environmental actions and how they influence the durability of reinforced concrete structures.

INTRODUCTION

Chloride induced reinforcement corrosion is one of the major reasons for reduction of the service life for reinforced concrete structures. Concrete structures are exposed to chlorides along roads where thaw-salts are spread or in or close to oceans with water that contain chlorides. The surfaces of the concrete structures are either directly exposed to chloride contaminated water, snow-slush, splash or aerosols (spray). The splash and aerosols are produced either by the drainage system of vehicle tyres or from breaking waves. The transport of chlorides into the concrete takes place principally in three different ways, [1]: (i) permeation of salt solution, (ii) capillary absorption and (iii) diffusion of free chloride ions. A common parameter for all these three transport mechanisms is that they require a certain level of moisture in the pore-system of the concrete. When the capillary pores are relatively dry, absorption dominates, and when they are relatively saturated, diffusion dominates. Thus to get a complete understanding of the chloride conditions it is required to measure both the chloride ingress and the moisture conditions in the concrete.

The transport and distribution of chlorides in concrete is very much a function of the environmental actions. The environmental actions on a concrete structure can be described as the temperature, moisture and chloride conditions and the concentration of carbon dioxide at the surface of the structure. The temperature and moisture conditions can be expressed as equivalent surface temperatures and equivalent surface humidities and time of wetness.

During the autumns of 1998 and 1999 seven reinforced concrete bridges in and around Göteborg on the west coast of Sweden were studied with respect to environmental actions and response. The investigations were made possible with funds from the Swedish National Road Administration, Vägverket. In the following a short description of some of the results from the investigation are presented – for more comprehensive descriptions see [2] or [3].

FIELD STUDIES

The field studies were made in November and December of 1998 and 1999. Seven reinforced concrete bridges in and around Göteborg have been included in the study. The columns and side-beams on the bridges have been examined for chloride penetration on selected positions. Cores have been drilled from the concrete for further analysis in the laboratory.

Before cores can be taken from a bridge the sampling locations have to be determined. The sampling locations are normally chosen based on the following:

- **Security**. The security involves the security for the staff involved in the sampling, e.g. work close to the traffic and at heights, and the safety for the structure, e.g. protection of the reinforcement. With structures in use cores must be taken in such way that the reinforcement is not unnecessarily damaged.

- **Availability**. The sampling locations must be available for the sampling equipment, e.g. drilling machine and cooling system. Samplings close to the road may require that the road is partly or completely closed and samplings at high heights may require a lift etc.

- **Desires from the owner**. The owner of the structure may have special desires of where the cores should be taken.

- **Material**. Cores should not be taken from inhomogeneities in the concrete, e.g. damages, graffiti and where previous samples have been taken.

After the cores have been taken from the structure they have been transported to and stored in the laboratory before the analysis. The storage is made in a careful way so the samples are not influenced or changed, e.g. due to redistribution of moisture and chlorides. To avoid these kinds of effects the time for transport and storage is kept as short as possible. Before the cores are analysed they are prepared by means of profile grinding. The diameters of the cores have been 50 mm.

Examined Bridges

The examined bridges have been chosen in such a way that they represent typical Swedish reinforced concrete bridges with an age of 25-35 years. All examined bridges have been designed and constructed according to valid Swedish concrete standards. The bridges are built with a K400 concrete, with a cement content of approximately 300-360 kg/m^3 and a w/b of approximately 0.45-0.50. In the following a short description of the examined bridges is given. A more extensive description can be found in [2] and [3]. The following bridges have been included in the study:

- **Bridge N 434**. Bridge, built in 1972, south of Göteborg, over the motorway E6 (four lanes with safety lanes on each side and exit- and entry-roads to the motorway), where E6 crosses the road between Kungsbacka and Onsala. Two columns and one side-beam have been examined. The motorway has much traffic and the speeds are high (>100 km/h) and the local road has fairly much traffic and the speeds are fairly high (60-70 km/h).

- **Bridge O 670**. Motorway bridge, built in 1967, north of Göteborg, where the motorway E6 (four lanes with safety lanes on each side) crosses the river Nordre Älv. The bridge is one of the largest bridges in Sweden, regarding the number of supporting columns and the area of the bridge-deck. One side-beam and the underside of the bridge slab (close to a joint between two slabs) have been examined. The motorway has a lot of traffic and the speeds are high (>100 km/h).

- **Bridge O 707**. Motorway bridge, built in 1989, placed in the centre of Göteborg, where an exit road from the motorway E6 crosses a municipal road. The exit-road has two lanes and no safety lanes. The bridge is constructed as a 180°-curve over the municipal road and a parking place. One side-beam and one column have been examined. The exit road has little traffic and the speeds are low (20-30 km/h) and the municipal road has also little traffic and the speeds are low (30-40 km/h).

- **Bridge O 762**. Motorway bridge, built in 1974, east of Göteborg, where the motorway Rv40 (five lanes with safety lanes on each side) crosses the road between Landvetter and Partille. One side-beam and one column have been examined. The motorway has relatively much traffic and the speeds are relatively high (90-100 km/h) and the local road has fairly much traffic and the speeds are fairly high (60-70 km/h).

- **Bridge O 832**. Motorway bridge, built in 1972, east of Göteborg, where two municipal roads crosses the motorway E20 (six lanes with safety lanes on each side). The bridge can be divided into three principal parts: two bridges that cross the motorway E20 and one bridge running parallel with the motorway. One side-beam and one column have been examined. The motorway has much traffic and the speeds are high (90-100 km/h). The local roads have fairly much traffic and the speeds are not so high (30-40 km/h).

- **Bridge O 951**. Motorway bridge, built in 1972, south of Göteborg, where a local road crosses the motorway E6 (four lanes with safety lanes on each side). One column has been examined. The motorway has much traffic and the speeds are often high (>100 km/h).

- **Bridge O 978**. Motorway bridge, built in 1974, east of Göteborg, where a local road crosses the motorway Rv40 (four lanes and safety lanes on each side). One column and one side-beam have been examined. The motorway has much traffic and the speeds are often high (100 km/h) and the local road has little traffic with low speeds (30-40 km/h).

The columns on bridge N 434, O 951 and O 978 have been examined on three different heights and in two (side column) and four (middle columns) directions towards the traffic.

Environmental Actions

The temperature and moisture conditions for the examined bridges have been assumed equal to the regional climate in Göteborg. In Table 1 the monthly extremes for Göteborg are given for minimum and maximum monthly means. However, the exposure environment varies between the bridges due to variations in the road environment. The road environment for an individual bridge follows from the amount and speed of the traffic, the amount of de-icing salt spread on the road number, the number of occasions of spreading during a season and the distance between the roadways and the structural parts.

The yearly amount of de-icing salt spread on the roads at the examined bridges varies between 2.3-3.1 kg/m^2. If 20 g/m^2 of de-icing salt is spread each time this means that de-icing salt is spread between 115 and 155 times each season! The distance between the road surface and the examined columns and side-beams is between 2.0 and 4.0 m.

Table 1 Environmental actions (regional climate) for Göteborg. Monthly means values during the period 1961-1990 at the meteorological station Säve in Göteborg. Data from [4]

METEOROLOGICAL DATA	MIN	MEAN	MAX
Air temperature – February[1]	-7.4°C	-1.7°C	+4.7°C
Air temperature – July[1]	+14.4°C	+16.2°C	+18.3°C
Air humidity, RH - January [1]	76%	87%	92%
Air humidity, RH - June [1]	61%	76%	82%
Precipitation – February	1 mm rain	41 mm rain	131 mm rain
Precipitation - November	23 mm rain	85 mm rain	171 mm rain
Solar radiation - December[2]	4 W/m^2	10 W/m^2	19 W/m^2
Solar radiation - June[2]	133 W/m^2	232 W/m^2	328 W/m^2

[1] Min and Max represent 5%- and 95%-fractiles respectively
[2] Min and Max represent 10%- and 90%-fractiles respectively

LABORATORY STUDIES

The concrete in the bridges has been analysed for chloride and calcium content at different depths. In the following the method used to determine the chloride and calcium content is briefly described. A more extensive description of the method is given in [3].

Chloride Conditions

The chloride conditions in the concrete samples have been determined as chloride penetration profiles, where the quotient between the chloride and calcium contents at different depths is given. The reason to analyse both the content of chloride and calcium is that the chloride ions are transported and bound only in the cement paste. If only the chloride content, relative the sample-weight, is measured the amount of aggregate in the sample will influence the result. However, if the chloride content is related to the binder content the effects from variations in the aggregate will be minimised. The method does not work if the aggregate contains calcium, e.g. limestone, but since the aggregate in Sweden only contains small amounts of calcium, this means that the analysed calcium content can be used to estimate the binder content.

The chloride and calcium contents have been determined from powder samples taken from the cores. The powder samples have been taken from each core by means of profile grinding and have been analysed for acid-soluble chlorides and calcium. The calcium and chloride content in the samples have been determined with potentiometric titration with calcium and chloride selective electrodes.

RESULTS

The results from the study are presented as chloride penetration profiles, where the quotient between chloride and calcium content at different depths is given. It should be observed that all results give the state at the time for sampling. The chloride penetration in a road environment varies over the year, which means that results from the analyses depend on the time for sampling, [6].

The chloride conditions are determined as chloride penetration profiles. All chloride penetration profiles are analysed from cores taken in November or December, i.e. before or in the beginning of the season when de-icing salts are applied. In Figure 1 a selection of chloride penetration profiles from all the examined bridges are presented. The first index shows if the profile comes from a column (C), side-beam (SB) or the underside of the bridge-slab (Ö), the second index shows which bridge the profile comes from and the third index indicates the name of the profile. A complete presentation of the results from the study is given in [2] and [3].

The columns on bridge N 434, O 951 and O 978 have been examined on three different heights over the roadways and in two or four directions towards the traffic. A selection of the chloride penetration profiles from these bridges is presented in Figure 2. The profiles have been selected in such way that the profiles with highest and lowest chloride penetration on each bridge are shown.

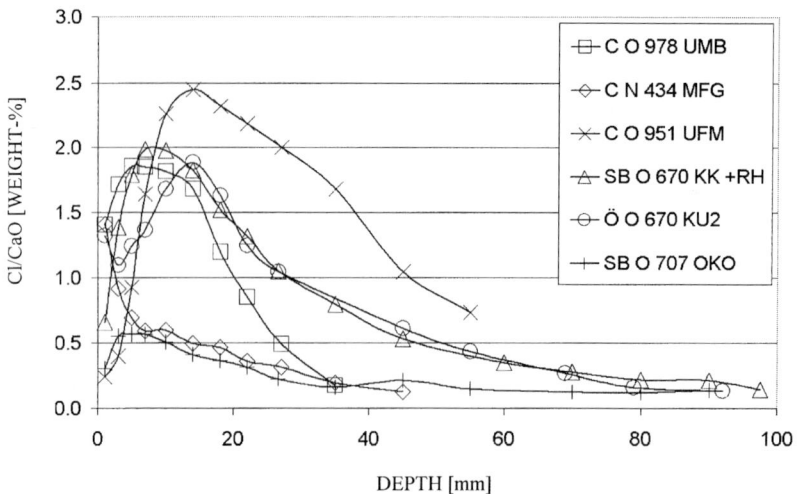

Figure 1   A selection of chloride profiles from all bridges examined in the study

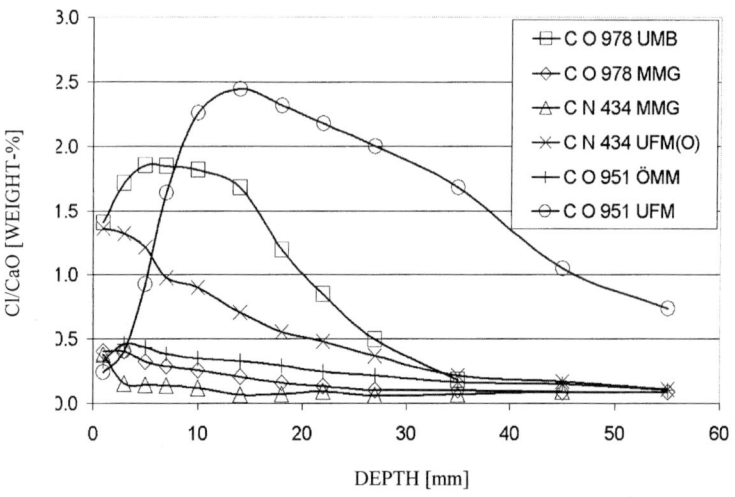

Figure 2   A selection of chloride profiles from the columns on bridge N 434,
O 951 and O 978

The chloride profiles from the examined bridges have been evaluated with curve fitting to the error-function solution of Fick's second law. The curve fitting has been done in accordance with a new method, proposed in [6], where the profile is divided into an outer convection zone, where the profile does not fit to the error-function, and an inner diffusion zone, where the profile fits to the error-function. The curve-fitting has resulted in four parameters: an apparent diffusion coefficient, D_{F2}, an apparent surface chloride concentration, C_{sa}, the thickness of a convection zone, x_c, and a surface concentration of chlorides for the diffusion-zone, C_{sc}. The results from the curve-fitting are presented in [2] and [3].

DISCUSSION AND ANALYSIS

Large variations have been observed in the chloride penetration profiles depending on the height over the roadway and orientation towards the traffic. The variations in chloride penetration in different directions on one column on bridge O 978 are shown in Figure 3a and 3b, where chloride profiles from surfaces facing towards and from the traffic from Borås respectively are presented.

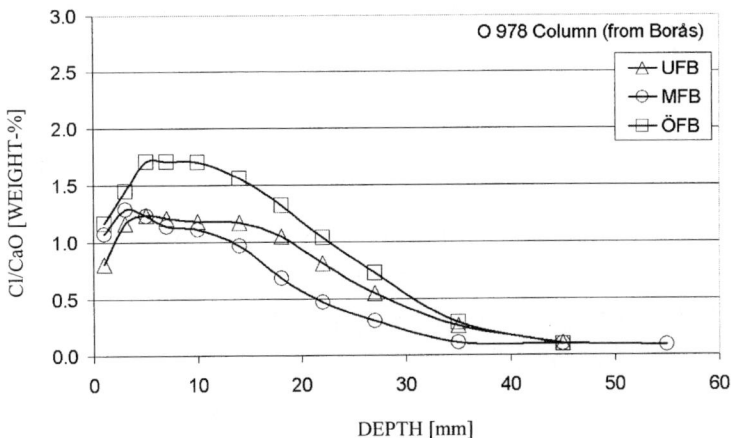

Figure 3a   Chloride profiles from the surface facing towards the traffic from Borås

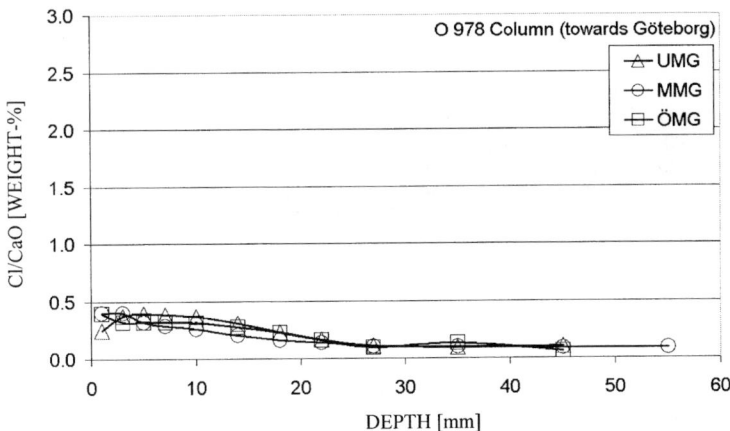

Figure 3b   Chloride profiles from the surface facing from the traffic from Borås

An explanation to these results can be that at certain wind directions airborne chlorides, i.e. aerosols from the road, will follow the air-stream and be deposited on the lee-side (direction FB) of the column. These chlorides will not be washed away but instead each time with "right" wind-direction the chloride concentration will increase.

In direction MG chlorides are constantly washed away on all heights over the roadway, resulting in lower chloride penetration. A schematic pattern for the wind-streams around bridge O 978 is shown in Figure 4.

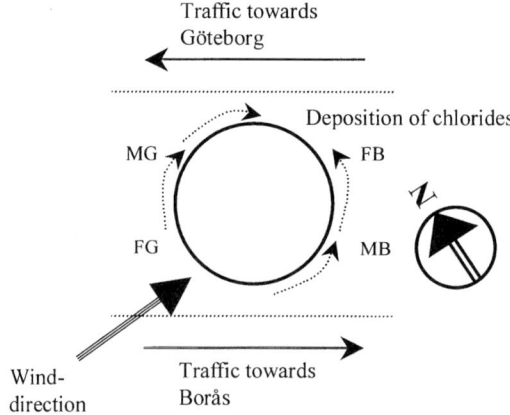

Figure 4   Schematic pattern for the wind-streams around bridge O 978

De-icing salts are normally applied during the nighttime or early in the morning. Thus, the surface on a concrete structure exposed to splash from the morning commuter-traffic is also the surface with the highest chloride exposure. This can be observed on bridge O 951 where the surface orientated towards the traffic to Göteborg (index FM) has larger chloride penetration compared to the other surfaces, see Figure 5.

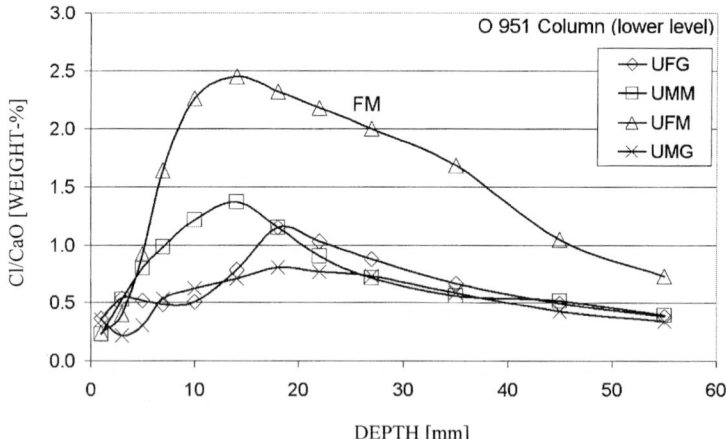

Figure 5   Chloride profiles from the lower level on bridge O 951

The effect from the orientation of the examined surface (horizontal/vertical) has been investigated on the side-beams on bridge O 978 and N 434. The results are shown in Figure 6, where filled symbols represent chloride profiles from horizontal surfaces (index H) and non-filled symbols represent vertical surfaces (index V).

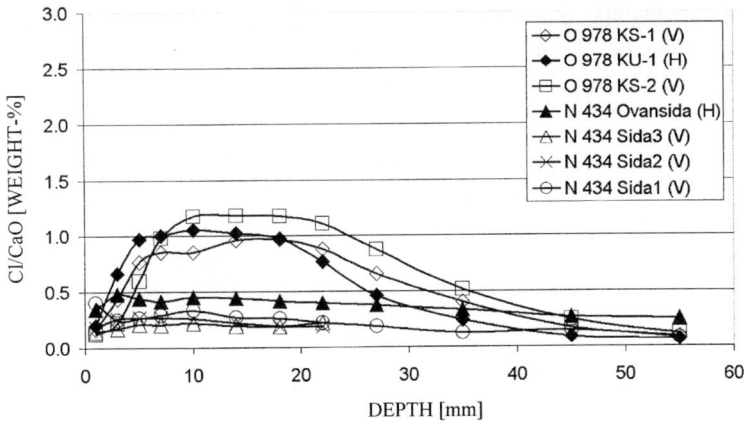

Figure 6   Chloride profiles from side-beams examined on bridge O 978 and N 434

The result presented in Figure 6 show that there is no significant difference between chloride penetration in horizontal and vertical surfaces on the side-beam. However the differences in chloride penetration between the examined bridges are significant – where the chloride penetration into bridge N 434 is significantly lower than into bridge O 978. Possible explanations for this are variations in the exposure environments and the material properties, e.g. effects from surface treatments.

The chloride penetration profiles from the columns and side-beams have different shapes, cf. Figure 2 and Figure 6, where the profiles from side-beams have a larger outer convection zone. This indicates that the exposure for chlorides takes place with different mechanisms on columns and side-beams, where columns are mainly exposed to aerosols while side-beams are mainly exposed to direct splash.

CONCLUSIONS

The objective of the presented study was to get information about the physical processes that govern degradation of concrete and the response from the concrete. The following conclusions have been drawn from the results:

- **Each bridge must be treated separately**. The results show that there are large variations between the examined bridges, especially concerning the chloride penetration. Thus to avoid large variations, when data on chloride penetration from different bridges are analysed, each bridge must be treated separately.

- **Conditions for chloride penetration**. The chloride penetration into a concrete structure is mainly a function of the exposure conditions but is also influenced by the concrete properties. Differences in exposure mechanisms can be identified for the chloride penetration into side-beams (mainly exposed to splash) and columns (mainly exposed to aerosols).

- **Variations in chloride penetration into one structural element**. The chloride penetration is influenced from the orientation direction towards the traffic, the height over the roadway and the microclimate (surface climate), e.g. wind-streams around the structure and the shelter against combined precipitation and wind.

Finally it should be pointed out that together with the workmanship, the maintenance and the microclimate have a decisive influence on the durability of the concrete in the bridge. Thus, when an assessment of the condition of a bridge is made, it is important to identify the parts of the bridge where the microclimate is most severe and where possible defects due to lack of workmanship and maintenance may adversely influence the durability.

REFERENCES

1. BASHEER, L, KROPP, J, CLELAND, D J, Assessment of the durability of concrete from its permeation properties: a review, Construction and Building Materials, 2001, No 15, p 93-103.

2. LINDVALL, A, ANDERSEN, A, Undersökning av kloridinträngning, armeringskorrosion, frysning och fukttillstånd på sju brokonstruktioner exponerade för tösalter (Investigation of chloride ingress, reinforcement corrosion, frost and moisture conditions on seven bridge-structures exposed to thaw-salts), Publication P-00:8, Department of Building Materials, Chalmers University of Technology, Göteborg, 2000, p 141 (in Swedish).

3. LINDVALL, A, Environmental Actions and Response - Reinforced concrete Structures exposed in Road and Marine Environments, Publication P-01:3, Department of Building Materials, Chalmers University of Technology, Göteborg, 2001, p 313.

4. HARDERUP, E, Klimatdata för fuktberäkningar – Väderdata från tio meteorologiska mätstationer i Sverige (Climate data for moisture calculations – Weather data from ten meteorological stations in Sweden), Report 3025, Department of Building Technique, Lund University, Lund, 1995, p 50.

5. AASHTO, Standard Method of Sampling and Testing for Total Chloride Ion in Concrete and Concrete Raw Materials, American Association of State Highway and Transportation Officials, Designation: T-260-84, Washington, 1984, p 12.

6. NILSSON, L-O, ANDERSEN, A, LUPING, T, UTGENNANT, P, Chloride ingress data from field exposure in a Swedish road environment, Publication P-00:5, Department of Buildings Materials, Chalmers University of Technology, Göteborg, 2000, p 29.

PERFORMANCE, REPAIR NEEDS AND RENOVATION OF PLAIN CONCRETE PAVEMENT OF AN INTERNATIONAL AIRPORT

S Z Bosunia

A M Hoque

Bangladesh University of Engineering & Technology

Bangladesh

ABSTRACT. The runway and the traffic areas of the Zia International Airport (ZIA) at Dhaka, Bangladesh were constructed with plain concrete during 1968-69. Due to non-completion of ancillary structures and other facilities the airport was not opened to routine air traffic till 1980. Subsequently the airport was opened to wide body aircraft in the mid-eighties though the runway and other areas were not initially designed for their use. The slab panels of the runway developed cracks even before the airport was opened to traffic. Such cracks became wider and propagated randomly with the operation of air traffic. The poor performance of the runway pavement demanded immediate repair and renovation works to be taken up in the early nineties since that was the only operational runway at the International Airport in Bangladesh. A crack survey revealed extensive cracks ranging from a few millimetres to more than 50 mm wide. In order to determine the repair needs, sub-grade soil characteristics, testing of concrete core samples taken from the runway, detail classification and condition of cracks were investigated. The methodology to renovate the entire runway and other air traffic areas included removing the patch repair works executed earlier, repairing the cracks of various widths, placing high density fabric over the cracks, joints and repaired areas and finally constructing the hot-mix asphalt overlay in 2-3 layers on the entire pavement area. The repair works and overlay constructions were carried out at night keeping the runway operational during the day. Since the completion of the repair and renovation works in 1996 the runway has been in service and no sign of distress has been observed.

Keywords: Performance, Repair, Airport pavement, Renovation.

Professor Shamim Bosunia is a Professor of Civil Engineering at Bangladesh University of Engineering and Technology (BUET), Dhaka. He obtained a PhD degree from University of Strathclyde, Glasgow in 1979. His research interests include cement and concrete technology, repair and rehabilitation of concrete structures and low cost construction.

Professor Alamgir Hoque is a Professor of Civil Engineering at Bangladesh University of Engineering and Technology (BUET), Dhaka. He obtained a PhD degree from University of Leeds, UK in 1980. His research interests include transport infrastructure planning and design, construction materials and management.

INTRODUCTION

The runway pavement at Zia International Airport (ZIA) in Dhaka, Bangladesh is a plain cement concrete (PCC) runway.   The runway was designed in early sixties and the construction was started in 1965 and completed in 1968-69.   The completion of the entire project was delayed due to several interruptions such as the liberation war of Bangladesh in 1971.   The runway and the other traffic area initially designed for middle sized aircraft were opened for routine service in 1980.   Subsequently the runway was opened for the operations of heavy and wide-bodied aircraft such as DC10, B747 in the mid-eighties.

The subgrade of the pavement was constructed by placing local soils and the total thickness of the compacted fill varied between 900 mm to 4200 mm.   A 150 mm thick plain concrete sub base having a proportion of 1:3:6 by volume (cement: sand: coarse aggregate) was provided.   The concrete runway is 3200 m long with 274 m long stopway (overrun) strips at both ends.   The runway width is 45.6 m with 7.6 m paved shoulders on each side.   The size of runway slabs at the two ends (152.3 m each) is 7.6×7.6m and 7.6×6.1 m in the interior. Taxiways are formed by 7.6×7.6 m slabs.   The edge slabs along both sides of the runway have a longitudinal contraction joint that divides them to a 3.8 m width.   The thickness of the concrete slab along the centre line of the runway is 300 mm and that for the edge is 250 mm.

The plain concrete runway pavement showed cracking even prior to the application of air traffic.   Subsequently many cracks developed on the runway pavement, taxiway and aprons. These cracks increased in number and in size progressively in the course of repetition of aircraft loads.   The Civil Engineering Department of Bangladesh University of Engineering and Technology (BUET), Dhaka was engaged by the Civil Aviation Authority of Bangladesh (CAAB) to provide the consulting services for the renovation of the pavement.

The consulting services required survey and documentation of all the cracks of the entire pavement, determination of the characteristics and condition of the sub-grade soil, evaluation of the structural safety of the pavement, recommendation of the methodology to repair the cracks.

The services also included the preparation of the technical specifications including documents for international tendering for the overlay construction only during night times. The services also included the total supervision of the renovation works of the runway pavement.   This paper presents the results of the investigations and detailed methodology developed for the repair and rehabilitation of the runway pavement.

CRACKS IN RUNWAY PAVEMENT

The plain cement concrete slabs of the runway developed severe cracks in the form of longitudinal linear cracks spread over the entire length of the runway.   Most of the slab panels near the runway centre line underwent multiple cracks and some of the panels near the touchdown and take off areas of the runway were wrecked to pieces.   Like the runway slabs the aprons and the taxiway also developed cracks, however, the intensity of such cracking both in number and width was much less than the cracks in the runway slabs.

Crack survey

To assess the repair needs an extensive survey of the existing cracks was carried out in 1993 and documented. The cracks were classified into four categories according to their width. A typical account of cracked pavement is presented in Figure 1.

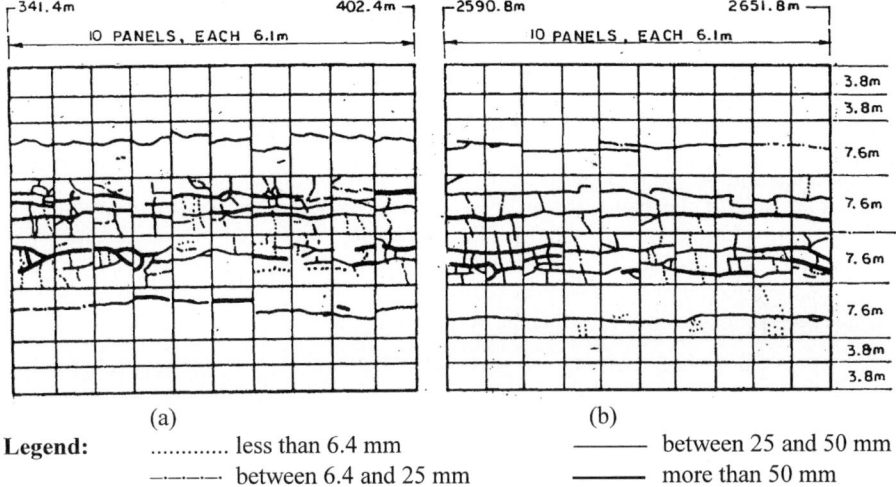

Legend: .... less than 6.4 mm   ——— between 25 and 50 mm
                   —·—·— between 6.4 and 25 mm   ——— more than 50 mm

Figure 1   Typical portions of the slabs at the touch-down and take-off areas of runway showing cracks of varying widths (a) North end (b) South end

Cracks of width less than 6.4 mm

Cracks of width less than 6.4 mm, well defined and visible were categorized in this type. Concrete adjacent to these cracks was found to be in good condition.   Cracks of such category had not been repaired previously.

Cracks of width between 6.4 mm to 25 mm

Cracks in this category were very well defined and many such cracks were repaired without cutting their sides.

Cracks of width between 25 mm and 50 mm

In this category the width refers either to the actual crack width in the pavement, or to the state of the crack at the time of the inspection after spalling of concrete had taken place from the adjacent areas. The earlier repair works for such cracks had been preceded by cutting 25 mm wide grooves and subsequent filling up the grooves with epoxy based joint sealant.

Cracks more than 50 mm wide

In this category the cracks were repaired earlier by the CAAB after cutting 150 mm wide and 75 mm deep grooves and then filling with asphalt concrete to the level of the pavement. The cases where two minor cracks were formed of about 50 mm spacing with the concrete in between weathered were also grouped under this category.

During the field survey it was observed that about one third of the joint seals of transverse and longitudinal joints (1524 m) were in poor condition and required resealing before the placing of the overlay.

SOIL INVESTIGATION PROGRAMME

A soil investigation was carried out to determine the characteristics and existing condition of the sub-grade soil of the pavement in order to assess whether the origin of cracks was associated with sub-grade soil properties. The investigation included the identification and classification of the sub-grade soil, determination of shrinkage and swelling characteristics of the same and evaluation of the condition of compaction. The strength properties of the sub-grade soil were estimated for the evaluation of the condition of runway pavement.

The sub-grade soil was found to be typically silty clay soils containing little to moderate amount of sand and of low to medium plasticity.   The shrinkage characteristics were evaluated in terms of shrinkage limit and linear shrinkage that varied from 14% to 17% and 7% to 13% respectively. Moisture density relationships and in-situ density of the subgrade soil were also determined. Compaction of the sub-grade soil and in-situ density test results showed that the sub-grade soil was in a state of fairly good degree of compaction.

The average value of linear shrinkage of the sub-grade soil was found to be moderate indicating a moderately low degree of shrinking. Swelling tests on disturbed and laboratory compacted samples indicated very low values of free swell and swelling pressure. The low shrinking and expansion of the sub-grade soil indicated that the development of cracks was not due to expansive nature of the underlying sub-grade soil. The underlying soil was also found to be very well compacted. As such the origin of the cracks in the concrete pavement could not be related with sub-grade soil characteristics and existing condition.

STRENGTH OF CONCRETE IN PAVEMENT

Before the recommendation of the repair methodology for the cracked pavement of the runway an investigation to evaluate the adequacy of concrete strength of the cracked pavement was needed. The strength of concrete in the runway pavement (both for the pavement slab and the lean concrete sub-base) was assessed by taking concrete cores at various locations from the main runway as well as from the taxiway. This evaluation was also supplemented by the strength evaluation of concrete using Schmidt hammer tests. Concrete core samples of 100 mm nominal diameter were collected to a depth of up to 300 mm covering the full depth of the pavement slab. Prior to compression testing, the cylinders were sized to have an approximate height to diameter ratio of about 2. The core samples from the runway showed an average compressive strength of 25.5 N/mm^2 and those from the taxiway showed an average value of 28.3 N/mm^2.

In the absence of appropriate flexural strength data, estimates of the flexural strength of the pavement concrete were obtained using the relationship of Walker and Boelm (1960). These estimates of flexural strength were found to be 3.7 N/mm^2 with a standard deviation of 0.34 N/mm^2 and 4.0 N/mm^2 with a standard deviation of 0.2 N/mm^2 for the pavement in runway and in taxiway respectively.

The condition of concrete in the pavement was also evaluated using Schmidt hammer test on various locations. The average rebound numbers, when converted to equivalent cylinder strength using a calibrated relationship indicated an average strength of 36.2 N/mm^2 with a standard deviation of 1.1 N/mm^2 in the runway pavement. For the sub-base concrete, where brick chips were used as coarse aggregate, the compressive strength of the core samples were found be about 24.82 N/mm^2 with a standard deviation of 0.17 N/mm^2. Such strength was found to be adequate for the sub-base concrete.

Flexural strengths, estimated from the compressive tests on concrete core samples from the existing runway pavement were found to be adequate. Considering the soil condition beneath the runway pavement and the adequacy of the concrete strength of the pavement, repair needs for the pavement were recommended. The needs included several preparations including the repair of all types of cracks on the pavement and finally the laying of the 200 mm thick (maximum) hot-mix asphalt overlay on the entire pavement areas of the runway.

PREPARATION OF PAVEMENT FOR OVERLAY

Careful and thorough preparations of the existing pavement prior to the laying of overlays were essential for good construction and effective overlay performance. The following preparations were made besides the repair of cracks of the concrete pavement. The repair methodology developed for the cracks of the pavement is given in a separate section.

Removal of rubber and paint markings

The rubber deposits and the paint markings on the pavement were removed using high pressure water jets, chemicals, high velocity particle impact and by mechanical grinding following relevant Federal Aviation Administration Advisory Circular (FAA 1978).

Seating slabs using heavy rollers

Heavy rollers having a minimum weight of 50 tons moving at a speed of 4 to 8 km per hour were used for seating of the slabs up to 15.25 m on either side of the centre line of the runway. Where movement of the slab could be noticed, during the movement of the rollers, the slab was under-sealed using asphalt meeting the requirements of ASTM Specification D3141 following the specification for Under-sealing Portland Cement Concrete Pavements with Asphalt (CL-13) of the Asphalt Institute, USA.

Repair of spalled areas

The spalled areas along the joints and elsewhere on the runway were removed by sawing the slab approximately 25 mm back of the spalled concrete and to sufficient depth (min 50 mm) until sound concrete was reached. The surface was then cleaned and dried. Emulsified asphalt tack coat (slow setting CSS-1 or CSS-1 h) was then applied on the surface and edges.

The areas were then back filled with hot asphalt mixture. The mixture was then satisfactorily compacted with hand operated vibrating roller. In case the compacted thickness of the fill was more than 75 mm, the fill was placed and compacted in layers.

Repair of patched areas

In the past years, the CAAB as a part of its routine maintenance had applied asphalt concrete patches for about 930 m^2 of slab panels. The thickness of these patches varied between 25 to 75 mm. Considering the poor condition of these patches the existing bituminous patch works were removed totally. If cracks were found underneath, they were repaired following appropriate methods specified later. A new patch was recommended to be placed in advance of the overlay when the average thickness of the patch was more than 35 mm.   Such patchwork was applied over an area of about 465 m^2.

The steps for applying the patchwork included (a) cleaning of the surface (b) application of emulsified asphalt tack coat (CSS-1 or CSS-1h) (c) placement of hot asphalt mix, similar to that specified for the surface course of the overlay on the tack coated surface (d) compaction of the mixture to 98 percent of laboratory density using rollers and (d) checking of the level of the compacted patch using a 5 m straightedge or a stringline.

Repair of corner breaks

These diagonal cracks forming a triangle with a longitudinal edge, joint, a transverse joint or crack were repaired before the overlay operation. After marking such broken corners those were removed using pavement saw/cutter. The sub-base was cleaned of all debris and the areas were patched with dense graded asphalt concrete that was placed in layers and compacted using rollers.

Resealing of joints

The crack survey revealed the poor condition of both longitudinal and transverse joints of the runway pavement. The joint seals were damaged and protruded above the pavement surface in many places. It was estimated that about one third of such joints (15240 m in total) were badly in need of resealing and repair. Such joints were cleaned, resealed and repaired before placing the overlay. The following methods were used for cleaning and resealing the joints.

1. The old seals were ploughed out to a depth of 38 mm at least.
2. The vertical faces of the joints were cleaned to remove foreign materials and the excess old seal materials from the pavement surface at least 38 mm on each side of joint was also cleaned.
3. Vertical faces of joints and the pavement surfaces were sand blasted at least 38 mm on each side of the joint. Hand tools were used to remove any traces of old seal that might be left. The joints were finally cleaned with compressed air.
4. A concrete joint sealer satisfying the requirements of ASTM D1191 was placed into the joint. The outer ends of transverse joints were dammed to prevent sealing material from running out on to the shoulder.
5. A tack coat of hot AC-20, satisfying the requirements of ASTM D3381, was uniformly applied on the joint areas and a layer of high density fabric, a stress-relief interlayer material consisting of high-density, heavy-duty mastic between two layers of rugged polyester fabric (McAdams, 1987) of width of 300 mm was placed on the tack coat.

REPAIR METHODOLOGY FOR CRACKS IN PAVEMENT

A repair methodology for the repair of cracks of different categories was developed and recommended. The entire repair work was carried out at nights and the runway, being the only runway of the airport, was kept operational for air traffic during the days.

Repair of cracks of width less than 6.4 mm

Field survey revealed the presence of about 3353 m of cracks in this category. Of these about 2743 m of cracks were not wide enough to receive any sealing material with ease. About 610 m of wider cracks in this category were sealed. The cracks were repaired by cleaning the pavement surface surrounding the cracks of all dirt, dust and loose material, using a wire brush, stiff bristled brooms and compressed air. The cracks were then filled with hot asphalt AC-20 by pressure injection method.

Repair of cracks of width between 6.4 mm and 25 mm

There was about 3657 m of cracks in this category. The vertical faces of the cracks to a depth of at least 25 mm and pavement surface at least 25 mm to each side of the cracks were sand blasted. The cracks were then cleaned of loose debris, dirt, previously placed filler material and vegetation using wire brushes, shovels and compressed air. The cracks were filled with a rubberized asphalt that met Federal Specification S-SS-1401 and satisfactory for local environmental conditions. Concrete joint sealer of the hot-poured elastic type (ASTM 1190) meeting the requirements of ASTM D1191 was used. A tack coat of hot AC-20 was then applied followed by placing a 500 mm wide high density fabric. The width of the asphalt tack coat application was the fabric width (500 mm) plus 50 mm on each side and was applied no further in advance of overlay material placement. Fabric was sized and cut in such a way that it overlapped the cracks to be repaired with a minimum of 300 mm at each end. The fabric also overlapped a minimum of 63 mm in the direction of paving operations. Typical crack repair and placement details of fabric are shown in Figure 2.

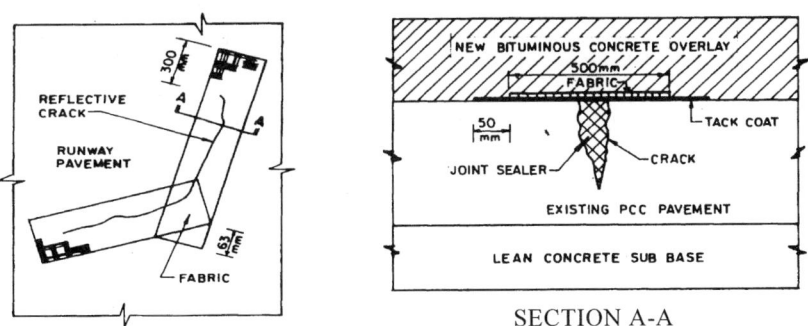

SECTION A-A

Figure 2   Typical repair method of cracks of width between 6.4 mm
and 25 mm and placement detail of fabric

Repair of cracks of width greater than 25 mm

Three types of cracks of this category were identified and the repair of all such types were recommended accordingly.

Type 1

Type 1 cracks were single non-interconnected cracks which were relatively straight, and where cutting of grooves could be done without causing damage in the adjoining areas of the cracks. Total length of cracks of this type was about 6096 m. Such cracks were repaired using the following procedure.

(a)   The area was marked taking at least 25 mm from the edge of the cracks.
(b)   Trenches 300 mm wide and 75 mm deep were formed, as shown in Figure 3, using concrete cutting wheel/machine.
(c)   The debris and milled materials were removed using sweepers and then the dust was blown by an air compressor.
(d)   Crack/cavity at the bottom of the trench, when present, was filled up either by a liquid sealer in case of cracks of width 6.4 mm to 25 mm or by a hot-mix asphalt for large cracks and cavities. A hand tamper was used to compact the hot-mix asphalt in the bottom of the trench.
(e)   A tack coat of hot AC-20 was applied on the bottom surface and the edges of the trench.
(f)   The high-density fabric was placed.
(g)   Additional tack coat of hot AC-20 was applied on top of the fabric.
(h)   The trench was then filled with hot-mix asphalt, similar to that specified for surface mix, and compacted to 98% of laboratory density using a roller.
(i)   Another tack coat of hot AC-20 was applied on the trench area and another layer of fabric of width of 600 mm, was placed.
(j)   Hot-mix asphalt overlay was placed following the placement at the second layer of fabric. Prior to hot-mix overlay, the top of the fabric was tack coated over along with the existing surface.

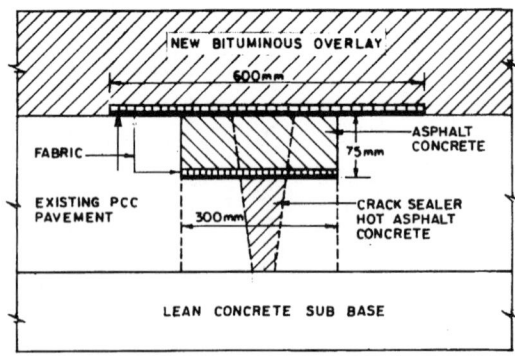

Figure 3   Repair of wide cracks (more than 25 mm in width, Type 1)

Type 2

The larger cracks, for which 150 mm wide and 75 mm deep trenches were cut and filled with asphalt concrete previously by the CAAB were included in this type. Survey revealed about 2800 m of such cracks were repaired by the following methods.

(a)   The details specified for the repair of Type 1 cracks were used for the cracks, where 300 mm wide trenches could be cut without damaging the adjacent areas.

(b)   Where trench cutting was not possible without damaging the concrete in the adjoining areas, the following steps were adopted for the repair works:

    (i)   The existing asphalt mix in the trench was ploughed out.

    (ii)   The vertical edges and bottom of the trench were cleaned and all foreign materials were removed.

    (iii)   If no crack/cavity was found at the bottom of the trench, the trench was filled up with asphalt concrete as shown in Figure 4(a). Tack coat of hot AC-20 was uniformly applied on the edge and the bottom of the trench before placement of the hot-mix asphalt. The mix was compacted using vibratory rollers.

    (iv)   If cavities were found at the bottom of the trench, those were filled up with a surface course of hot-mix asphalt and compacted to the level of the bottom of the trench. The mixture was tamped appropriately in place as shown in Figure 4(b). Tack coat of hot AC-20 was uniformly applied on the vertical edge and bottom of the trench and the trench was filled with hot-mix asphalt and compacted to the level of the pavement slab.

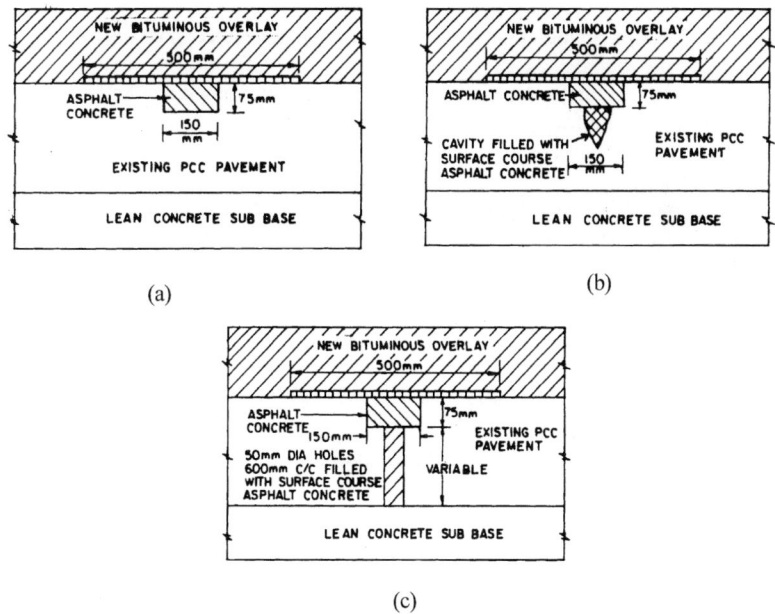

(a)

(b)

(c)

Figure 4   Repair of wide cracks (more than 25 mm in width, Type 2) (a) cracks of no cavity at bottom of trench; (b) cracks with cavity at bottom of trench; (c) wide cracks at bottom of trench (v)　If cracks were found at the bottom of the trench, 50 mm diameter holes were drilled along the crack 600 mm centre to centre up to the bottom of the slab. Using an air compressor with hose and nozzle attachment, the dust from the holes was blown out. An asphalt (AC-40) meeting the requirements of ASTM D3381 was heated to a safe temperature to make it sufficiently fluid suitable for pumping through the holes. After the asphalt had hardened the holes were filled with an asphalt surface course mixture, thoroughly tamped in place. Tack coat of hot AC-20 was uniformly applied on the vertical edge and bottom of the trench and the trench was filled with hot-mix asphalt and compacted to the level of pavement slab.

(vi)　A tack coat of hot AC-20 was uniformly applied on the trench area followed by placement of fabric of width 500 mm according to the procedure shown in Figure 4(c).

(vii)　Hot mix asphalt overlay was then placed after a uniform application of tack coat of hot AC-20 on the fabric.

Type 3

Crack survey revealed the presence of many random and interconnected wide cracks where trench cutting similar to that for Type 1 cracks was not feasible and any attempt to groove cutting was likely to cause serious damage to the runway slab. A total length of about 6096 m of cracks of this type was identified. The following repair methodologies were used for such cracks.

(a)　Holes, 50 mm diameter were drilled through the entire slab on 600 mm centre to centre along the crack.

(b)　Debris from the holes was removed and using compressed air the holes were cleaned of dust, loose and foreign materials.

(c)　The cracks and holes were filled up using materials as mentioned in step (iv to v) for Type 2 cracks.

(d)　A tack coat of hot AC-20 was then uniformly applied and high-density fabric of width 500 mm was placed following standard procedures as shown in Figure 5.

(e)　Hot mix asphalt was then placed.

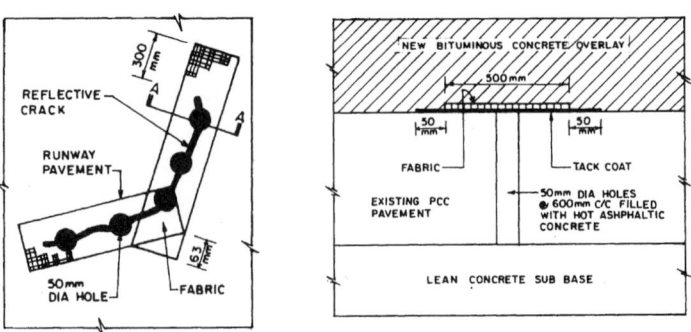

Figure 5　Repair of wide cracks (more than 25 mm in width, Type 3)

CONCLUSIONS

A crack survey revealed extensive cracks from less than 6.4 mm to more than 50 mm on the entire concrete pavement of the runway at the Zia International Airport, Dhaka, Bangladesh. The repair need was necessitated for structural strengthening of the runway pavement. Repair methodology was developed for each category of crack. High-density stress relief fabric was laid over the repaired cracks, and over the longitudinal and transverse joints of the runway slab. The repair work was followed by the application of 200 mm thick flexible (bituminous) overlay on the pavement. The entire repair works and the construction of the hot-mix asphalt overlay were carried out at night and the runway, being the only runway of the airport, was kept operational for air traffic during the day. The overall performance of the work has been reported to be satisfactory and the runway has been in service including operation of wide-bodied aircraft.

REFERENCES

1.   ANNUAL BOOK OF ASTM STANDARDS, vol 04.03, "Road and Paving Materials Traveled Surface Characteristics", ASTM, 1986.

2.   ASPHALT IN PAVEMENT MAINTENANCE, Manual Series no 16, The Asphalt Institute, 1983.

3.   McADAMS, R R, "Use of Fabrics/Mastics to Control Reflective Cracking in Bituminous Overlays : State-of-the-Art'. Proceedings of the 10th Annual FAA/Penn Slate Airport Paving Conference, Hershey, Pennsylvania, USA, 1987.

4.   FEDERAL AVIATION ADMINISTRATION (FAA), "Advisory Circular 150/5320-12", United Sates Department of Transportation, Washington, 1978.

5.   WALKER, S., AND BOELM, D L, Effects of Aggregates size on properties on concrete, ACI J. 32(3), 283-298, 1960.

THEME TWO:

ASSESSMENT AND REPAIR TECHNIQUES

DOCUMENTATION OF ELECTROCHEMICAL MAINTENANCE METHODS

O Vennesland

Norwegian University of Science and Technology

Norway

ABSTRACT. The electrochemical methods used for corrosion protection of concrete structures are cathodic protection, electrochemical realkalisation and electrochemical chloride removal. By these electrochemical methods the protective properties of the concrete are either reinstated (chloride removal, realkalisation), or the potential of the reinforcement is changed to a value that brings back passivity to the reinforcement (cathodic protection).

The execution of the processes should be controlled by current measurements (all processes) or by potential shifts (cathodic protection).

For electrochemical chloride removal the result should be documented by chloride analyses of the concrete before and after treatment. For electrochemical realkalisation the result should be documented by the pH value of the concrete pore water. For both chloride removal and realkalisation the current charge per unit area should be documented.

Keywords: Reinforcing steel corrosion, Electrochemical maintenance methods, Documentation.

Professor O Vennesland, is Professor at the Department of Structural Engineering at the Norwegian University of Science and Technology in Trondheim, Norway. His research focuses mainly on corrosion of reinforcing steel, concrete durability and repair and maintenance of concrete structures.

THE PROCESSES

General

The processes are reviewed in the final report of COST 509 [1]. All the three processes are generally based on applying a potential to the reinforcing steel. The potential is applied by means of an external electrode as shown in Figure 1.

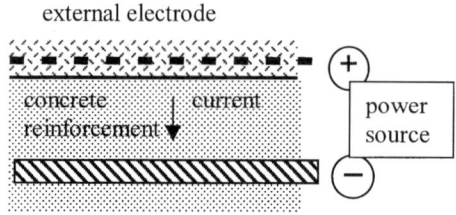

Figure 1   Set-up for electrochemical maintenance methods [2]

For electrochemical removal and electrochemical realkalisation the current – and subsequently the polarisation is high – while for cathodic protection the current is two orders of magnitude less (Table 1).

Table 1 Main differences between electrochemical maintenance methods [2]

METHOD	DURATION	TYPICAL CURRENT DENSITY
Cathodic protection	Permanent	10 mA/m^2
Realkalisation	Days to weeks	1 A/m^2
Chloride removal	Weeks to months	1 A/m^2

Cathodic Protection

Cathodic protection is based on changing the potential of the reinforcing steel to more negative values, reducing potential differences between anodic and cathodic sites and thereby reducing the corrosion current to negligible values. A CP system protects the steel as long as the current is flowing, that is, as long as the system is functioning properly. Many practical projects have shown that cathodic protection is an alternative to conventional repair and in many cases CP is the best choice of maintenance method for structures suffering reinforcement corrosion, in particular when the corrosion is due to the presence of chloride.

Electrochemical Chloride Removal

The principle of electrochemical chloride removal is to apply a direct current between the reinforcement, which is made to act as a cathode, and an anode that is placed temporarily on the outer surface of the concrete. The anode is normally an activated titanium wire mesh. Sometimes the anode is a reinforcing steel net that corrodes during the process.

The anode is surrounded by an electrolyte, which is an aqueous solution of calcium hydroxide or water in a paper pulp mash. The negatively charged chloride ions in the concrete migrate away from the reinforcement towards the positively charged anode. After the process is finished, the anode and the electrolyte with the incorporated chloride ions are removed from the structure.

Electrochemical Realkalisation

By electrochemical realkalisation the pore system of the concrete becomes alkaline. The pH around the steel is increased and the passivating properties of the concrete are restored. The technique involves passing a current through the concrete to the reinforcement by means of an externally applied anode that is attached to the concrete surface – as for chloride removal. For realkalisation the electrolyte is a 0.6 molar solution of sodium or potassium carbonate

The development of realkalisation is schematically shown in Figure 2. The Figure 2a shows the extent of carbonation before treatment. The Figure 2b shows the progress of alkaline material after a short duration: around the reinforcement (due to electrolysis) and from the surface into the concrete (due to absorption, electro-osmosis and/or diffusion). The right hand Figure 2c shows a more advanced state of realkalisation, where the alkaline zone around the steel has become continuous with the alkaline material penetrating from the surface.

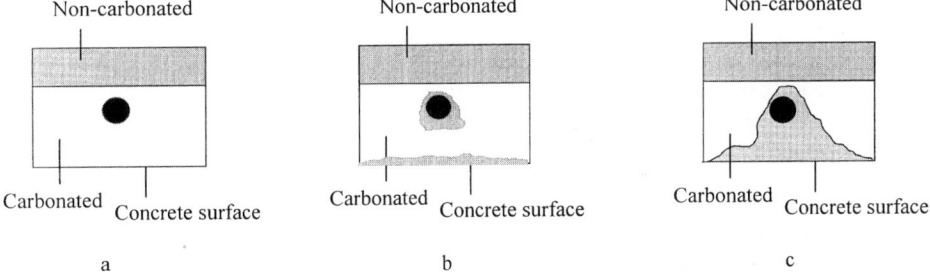

Figure 2   Development of realkalisation; left untreated, middle after a short treatment,
right after longer treatment [2]

Realkalisation of concrete made with blast furnace slag cement seems much more difficult than realkalisation of concrete with Portland cement [2]. In one case, after 14 days of realkalisation, nearly total realkalisation was observed in case of the use of Portland cement, as shown by phenolphthalein colouring purple. However, after 28 days of treatment, in case of the use of blast furnace slag cement, only a small "pink" ring around the rebar was observed after spraying a phenolphtalein pH indicator. Similar effects were found with fly ash cement concrete.

It appears that concrete made with slag cement or fly ash cement needs considerably more electric charge than Portland cement concrete before it has reached a high pH over a substantial part of the cross section. Consequently, the treatment time is longer for structures made with slag or fly ash cements.

PRE-CONTROL

Before any decision is made regarding the repair and/or maintenance method for a structure, an investigation clarifying the structural condition should always be made. This is even more important if electrochemical methods are considered. The assessment of the structure should include a general survey, identifying the presence of structural cracks, deformations and other visual defects. If such defects are present to a significant level, the treatment should be reconsidered and structural repairs carried out. Where structural repairs are not necessary, the inspection should focus on the preparation for electrochemical treatment. The following items should be measured over representative parts of all areas (or types of concrete elements) to be treated:

- Concrete cover to the steel.
- Chloride content and distribution.
- Depth of carbonation.
- Electrical continuity of the reinforcement.
- Electrical continuity of the concrete.
- Presence of potentially alkali-reactive aggregates.
- Presence of pre-stressing steel.

PROCESS CONTROL

General

The current and voltage should be measured on a daily basis and, if necessary adjustments shall be made according to pre-set values.

Cathodic Protection

The protection offered by a CP system must be controlled at regular intervals. Measuring the so-called depolarisation potential controls the performance of an installed cathodic protection system. The depolarisation potential is the potential decay that is measured by interrupting the protection current and monitoring the change of the steel potential against a reference electrode over a period of 24 hours (normally). This potential change is recorded at many representative points in the concrete structure using embedded reference electrodes.The principle is shown in Figure 3

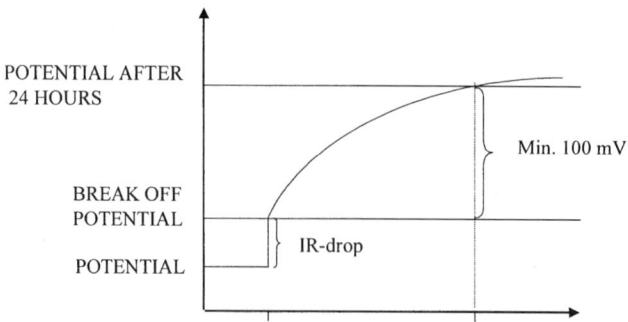

Figure 3   Depolarisation at a cathodically protected concrete structure

Electrochemical Chloride Removal

The efficiency of an electrochemical chloride removal project is controlled by analyses of the chloride content of the concrete. The specimens for chloride analysis should be taken as closely as possible to corresponding analyses before the treatment started (as a part of the pre-control). When the chloride analyses are based on drilled dust the bore diameter should be at least 16 mm. After drilling or coring the holes must be filled with mortar.

For process control any analysing method – to the contractor's preference – may be used. When typical field methods are used it is advised that the results are controlled by the method that shall be used for final documentation.

Electrochemical Realkalisation

The efficiency of an electrochemical realkalisation is controlled by the pore water pH of the concrete. As process control the pH is normally controlled by the phenolphthalein-test.

However, it has to be pointed out that the phenolphthalein indicator turns from clear to pink as the pH rises over about 9. This may leave the steel in a non-passivated condition. Thymolphtalein, however, has a colour change at a pH close to 10. Pollet and Dieryck [3] therefore advise to use thymolphtalein and not phenolphthalein. To be visible on concrete, the thymolphthalein solution has to contain 1% thymolphtalein in a solution of 70 % ethyl alcohol. The specimens for testing should be taken both above reinforcing steels and between steels.

FINAL CONTROL

Electrochemical Chloride Removal

After the application of chloride extraction, non-destructive measurements may be carried out, for controlling the durability of this treatment, several months or years after its decommissioning. Such further controls (types of measurements, locations of checking points, etc.) shall be defined jointly by the contractor and the structure owner [4].
Specimens for chloride analyses shall be taken above and close to the reinforcing steels in a pre-arranged set-up. Location of specimens depends upon the reinforcement (amount, centre distance and cover thickness). The locations should be marked before the process starts – e.g. by plastic markers – corresponding to the locations for determination of the chloride content during pre-control (structural assessment). Alternatively the locations might be determined and agreed upon by localising the reinforcement by use of a cover meter.

While during the process any analysing might control the chloride reduction method the final documentation shall be made by using generally accepted methods by renowned laboratories. Preferable the national standardised method(s) should be used.

The advises in the Norwegian Technical Specifications [5] are:

1. For documentation of the chloride content one set of specimens shall be taken from each $50 \, m^2$ concrete surface and at least 2 sets of specimens from each treatment area. Each set of specimens shall give results from 3 locations.

2. At each location 2 individual specimens shall taken – consisting of two drilled cores with minimum diameter 25 mm.

3. One core shall be drilled above and in to the reinforcement; the other core shall be drilled close to and to a depth of the rear of the reinforcement.

4. The chloride removal current shall be measured within each section.

Sometimes it is advisable to record the development of the electrochemical potential after finishing the process – either by using embedded reference electrodes or by surface potential measurements. Such measurements must be made after a long resting period (e.g. one year) in order to present safe values [3]. This is because of the strong polarisation of the reinforcing steels during the process.

Electrochemical Realkalisation

Phenolphthalein is a widely used pH indicator for the measurement of the carbonation depth. As earlier discussed it has to be pointed out that the phenolphthalein indicator turns from clear to pink as the pH rises over about 9 while thymolphtalein has a colour change at a pH close to 10. Realkalisation is linked with the current density and with the total charge flow. Contractors applying this method operate with an empirical value between 200 to 450 Ah/m² [2]. These values are obtained with treatment duration between 8 and 18 days with a current density of 1 about A/m² of steel.

When sodium carbonate is used as electrolyte the determination of sodium content in depth profiles might be used as additional control of the treatment efficiency (and correspondingly potassium analyses when the electrolyte is potassium carbonate). As the sodium and potassium content can vary strongly in concrete depending among others on the cement type, the results of the measurements should always be compared to measurements before realkalisation.

ACCEPTANCE CRITERIA

General

The acceptance criteria of the final results shall be determined and agreed jointly by the structure owner and the applicator of the process.

Cathodic Protection

It is generally agreed that the effect of cathodic protection shall be controlled by the depolarisation (100 mV) within 24 hours [6].

The questions still are:

- How many reference electrodes?
- Where shall the electrodes be placed?
- What percentage of electrodes shall achieve the requirements?
- When - after start of the process - shall the requirement be valid?

These questions are commented in Table 3.

Electrochemical Chloride Removal

The criteria usually concern the content of chloride in contact with reinforcing steel. The requirement to remaining chloride content as well as percentage of single measurements that may be below the requirement should be agreed upon. In addition the allowable deviation of single measurement should be given.

When the amount of current shall be used as method of documentation the minimum must be given.

Electrochemical Realkalisation

Phenolphthalein is a widely used pH indicator for the measurement of the carbonation depth. As earlier discussed it has to be pointed out that the phenolphthalein indicator turns from clear to pink as the pH rises over about 9 while thymolphtalein has a colour change at a pH close to 10.

The determination of sodium content in depth profiles is often used as additional control of the treatment efficiency. As the sodium content can vary strongly in concrete depending among others on the cement type, the results of the measurements should always be compared to measurements before realkalisation. When the amount of current shall be used as method of documentation the minimum must be given.

Table 3   Questions regarding cathodic protection

QUESTION	AUTHOR'S OPINION
How many reference electrodes?	No number can be presented. It depends upon the type of structure and upon the anode system. The number of reference electrodes should be as low as possible.
Where shall the electrodes be placed?	Generally where important information is obtained: Close to – and far from – reinforcing steel. Close to – and far from – current distributors. In areas with high chloride content as well as in areas with low chloride content.
What percentage of electrodes shall achieve the requirements?	We know that reference electrodes fail. In the author's opinion 80 % is a sensible figure. The figure depends, however, on where the electrodes are placed. It is, e.g. much simpler to achieve the criteria close to reinforcing steel than far from the steel.
When – after start of the process - shall the requirement be valid?	The requirements should never be enforced immediately after start of the project. In the author's opinion the requirements should be valid one year after start of the project.

CONCLUSIONS

The acceptance criteria of the final results shall be determined and agreed jointly by the structure owner and the applicator of the process.

For cathodic protection it should be agreed upon the number of reference electrodes, locations of the electrodes and the percentage of electrodes to achieve the requirements. In addition to the depolarisation criteria is should be agreed when this criteria shall be effected.

The requirement to remaining chloride content as well %-age of single measurements that may be below the requirement should be agreed upon. In addition the allowable deviation of single measurement should be given.

When amount of current shall be used as method of documentation the minimum charge must be given.

Phenolphthalein is a widely used pH indicator for the measurement of the realkalisation effect. Using thymolphtalein might be better as thymolphtalein has a colour change at a pH close to 10 while the phenolphthalein indicator turns from clear to pink as the pH rises over about 9.

The determination of sodium or potassium content in depth profiles is advised as additional control of the treatment efficiency. As both the sodium and potassium content can vary strongly in concrete depending among others on the cement type, the results of the measurements should always be compared to measurements before realkalisation.

REFERENCES

1. COST 509, 1997, Corrosion and protection of metals in contact with concrete, Final report, Eds. Cox, R N, Cigna, R, Vennesland, O, Valente, T, European Commission, Directorate General Science, Research and Development, Brussels, EUR 17608 EN, ISBN 92-828-0252-3, p 148.

2. COST 521, To be published, Corrosion of Steel in Reinforced Concrete Structures, Final report, European Commission, Directorate General Science, Research and Development, Brussels.

3. POLLET, V, DIERYCK, V, 2000, Re-alkalisation: specification for the treatment application and acceptance criteria, Annual Progress Report: 1999-2000, COST 521 Workshop, Belfast.

4. ELSENER, B, MOLINA, M., BÖHNI, H., Electrochemical Removal of Chlorides from Reinforced Concrete Structures, Corrosion Science, Vol 35, No 5-8, 1993, p 1563-1570.

5. Repairing Concrete Structures, Norwegian Technical Specifications, Association of Consulting Engineers, Norway, ISBN 82-91510-03-2, 1995.

6. CEN, 2000, Cathodic protection of steel in concrete, EN 12696-1.

ACCELERATED EXAMINATION OF LONG-STANDING REINFORCED CONCRETE VIADUCT

S M Skorobogatov   B P Pasynkov   A V Chernyavskiy

A V Kurshpel   A K Yagofarov

Urals State Univeristy of Railway Transport

A M Mukhametshin

Russian Academy

Russia

ABSTRACT. The paper is devoted to elaboration of technique of a complex investigation of an extensive and massive long-standing viaduct at passing the peak of the exploitation term of 80 years under the urgent condition and small accessibility. The seismometry profiling resulted from not only integral assessment but data on weak points for the further detailed, local examination by ultrasound and sclerometry which gives materials for checking calculation. Seismotomography is predesignated for massive structural elements and resulted in the pattern of isolines of equal speeds of longitudinal sound waves. This allows to gain the mean actual strength of concrete independently of the destroyed superficial layers. Analysis of the ratio of speeds of the transverse and longitudinal waves can be used for determination of the crack texture of concrete and presents the basis of elaboration of a new method.

Keywords: Seismometry profiling, Seismotomography, Concrete strength, Cracks, Long-standing capacity, Heterogeneity of properties.

Professor S M Skorobogatov, is Doctor of Sciences, Honoured Science Worker of Russia, Corr.-member of Academy RAASN. He works at the Urals State University of Railway Transport. His research interests lie in the field of reinforced concrete structures. Over the past decade his current research is directed towards the development of the theory of technical and natural catastrophes.

Professor B P Pasynkov, is a Research/Teaching Fellow. His research interests includes dynamics of building structures.

Mr A V Chernyavskiy, is a reader and Dean of Civil Engineering Department. His main scientific interest is geodesy.

Doctor A M Mukhametshin, is Director of the laboratory of mining geophysics and seismometry of the Institute of mining of Russian Academy. His research is concerned with seismometry and seismotomography.

Mr A V Kurshpel, is a practising civil engineer.

Mr A Kh Yagofarov, is a practising civil engineer.

INTRODUCTION

The 21th century will certain be a century of examination, reconstruction and rejuvenation of long-standing concrete structures and buildings. The case is that concrete is a historically young construction material and its average service life in the open air is approximately 60 years. Durability of concrete was usually elucidated with its frost-resistance and erosion and not with its carbonation. After passing the peak of the service life, an increasing of fracture in concrete and in reinforced bars took place. That is why, prior to examination of a long-standing building there appears to be a challenge of the acceleration of investigation procedure in the urgent condition. Examined viaduct is a sixty span vaulted bridge over the river of Duck with the common length of about 200 m and 30 m in height (See Figure 1). It was built in 1916-1919 and its age has exceeded the normal term of seventy years. The bridge was designed by famous professor G P Perederi and described in detail in his textbook [1].

STRATEGY OF ACCELERATION OF EXAMINATION

From the standpoint of the accelerated strategy of examination, the methods of research can be classified as follows:

1. Preliminary methods of inspection and acquaintance included seeking information, short-range search, analysis of drawing-and-designing documentation, making new drawings, drawing up charts of cracks and defects, geodesy survey, establishing a graduation curve of "sound speed-concrete strength" and elaboration of scheme for the further seismological profiling and seismography.

2. Integral methods of spatial, dimensional investigation included seismology profiling for extensive structures and seismography for massive structural elements. Thereafter the scheme of the further local investigation was elaborated.

3. Local methods of investigation of selected separated points in the extensive structures included ultra-sound and sclerometry.

4. Methods that use the integral rigidity of structural elements to verify adequacy of measured strength of concrete in buildings.

These methods require testing under static and dynamic loads as well as substantiation of a structural diagram of the viaduct. The third group also includes electrochemical analysis for determining pH value of cement stone, presence of electrical surface-leakage and stray earth current, an estimate of erosion and carbonation of concrete and corrosion of reinforcement.

The previous (5-6 years earlier) examination of the viaduct resulted in not only integral characteristics (values) of rigidity which could not only determine deflection (sagging) but find out the weaker points which could lead to failure of structural elements. Only the last complex examination confirmed the presence of separate weak points in the fifth and sixth vaults of the viaduct omitted at the previous examination. In the lower surfaces of all vaults the protective concrete covers were fractured. Most of the longitudinal reinforcement bars were revealed, covered with corrosion up to 1 mm depth. On the first riverside abutment one of the bars was destroyed completely.

CLASSIFICATION OF CRACKS AND DEFECTS IN STRUCTURAL ELEMENTS

The main determinations from the well-known Code [2], Instructions [3.4] and other Literature [5] were drawn on to make clarity of analysis of fractures and defects.

Results of the analysis are displayed in the classification below.

Force transversal visible cracks with the width $a_{crc} \geq 50$ mkm appeared from bending were observed very seldom. This fact could be explained by excessive great reserve in cross section designed by the old Code at the beginning of the 20th century.

Invisible cracks with the crack width $a_{crc} < 50$ mkm were predicted from the seismological profiling.

Manufacturing and settling cracks resulted from the non-uniform settlement of concrete mortar in the process of its laying and roading, low-quality curing of concrete. These cracks were observed everywhere, especially in the support places of the intermediate vertical columns over the vaults. The cracks were through and their width was equal from 0.1 to 0.5 mm.

Temperature-shrinkage cracks resulted from the inhomogeneous temperature deformation because of the surrounding air, shrinkage and the inhomogeneous temperature deformation. These cracks looked like an orthogonal, right-angled net with crack step of 300 to 350 mm. Their widths varied within the limits of 0.1 to 0.2 mm. In accordance with the crack hierarchy suggested by one of the authors [5], greater cracks had a step approximately of 1 m. These crack nets were peculiar to the riverside abutments.

Corrosion longitudinal cracks were placed along with longitudinal reinforcement bars in the concrete cover. Such cracks appeared as the consequence of increasing width of the corrosion product 2 to 2.5 times as much as metal. The corrosion products were pressing out the concrete cover and breaking it off. The first main reason of the corrosion was a small concrete cover assumed its manufacture. The cover thickness was decreased up to 0.5 cm instead of 3 cm. The second main reason of the corrosion was the constant moister content of the superficial layers from too short spillway tubes left by the ancient builders.

The classification of other concrete defects included caverns, cavities, split-offs, places of leaching resulted from manufacturing reasons and climate factors. The analysis of cracks and defects of the old building induces two more limit states in calculating of long-standing structures: erosion of concrete and corrosion of reinforcement.

DEGREE OF CORROSIVE WEAR AND EROSION OF CONCRETE

The corrosive wear of reinforcement bars can be determined through the width of longitudinal superficial cracks lined up along with the bars. A schematic diagram of the external manifestation of corrosion products on the concrete cover was adopted from works by Dr A I Vasilyev. There is a simple formula, changed by the first author, for determination of a corrosion value through the width of superficial cracks.

The erosion of the superficial layers of concrete was determined with a special method by Dr A G Mokhov. The sampling of old concrete from the structural elements of the viaduct resulted in the absence of hydro-carbonate and carbonization of concrete. This unexpected result can be explained by the presence of the pure air in the environment.

"SOUND-SPEED – CONCRETE STRENGTH"

Before the plotting of the graduation curve there were two difficulties in preparatory work. The first one consisted in the necessity of verification of adequacy in values of the speeds of sound and ultrasound waves. If the adequacy existed, there would be the opportunity of using the vast literature on ultrasound oscillations. The second one is the absence of check samples. Sampling, withdrawal of audit samples from existent, real buildings is a very difficult technical and methodical problem. The weakening of the cross section of structural elements is dangerous and hardly probable.

In the example of the viaduct the audit samples were sawn from the less stressed parts of the vaults. The form and dimensions of the samples were approaching a prism with the size of 100x100x600 mm. The number of audit samples was limited. That is why the investigators tried to get maximum information from one sample. All the samples were used for sclerometry, ultrasound and mechanical test on bending, splitting and compression.

Tensile strength was taken from bending and especially from splitting samples. Samples left after splitting were used for compression test through rubber layer pads. In doing so, the tests resulted in the prism (axial) compression because fracture cracks were vertical and typical for prism failure.

At the age of 80ty the old concrete showed concrete strength 1.5 to 2.0 times as much as modern concrete. We mean that the tensile strength is taken relatively the compression strength (R_{bt}/R_b). The old concrete is more porous, brittle and that is why more dangerous. Particular attention must be given to the heterogeneity of properties of old concrete which was very high. This sharply reduces the value of the design strength used in calculating of load-carrying capacity for a building.

The experimental data resulted from these tests had been successfully applied in plotting the graduation curve of "sound-speed concrete strength". The conversion coefficient appeared to be 1.5 to 2.0 times greater than for modern concrete.

There will be new challenges connected with the study of unknown physical and mechanical properties of old concrete. This is necessary for the calculation of the reserve of serviceability of an old building (see the other report of the first author of this paper in the proceedings or book [5]).

TECHNIQUE OF ACCELERATED INVESTIGATION

The searching of weak points in structural elements of the viaduct had been motivated by determination of value of the design strength for testing and checking calculation. Such a searching is a very difficult technical and economical problem if researchers use traditional local methods.

To examine an extensive structure and massive structural element, it is expedient to use seismometry profiling created by Dr A M Mukhametshin and seismotomography by eng. V V Bodin.

The positive results of examining the extensive viaduct and its massive concrete piers make it possible to recommend the following two methods:

1. Seismometry profiling along an extensive structure with measuring values of speeds of the longitudinal and transverse waves.

2. Seismomonography of cross sections of massive elements with plotting a pattern of isolines of equal speeds of the longitudinal and transverse waves.

On the whole, the continuous seismometry profiling gives only an integral value of rigidity of a cross section and hence deflections. If the researcher installs seismo-receivers along an extensive structure with a step of 4 to 5 m, then there will be information about weak points along the axis of the vaults. This results in the data for checking calculation.

DETERMINATION OF STRENGTH OF CONCRETE

By values of the longitudinal and transverse waves, all the vaults can be divided into two groups. The first group includes first three vaults with the speed longitudinal waves $V_p = 4.005\,\text{km}/\text{sec} \pm 0.048$ and transverse waves $V_s = 2.488\,\text{km}/\text{sec} \pm 0.039$.   The other vaults of the second group are with $V_p = 3.711\,\text{km}/\text{sec} \pm 0.060$. The reducing of speeds of the longitudinal waves is small, i.e. 8%, but in accordance with the graduation curve the reducing in concrete strength is great enough, i.e. 33%. The surprising thing is that the relation of speeds $V_s/V_p = 0.621$ and 0.559 almost coincides with the second parameter point by Dr. O.Ya.Berg.

The second parameter point corresponds to the stress in concrete when the avalanche process of appearance and growth of microcracks appears, when the cracks merge together and convert into mezocracks irreversible in time.
Reasonable high relationship of $V_s/V_p = 0.621$ for average concrete strength means that the transverse waves don't meet longitudinal mezocracks on their way. Above the value of $V_s/V_p = 0.621$ we have transverse mezocracks and they are placed on the way of the longitudinal waves.

Such an attempt with the relationship of the transverse and longitudinal waves V_s/V_p presents the basis for elaboration of a new method of examination.

VERIFICATION OF RESULTS TAKEN FROM NONDESTRUCTIVE (SEISMOMETRY) METHOD

To get the concrete strength directly from the static test, the researchers executed the special static calculation. They were directed at the choice and substantiation of the structural model of the viaduct with the high intermediate piers.

The traditional model of arch with fixed end in the abutments appeared to give very small values of deflection. Taking into consideration that the terminal rigidity of piers, we have the structural model resulted in surprising coincidence of theoretical and experimental data under load of 1,200 kN.

It is interesting to note that the average concrete strength $R = 18.2\,MPa$ from the test complex coincided with the design strength of 18.5 MPa, put in common practice for similar viaduct [1].

Results of the dynamical test under load of 120 kN gave comparative estimation of vaults in the viaduct (See Figure 1).

The frequency of free vibrations appeared to be minimum for the fifth vault. The reason of this fact lies in a lessened value of rigidity in cross sections because of the presence of the transverse cracks, which were revealed by the previous seismometer profiling.

The pattern of the amplitudes of free vibrations was very symptomatic. In the character of changing, the pattern coincides with the diagram of deflection got through value of rigidity. The pattern also coincides with the diagram of concrete strength gained from the seismometry. On the whole, the pattern of free vibration confirmed that the fifth and the sixth vaults were in the worst condition.

The pattern of the amplitudes of forced vibrations and the logarithmic decrement of damped vibration happened to be little information bearing.

The pattern of the isolines of equal speeds of the longitudinal waves for massive element – riverside abutment was gained from seismometry method. This pattern showed that the average concrete strength was higher than 10 MPa. This result was of great importance for determination of the load-carrying capacity because the superficial layer of concrete was destroyed with the depth of approximately 20 to 25 cm and the last fact could be taken for a wrong estimation.

CONCLUSIONS

1. The complex of investigations, including the seismometry profiling for extent structures and seismotomography for massive elements, makes it possible to solve the problem of accelerated examination of long-standing buildings under the urgent condition of exploitation.

2. Seismotomography methods for massive elements enable to obtain the pattern of isolines of equal speeds of the longitudinal waves and the average concrete strength without regard to the strength in the superficial layers.

3. The analysis of the relationship of speeds of the transverse and longitudinal waves permits to obtain not only the concrete strength but also the crack texture of concrete.

4. Analysis of the stage of a long-standing structure demands to take into consideration two more limit states of a structural element in addition to the Building Code: erosion of concrete and corrosion of reinforcement.

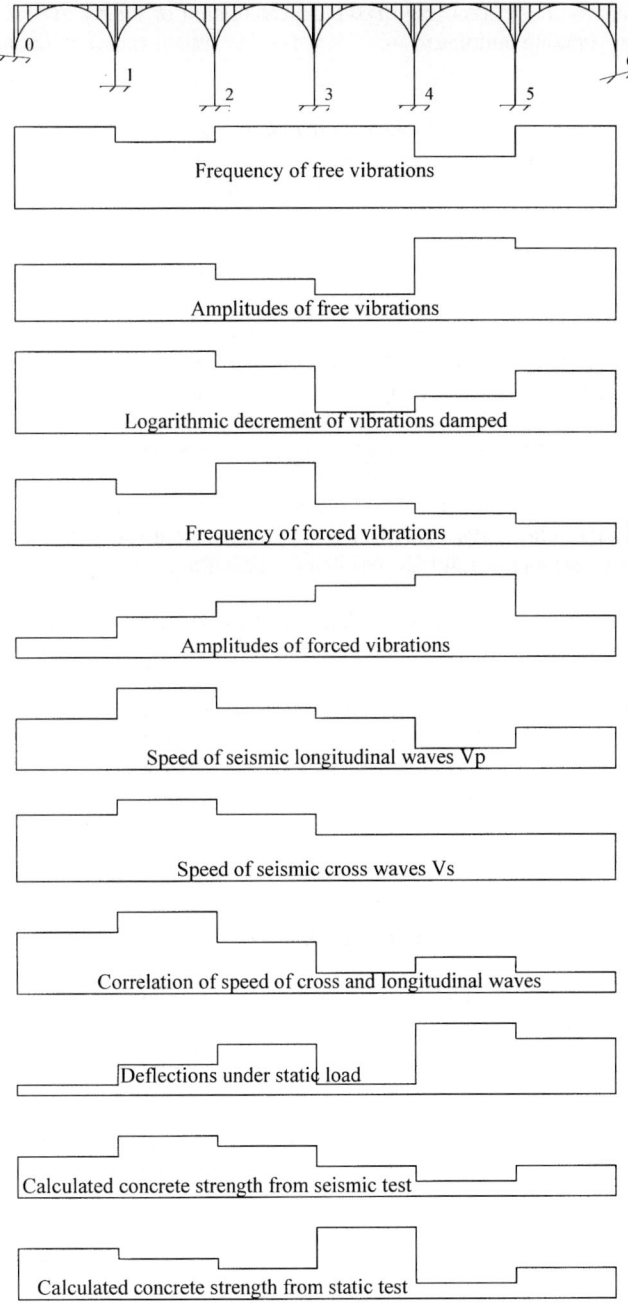

Figure 1   Generalized parameters of the concrete strength and the flexural rigidity for vaults
of the viaduct

5.   Study of peculiarities of old concrete, including tension and compression strength, heterogeneity, is of great importance for calculation of the reserve of serviceability of a long-standing building.

REFERENCES

1   PEDERERI, G P.   The course of reinforced concrete bridges.   Moscow, Gostransizdat, 1931, 512 pp.

2   BRIDGES AND TUBES.   Rules of examinations and tests, Building Code SniP 3.06.07-86, Moscow, Gosstroi, 1987, 41 pp.

3   INSTRUCTIONS OF MAINTENANCE OF ARTIFICIAL BUILDINGS.   Moscow, Transport, 1999, 108 pp.

4   INSTRUCTIONS ON CARRYING OUT INSPECTION OF BRIDGES AND TUBES IN HIGHWAYS. VSN-4-81, Moscow, Minavtodor of Russia, 1990, 36 pp.

5   SKOROBOGATOV, S M. The principle of infrmation entropy in fracture mechanics of buildings and rock seams. Ekaterinburg, URGUPS, 2000, 420 pp.

USE OF INTERMITTENT VIBRATION MEASUREMENTS FOR DETERMINING THE INTEGRITY OF CONCRETE STRUCTURES

N R Short   O T Owolawi

M G Wood   J E T Penny

J A Purkiss

Aston University

United Kingdom

ABSTRACT. Operation of plant containing safety-critical concrete structures requires early warning of damage initiation so that remedial action can be taken well in advance of loss in serviceability.   Whilst crack initiation and propagation is difficult to detect using conventional means, vibration measurements are thought to be sufficiently sensitive to detect and monitor damage evolution.   Initial investigations involved testing lightly reinforced concrete beams fabricated with a range of constituents.   Vibration models describing the natural frequencies, damping characteristics, and mode shapes of undamaged beams were established.   The beams were then statically loaded in discreet steps and changes in the vibration parameters monitored at each step.   This paper describes the method of modal analysis, presents initial results obtained, particularly in relation to aggregate type, and assesses the potential for its application to reinforced concrete structures in the field.

Keywords: Concrete, Structures, Deterioration, Testing, NDE, Vibration, Integrity.

Dr N Short, is a Senior Lecturer in Engineering Materials, School of Engineering and Applied Science, Aston University, Birmingham, UK.   Current research interests are concerned with the durability of construction materials.

Mr O Owolawi, is a Research Student, School of Engineering and Applied Science, Aston University, Birmingham, UK. He has a background in mechanical engineering and is currently undertaking the research reported in this paper.

Dr J Penny, is Senior Lecturer in Mechanical Engineering School of Engineering and Applied Science, Aston University, Birmingham, UK.

Dr M Wood, is an Independent Consultant and Honorary Visiting Fellow at Aston University. Research interests include the development of techniques for recognising and locating changes in state of heterogeneous materials.

Dr J Purkiss, is a Lecturer in Structural Engineering, School of Engineering and Applied Science, Aston University, Birmingham, UK. He is concerned with general structural degradation, specialising in the design of structures, especially fire safety design.

INTRODUCTION

The operation of plant containing safety-critical concrete structures, such as those found in nuclear power plants requires early warning of damage initiation so that remedial action can be taken well in advance of loss in serviceability [1-4]. Crack initiation and propagation may occur at loads far below those for actual structural failure and in the early stages, is impossible to detect using visual inspection and other conventional means [1-3]. However, vibration measurements are thought to be sufficiently sensitive to detect and monitor damage evolution [5-9], even when the micro-cracks are located well within the structure. In addition, such tests are non-destructive and may be carried out with minimum disruption to the function of the structure.

One of the approaches of this type of damage assessment is to use changes in modal parameters. This involves first determining a mathematical model of the vibration characteristics of a component or structure, the data for this model being derived from experimental investigations. The model describes the natural frequencies, damping characteristics and mode shapes during vibration. Thus subsequent monitoring of these parameters should enable any changes in structural integrity to be detected.

Prior to the application of this technique to real structures it is important to establish its sensitivity to damage induced in simple concrete structures under carefully controlled conditions. Thus initial investigations have involved testing lightly reinforced concrete beams fabricated with siliceous, limestone, granite and Lytag aggregates. Vibration models, using undamaged data, have then been established for these beams. The influence of factors such as (a) method of beam support, (b) nature of the excitation device and its position, (c) the detection device and its position, on the reproducibility of results has been established. The beams were then statically loaded and the changes in modal response determined.

The vibration theory behind this approach has been reported elsewhere [10] and this paper concentrates on preliminary results looking specifically at the influence of aggregate type on the nature of response.

EXPERIMENTAL

Specimen Fabrication

The test specimens were lightly reinforced concrete beams, which contained two longitudinal D8 or D16 bars. The beam dimensions and position of the reinforcement are shown in Figures 1(a) & (b). The concrete was made using a normal Portland cement, medium zone quartz sand and either siliceous (beams 1A, 1B), limestone (beam 2A), Lytag (beam 3A) or granite (beams 4A, 4B) coarse aggregates.

The mix was 1:2:4 cement/sand/coarse aggregate with a 0.5 w/c. In addition three cubes were cast to determine the concrete strength. Eleven threaded pins were embedded in the concrete to attach the accelerometer and act as excitation points, Figure 1(b).

Loading Procedure

The beam was mounted in a test frame, supported 50 mm from each end and loaded at the third points of the clear span. This procedure was used in order to gradually introduce crack-damage. The use of constant load measurements does not imply constant increments of deflection or strain and thus may distort any relationships. The load was applied by a hydraulic ram in a series of discrete steps until the beam failed (Table 1). Visual observations were made of the extent of cracking at each step in the load cycle.

Table 1   Load increments

Load Step	0	1	2	3	4	5	6	7	8	9	10	11	12
Load (kN)	0	6.2	9.3	12.5	15.6	18.7	21.8	24.9	28.0	31.1	34.3	37.4	40.5

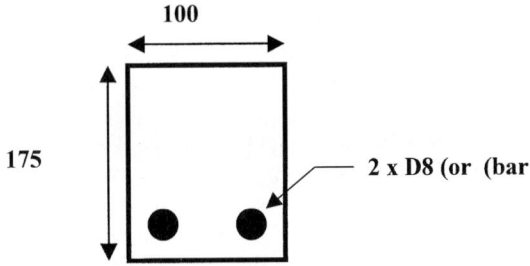

Figure 1(a)   Cross-section of the beam

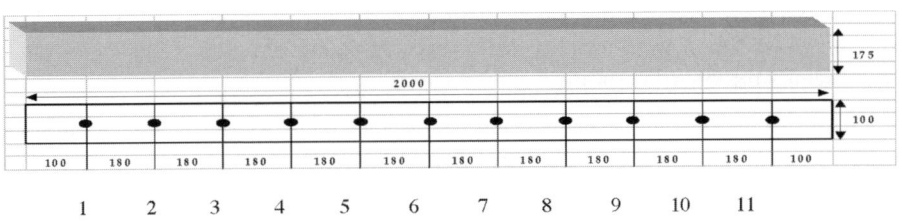

Figure 1 (b)   Location of vibration pins on the beam
(All dimensions in mm)

Vibration – Excitation and Sensing

A B&K modal hammer was used to excite the beam and the response was measured using a piezo-electric accelerometer. Amplified signals from the accelerometer and the force transducer in the hammer were logged and analysed using a Data Physics Ace spectrum analyser plugged into a laptop computer. Experience was gained using the modal hammer so that the hammer strikes were consistent in quality. A number of methods were used to support the beam during the vibration testing including steel, rubber and timber supports. Results obtained using these supports were difficult to interpret.

The most satisfactory arrangement was to suspend the to beam in a harness at its mid point. This simulated a free-free boundary condition.

The accelerometer was first attached to point 1 on the beam (see Figure 1(b)) and the frequency response was obtained by exciting with beam using hammer strikes at points 1-11 in turn. The accelerometer was then moved to point 2 and the response was again obtained by exciting the beam at points 1-11. The accelerometer was then moved to points 3, 4 etc. until it had been moved to all 11 points. It should be noted that there is now an 11 x 11 set of data for analysis.

The beam was then placed in the loading rig and loaded. After the first loading step 1, the beam was removed from the rig, suspended and sets of frequency responses were obtained for all the points. This procedure was repeated for load steps 2-12 (or until failure) and the incidence of damage noted after each step.

RESULTS AND DISCUSSION

When the accelerometer was at point 1 and the hammer excited the beam at point 2 the frequency response was such that several resonant frequencies, modes 1, 2, 3, and 4 at about 200, 400, 800 and 1200 Hz respectively, were evident. The response reflects the global stiffness and mass properties of the beam material. Similar, but not identical, responses were obtained for the other relative positions on the same beam. As the load on the beam increases changes in the magnitude of response and resonant frequency were observed.

Shifts in mode frequency with load step, for six different beams, are shown in Figure 2. With each of the beams a general trend is apparent in that mode frequency decreases as the load on the beam increases. This decrease may be attributed to increasing damage in the beam as the load on it increases. For a given beam, normalising shows that the trends for the individual mode frequencies are similar with the exception that mode frequency 1 reaches a minimum and then starts to increase. Incipient failure is indicated by a sudden drop in normalised frequency for beams 1A, 1B and 2A which failed in bending whilst this is not evident in the other beams which failed in shear.

Averaging the four normalised mode frequencies facilitates comparison between beams of the same aggregate (reproducibility of results) and beams made with different aggregates (influence of aggregate). These averages are plotted against load step in Figure 3. It can be seen that the lines for beams 1A & 1B are almost identical until load step 6 and the lines for beams 4A & 4B almost identical until load step 7. This suggests that results are reproducible for beams made with the same aggregate until incipient failure when a divergence occurs. Comparing the trends for the different aggregates it would appear that there are noticeable differences, particularly between load steps 2 and 5, which is of most interest structurally. This is probably a result of different damage patterns for beams failing in shear compared to bending rather than the different type of aggregate.

To extract as much information as possible from the data a different analytical approach has been tried whereby a direct comparison is made between any two points on the beam. Thus the data when the accelerometer is e.g. on 1 with the hammer striking 2, is compared to the data when the accelerometer is on 2 with the hammer striking 1. Theoretically these should be the same but subtle differences are evident.

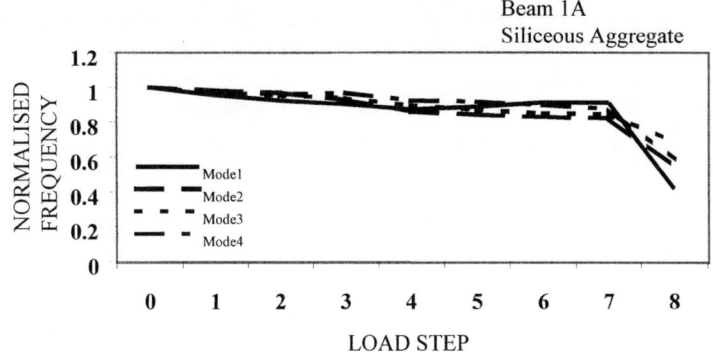

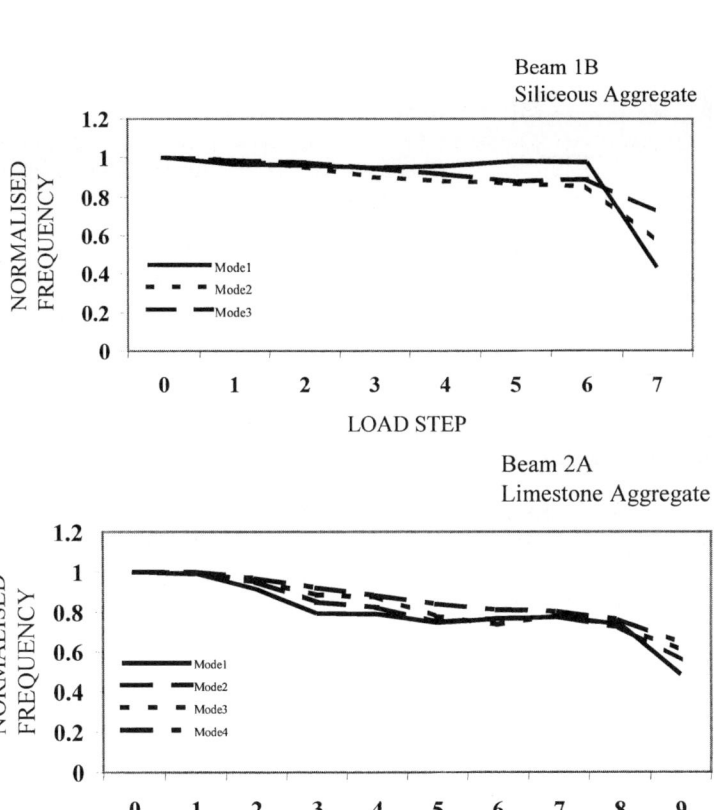

Figure 2   Change in mode frequency with load

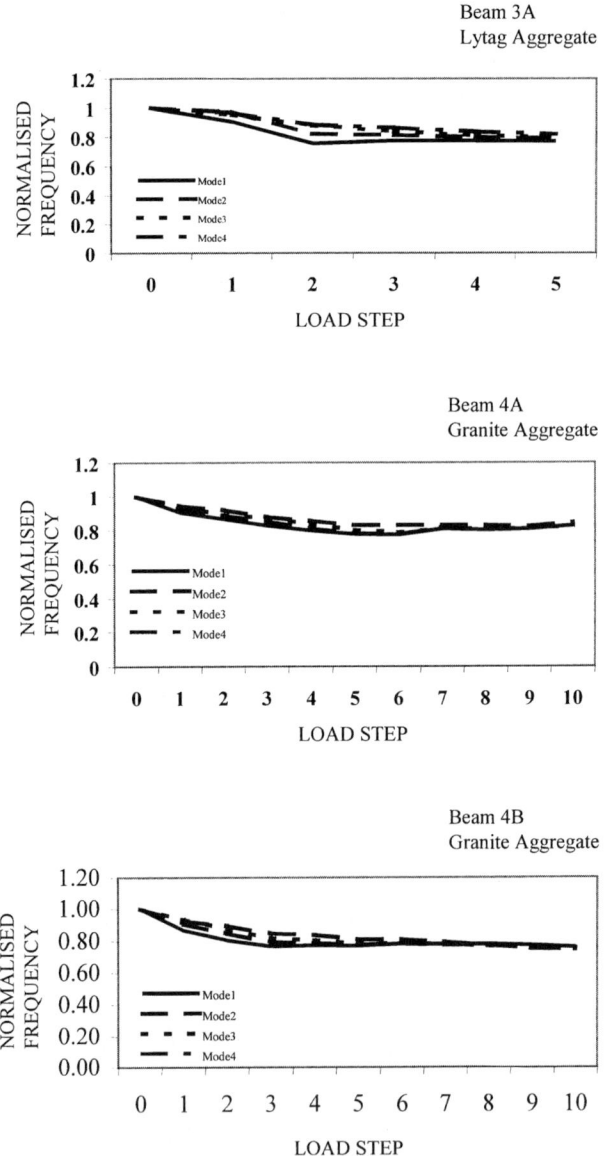

Figure 2 (continued)   Change in mode frequency with load

At low loads differences in the magnitude rather than the frequencies of the resonance are found whilst at higher loads there are evident shifts in frequency as well as magnitude. The sum of the two responses gives the difference of reciprocity (DoR) and these may then be compared at different load steps.

Figures 4 (a)-(d) show the DoR between points 1 and 2 at load steps 0 and 5, for beams made with four different types of aggregate. Examination of these Figures (and those obtained for determinations between other points on the beams) suggests that there are significant changes in magnitude and position of the resonant peaks when comparing load steps 0 and 5 and that these changes may be related directly to the levels of damage induced in a test-beam.

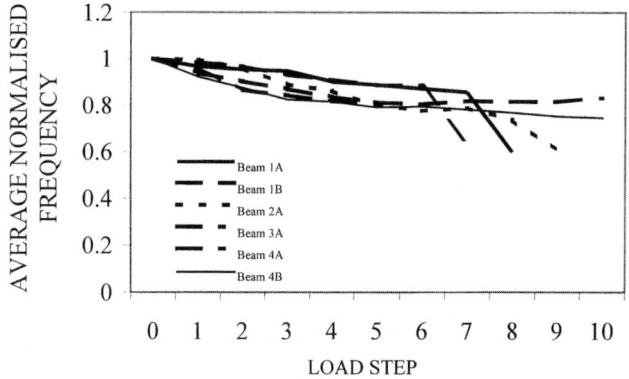

Figure 3   Change in averaged mode frequency with load

These DoR data have been manipulated and then assessed as a series of mathematical relationships ranging from simple arithmetic comparisons to averaging techniques such as Root Mean Square (rms) with the result that, within the bounds of the available data, the rms representation shows much promise.

The significance of DoRs measured over a range of structural positions has yet to be addressed fully, however, there appears to be a correlation between the reciprocal positions of measurement and the normalized DoR (NdoR) levels recorded (for a limestone aggregate beam), see Figure 5.

CONCLUDING REMARKS

Investigations have been initiated to ascertain the viability of using vibration measurements for integrity assessment of safety-critical structures. Preliminary experiments focused on determining the most suitable method of beam support and the influence of the nature / position of the excitation and detection devices. Changes in mode frequency observed were related to the level and pattern of damage induced in the beam.

The type of aggregate used was not very significant. Methods of presenting results in terms of difference of reciprocity and rms DoR look useful as way to locate the position of damage as it evolves. The results suggests that the method has promise although considerably more work needs to be done to first confirm this and then see if it can be applied to real structures.

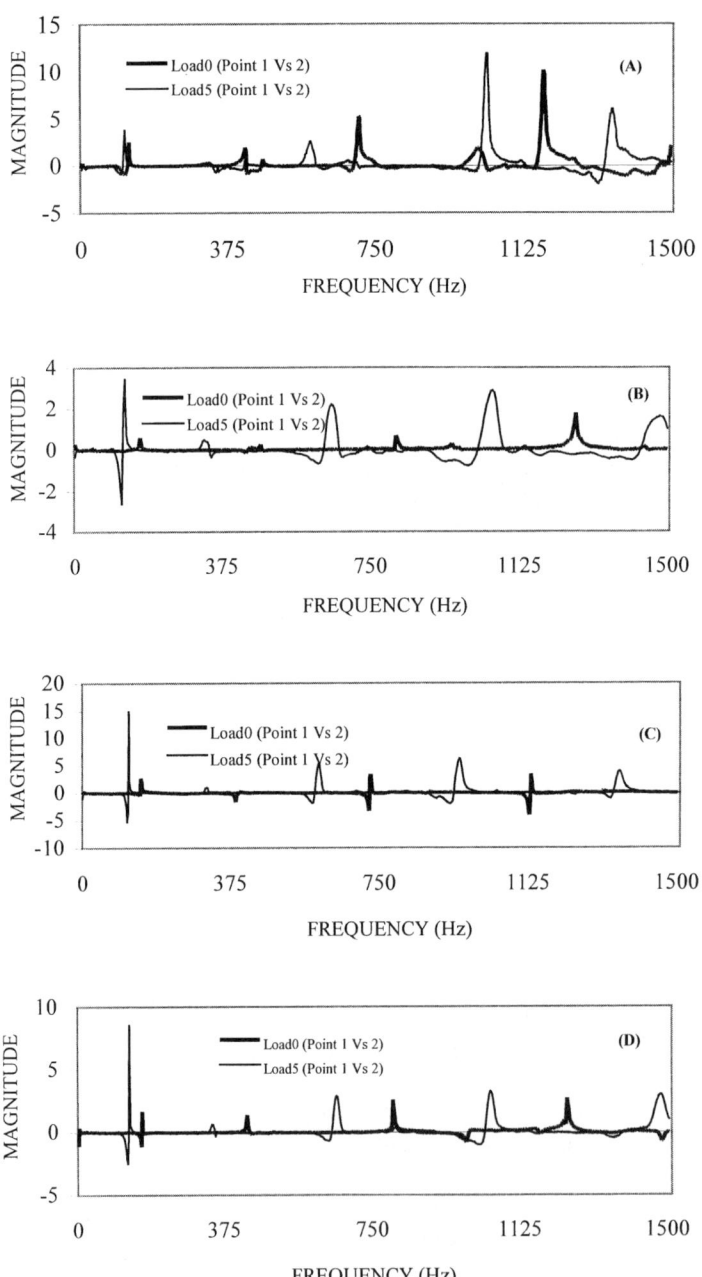

Figure 4   Difference in reciprocity for beams made with: (A) siliceous;
(B) limestone; (C) Lytag; (D) granite aggregates

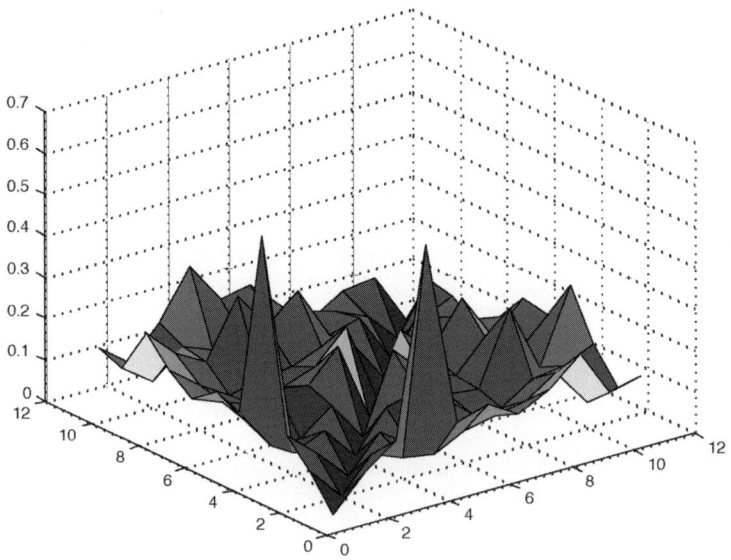

Figure 5(a)   Undamaged beam (Vertical Units rms NDoR)

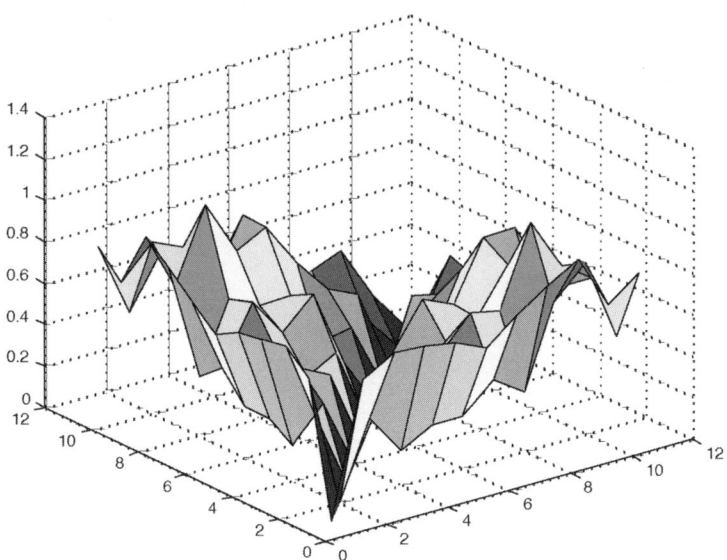

Figure 5(b)   Damaged beam (Vertical Units rms NDoR)

ACKNOWLEDGEMENTS

Support from EPSRC (Total Technology) and HSE is gratefully acknowledged.

REFERENCES

1. NUCLEAR ENERGY AGENCY, (1998), Development priorities for non-destructive examination of concrete structures in nuclear plant, OECD, Issy-les Moulineaux, France, NEA/CSNI/R(98)6.

2. RUDZINSKY, J, BONDARYK, J, CONTI, M, (1999), Feasibility of high frequency acoustic imaging for inspection of containments: Phase II. Office of Nuclear Regulatory Research, US, ORNL/NRC/LTR-99/11, ETC RPT:100546(U).

3. NAUS, D J, (1999), Considerations for use in managing the aging of nuclear power plant concrete structures, RILEM Technical Committee 160-MLN 'Methodology for life prediction of concrete structures in nuclear power plants', Report 19.

4. SMITH, L M, (1996), In-service monitoring of nuclear-safety-related structures, The Structural Engineer, Vol 74, p 210-211.

5. WOOD, M G, (1992), Damage analysis of bridge structures using vibrational techniques, PhD Thesis, Aston University.

6. WOOD, M G, BAILEY, M, FRISWELL, M I, PENNY, J E T, PURKISS, J A, (1991), Damage location in reinforced concrete beams using vibration responses. Proceedings of the 9th International Modal Analysis Conference, Florence, Union College/Society for Experimental Mechanics, p 197-203.

7. WOOD M G, FRISWELL, M I, PENNY, J E T, (1992), Exciting large structures using a 'bolt-gun'. Proceedings of the 10th International Modal Analysis Conference, San Diego, Union College / Society for Experimental Mechanics, p 204-210.

8. ABEELE, K V D, VISSCHER, J D, Damage assessment in reinforced concrete using spectral and temporal nonlinear vibration techniques, Cement and Concrete Research, Vol 30, 2000, p 1453-1464.

9. PEARSON, S R, OWEN, J S, CHOO, B S, The use of vibration signatures to detect flexural cracking in reinforced concrete bridge decks, Key Engineering Materials, Vol 204-205, 2001, p 17-26.

10. OWOLAWI, O T, WOOD, M G, PENNY, J E T, PURKISS, J A, SHORT, N R, Vibration measurements for determining the integrity of concrete structures. Concrete Communication Conference 2001, Proc. 11th Annual BCA/Concrete Society Conference on Higher Education and the Concrete Industry, UMIST Manchester July 2001, British Cement Association, p 159-168.

STATISTICAL NONLINEAR ANALYSIS OF CONCRETE STRUCTURES

M Vořechovský   V Veselý   R Rusina
Brno University of Technology
V Červenka   R Pukl
Červenka Consulting Company
Czech Republic

ABSTRACT.   This paper presents some results of prestressed concrete beams using statistical nonlinear fracture mechanics analysis. The program ATENA represents computer simulations of concrete failure. The fracture of concrete is modelled by the crack band theory based on fracture energy-related softening. Statistical variability of material parameters is introduced in a form of small-sample Monte Carlo simulation Latin hypercube sampling. The approach is demonstrated on bending failure of prestressed concrete beams. Numerical examples are focused on comparison of both deterministic and probabilistic simulation results with experimental data of frequently used Australian bridges.

Keywords: Nonlinear fracture mechanics, Software, Statistics, Latin hypercube sampling.

Mr M Vořechovský, is a PhD student at the Brno University of Technology, Brno, Czech Republic. He completed his diploma thesis on main research interest including stochastic computational mechanics, random fields, fracture mechanics, software development.

Dr R Pukl, is a Project Manager in Červenka Consulting Company, Prague, Czech Republic. His research interests include fracture mechanics, finite element method, and computer graphics. Co-author of the program system SBETA/ATENA. Numerical simulation of concrete structures and fastenings, software development. Member of FraMCoS.

Mr V Veselý, is a PhD student at the Brno University of Technology, Brno, Czech Republic. His research interests include fracture mechanics, FE modelling.

Dr V Červenka, is Owner and Director of the Červenka Consulting company, Prague, Czech Republic. His research interests include: internationally recognized expert in mechanics of concrete structures. Co-author of the program system SBETA/ATENA. Member of the international associations FraMCoS, fib and IABSE (WC 3 "Concrete Structures"), and national engineering associations Czech concrete society and Czech society for mechanics.

Mr R Rusina, is an external PhD student at the Brno University of Technology, Brno, Czech Republic. His research interests include FE modelling, random fields, software development.

INTRODUCTION

The aim of this paper is to present some results of the statistical simulation of nonlinear prestressed beams using the program SBETA/ATENA. SBETA is well-known software for nonlinear finite element simulation of concrete and reinforced concrete structures. It employs nonlinear fracture mechanics and energy based concept of concrete fracture [1], [2]. Recently introduced software ATENA is an extension of SBETA. ATENA is conceptually object-oriented and based on a MS-Windows environment [3]. It includes all material models developed and verified in SBETA. In addition, a wide range of extensions in the ATENA analytical core is available. The user-friendly graphical environment offers excellent support to the user during all stages of the nonlinear analysis. ATENA software was used for the nonlinear analysis of prestressed concrete beams presented in this paper.

Objective failure modelling of concrete structures with significant nonlinear effect should use the tools of fracture mechanics. Such modelling usually remains at deterministic level. The aim is to utilize efficient methods of reliability engineering for statistical analysis of structures by nonlinear fracture mechanics. "Randomisation" of fracture mechanics problems will be based on ATENA software. Statistical simulation of Monte Carlo type will enable so called statistical analysis - estimations of statistical characteristics of structural response (ultimate load, stress, deflection, crack width etc.). The utilization of results is supposed in the following areas: 1) Qualitativelly higher level of structural modelling; 2) The solution of inconsistency of design using partial safety factors according to EUROCODES for nonlinear problems.

NONLINEAR FRACTURE MECHANICS: SOFTWARE ATENA

Fracture is one of the most important features of concrete behaviour with significant nonlinear effect. Sufficiently precise and efficient methods for numerical analysis of reinforced concrete structures are the objects of research of last decades. First practical achievements in the nonlinear analysis of concrete structures were reached in the late 1960s by Rashid (1968) and Červenka & Gerstle (1971-1972). They evolved in a numerical tool for simulation of the real behaviour of concrete structures. A complete response of a structure to the given imposed loading can be obtained by such analysis including stages of crack propagation in the pre-peak serviceability state, the failure load and failure mode, and the post-peak behaviour.

In the numerical investigations, the ductile behaviour can be covered by plasticity, while the brittle behaviour corresponds to the linear elastic fracture mechanics (LEFM). For the transtion nonlinear fracture mechanics (NLFM) with softening based on fracture energy can be effectively used. NLFM is suitable for analysis of quasi-brittle materials with certain toughness like concrete. The plastic and the brittle behaviour can be treated as limit situations. There is still a gap between research achievements and engineering practice in modelling of failure behaviour of concrete structures. The academic community is concerned with many different interesting topics like damage localization, crack formation, size effect, strain softening etc. On the other hand the industry and design practice provides the motivation for purpose minded community concerned mainly with efficiency aspects like implementation of efficient material models, solution strategies, discretization and interpretation of results. Commercial software ATENA [2] definitely belongs to this category.

Post-test analysis was also carried out based on measured material properties. The analysis was carried out by nonlinear finite elements (NLFEA) using the software ATENA to determine the ultimate load capacity, the load displacement response, cracking patterns and strains in the prestressing wires. The constitutive model used in ATENA reflects all the essential features of concrete behaviour, namely cracking in tension. It is based on nonlinear damage and failure functions in plane stress state [3]. A smeared crack approach simulates discrete cracks occurring in real concrete structures by strain localisation in a continuous displacement field. Concrete fracture is covered by nonlinear fracture mechanics based on fracture energy [1], [3]. Exponential softening law derived experimentally by Hordijk (1991) is used. Objectivity of the finite element solution is assured by crack band approach - the descending branch of the stress-strain relationship is adjusted according to the finite element size and mesh orientation.

SMALL-SAMPLE MONTE CARLO SIMULATION

The paper shows the possibility of simple "randomisation" of nonlinear analysis of concrete structures. As the scatter of some basic input parameters of computational model (tensile/compressive strength of concrete, fracture energy etc.) can be large, it is desirable to consider them to be random variables. Then applying a suitable type of Monte Carlo simulation technique, random variables can be randomly generated under their probability distribution functions. Consequently, the particular problem is repeatedly solved and statistical characteristics of structural response can be obtained and assessed. As nonlinear analysis is computationally intensive, a suitable technique of statistical Monte Carlo simulation should be utilized. The Latin hypercube sampling technique appeared to be a very efficient technique in this context because it requires rather small number of simulations for accurate results [4].

Such an approach will be shown on problem of simple supported prestressed concrete beams. The paper presents first selected deterministic simulation results compared to set of experimental data. Nonlinear fracture mechanics as applied in ATENA resulted in a good agreement with these test data. Second, using a probabilistic approach also the variability of resistance could be obtained with reasonable computational effort.

The aim of ATENA statistical nonlinear analysis is to obtain the estimation of the structural response statistics (stresses, deflections, failure load etc.). Procedure can be itemized as follows:

- Uncertainties are modelled as random variables described by theirs probability distribution functions (PDF). The optimal case is if all random parameters are measured and real data exist. Then statistical assessment of this experimental data (e.g. data on strength of concrete or loading) should be done resulting in selection the most appropriate PDF (e.g. Gaussian, lognormal, Weibull, etc.). The result of this step is the set of input parameters for ATENA computational model – random variables described by mean value, variance and other statistical parameters (generally by PDF).
- Random input parameters are generated according to their PDF using Monte Carlo LHS type simulation. Statistical correlation between parameters can be imposed [5].

Previous step is repeated N-times (N is the number of simulations used). Generated realizations of random parameters are then used as inputs for ATENA computational model. The complex nonlinear solution is performed and results (response) are saved. At the end of the whole simulation process the resulting set of structural responses is statistically evaluated.

The results are mainly mean value, variance, coefficient of skewness, histograms, empirical cumulative probability density function of structural response [6], [7]. Here we can also perform some types of sensitivity analyses, it means we can observe dependence of response variables on each input variable.

STRENGTH ASSESSMENT OF PRESTRESSED CONCRETE BEAMS

Theoretical assessments were made of a number of precast prestressed concrete beam units. The units reviewed were generally constructed during the period 1950 to 1976 and designed for MS18 loading according to Australian Design Code. This paper reports on the theoretical assessment and ultimate load testing of a number of 1958 series prestressed rectangular beam units. These were the first standard precast prestressed concrete beams components used for bridges in Victoria, NS Wales in Australia. The prestressed rectangular beam units were introduced to complement the precast reinforced concrete U-Slabs. The use of prestress enabled a shallower unit to be used for a given span. These were developed for bridging over irrigation channels, where minimising structural depth was important to maximise the waterway area and reduce the probability of trapping debris without having to raise road grade lines and bridge lengths. There are approximately 1100 of these bridges on the Vic Roads arterial road network, and a similar number on the local road network. In addition, Victorian rural water authorities have a number of these bridges that are not on the public road system [8].

Three 4.5 m long prestressed rectangular beam units were obtained from old but unused stock located in storage. The theoretical analysis and testing of those units is described below. Results of experimental data and data describing beams are taken from the paper [8] and for complexity of our numerical modelling and comparison purposes some details of experimental testing are based on the same paper.

Experimental testing

During December 1999, three 4.5 m (15 foot) Country Roads Board beams became available for testing. These units were located at the Vic Roads depot in Benalla and had been manufactured by the State Rivers and Water Supply Commission. Three 4.5 m beams were tested, designated 4.5-1, 4.5-2, and 4.5-3. Geometry and reinforcing details are given in Figure 1.

After testing, concrete cores and samples of prestressing wire were retrieved from the beams and tested to determine the material properties. The concrete strength results had some scatter both within and between beams [8].

All beams were tested as simply supported units subjected to four-point bending loading with total span 4.2 m and bending span 1.524 m.

This arrangement approximates the loading from a dual axle. Results were modified to allow for 1.2 m spacing on road and design vehicles. The beams were tested using load control at a quasi-static rate of loading, typically 4 kN/minute. Loading was halted several times during the testing to visually record the extent of cracking in the beams. All beams were loaded to failure.

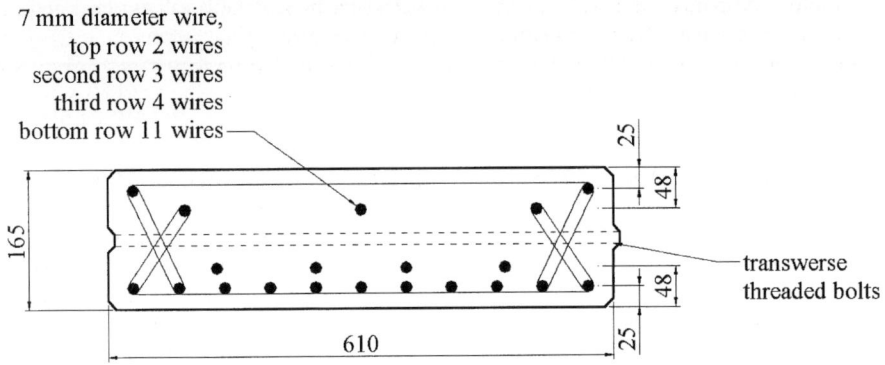

7 mm diameter wire,
top row 2 wires
second row 3 wires
third row 4 wires
bottom row 11 wires

transverse
threaded bolts

Figure 1   Cross section for 4.5m prestressed beams (in mm)

Load deflection plots for the 4.5 m units are given in Figure 4.   There is good agreement between the three beams that were tested in each series.   The ultimate loads were 82 kN, 82 kN and 86 kN.   Cracking was visually observed at a jack load of approximately 50 kN.   At failure cracking had developed throughout the constant moment region.   Failure in every case was by yielding of the steel followed by crushing of the concrete [8].   In order to obtain an estimate of the level of prestress that existed in the beams prior to testing, one wire was de-stressed by carefully breaking out the concrete around the wire, while the strain in the wire was measured.   A strain gauge was fitted to a wire in beam 4.5-I at 1000 mm from the end of the beam.   Concrete was broken away to expose the wire, working progressively from the end of the beam.   Using a value of 200 GPa for the elastic modulus of the wire, the prestress in the wire was 1050 MPa.

NONLINEAR ANALYSIS OF THE BRIDGE BEAMS

Each beam was modelled with layers of plane stress elements to represent concrete and line elements to represent prestressing wires.   Assumed or measured material properties of the prestressing wires (Figure 2), including strain hardening, were incorporated in the model.   2D plane stress state of concrete and biaxial failure load is modelled by equivalent 1D stress-strain relationships [3].   In tension, concrete was assumed to be elastic up to its tensile strength followed by an exponential descending response to represent tension softening (Figure 2).   For the concrete in compression, softening behaviour (uniaxial effective concrete compressive stress-strain curve) was assumed to account for the descending part of the stress-strain relationship for concrete (Figure 2, CEB-FIP Model Code MC-1990).   To account for the ability of cracks to transfer shear stresses by aggregate interlock, reduced material shear stiffness, known as shear retention, is used.

The beams were loaded incrementally to failure under load control.   It was not used displacement driven loading to enable unsymmetrical failure.   In fact, response was symmetric in the major part of pre-peak load-deflection curve.   Prestressing was introduced in ten load steps using Newton-Raphson method with elastic stiffness matrix, and the same solution method was used for approximately first four point loading steps (until the structure was elastic).

Table 1 Values of material properties for 4.5m specimen, default values generated automatically as a function of cylinder strength f_{cu} (right column)

PARAMETER		VALUE	UNIT	FORMULA, DEFAULT VALUES
Cylinder strength	f_{cu}	50	MPa	input value
Initial elastic modulus	E_c	3.695E+04	MPa	$(6000-15.5 f_{cu}) \, sqrt \, (R_{cu})$
Poisson's ratio	v	0.200	-	0.2
Tensile strength	f_t	3.257E+00	MPa	$0.24 f_{cu}^{2/3}$
Compressive strength	f_c	-4.250E+01	MPa	$-0.85 f_{cu}$
Type of tension softening				Exponential, based on G_f
Specific fracture energy	G_f	81.43	N/m	$2 f_t^{ef}$
Crack model		fixed		
Compressive strain at compr. strength	e_c	-2.301E-03	-	-2.301E-03
Reduction of compr. strength due to cracks	c	0.800	-	0.8
Type of compression softening		Crush Band	-	Crush Band
Critical compressive displacement	w_d	-7.0000E-04	m	-0.0005
Shear retention factor			variable	
Tension-compression interaction			linear	

Next steps (pre-peak and post-peak) were performed using Arc-Length method with tangent stiffness matrix. Monitored values were mainly: applied forces (left and right), reactions (left and right), deflections under forces (left and right) and deflection in the middle of the beam.

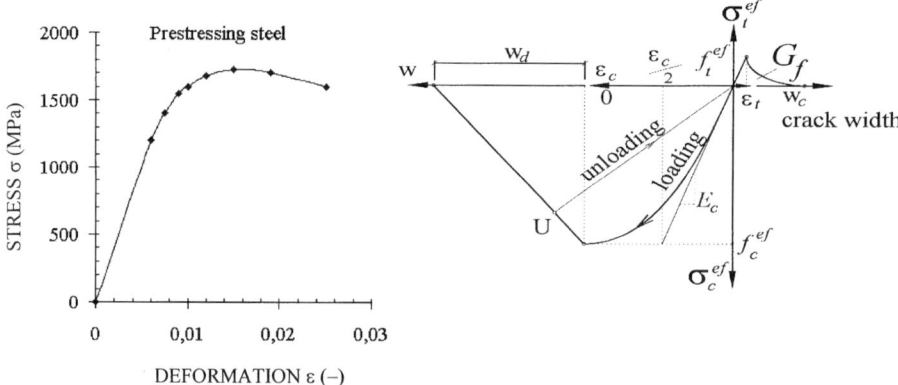

Figure 2 Stress-strain diagram of prestressing steel (left), constitutive law of ATENA concrete model (right)

The load-displacement curve, which resulted from the analysis, is shown in Figure 4, mean. It is evident from this plot that the ultimate strength, initial stiffness and mode of failure are well predicted by the NLFEA models. Gradual development of cracking is showed in Figure 3. There is also comparison of σ_x (longitudinal stress) distribution at the middle of the beam and sketched FE-mesh. Failure was in each simulation by crushing of concrete.

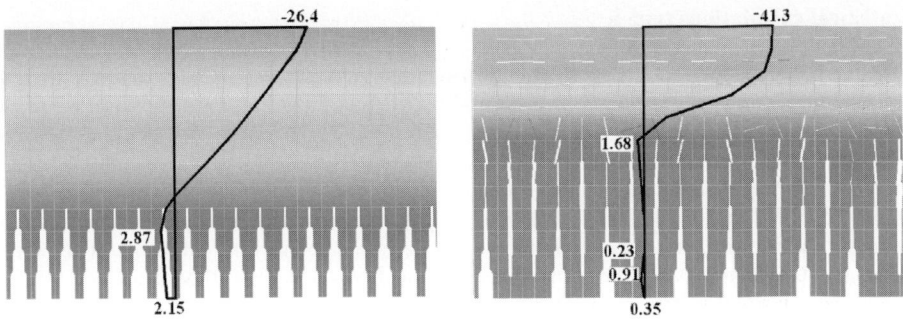

Figure 3   Left: load step 20, [F=55 kN, deflection 21 mm], max. crack width $8\cdot10^{-3}$ mm
Right: load step 150, [F=81.9 kN, defl. 115 mm], max. crack width 0.3 mm
Cracking in tension, crushing under compression

Engineering codes require that structures be designed to be as ductile as possible so that adequate warning of incipient collapse is given by large deflections and crack widths. Ductility is also required to allow moment and load redistribution in redundant structures. Although each individual prestressed rectangular beam is a statically determinant element, when combined with adjacent beams, the ability of individual beams to sustain large deflections before failure allows significant transverse redistribution of load. The Codes follows the classical theoretical approach to try and ensure that engineers design ductile structures, by penalising sections that fail by concrete crushing rather than by yielding of reinforcement. It can be found several reasons to require ductility in concrete structures: to assist in providing robustness, to give warning of incipient collapse, to enable moments in indeterminate structures to redistribute themselves, to enable energy to be absorbed without collapse during an earthquake. For the prestressed concrete rectangular beams discussed in the paper only first two reasons are applicable.

The following can be observed from the experimental testing and NLFE analysis: cracking was visually observed at less than 65% of the ultimate load in each case, the load deflection curves (Figure 4) show that the beams exhibit a strain hardening behaviour, for the 4.5 m beam, deflection at failure > 90 mm, i.e. greater than span/50.

STATISTICAL CALCULATIONS

Two different types of statistical assessment were performed. These two types are described in the following sections.   In case of the design application, according to most current standards, the material properties for calculation of structural resistance (failure load) are considered by minimal values with applied partial safety factors.   The resulting maximum load can be directly compared with the design loads.   More appropriate approach would be to consider the average material properties in nonlinear analysis and to apply a safety factor on the resulting integral response variable (force, moment).   However, this safety format is not yet fully established.   In cases of the simulation of real behaviour, the parameters should be chosen as close as possible to the properties of real materials.   The best way is to determine these properties from mechanical tests on material sample specimens (in consideration with size effect).

Statistical calculations type I

LHS methodology was used here to provide randomness of concrete. Only concrete strength under compression was assumed as a normally distributed random variable with the mean value f_{cu} = 50 MPa and COV = 0.12 (12% variability). Eight realizations of concrete were used to calculate response of the beam. Other parameters of the ATENA constitutive model for concrete were automatically generated using the default formulas given in the right column of the Table 1, [3]. In such a case, only the cube strength of concrete f_{cu} (nominal strength) is specified and the remaining parameters are calculated as functions of the cube strength. The formulas for these functions are based on the CEB-FIP Model Code 90 and other research sources, here used [3].

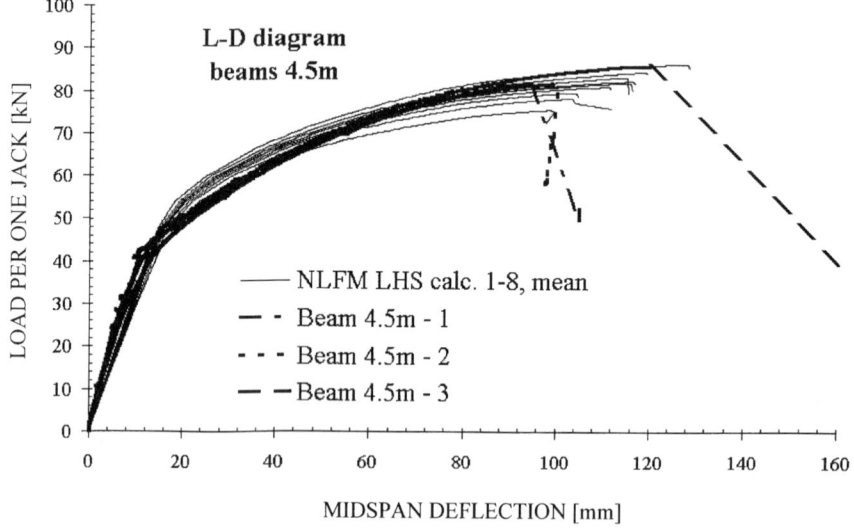

Figure 4   Results of eight LHS calculations – comparison with the three experiments

The eight beams were calculated to the peak value of response and eight values of the strength has mean value 81.378 kN, standard deviation 3.405, COV 0.04, standard skewness - 0.250 and standard kurtosis 2.168. Considering normal probability distribution of the response, we obtain 1% percentile (design value according to the EC) equals to 73.91kN, it means, that probability that strength of our beam is lower than 73.91kN is 1% (0.01). The so-called characteristic value (5%percentile) is 76.10 kN.

Statistical calculations type II

Seventeen different compositions of random concrete by means of LHS method again were used in this type of calculations. Random parameters were as summarised in Table 2. Statistical correlation according to Table 2 was imposed, [5]. The 17 beams were calculated to the peak value of response and 17 values of the strength has mean value 80.194 kN, standard deviation 3.3208, COV 0.041, standard skewness -0.572 and standard kurtosis 2.999.

Table 2   Correlation matrix

	E	w_d	$-f_c$
E	1	0	0.75
w_d	0	1	0
$-f_c$	0.75	0	1

Considering normal probability distribution of the response, we obtain 1% percentile (design value according to EC) equals to 72.47kN, it means, that probability that strength of our beam is lower than 72.47kN is 1%. The so-called characteristic value (5%percentile) is 74.73 kN.

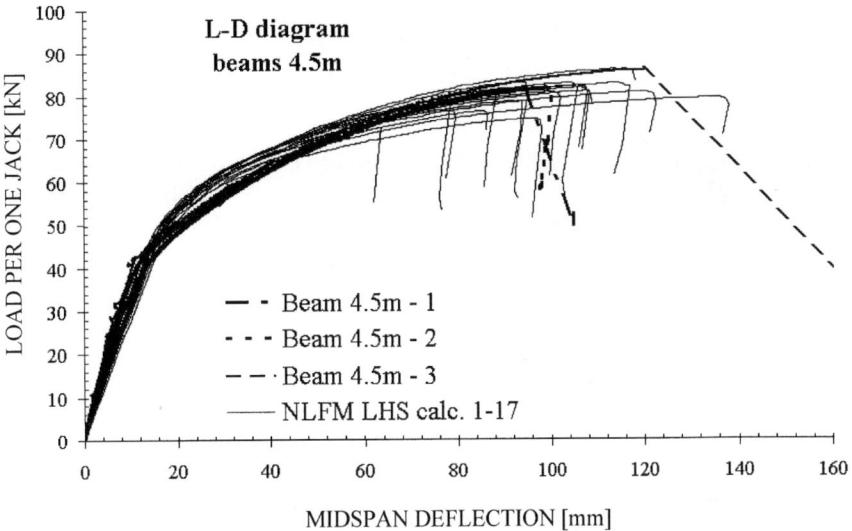

Figure 5   Results of 17 LHS calculations – comparison with the three experiments

CONCLUSIONS

Ability of ATENA software to capture behaviour of prestressed beams is showed. The idea of "randomisation" of this software has been conceived: It is based on small-sample Monte Carlo approach Latin hypercube sampling which gives accurate estimates of statistical characteristics of response using only a small number of simulations. The feasibility of statistical approach is shown using numerical examples. Statistics of some calculated data are presented. The approach is general and can be applied for basic statistical analysis of computationally intensive (fracture mechanics) problems. The aim of stochastic calculation could be the estimation of reliability of the structure using statistical characteristics of response. It is higher level than deterministic calculation only. In cases when an engineer does not know the parameters of the structure accurately, he can calculate the structure several times and compare modes of failure. Sensitivity analysis will help him to estimate which parameters plays main role in the structure and should be measured well.

ACKNOWLEDGMENTS

The results presented in this paper are related to research topics supported by grants of Grant Agency of Czech Republic (GAČR) No. 103/00/0603, project CEZ: J22/98:261100007. The financial support is greatly appreciated.

REFERENCES

1. MARGOLDOVÁ, J, ČERVENKA, V, PUKL, R, Applied brittle analysis, Concrete Engineering International, 1998, Vol 8, No 2, p 65-69.

2. ČERVENKA, V, Simulating a response, Concrete Engineering International, 2000, Vol 4, No 4, p 45-49.

3. ČERVENKA, V, PUKL, R, ATENA – Computer Program for Nonlinear Finite element Analysis of Reinforced Concrete Structures. Program documentation. 2000, Prague, Czech Republic: Červenka Consulting.

4. NOVÁK, D, TEPLÝ, B, KERŠNER, Z, The role of Latin Hypercube Sampling method in reliability engineering. In Shiraishi, N, Shinozuka, M, Wen, Y K, (eds), Proceedings of ICOSSAR-97, Kyoto, Japan, 1998, Rotterdam: Balkema, p 403-409.

5. VOŘECHOVSKÝ, M, RUSINA, R, NOVÁK, D, Correlated Random Variables in Probabilistic Simulation, Proceedings of 4th Ph.D. Symposium in Civil Engineering, Munich, Germany, September 2002.

6. KALA, Z, NOVÁK, D, VOŘECHOVSKÝ, M, Probabilistic nonlinear analysis of steel frames focused on reliability design concept of Eurocodes, Proceedings of ICOSSAR 2001, 8th International Conference on Structural Safety and Reliability, 2001, USA.

7. NOVÁK, D, VOŘECHOVSKÝ, M, PUKL, R, ČERVENKA, V, Statistical nonlinear fracture analysis: Size effect of concrete beams, International Conference on Fracture Mechanics of Concrete Structures FraMCos 4, Cachan, France, Balkema, Lisse, 2001, The Netherlands, p 823-830.

8. TAPLIN, G, AL-MAHAIDI, R, BOULLY, G, KWEI, S, AUSTROADs, Proceedings of the bridge conference, Vol 3, Adelaide, South Australia, December 2000.

THE EFFECT OF PORE SIZE ON ION MIGRATION IN CONCRETE DURING ELECTROCHEMICAL CHLORIDE EXTRACTION

M Siegwart   **B J McFarland**

J F Lyness   **W Cousins**

University of Ulster

United Kingdom

ABSTRACT. Electrochemical chloride extraction (ECE) is the removal of chlorides from the vicinity of the reinforcement leaving sound concrete intact. In a short period, 3 to 5 weeks, a high current, relative to other applications such as cathodic prevention or protection, is applied to the structure. Concrete shows a difference in pore size and pore size distribution before and after ECE treatment. Furthermore during ECE applications an increase in concrete resistance can be observed. An experimental study was conducted to investigate the influence of pore size on the resistance of artificial concrete pore liquid. The results showed that the resistance increases with decreasing pore size if a certain limit is exceeded. This limit appears to be related to the ion species and the concentration of the ions in the solution.

Keywords: Electrochemical chloride extraction, Pore size, Ion redistribution, Migration, Concrete resistivity

M Siegwart obtained a B.Eng.(Hons) in Civil Engineering from the University of Ulster. He is currently a postgraduate researcher at the University of Ulster with an interest in electrochemical chloride extraction and related areas.

B J McFarland, MICE, MICorr, MIEI, specialist engineer NACE, B.Eng.(Hons) Civil Engineering (Aston University). He is an independent consultant specialising in all aspects of reinforced concrete inspection, testing, assessment and repair.

J F Lyness, MICE, MIStructE, obtained a Ph.D. in civil engineering from University College Swansea, Wales. He is currently Reader in Civil Engineering at the University of Ulster with an interest in numerical methods in engineering.

W Cousins, CEng, MICE, obtained a PhD in Civil Engineering from the University of Ulster. He is currently a Lecturer in structures and materials at the University of Ulster.

INTRODUCTION

The two main causes of reinforcement corrosion are chloride and carbonated concrete. Electrochemical chloride extraction and its sister technique, re-alkalisation, are used to restore chloride contaminated or carbonated concrete. Both processes take advantage of the fact that migration of ions under an electrical field in concrete is much faster than diffusion. The chloride contamination through diffusion can thus be reversed in a short period of time. Migration and diffusion of a charged species such as chloride in concrete is dependent on a combination of factors. One of them might be the pore structure, because it is altered during electrochemical chloride extraction. It is difficult to examine the contribution of one single factor, such as the pore size, on ion migration in concrete, because concrete is a heterogeneous, reactive environment. Therefore, studies were conducted on thin membranes with known pore size in artificial concrete pore water and its components.

CHLORIDE CONTAMINATION THROUHGH DIFFUSION

Chloride ions diffuse through concrete towards the reinforcement when the structure comes in contact with e.g. de-icing salts. During diffusion the cement binds chloride as Friedel's salt [1]. Fundamental equations of diffusion are Fick's 1st and 2nd laws. When Fick's law of diffusion is applied to ions in concrete it only applies to the diffusion of undissociated molecules, not ions such as chloride. Fick's 1st law applies only for steady state diffusion e.g. where the chlorides were cast in. The transport process under Fick's law is a simple diffusion process and cement is not an inert or fully open medium to diffusants (i.e. water). Therefore, the diffusion equations need to be modified. Only the pore characteristics allow the diffusants to pass through the sample and there are interactions between the diffusants and the cement.

The formation of Friedel's salt and the adsorption of sodium ions by CSH will reduce the effective concentration gradients. The diffusion coefficient becomes an apparent diffusion coefficient for cement [2]. For ionic materials the influence of the electronic double layer and the fact that concrete reacts with the diffusants needs to be taken into consideration [3]. The concentration gradient of ions across a permeable barrier is generally accompanied by an electrical potential gradient. This electrical potential gradient hinders the diffusion of ions across the barrier and a membrane potential is formed [4].

CHLORIDE EXTRACTION THROUGH MIGRATION

Chatterji [5] suggested that ion transport through cement based material can properly be described only by using a combination of Nernst-Planck and the electrical double layer diffusion processes. Movements of positive and negative ions are coupled, which would explain the high activation energy of chloride diffusion. The high activation energy of chloride migration processes in concrete is considered to be an indication of other linked processes. These processes are the transport of positive and negative ions, the dissolution of cement hydrates and the migration of calcium and hydroxide in opposite directions, the maintenance of calcium in the double layer and the linked formation of Friedel's salt.

For a steady state diffusion there is a solution due to Planck. All ions present in the pore solution that have been generated at the electrodes will move towards the counter pole.

This movement has to follow the mass transport law for electrolytes (Nernst-Planck), which states that the total net flow is the addition of diffusion, migration and convection. The Nernst-Planck equation is expressed as follows:

$$J_j(x) = -D_j \frac{\partial C_j(x)}{\partial x} - \frac{z_j F}{RT} D_j C_j \frac{\partial \phi(x)}{\partial x} + C_j v(x) \qquad (2)$$

Migration takes into account the movements due to the application of an electrical field. The migration term is also named the Nernst-Einstein equation. The diffusion coefficient in the Nernst-Einstein equation is a function of the electrolyte conductivity. In an ionic solution, a electrostatic field due to the other ions in the solution restricts the movements of ions. This restriction becomes particularly important when the diffusion occurs through a permeable barrier such as concrete across which there is a concentration gradient.

CONCRETE RESISTIVITY

The ions in the pore water transport electrical current, hence the resistivity is a measure of the continuity of the pore solution through which chlorides migrate [6,7]. Andrade [8] and Bamforth[9] suggest measuring the resistivity of concrete as an indication of chloride permeability. The use of conductivity (reciprocal of resistivity) of concrete for measurements of chloride permeability has also been suggested [10]. Dry concrete acts as electrical insulator; its resistivity (oven-dry) is approximately $109\Omega m$, but moist concrete acts as conductor with a resistivity of approximately 50 Ωm [11].

Resistivity may be obtained by dividing the IR voltage drop across the specimen by the average current density and the specimen thickness [12]. Andrade [13] proposed a mathematical formula for the calculation of the chloride diffusivity in concrete related to its resistivity. The resistivity of blended cement is higher than resistivity of OPC concrete [14], because of smaller pores. For the same reason carbonation increases the electrical resistivity and water adsorption of concrete [15].

EXPERIMENTAL PROCEDURE

Concrete specimens were subjected to ECE treatment for a period of 40 days. The concrete pore size distribution was analysed prior to and after the ECE treatment at various depths. It was noted that the pore size and distribution had changed due to ECE. An experimental study was then conducted to investigate the influence of pore size on migration of ions during ECE. The diffusion and migration of ions in concrete is a complex interaction of a multitude of factors such as concentration of migrating species in different layers of the concrete, physical properties of the concrete, etc. It is thus not possible to investigate the influence of one single parameter on ion migration using real concrete. The composition of concrete pore water and the range of concentrations of individual species can be reproduced for laboratory investigation [16]. The concentrations of ions used in this investigation are given in Table 1.

Table 1   Concentration of Ions in Artificial Concrete Pore Solution

SPECIES	CONCENTRATION [g/l]
Potassium hydroxide (KOH)	2.0
Sodium hydroxide (NaOH)	0.3
Calcium hydroxide (Ca(OH)$_2$)	0.1
Sodium chloride (NaCl)	2.5
Artificial concrete pore water	Sum of above components

The pore size of concrete and cement paste is reported to be between 0.003 and 1.000 μm for young cement pastes and between 0.003 and 0.1 μm for mature cement pastes [17]. Instead of concrete a material with known pore size - *Whatman*[®] *Anopore* filter membranes with 0.02, 0.10 and 0.20 μm pore diameter - were used to investigate the influence of pore size on migration. The experimental set-up is shown in Figure 1. An impermeable Perspex[©] membrane with a ∅35mm hole in the centre was glued into a crystallisation dish to form two compartments and the hole was covered with a permeable filter membrane and sealed. Migration of ions was therefore possible only through the membrane. Two platinum electrodes with an electrode area of 1cm^2 were placed at a distance of 2cm from each other, separated only by the membrane or plain solution in case of the measurements without membrane.

The filter membrane is very thin compared to concrete cover, but one thin membrane can represent concrete, because during impedance the migration of species is reversed at high frequencies. One membrane (filter membrane) does, therefore, have the same effect as multiple membranes (or concrete).

Figure 1   Experimental set up

The electrical resistance of the system reflects the reactions of an electrode system. The resistance of the solution (R_s), the evolving double layer on the electrodes (R_p) and the capacitance of the electrode (C) give the total resistance or impedance [18]. The capacitance is constant, but the solution resistance and the resistance of the double layer depend on the concentration of the ions in the solution. Impedance Z and resistance are obtained the same at high frequencies when reactions at the electrode do not take place.

The impedance measurements of the influence of membranes in artificial concrete pore solution were conducted with a Schlumberger SI 1260 Impedance/Gain Phase Analyser and SI 286 Electrochemical Interface.

EXPERIMENTAL RESULTS AND IMPLICATIONS

The impedance measurements showed, for different electrolytes, an increase in resistance when a membrane was introduced between the electrodes. This effect was observed at low and at high frequencies.

The results of the impedance measurements of the artificial concrete pore solution are displayed in Figure 2. The resistance significantly increases below a certain frequency. This can be attributed to the build up of a double layer and the occupation of free sites on both electrode surfaces by the adsorption and subsequent reactions of ions. This effect is even more significant during ECE as a direct current is applied.

This results in an instantaneous formation of a reactive layer on the surface with a corresponding increase in resistance. Therefore, resistance/resistivity measurements under a direct current source always contain an element of error due to the build up of the double layer and do not truly reflect the migration properties of ions in the concrete.

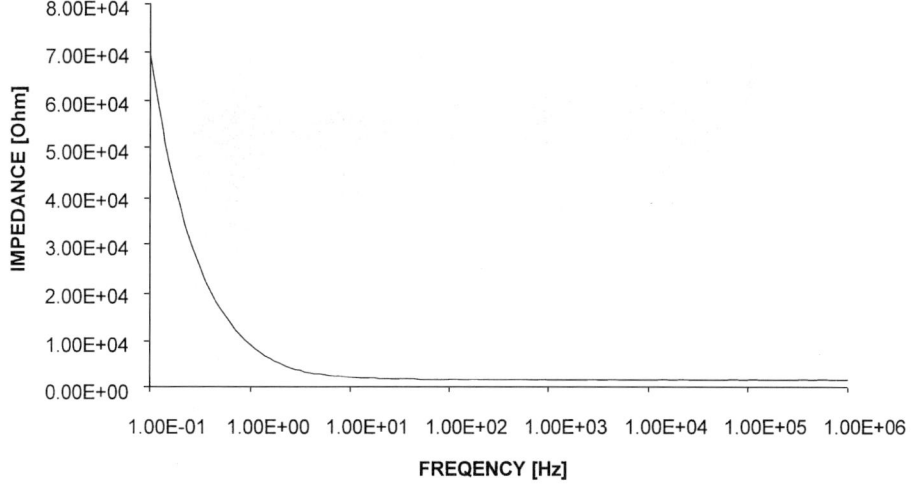

Figure 2   Results of Impendence measurement for artifical concrete pore solutions with 0.02μm membrane

During ECE ions are removed from the bulk concrete. Hydroxide and chloride form water and chlorine at the anode and cations precipitate at the pore walls near the cathode once their maximum solubility is exceeded. Thus these ions do not take part in the migration process and fewer ions carry the current, which results in increasing resistivity during ECE.

Concrete acts a membrane and due to the build up of a concentration gradient across the membrane, i.e. the reinforcement and the anode, a potential gradient develops, which is caused by the non-uniform distribution of charged species. This potential is called the membrane potential and contributes to the increase of the resistance and was observed where a DC current was applied between two electrodes separated by membranes.

Two effects can be observed, first, the initial resistance of the system is different due to different pore size of the membranes and second, with time the resistance increases slightly.

To analyse the influence of the parameter of pore size it is important to use only the impedance measurements that represent the resistance of the solution. This is equivalent to the part of the impedance, which is parallel to the horizontal in Figure 2. The relative increase of resistances (percent) of the different solutions is plotted against the pore sizes in Figure 3, the resistance of the solution without membrane was thereby taken as 100%. It can be seen that with decreasing pore size the resistance of the system increases.

A linear relationship best describes the relationship between pore size and resistance for the pore solution and calcium hydroxide and for the remaining three solutions a logarithmic relationship gives a good approximation of the influence of pore size on resistance.

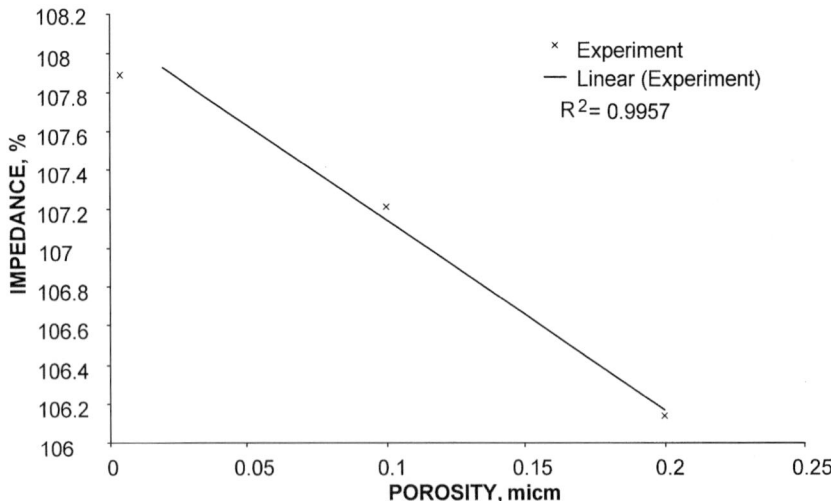

Figure 3   Relationship between impedance and pore size of membranes

The impedance measurements are based on four points on the vertical, but only for three points, the membrane, the pore sizes (horizontal) are known – not for the measurements without membrane. To find the pore size where migration remains unhindered, the graph has to be extrapolated to reach the horizontal.

The results of this analysis are shown in Figure 4 and Figure 5. The resistance remains unhindered for a pore size of larger than 4 μm for NaCl and larger than 10 μm for sodium hydroxide. It is not possible to isolate and investigate the influence of one type of ion because a cation is always accompanied by an anion to satisfy the electroneutrality condition. It appears that the coupled migration of sodium and hydroxide requires more space than the coupled migration of sodium and chloride.

Figure 4 shows to be a logarithmic plot, but the data in Figure 5 is plotted on a normal scale. It becomes apparent that the logarithmic graph gives a better fit when the data is extrapolated, because the curve approximates better to the known values of pore sizes.

The logarithmic relationship can, however not be used to find the minimum pore size, because it would give unrealistically high values for minimum pore sizes (100 μm for calcium hydroxide and 1000 μm for pore solution).

If a linear relationship is used for extrapolation a minimum pore size of 10.41 μm for calcium hydroxide and 6.48 μm for plain pore solution is obtained. Potassium hydroxide can migrate undisturbed through pores with a pore size of 1 μm whereas calcium hydroxide requires a much wider space although its concentration was much less compared to potassium hydroxide.

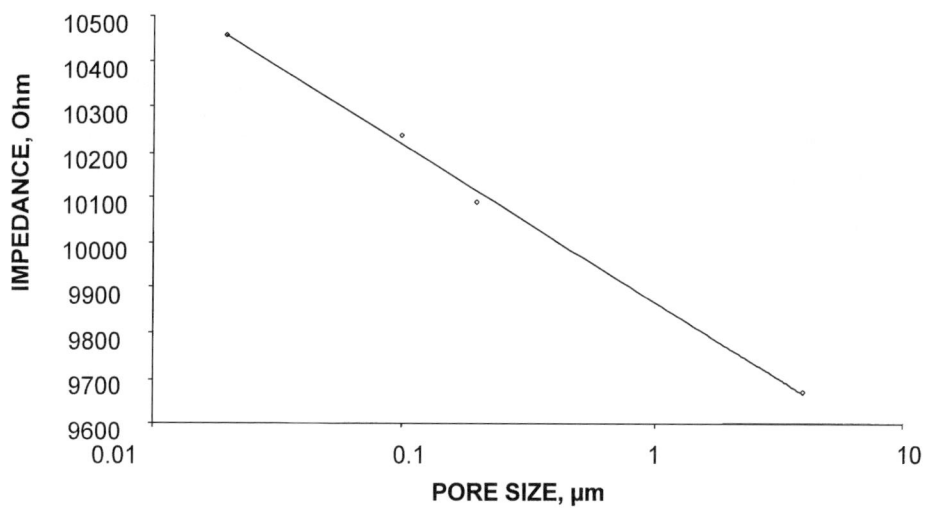

Figure 4   Relationship between impedance and pore size in sodium chloride solution

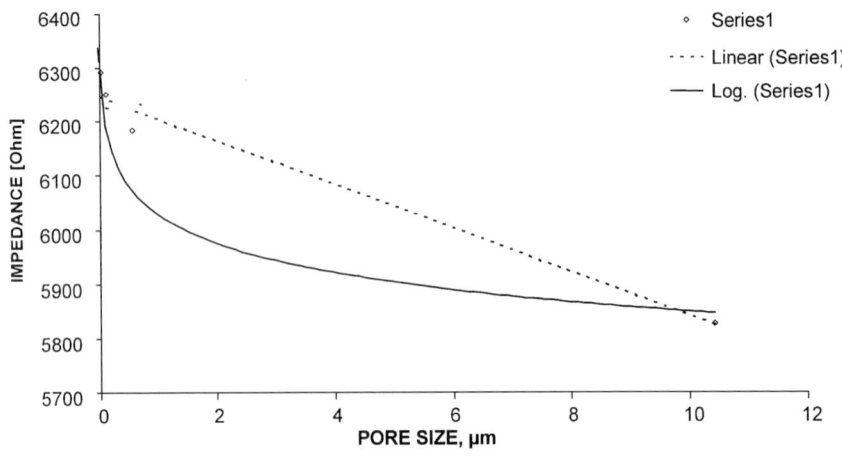

Figure 5 Relationship between impedance and pore size of membrane in pore solution

CONCLUSIONS

From the study a number of conclusions can be drawn.

- The diffusion of charged species such as chloride or hydroxide and their corresponding cations are related to their electrical properties, interactions with the diffusive medium and interactions with each other.

- During ECE the pore size and the pore distribution of concrete change.

- Amongst other factors, the change in pore size, if it falls below a certain limit, will influence the migration properties of the species and change the resistance of the system.

- However, this limit is different for different ions and it varies with concentration. Further work is recommended to consider the effects of changing pore size for the migration of ions in concrete.

ACKNOWLEDGEMENTS

The authors are in dept of Dr B. Eggins from the Department of Chemistry at the University of Ulster for his advice. Furthermore, the first author would like to express his special thanks to A. Richardot from the Northern Ireland Bio-Engineering Centre (NIBEC) for his technical assistance during the impedance measurements.

REFERENCES

1.   SURYAVANSHI A.K., SCANTLEBURY J.D., LYON S.B., Mechanism of Friedel's salt formation in cements rich in C3A, Cement and Concrete Research, 26(5), 1996, pp717-727.

2. CHATTERJI S., Transportation of ions through cement based materials. Part 2 adoption of the fundamental equations and relevant comments, Cement and Concrete Research, 24 (6), 1994, pp 1010-1014.

3. CHATTERJI S., On the applicability of Fick's second law to chloride ion migration through portland cement concrete, Cement and Concrete Research, 25 (2), 1995, pp299-303.

4. BUENFIELD N.R., GLASS G.K. HASSANEIN A.M., ZHANG J., Chloride transport in concrete subjected to electric field, Journal of Materials in Civil Engineering, 10 (4), 1998, pp220-228.

5. CHATTERJI S., Transportation of ions through cement based materials. Part 3 experimental evidence for the basic equations and some important deductions, Cement and Concrete Research, 24 (7), 1994, pp 1229-1236.

6. FELDMAN R.F., CHAN G.W., BROUSSEAU R.J., TUMIDAJSKI P.J., Investigation of the rapid chloride permeability test, ACI Materials Journal, 91 (2), 1994, pp. 246-255.

7. FELDMAN R.F., PRUDENCIO L.R., CHAN G.W., Rapid chloride permeability test on blended cement and other concretes: correlations between charge, initial current and conductivity, Construction and Building Materials, 13, 1999, pp149-154.

8. ANDRADE C., Calculation of chloride diffusion coefficients in concrete from ionic migration measurements, Cement and Concrete Research, 23, 1993, pp724-742.

9. BAMFORTH P.B., The derivation of input data for modelling chloride ingress from eight-year UK coastal exposure trials, Magazine of Concrete Research, 51 (2), 1999, pp87-96.

10. XINGYING Lu, Application of the Nernst-Einstein equation to concrete, Cement and Concrete Research, 27 (2), 1997, pp293-302.

11. BUENFIELD N.R., NEWMAN J.B., Examination of three methods for studying ion diffusion in cement pastes, mortars and concrete, Materials and Structures, 20, 1987, pp3-10.

12. HASSANEIN A.M., GLASS G.K. BUENFIELD N.R., Effect of intermitted cathodic protection on chloride and hydroxyl concentration profiles in reinforced concrete, Corrosion Reviews, 17 (5), 1999, pp423-441.

13. ANDRADE C., SANJUÁN M.A., RECUERO A., RIO O., Calculation of chloride diffusivity in concrete from migration experiments, in non steady-state conditions, Cement and Concrete Research, 24 (7), 1994, pp1214-1228.

14. BAWEJA D., ROPER H SIRIVIVATNANON V., Chloride induced steel corrosion in concrete: Part 1 – corrosion rates, corrosion activity, and attack areas, ACI Material Journal, 95 (3), 1998.

15. Al-KADHIMI T.K.H., BANFIL P.F.G., MILLARD S.G., BUNGEY J.H., An accelerated carbonation procedure for studies on concrete, Advances in Cement Research, 8 (30), 1996, pp47-59.

16. BREIT W., Untersuchungen zum kritischen korrosionsauslösenden Chloridgehalt für Stahl in Beton, PhD thesis, RTWH Aachen, 1997.

17. TAYLOR H.F.W., Cement Chemistry, 2nd Edit, Thomas Telford, 1998.

18. BARD A.J., FAULKNER L.R., Electrochemical methods, John Wiley & Sons, 1980.

INFLUENCE OF SPECIMEN SIZE ON MEASURED DIRECT TENSILE STRENGTH OF CONCRETE

A K H Kwan

P K K Lee

W Zheng

University of Hong Kong

China

ABSTRACT. The influence of specimen size on the measured direct tensile strength of concrete was investigated. In the investigation, two similar direct tension test setups were employed to measure the direct tensile strengths of concrete prisms of four different sizes. The test results revealed that the specimen size effect in the direct tension test was significant. Generally, the measured direct tensile strength decreased as the length, cross-sectional area, or volume of the concrete specimen increased. On the other hand, the within-test error of the direct tension test decreased when the length decreased or the cross-sectional area increased. Based on the test results, regression equations correlating the direct tensile strengths measured from specimens of difference sizes with the corresponding cube compressive strength have been derived. The regression analysis showed that the correlation between direct tensile strength and cube compressive strength varied with the size of the specimen used in direct tensile strength measurement.

Keywords: Tension test, Tensile strength, Size effect.

A K H Kwan is an Associate Dean of Engineering at The University of Hong Kong. He obtained his doctorate from The University of Hong Kong and has acquired many years of practical experience before returning to the academia. His research interests include concrete technology, tall building structures and earthquake resistant structures.

P K K Lee is the Head of Department of Civil Engineering, The University of Hong Kong. He has been teaching and researching as an academic and practicing as a specialist consultant for more than thirty years. His research covers many aspects of construction including piled foundation, rock engineering, concrete technology and structural health monitoring.

W Zheng is a Ph.D. student studying at Department of Civil Engineering, The University of Hong Kong. He earned his BEng from Tongji University, Shanghai, China. His research interests include static and dynamic testing methods for concrete properties as well as high-performance concrete.

INTRODUCTION

Progress in concrete technology has led to growing concern about the tensile behaviour of concrete. Research has revealed that if the no-tension design is used in the construction of plain concrete structures, such as dams and retaining walls, the factors of safety cannot be guaranteed to have the specific values and that therefore the tensile strength of concrete cannot be neglected [1]. Moreover, for crack control during casting and curing of concrete, the stress/strain capacities of the concrete in tension need to be known [2]. In both regards, knowledge of the tensile behaviour of concrete is important.

The tensile strength of concrete is usually measured by the flexural test, the splitting tension test and the direct tension test, which yields the flexural tensile strength, the splitting tensile strength and the direct tensile strength respectively. However, the tensile strength values obtained vary with the test method used. Among the existing test methods, it is believed that the direct tension test method is the best in producing realistic measurement of the uniaxial tensile strength of concrete [3].

It is well known that the size of the specimen used could substantially affect the strength results [4-7]. Based on Griffith theory, several theoretical expressions for estimating the effect of specimen size on the measured strength of concrete have been developed. These include the "weakest link principle" [5], the "multifractal scaling law" [6] and "Bazant's size effect law" [7]. They have been applied successfully to quantify the size effect in the flexural test and the splitting tension test [8-10].

However, the influence of specimen size on the measurement of direct tensile strength has not been well investigated. This is due partly to the generally larger difficulty of measuring the direct tensile strength and partly to the absence of a well-established direct tension test method [3,4]. It is the authors' opinion that the most reliable way of applying direct tension to concrete without inducing secondary stresses is to pull steel plates glued to the ends of the concrete specimen [11-13].

Adopting this way of applying direct tension, a direct tension test method that is applicable to prismatic concrete specimen cast either horizontally or vertically has been developed. As a preliminary study on the tensile behaviour of concrete, this newly developed test method has been employed to measure the direct tensile strength of concrete prisms of four different sizes. The results so obtained on the influence of specimen size on the measured direct tensile strength of concrete are presented herein.

DIRECT TENSION TEST SETUPS

A new direct tension test method that employs bonded steel end plates to apply tension load to the concrete specimen has been developed recently by the authors [3]. Each set of end plates attached to one end of the concrete specimen consists of two steel plates: an inner plate and an outer plate. The inner plate is glued to the specimen by epoxy while the outer plate is connected to the inner plate by bolts. The outer plate has a pulling rod attached to it at the centre for gripping by the testing machine (the outer plate and the pulling rod attached to it are actually machined from a single piece of steel).

Tension load is applied to the pulling rod of the outer plate and transmitted to the inner plate through the bolts, which help to spread the load uniformly across the inner plate. A sketch of the test setup designed for prismatic specimens with 100 mm × 100 mm section has been presented in an earlier paper [3].

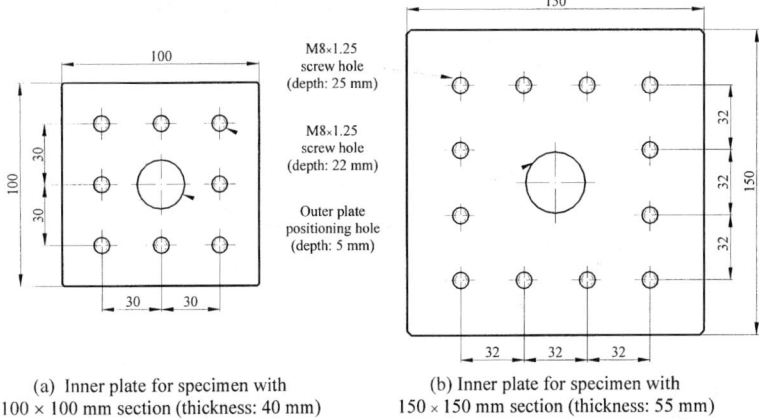

(a) Inner plate for specimen with
100 × 100 mm section (thickness: 40 mm)

(b) Inner plate for specimen with
150 × 150 mm section (thickness: 55 mm)

Figure 1   Inner plates used in the two direct tension test setups

In order to study the specimen size effect, a similar test setup for prismatic specimens with 150 mm × 150 mm section has been designed and fabricated. The major differences between the two setups are the sizes of the steel end plates and the arrangement of the bolts connecting the inner and outer plates together. In each setup, the inner and outer plates have the same cross-sectional sizes as the specimen to be tested. In the setup for specimens with 100 mm × 100 mm section (Setup A), the thicknesses of the inner and outer plates are 40 mm and 20 mm respectively, while in the setup for specimens with 150 mm × 150 mm section (Setup B), the thicknesses of the inner and outer plates are 55 mm and 25 mm respectively. Regarding the arrangement of the bolts, the numbers of bolts used in Setup A and Setup B are 8 and 12 respectively. All bolts are M8 × 1.25 bolts. Figure 1 shows the actual arrangement of the bolts in each setup. Both setups can be used for specimens with different lengths. In Setup A, the length of the specimen is adjustable between 150 and 500 mm, whereas in Setup B, the length of the specimen is adjustable between 250 and 750 mm.

Three-dimensional finite element analysis of the two tension test setups has been carried out to check whether the tensile stresses applied to the specimen are truly uniaxial and uniform. Several different specimen lengths, from 2.0 to 5.0 times the width, have been used in the analysis. Almost identical stress/strain results were obtained for the steel end plates and the epoxy layer regardless of the specimen length. Hence, any specimen length longer than 2.0 times the width has little effect on the stress distribution near the specimen ends. Even at the end plate-specimen interface, the tensile stresses are within 100±2% of the average stress over 87% and 78% of the sectional area in Setup A and Setup B respectively. The variation of tensile stress within a section in the specimen diminishes rapidly with the distance from the end plate-specimen interface. At a distance of 0.5 times the width from the specimen end, the variation becomes less than 3%. Moreover, throughout the whole length of the specimen, the tensile stresses are almost purely uniaxial.

EXPERIMENTAL PROGRAMME

To cover typical concretes of different strengths, three concrete mix designs were adopted. From each designed mix, three batches of concrete were cast, each for testing at ages of 3-, 7- and 28-day respectively. The physical properties of the raw materials used are presented in Table 1 while the mix proportions of the concrete mixes are given in Table 2.

Table 1   Physical properties of raw materials

MATERIAL	BRAND OR SPECIFICATION	SPECIFIC GRAVITY	MAJOR CHARACTERISTICS
Cement	"Green Island" ordinary Portland cement	3.14	Specific surface: 433 m^2/kg
Fine aggregate	Crushed granite	2.65	Grading complied with BS 882: 1992
Coarse aggregate	10 mm and 20 mm crushed granite	2.65	Grading complied with BS 882: 1992
Admixtures	"Daracem 100" superplasticizer	1.20 (liquid)	Complied with BS 5075: Part 3: 1985

Table 2   Details of mix proportions

MIX NO	CEMENT CONTENT (kg/m^3)	WATER/CEMENT RATIO	FINE TO TOTAL AGGREGATES RATIO	SUPERPLASTICIZER (litre/m^3)
1	265	0.80	0.45	0
2	320	0.66	0.37	0
3	360	0.56	0.37	0.95

Out of each batch of concrete, three 150 mm cubes, three $100 \times 100 \times 300$ mm prisms, three $100 \times 100 \times 500$ mm prisms, three $150 \times 150 \times 450$ mm prisms and three $150 \times 150 \times 750$ mm prisms were cast. Steel moulds were used for casting. After casting, the unmoulded surfaces were covered by polythene sheets until the steel moulds were removed at the age of 1-day. Then, the specimens were water cured at 20±1 $^\circ$C until the time of testing.

At the designated test age, the cubes and prisms were tested for cube compressive and direct tensile strengths respectively. From each test group of three specimens, the test results were averaged to give the mean cube compressive/direct tensile strength of the concrete. The coefficient of variation of the test group was also calculated to give the within-test variation of the test.

If, within one test group, none of the individual test results deviated from the mean of three test results by more than 15%, the mean was taken as the final result. In the case of one individual test result deviated from the mean by more than 15%, the median of the three test results was taken as the final result. In the case of two or more individual test results deviating from the mean by more than 15%, the test results of the whole test group were discarded.

Where applicable, the tests were carried out in accordance with relevant British Standards. However, the direct tension tests were conducted using the method developed by the authors, as depicted below. For the direct tension tests, a 1000 kN computer controlled hydraulic testing machine was employed. An electronic load cell, connected to the test assembly in sequence, was used for load measurement.

The tension load was applied gradually at a constant rate of 0.015 MPa/s, which was well within the range of loading rate recommended by US Bureau of Reclamation [13], until the concrete prism failed. After then, the maximum applied tension load and the fracture location were recorded, and the fracture surfaces were inspected for presence of any foreign matter or weathered rock aggregate particles that might have caused unreasonably low tensile strength results.

RESULTS AND DISCUSSIONS

The direct tensile strength results are listed in Table 3 and plotted against the corresponding cube strength results in Figure 2. Assuming that the direct tensile strengths may be correlated to the cube compressive strength by square root functions, as in the following form:

$$f_t = k \, (f_{cu})^{1/2} \qquad\qquad\qquad (1)$$

in which k is an unknown coefficient to be determined, regression equations for the test results have been obtained. These equations are plotted in Figure 2 as best-fit lines alongside the data points to show how they fit the test results. The regression equations, together with their respective correlation coefficients R are printed by the side of the best-fit lines.

From these equations, it can be seen that the k-values for $100 \times 100 \times 300$ mm prisms, $100 \times 100 \times 500$ mm prisms, $150 \times 150 \times 450$ mm prisms and $150 \times 150 \times 750$ mm prisms are 0.49, 0.46, 0.47 and 0.43 respectively. The variation of k-value with specimen size reveals the influence of specimen size on the measured direct tensile strength.

Table 4 presents the coefficients of variation calculated from the individual results of the direct tension tests. The coefficients of variation, which may be interpreted as the within-test variations, vary from one set of test results to another. On the whole, averaged values of 7.7%, 11.4%, 6.3% and 9.8% have been obtained for the $100 \times 100 \times 300$ mm prisms, $100 \times 100 \times 500$ mm prisms, $150 \times 150 \times 450$ mm prisms and $150 \times 150 \times 750$ mm prisms respectively.

These results indicate that the wider and shorter specimens yielded smaller coefficients of variation. In other words, the within-test variation of the direct tension test reduces as the specimen width increases or the specimen length decreases.

Table 3   Direct tensile strength results

MIX NO.	AGE (DAYS)	CUBE STRENGTH (MPa)	DIRECT TENSILE STRENGTH (MPa)			
			100×100×300 prism	100×100×500 prism	150×150×450 prism	150×150×750 prism
1	3	11.1	1.41	1.37	1.34	1.41
	7	19.4	2.01	2.19	2.07	1.95
	28	28.9	3.38	2.81	2.74	2.65
2	3	14.0	1.77	1.79	1.68	1.41
	7	24.0	2.61	2.47	2.39	2.41
	28	35.9	3.03	2.73	2.83	2.87
3	3	22.6	2.08	1.92	2.20	2.07
	7	31.7	2.71	2.70	2.85	2.14
	28	48.1	2.98	2.74	2.99	2.69

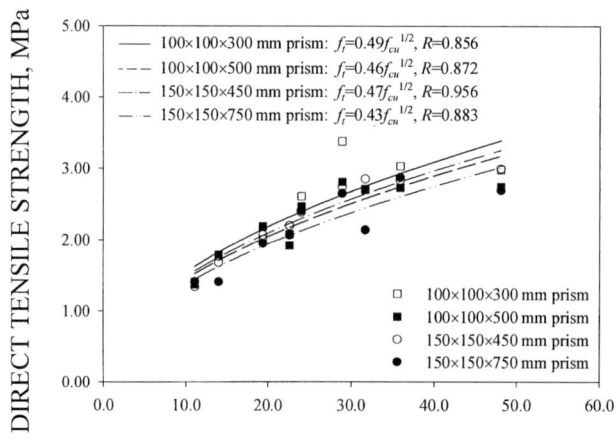

Figure 2   Relations between direct tensile strengths and cube compressive strength

Influence of Volume of Specimen

From the k-values of the regression equations, it can be seen that as the specimen size increases, the direct tensile strength decreases. On average, the direct tensile strengths of the 100 x 100 × 500 mm prisms, the 150 × 150 × 450 mm prisms and the 150 × 150 × 750 mm prisms are respectively 6.1%, 4.1% and 12.2% lower than that of the 100 × 100 × 300 mm prisms.

Table 4 Coefficient of variation of direct tension test

MIX NO.	AGE (DAYS)	COEFFICIENT OF VARIATION OF DIRECT TENSION TEST (%)			
		100×100×300 prism	100×100×500 prism	150×150×450 prism	150×150×750 prism
1	3	5.0	10.5	1.5	31.0
	7	17.1	12.9	10.9	5.2
	28	6.3	6.1	14.1	1.8
2	3	9.6	7.1	9.9	23.9
	7	8.2	15.4	4.3	9.8
	28	4.3	23.0	5.6	2.4
3	3	6.2	3.0	2.1	3.5
	7	3.7	16.0	1.9	6.2
	28	8.5	8.8	6.5	4.2

The k-values are plotted against the volume of the specimen in Figure 3, where a linear trendline is also drawn to show how the volume of the specimen influences the measured direct tensile strength at the same compressive strength level. A quantitative relation between the k-value and the volume of specimen cannot be derived yet because the amount of experimental data available is still not sufficient.

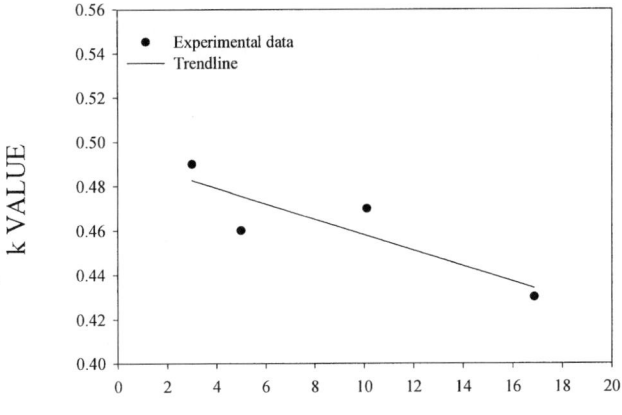

VOLUME OF CONCRETE SPECIMEN, 10^6 mm^3

Figure 3 Variation of k-value with volume of specimen

Influences of Cross-Sectional Area and Length of Specimen

Putting the test results of the $100 \times 100 \times 300$ mm prisms and the $100 \times 100 \times 500$ mm prisms together for regression analysis, the regression equation for prisms with a cross-section area of 10,000 mm^2 (100 mm $\times$ 100 mm) is obtained as:

$$f_t = 0.47 \, (f_{cu})^{1/2} \tag{2}$$

The correlation coefficient of this equation is 0.853. Similarly, putting the test results of the $150 \times 150 \times 450$ mm prisms and $150 \times 150 \times 750$ mm prisms together, the regression equation for prisms with a cross-section area of 22,500 mm^2 (150 mm $\times$ 150 mm) is obtained as:

$$f_t = 0.45 \, (f_{cu})^{1/2} \tag{3}$$

The correlation coefficient of this equation is 0.910. Comparing Equations (2) and (3), it can be seen that the k-value for prisms with a cross-section of 150 mm $\times$ 150 mm is about 4% lower than that for prisms with a cross-section of 100 mm $\times$ 100 mm. This reveals that the measured direct tensile strength decreases as the cross-sectional area increases.

Putting the test results of the $100 \times 100 \times 300$ mm prisms and the $150 \times 150 \times 450$ mm prisms together for regression analysis, the regression equation for prisms with a length/width ratio of 3.0 is obtained as:

$$f_t = 0.48 \, (f_{cu})^{1/2} \tag{4}$$

This equation has a correlation coefficient of 0.897. On the other hand, putting the test results of the $100 \times 100 \times 500$ mm prisms and the $150 \times 150 \times 750$ mm prisms together for regression analysis, the regression equation for prisms with a length/width ratio of 5.0 is obtained as:

$$f_t = 0.45 \, (f_{cu})^{1/2} \tag{5}$$

The corresponding correlation coefficient is 0.872. Comparing Equations (4) and (5), it can be seen that the k-value for prisms with a length/width ratio of 5.0 is about 6% lower than that for prisms with a length/width ratio of 3.0. Hence, it is evident that the measured direct tensile strength decreases as the length/width ratio increases.

Comparing the influences of the cross-sectional area and the length of the prismatic specimen, it appears that the influence of the length/width ratio is slightly larger than the influence of the cross-sectional area.

CONCLUSIONS

From the experimental results, it is evident that the specimen size has significant effect on the measurement of the direct tensile strength of concrete. The measured direct tensile strength decreases as the volume of the specimen increases. Specifically, the direct tensile strength decreases as either the cross-sectional area or the length/width ratio of the specimen increases. The dependence of the direct tensile strength on the cross-sectional area and the length of the specimen may be because the direct tensile strength is controlled by the weakest spot in the specimen and the chance of having weaker concrete is larger in a longer and wider specimen. Furthermore, the within-test variation of the direct tension test decreases as the cross-sectional area of the specimen increases or the length of the specimen decreases. This implies that the repeatability of the test could be improved by using shorter and wider specimens. Compared with prisms of other sizes, the 150×150×450 mm prisms yielded the lowest within-test variation. Hence, the most suitable specimen size for direct tension test of concrete appears to be 150 × 150 × 450 mm.

REFERENCES

1.   BAZANT, Z.P., Is no-tension design of concrete or rock structures always safe? Fracture analysis, Journal of Structural Engineering, ASCE, Vol 122, No 1, 1996, pp 2-10.

2.   FREIDIN, C., Effect of aggregate on shrinkage crack-resistance of steam cured concrete, Magazine of Concrete Research, Vol 53, No 2, 2001, pp 85-89.

3.   ZHENG, W., KWAN, A.K.H., LEE, P.K.K., Direct tension test of concrete, ACI Materials Journal, Vol 98, No 1, January-February 2001, pp 63-71.

4.   NEVILLE, A.M., Chapter 12: Testing of hardened concrete, Properties of Concrete (4th Edition), Addison Wesley Longman Limited, Essex, UK, 1995, pp 581-648.

5.   POPOVICS, S., Section 3.8: Other major factors affecting the strength of concrete, Strength and Related Properties of Concrete - A quantitative Approach, John Wiley & Sons Inc., New York, USA, 1999, pp 244-245.

6.   CARPINTERI, A., CHIAIA, B., FERRO, G., Scale dependence of tensile strength of concrete specimens: a multifractal approach, Magazine of Concrete Research, Vol 50, No 3, 1998, pp 237-246.

7.   BAZANT, Z.P., PLANAS, J., Fracture and Size Effect in Concrete and Other Quasibrittle Materials, CRC Press, Boston, USA, 1998, 616pp.

8.   BAZANT, Z.P., KAZEMI, M.T., HASEGAWA, T., MAZARS, J., Size effect in Brazilian split-cylinder tests - measurements and fracture analysis, ACI Materials Journal, Vol 88, No 3, May-Jun 1991, pp 325-332.

9.   BAZANT, Z.P., NOVAK, D., Proposal for standard test of modulus of rupture of concrete with its size dependence, ACI Materials Journal, Vol 98, No 1, January-February 2001, pp 79-87.

10.   ZHOU, F.P., BALENDRAN, R.V., JEARY, A.P., Size effect on flexural, splitting tensile, and torsional strengths of high-strength concrete, Cement and Concrete Research, Vol 28, No 12, 1998, pp 1725-1736.

11.   NORDTEST, Concrete, hardened: tensile strength of test specimens (Edition 2), Nordtest Method, NT Build 204, Finland, 1984, pp 1-4.

12.   RILEM, CPC 7: Direct tension of concrete specimens, 1975 (TC14-CPC), RILEM Technical Recommendations for the Testing and Use of Construction Materials, E&FN Spon, London, UK, 1994, pp 23-24.

13.   U.S. BUREAU OF RECLAMATION, Procedure for direct tensile strength, static modulus of elasticity, and Poisson's ratio of cylindrical concrete specimens in tension (USBR 4914-92), Concrete Manual, Part 2 (9th Edition), US Bureau of Reclamation, Denver, USA, 1992, pp 726-731.

BOND DETERIORATION OF REINFORCING STEEL IN CONCRETE DUE TO CORROSION

L Amleh

Ryerson University

M S Mirza

B B N Ahwazi

McGill University

Canada

ABSTRACT. The relative corrosion performance of two cementitious systems, consisting of normal Portland cement (NPC), with and without additional fly ash, is summarily reviewed. Tension tests were conducted on 20 tension specimens, 1000mm long and 125mm in diameter, made from NPC concrete (w/c ratio = 0.32) and another concrete mixture incorporating 58% of Sundance fly ash by mass of the cementitious materials (w/cm ratio = 0.32).

Direct current was applied for increasing time periods to the reinforcing bar embedded in tension specimens, immersed in a concentrated salt solution (5% NaCl by the weight of water). The reinforcing bar served as the anode while a bare steel bar, located in the bath, served as the cathode. The bond strength at the steel-concrete interface was noted to decrease with an increase in the level of corrosion, although with minor loss of cross-sectional area, the bond strength improved slightly. The fly ash concrete showed better bond response at higher levels of corrosion, when, the bond strength at the steel-concrete interface, was almost completely lost for the NPC concrete.

Keywords: Accelerated corrosion, Bond behaviour, Concrete, Corrosion products, Cracking behaviour, Fly ash, Reinforcing steel, Tension tests.

Professor L Amleh, is an Assistant Professor at Ryerson University, Toronto, Ontario, Canada, in the Department of Civil Engineering.

Professor M S Mirza, is a Professor at McGill University, Montreal, Quebec, Canada, in the Department of Civil Engineering and Applied Mechanics.

Mr B B N Ahwazi, graduated recently with an MEng degree from McGill University, Montreal, Quebec, Canada, in the Department of Civil Engineering and Applied Mechanics.

INTRODUCTION

Corrosion of reinforcing steel in concrete is a multi-billion-dollar problem, which affects construction, transportation and some other industries. Normally, the concrete alkaline environment protects the reinforcing steel against corrosion, however, when salts (chlorides or sulphates) penetrate the concrete and reach the rebars, they damage the passive layer on the steel bar, and if oxygen and water are available, corrosion normally commences. The corrosion products can occupy a volume up to seven times the original volume of iron in the steel, developing large pressures within the concrete, which can exceed the concrete tensile strength.

This, in turn, causes longitudinal and occasionally transverse cracking, and spalling of the concrete cover, thereby considerably reducing the load carrying capacity of the structural element and impairing its structural integrity. The corrosion products cause loss of adhesion, cohesion and friction between the concrete and the reinforcing steel even before the specimen is loaded. This breakdown of friction between the concrete and the rebar, except at low corrosion levels, and the change in the roughness of the bar surface result in loss of bond strength at the steel-concrete interface (Al-Subaimani et al [1], Amleh [2]).

Supplementary cementing materials, such as fly ash, silica fume and blast furnace slag are being used increasingly in present-day concretes to improve their overall performance, strength and durability through decreased permeability. Presently, very little data is available on the corrosion protection characteristics of the concretes made with supplementary cementing materials. Therefore, a comparative study was undertaken to study the corrosion protection characteristics of a fly ash concrete as compared with that of the NPC concrete, both with the same water/cement ratio at different levels of corrosion.

MATERIALS

Two different mixtures of concrete were studied in this program using tension specimens a concrete mixture incorporated 58% Sundance fly ash by mass of the total cementitious materials with a w/cm ratio of 0.32, and an NPC concrete mixture for comparison purposes, with the same w/c ratio of 0.32.   The details of the concrete mixes used are presented in Table 1; the concrete compressive strength, obtained from compression tests on 150 by 300mm standard cylinders, is also presented.

Table 1   Concrete mix proportions and compressive strength

MIX NO.	CEMENT TYPE	FLY ASH TYPE	FLY ASH CONTENT (%)	W/CM	QUANTITIES (kg/m³)						A.E.A** (mL/m³)	COMPRS-SIVE STRENGTH (MPa)
					Water*	Cement	Fly Ash	Fine agg.	Coarse agg.	Sp**		
C3	NPC	Sundance	58	0.32	121	157	218	720	108	6.0	558	66.6
C4	NPC	-	0	0.32	119	371	0	750	3	10.1	401	60.2

NPC-	Normal Portland cement	Sp**	Superplasticizer, naphthalene based
*	Including water in the superplasticizer	A.E.A**	Air-entraining admixture
agg.	Aggregates		

Locally available, 20M reinforcing steel bars, with a nominal diameter of 19.5mm, were used in the tests; they conformed to the ASTM Standard A615-72 (1972), with an average yield strength of 432 MPa. The CSA Type 10 or the ASTM Type I normal Portland cement was used for both mixtures. The Sundance fly ash samples tested were obtained from Alberta, Canada and had a CaO content of 12.5%, but its combined (SiO_2 + Al_2O3 + Fe_2O_3) content was higher than 70%, thus meeting the requirement of the CSA or the ASTM Class F fly ash, and it had a relatively high specific surface of 408 m^2/kg (Blaine). The fine aggregate was natural sand from the Ottawa region, while the coarse aggregate was local limestone with a maximum size of 19mm.

EXPERIMENTAL SET-UP FOR ACCELERATED CORROSION TESTING

Twenty concrete cylindrical specimens, 1000mm long by 125mm diameter, each reinforced symmetrically with a pre-weighed single 20M deformed reinforcing bar and with a concrete cover of approximately 52mm were cast and cured. Sixteen (16) specimens were subjected to accelerated corrosion in the special tanks, while 4 specimens served as the control specimens (2 per concrete mixture). The tanks contained 5% sodium chloride (NaCl) solution by the weight of water, which was changed every week to eliminate any change in the concentration of NaCl and pH of the solution.

A DC voltage (5 volts) was applied to the specimens to accelerate the corrosion process, with the reinforcing bar serving as the anode, and the bare steel bar installed in the tank acting as the cathode (Amleh [2]). The voltage was monitored on a daily basis and the readings of both the current and the potential were recorded every 48 hours using a SMART digital multimeter that read both the current and the voltage. Figure 1(a) shows a schematic accelerated corrosion set-up, while Figure 1(b) shows a typical tension specimen.

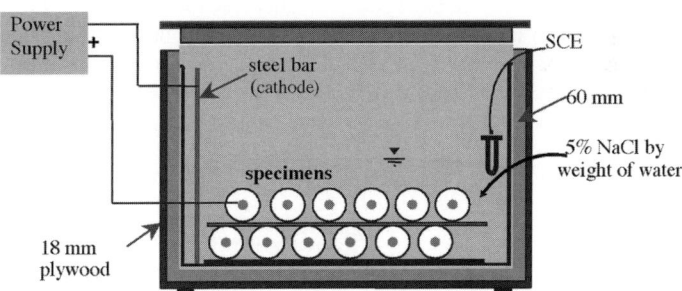

Figure 1(a)   Accelerated corrosion set-up

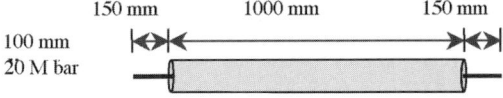

Figure 1(b) Typical tension specimen

Development of Various Corrosion Levels

The levels of corrosion attained in the various specimens varied from no corrosion (control specimens) to extensive corrosion. The width of the longitudinal cracks was used as the basis for ascertaining the level of corrosion attained; the current readings were also used to define the level of corrosion. The corrosion level was also verified by determining the chloride ion profile within the concrete cover and the chloride content at the rebar level. The corrosion level was confirmed by inspection of the steel rebar, retrieved from each specimen after completion of the test, and by measurement of the corrosion (pit) depth, and changes in the steel ultimate tensile resistance. The evaluation was normally carried out by determination of the weight loss after cleaning the bar with an acid (Amleh [3]).

EXPERIMENTAL PROGRAM

The tension tests were performed in the Structures Laboratory at McGill University using the MTS hydraulic, servo-controlled universal testing machine, with a capacity of 1000kN in tension. The machine was equipped with built-in linear voltage differential transducers (LVDT) for deformation measurements, which were connected to a PC workstation.

RESULTS - INFLUENCE OF CORROSION ON BOND BEHAVIOUR

Load-Elongation Response

The corrosion levels are not in the same sequence as the numbering of the specimens, however, the following sections deal with these specimens in an ascending order of the level of corrosion. The load-elongation responses of all of the NPC concrete specimens are shown in Figure 2, while those of the Sundance fly ash concrete specimens are shown in Figure 3, along with the response of the bare steel bar. The phenomenon underlying the transfer of the force from the steel bar to the concrete and vice-versa is best understood by comparing the measured load-elongation response with that of the bare bar. Similarly, the influence of the level of corrosion can be established by comparing the response of the corroded specimen with that of the uncorroded control specimens.

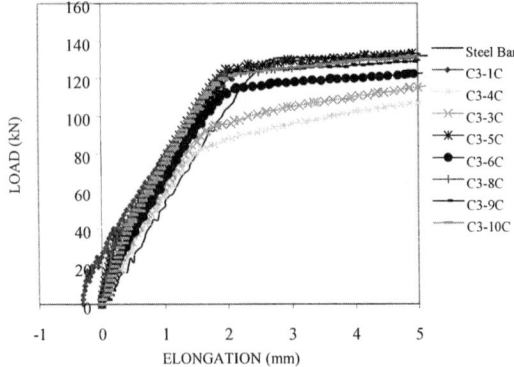

Figure 2   Load-elongation response for all of the specimens
(Normal Portland cement with w/c ratio = 0.32)

The control specimen elongation is negative prior to the load application because of the concrete shrinkage. Ignoring this member shortening would underestimate the tension stiffening effect (Bishoff [4]), who noted an average shrinkage strain of 0.3×10^{-3} for concretes made from the similar aggregates and cement as used in the present investigation. The corroded members were continually immersed in the NaCl solution, and hence there was no shrinkage.

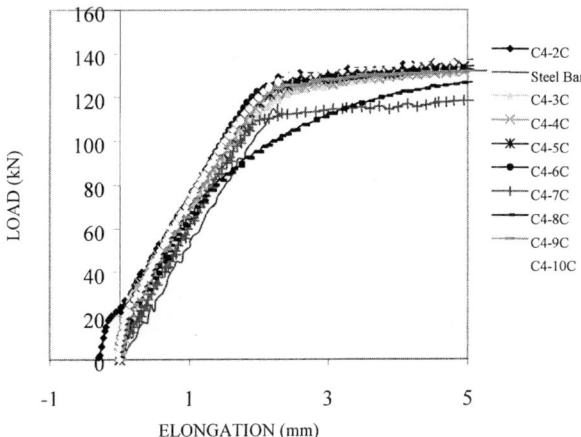

Figure 3   Load-elongation response for all of the specimens
(Sundance fly ash concrete)

Each load-elongation curve commenced basically with an uncracked elastic response, followed by a slight drop in the applied load at cracking, and subsequent stabilization of the cracks, and by yielding of steel reinforcement as the load was increased further. At this stage, the specimen was considered to have failed. All of the relevant experimental data (corrosion stage, yield load and the mass loss) are summarized in Table 2.

The load-elongation responses showed a gradual decrease in the load carrying capacity with an increase in the level of corrosion for both NPC and fly ash mixtures. This capacity reduction is attributable to the loss of bond, and to the reduction in the steel bar cross-sectional area, both as a direct result of corrosion.

Cracking Behaviour

The results for the number of transverse cracks and their maximum and average spacings (Table 2) show that as the level of corrosion increases, the crack spacing increases, showing increasing loss of bond at the steel – concrete interface. Figures 4 and 5 show the steel stress at the crack versus the crack width for NPC and Sundance fly ash specimens, respectively, showing that as the crack width increases with the level of corrosion, the bond at the steel-concrete interface deteriorates. Comparison of Figures 4 and 5 also shows better performance of the fly ash concrete specimens with smaller crack widths at the same steel stress level as compared with the NPC concrete specimens, demonstrating improved bond strength.

Goto [5], Houde and Mirza [6], and others observed that for the concrete specimen to crack, the force transferred from the steel bar to the concrete through the mechanical interlock between the bar ribs and the concrete must exceed the section tensile strength to cause it to crack. After one or more cracks have developed in the specimen, further cracking can occur only if enough force is transferred between the steel and the concrete over half of the distance between the two cracks (or a crack and a free end) to cause the concrete to crack further. The corrosion of the bar ribs has a significant influence on the loss of adhesion, cohesion and the mechanical interlocking with the concrete, resulting in the deterioration of bond at the rebar-concrete interface. Thus, the force is not transferred as efficiently from the steel bar into the concrete, leading to a larger spacing between the cracks in the corroded specimens.

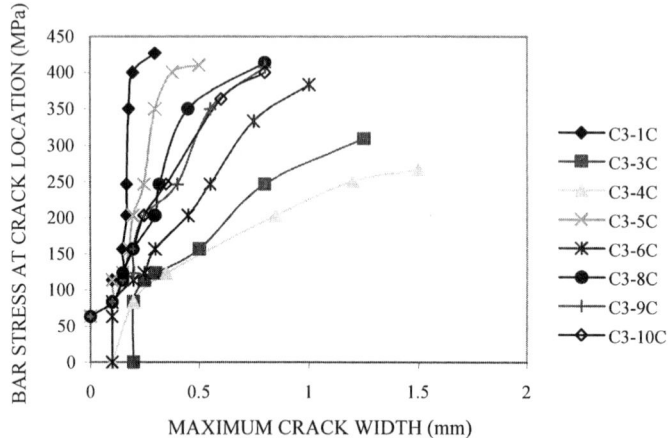

Figure 4   Variation of steel bar stress at crack location with the max. crack width for all of the specimens made with NPC with w/c = 0.32

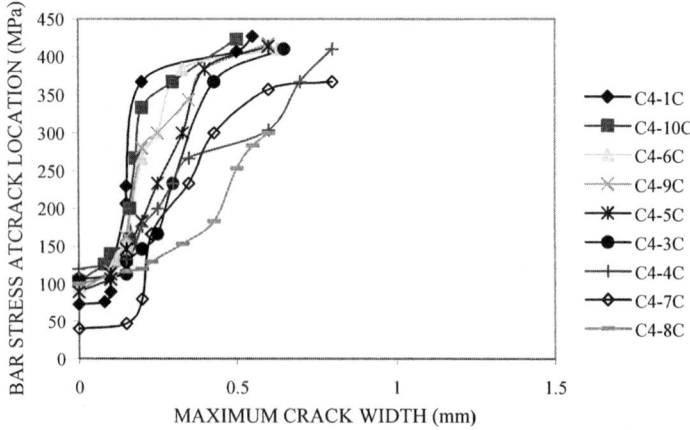

Figure 5   Variation of steel bar stress at crack location with the max. Crack width for all of the specimens made with Sundance fly ash concrete

Corrosion Percentage

After completion of the test, the bar was retrieved from each specimen for observations of its condition. The depths of the corrosion pits and the changes in the rebar diameter were measured to determine the loss of the cross-sectional area. After cleaning and brushing of the corrosion products, the weight loss was also determined, along with the cross-sectional areas at different locations.

Relative Bond Performance of Corroded Bars

Assuming the bond stress to be uniformly distributed over each half the length of the bar between the cracks, the bond stress is inversely proportional to the average crack spacing, and directly proportional to the bar diameter. All of the specimens are reinforced with the same size (20M) bars. Therefore, if the bond stress in the uncorroded specimens is considered to be 100%, the relative percentage of residual bond is calculated, based on the crack spacing. (Table 2 and Figure 6). It can be noted from this qualitative plot that the bond between the concrete and the reinforcing steel deteriorates increasingly as the level of corrosion increases.

Table 2   Test results for tension specimens

SPECIMEN	CORROSION STAGE	FIRST CRACKING LOAD	YIELD LOAD	ULTIMATE LOAD	REBAR MASS LOSS DUE TO CORROSION	NUMBER OF TRANSVERSE CRACKS	AVERAGE TRANSVERSE CRACK SPACING	MAXIMUM TRANSVERSE CRACK SPACING	NOMINAL BOND STRESS
		(kN)	(kN)	(kN)	(%)		(mm)	(mm)	(%)
				(NPC Concrete, w/c = 0.32)					
C3-1C	0	19	127	200	0	12	77	95	100
C3-5C	1	34	125	202	1.5	12	77	100	100
C3-8C	2	48	124	199	2.2	10	91	180	85
C3-10C	3	35	121	193	4.5	8	111	220	69
C3-9C	4	30	122	193	5.7	6	143	250	54
C3-6C	5	31	115	189	7.7	4	200	220	39
C3-3C	6	48	93	147	11.1	3	250	270	31
C3-4C	7	37	83	127	14.5	2	333	350	23
				(Sundance Fly Ash Concrete, w/cm = 0.32)					
C4-2C	0	21	128	200	0	10	91	110	100
C4-10C	1	31	127	198	0.9	11	83	112	110
C4-6C	2	29	124	196	1.0	9	100	140	91
C4-9C	3	27	126	197	1.5	9	100	135	91
C4-5C	4	27	125	197	1.75	8	111	150	82
C4-3C	5	32	124	195	2.0	8	111	153	82
C4-4C	6	36	124.5	196	2.5	7	125	158	73
C4-7C	7	22	110	173	3.35	7	125	190	73
C4-8C	8	30	90	128	4.5	6	143	150	64

For the NPC specimens, a 1.5 percent mass loss due to corrosion resulted in a 2 percent loss of the bond strength, while 14.5 percent mass loss due to corrosion, resulted in an 85 percent loss of the bond strength. This agrees well with the findings by Andrade et al [7], who noted that a reduction of 10 to 25 percent in the bar section in the critical zones of the structure will result in the depletion of its service life, while a reduction up to 5 percent, even with cracking and spalling, will indicate an early stage of deterioration with the remaining service life not as significantly influenced, provided that adequate repairs are undertaken urgently.

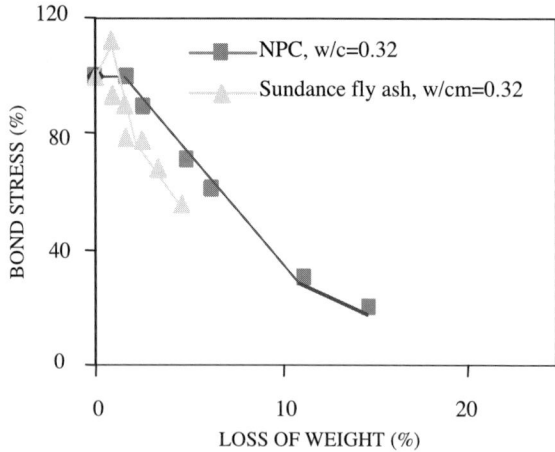

Figure 6   Relative bond stress versus level of corrosion

The qualitative plot of the relative bond stress of the corroded Sundance fly ash concrete specimens (Figure 6) shows that the bond between the concrete and the reinforcing steel increases by up to 10% up to a maximum bar mass loss of 0.9 percent, after which the bond deteriorates gradually with an increase in the level of corrosion. Mass losses of 1.5% and 4.5% result in relative bond strength losses of 10% and 55%, respectively. The most significant deterioration of bond in the postcracking corrosion stage occurs with a 14.5 percent mass loss in the NPC specimen, resulting in an 77 percent loss of bond strength, i.e. the bond between the steel and the concrete is almost completely lost.   Again, the performance of fly ash concrete specimens is superior than that of the NPC concrete specimens, with a much lower bond strength loss at higher levels of corrosion (Table 2).

The effect of the concrete permeability on the bond between the concrete and the reinforcing steel was not considered directly in this analysis. The results in the literature and those by Amleh [3] show that the permeability of the concrete mixture is a very important parameter which controls the ingress of chloride ions into the concrete.   Therefore, the concrete permeability must be considered an important factor in comparing the bond strength at the interface between the steel and the concrete in different concrete specimens and in different types of aggressive environments, at different levels of corrosion. This is in agreement with the results of the 192 pullout specimens, tested by Amleh [3], which showed that the permeability of the Sundance concrete mixture was about two orders lower that that of the NPC concrete (w/c ratio = 0.32).

CONCLUSIONS

1.   The tension specimens made from two different concretes – a normal Portland cement concrete with w/c ratios of 0.32, and the Sundance fly ash concrete with a w/cm ratio of 0.32 - showed generally similar behaviour tends, concrete cracking, loss of bond stiffness (slope of the bond stress-slip curve), followed by yielding of the reinforcing steel.

2. The corrosion of the reinforcing steel bar was generally uniform over the entire length of the bar, with concentrated or pitting corrosion at some locations. This pit is similar to a notch, which causes the steel rebar to behave in a brittle manner, as was noted in a few of the specimens, which failed suddenly in a very brittle manner.

3. As the level of corrosion of the steel rebar increased, the load-elongation curve showed a shift to the right, demonstrating a decrease in the concrete contribution due to the tension stiffening. When the corrosion level was high, the load-elongation response showed almost no concrete contribution by way of tension stiffening and it coincided with the load-deformation curve for the bare bar.

4. The tension specimens showed a gradual decrease in their maximum strength (almost linear) with an increase in the level of corrosion.

5. The performance of the fly ash concrete specimens in terms of cracking, crack width, and deterioration of bond was superior than that of the NPC concrete specimens.

ACKNOWLEDGEMENTS

The authors would like to acknowledge the NSERC Strategic Research Grant under which this and other related research programs were undertaken. The contributions of Natural Resources Canada/CANMET (Dr. V.M. Malhotra) and Hydro-Quebec (IREQ, Dr. J. Mirza) are gratefully acknowledged.

REFERENCES

1. AL-SUBAIMANI, G J, KALEEMULLAH, M, BASUNBUL, A, RASHEEDUZZAFAR, Influence of corrosion and cracking on bond behavior and strength of reinforced concrete members, ACI Structural Journal, Vol 87, No 2, 1990, p 220-230.

2. AMLEH, L, Bond behaviour between reinforcing steel and concrete, M.Eng. Thesis, McGill University, Montreal, Canada, 1996.

3. AMLEH, L, Bond deterioration of reinforcing steel in concrete due to corrosion, Ph.D. Thesis, McGill University, Montreal, Canada, 2000.

4. BISCHOFF, P H, Influence of shrinkage on tension stiffening of concrete, Proceedings of the 1995 Annual Conference of the Canadian Society for Civil Engineering, Ottawa, Ontario, p 433-442.

5. GOTO, Y, Cracks formed in concrete around deformed tension bars, ACI Journal, Proc., Vol 68, No 4, 1971, p 244-251.

6. HOUDE, J, MIRZA, M S, A study of bond stress - slip relations in reinforced Concrete, Structural Concrete Series No 72-8, McGill University, Montreal, Canada, 1972.

7. ANDRADE, C, ALONSO, M C, GONZALEZ, J A, An initial effort to use the corrosion rate measurements for estimating rebar durability, Corrosion rates of steel in concrete, Special Technical Publication 1065, American Society For Testing and Materials, Philadelphia, Pennsylvania, 1990, p 29-37.

8. AMLEH, L, MIRZA, M S, Corrosion deterioration of Dickson Bridge, Departmental Report, McGill University, Montreal, Canada, 2000.

9. ASTM A615-90, Specifications for deformed and plain billet-steel bars for concrete reinforcement, Book of ASTM Standards, Section 1 - Iron and Steel Products, Vol 01.04, Steel - Structural, Reinforcing Pressure Vessel, Railway, 1992, p 389-392.

10. CSA STANDARD G30.14-M83, Deformed steel wire for concrete reinforcement, Canadian Standards Association, Rexdale, Ontario, 1983.

EXPERIMENTAL INVESTIGATION OF THE FAILURE PATTERNS AND MECHANICAL PROPERTIES FOR PLAIN CONCRETE WITH CRACKS

L Zheng

University of Dundee

United Kingdom

S Wang

Zhejiang University

China

ABSTRACT. In this paper, the failure patterns and mechanical properties including full stress-strain (σ–ε) relationships for plain concrete with cracks which were artificially made in certain angles (α) from 0° to 90° to the principal stress were experimentally investigated in one-dimensional case. Three typical failure patterns were found in different α between stress and crack direction. This means that failure mechanism of the concrete is affected by the α values. It was found that the compressive strength (σ_o) and the ultimate strain (ε_o) of the concrete are functions of α, and the minimum σ_o and ε_o values were found at the α=30°. The damage evolution analysis also shows that the most dangerous cracks are those with α=30°. Concrete damages from microcracking process, therefore, depend not only on crack quantities developed but also on the cracking orientation to the principal stress. It was also found that the σ–ε relationships related to the α values especially in descending branch. Based on measured σ–ε curves and damage mechanics a modified Seanz σ–ε model was proposed.

Keywords: Crack, Failure pattern, Damage, Stress-strain, Model

Li Zheng is a research fellow in the Concrete Technology Unit in the Department of Civil Engineering at the University of Dundee, UK. He was an associate professor in the Department of Civil Engineering at Zhejiang University, PR China, prior to the current post. His research focuses mainly on concrete durability, mix optimisation and mathematical modelling.

Shuyu Wang is a professor in the Department of Civil Engineering at Zhejiang University, PR China. His research focuses mainly on concrete dam design, durability evaluation, optimisation and mathematical modelling.

INTRODUCTION

Cracks are one of the most common kinds of damage in concrete, which seriously affect the mechanical and durability properties of the concrete. It has been recognised that a number of microcracks have been developed in concrete prior to application of the load [1]. They are probably due to the inevitable differences in mechanical properties between the coarse aggregate and the hydrated cement paste, coupled with shrinkage or thermal movement. The cracking process in concrete is distinguished from cracking of other materials, such as metal and glass, in that it is not a sudden onset of new free surfaces but a continuous forming and connecting of microcracks [2]. The microcracking process can cause the stiffness degradation, which can be observed in concrete subjected to cyclic loading [3]. Since the forming of the microcracks in concrete is actually analogous to damaging of the material, in continuum damage mechanics, the damage or stiffness degradation can be modelled by defining the relationship between stresses and effective stresses [3-5]. In recent years, a lot of damage models for concrete, such as unilateral elastic damage models, elastoplastic damage models, plastic damage models, fracture damage models and endochronic damage models have been developed [5,6]. However, the verification for some of these models is still inadequate because the experimental studies are insufficient. Failure Patterns and their relations to mechanical properties have not been taken into account. Experiments about the influence of the angles α between crack direction and principal stress on the concrete failure patterns and damage evolution have not been investigated so far as suggested in the research literature.

In this paper, the experiments to investigate the influence of crack angle α to the principal stress on failure patterns, stress-strain curves and damage evolution of concrete were carried out. The cracking process and typical failure patterns were described. Experimental results on concrete mechanical properties and full stress-strain (σ–ε) curves with different α were reported. Based on mechanical damage theory, the Seanz σ–ε formula was modified to describe the full stress-strain relationships for concrete with cracks. Good agreement between the calculated and measured results indicates that this simply modified model can be used to predict the evolution of damage of the concrete with cracks.

EXPERIMENTAL PROGRAMME

Test Equipment

The test device is shown in Figure 1. The compression-testing machine with a capacity of 2,000 kN was used for loading and the two hydraulic jacks were used for rigidity support. The displacement rate of the jacks can be controlled in order to adjust the loading rate. Two PE (polyethylene) films and lubricant were used on two bearing surfaces of the specimen to reduce the effect of friction between the machine and the sample.

Design of the Specimens

The designed concrete 28-day cube strength is 25 N/mm^2 and the controlled slump is 30 to 50mm. The specimen code and mix proportion are given in Table 1. The measured 28-day mean cube strength was 27.2 N/mm^2.

The 200×200×70 mm slab specimens were used. Seven cracks of 20 mm in width were artificially made with following angles, α = 0, 15, 30, 45, 60, 75, and 90° respectively, immediately after the casting using special designed steel slices. That means crack quantity is about 175 /m². Twenty-four specimens, 3 samples for each α were cast.

Test Procedures

The specimens were cured in water (20°C) for 7 days and then cured in room dry condition till 28 days. Both electrical and mechanical strain gauges were used to measure the strains of the specimen. The specimen was installed as shown in Figure 1b. The compressometer was installed above the specimen to measure the load. The displacements between the machine platens were also monitored and the rate of the displacement was controlled by the two jacks. The full σ–ε curves were drawn automatically by an X–Y recorder.

(a) Full view of the testing device   (b) Details of the specimen installation

Figure 1   Device for testing the full stress-strain curves of the concrete with cracks

Table 1   Materials and Concrete Mix Proportion

SPECIMEN CODE	PW	P90	P75	P60	P45	P30	P15	P0
Angle Between Pre-made Cracks and Principal Stress α	without pre-made cracks	90°	75°	60°	45°	30°	15°	0°

Constituent Materials	Cement	Coarse Agg.	Fine Agg.	Water
	PC	Crushed	Natural Sand	Tap Water
	f_{ce}=42.5N/mm²	D_{max}=15mm	M_x=2.2	
Mix proportion	350 kg/m³	1325 kg/m³	500 kg/m³	225 kg/m³

EXPERIMENTAL RESULTS

Failure Patterns

Test results were summarised in Table 2. Three typical failure patterns of concrete with different angles between stress and pre-made cracks, α, were observed in the test under unilateral compressive stress, as shown in Figure 2. These failure patterns are defined as Type I) Stud compressive failure, II) Shearing failure, and III) Crack-through failure.

It was observed that about 42% of the specimens failed in Type I failure pattern. This failure pattern occurred for specimens with large α cracking conditions, i.e. α = 75° and 90° and for specimens without pre-made cracks (Figure 2a and 2b).

About 46% of the specimens failed in Type II failure pattern. For this type of failure, a V shape macro-crack can be seen on the specimen and the angle to the principal stress is about 20-30°. This failure pattern mainly occurred on the specimens with α between 30 – 60°, and sometimes occurred on the specimens with α = 75° and 90° and PW samples. However, this failure pattern was not observed on the specimens with α = 0° and 15°.

Type III, crack-through failure pattern, on which cracking was found along the direction of pre-made cracks and parallel to each other, was observed mainly on the specimens with α = 0°, 15° and 30°.

In unilateral loading case, the failure of concrete under the vertical compressive stress is mainly caused by the horizontal tensile strain which causes cracking parallel to the principal compressive stress [7]. This is the base to form the Type I failure pattern. When the cracking is resisted by an aggregate or separated by a pre-made crack, Shearing cracking may be found because the resistance in oblique slide may be less than that in vertical cracking.

Table 2   Summary of the Failure Patterns

SPECIMEN CODE	ANGLE α	SPECIMEN NUMBER	TYPE I NUMBER	TYPE II NUMBER	TYPE III NUMBER
PW	None	3	3	1	0
P90	90	3	2	1	0
P75	75	3	2	1	0
P60	60	3	1	3	0
P45	45	3	0	3	1
P30	30	3	0	2	2
P15	15	3	1	0	2
P0	0	3	1	0	2
Total Number Tested		24	10*	11*	7*
Percentage of Failure Patterns (%)		-	41.7	45.8	29.2

*   Note: some samples showed combined failure patterns and therefore the sum of these three failure patterns is larger than the number of the specimens.

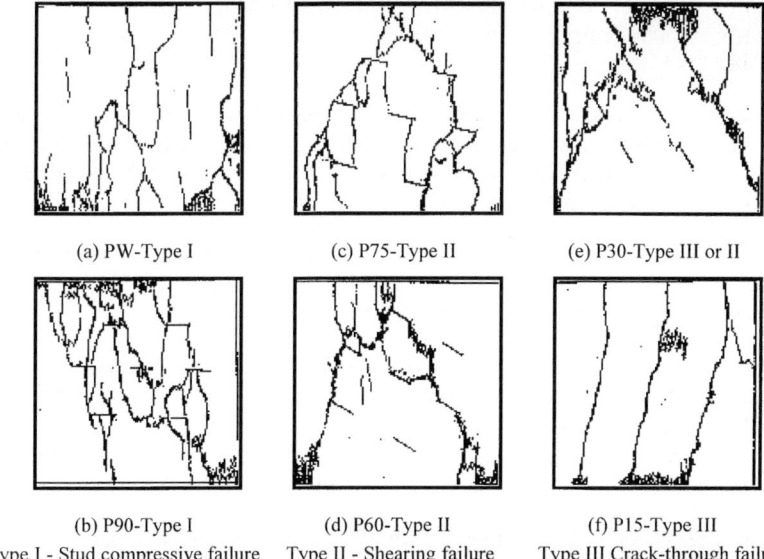

(a) PW-Type I	(c) P75-Type II	(e) P30-Type III or II
(b) P90-Type I	(d) P60-Type II	(f) P15-Type III
Type I - Stud compressive failure	Type II - Shearing failure	Type III Crack-through failure

Figure 2   Typical failure patterns of the specimens

When α of the pre-made crack to principal stress is about 20-30°, the direction of the crack and the oblique shearing are the same, and Type III failure pattern occurs. In this failure pattern, the number of cracks is less than that of the other two failure patterns.

Different failure patterns, which were found in different samples, reflect the different failure mechanisms, and hence result in different mechanical properties of the concrete. Such as in type III pattern which was mainly found in specimens with α=15-30°, the strength and ultimate strain decreased remarkably. (See Figure 3)

Mechanical Properties

The ultimate strain, ε_o, relative strength, σ_o/σ_{ow}, and residual strength, σ_u (when ε_u=0.01), varying with α are shown in Figure 3. It can be seen that the ultimate strain and relative strength are dependent on the pre-made crack angle α. They all decrease with the increase of α in the range of 0<α<30° and increase afterwards. It may be of interest to note that both minimum σ_o/σ_{ow} and ε_o occur at the same α = 30°. However, the residual strength varying with α does not show a significant minimum σ_u value at α = 30°. It can be thought that the residual strength for all specimens were similar. The mean residual strength is about 7.3 N/mm^2 for this set of the specimens while ε_u=0.01. The concrete maintains a residual strength σ_u because of the sliding friction between coarse aggregates or other non-continuous surfaces. It seems irrelevant to the pre-made cracks

Figure 4 shows initial modulus of elasticity, E_o, ultimate secant modulus, E_c (= σ_o/ε_o) and their ratio E_o/E_c varying with the pre-made crack angle α. It can be seen that there are no significant variations in E_o and E_c with the pre-made crack angle α. However, the E_o/E_c ratios were lower in the range of 15<α<45°. The lowest E_o/E_c range is approximately the same to that of minimum σ_o/σ_{ow} and ε_o.

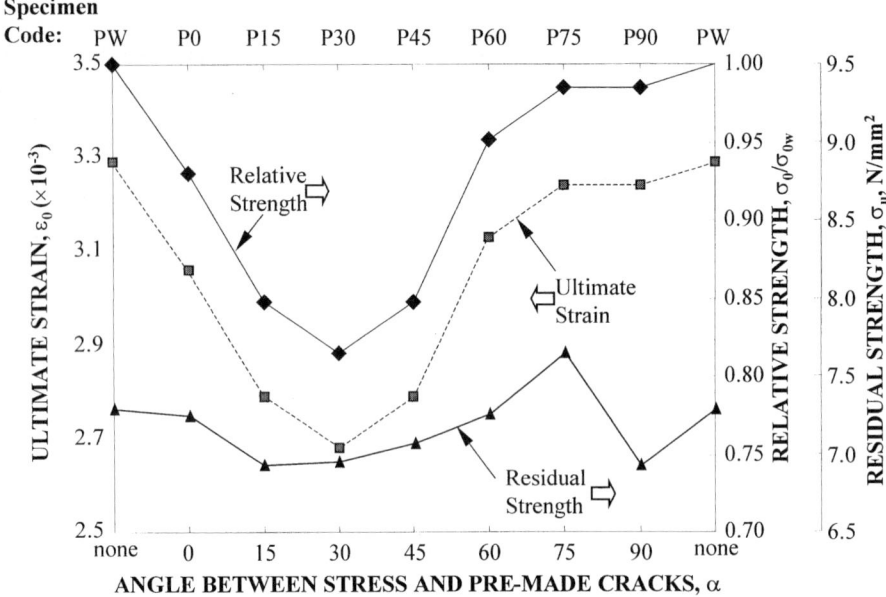

Figure 3   The relationships between ε_0, σ_0/σ_{0w}, σ_u and α

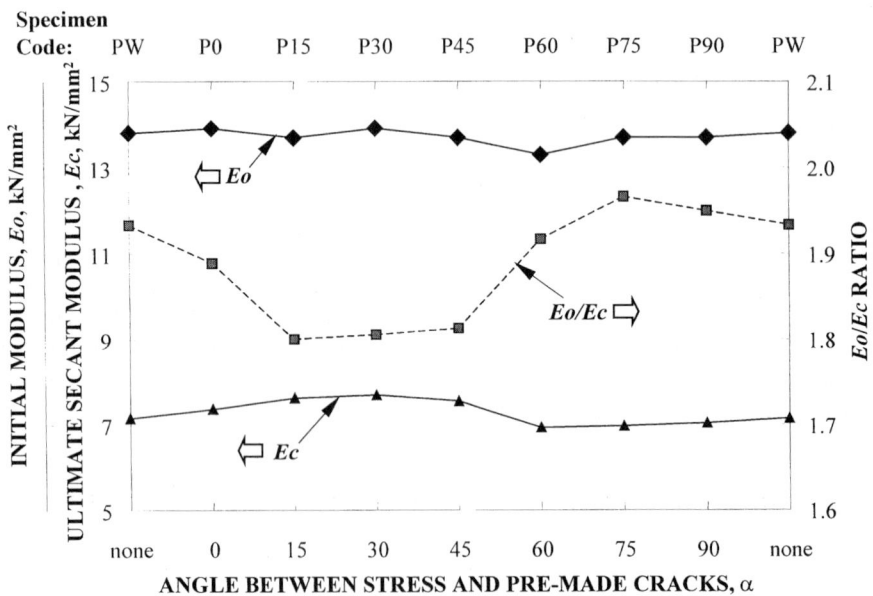

Figure 4   The relationships between E_o, E_c, E_o/E_c and α

Full Stress-Strain Curves

Figure 5a shows a typical stress-strain curve for intact concrete, i.e. specimen PW. In the Figure, point **a** is the elastic limit of the concrete and it is about 30~50% of the strength σ_0 and **oa** is approximately a straight line. The micro-cracks in the concrete start to extend when the stress exceeds **a**, however the crack development is stable, i.e. it increases with an increase in stress, when the stress is between **a** and **b**. Point **b** is called the critical stress point which is about 70~90% of σ_0. The crack development becomes unstable, i.e. it increases even without any further increase in stress, when the stress exceeds **b**. In the test, it can be seen that the cracks extend through each other in the concrete specimens and a great number of increases in strain (deformation). Point **c** is the maximum stress point. The cracks extending more quickly after point **c** and the stress starts to decrease, or called compressive stress softening, with the deformation increase. After point **d**, the stress decrease becomes lower and concrete maintains a residual strength σ_u.

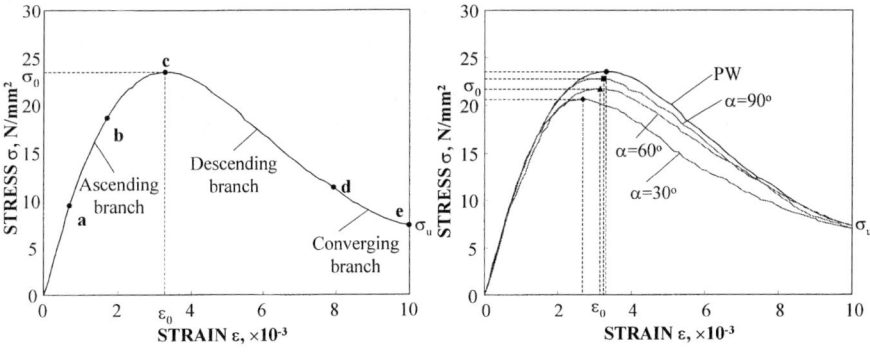

a Full stress-strain curve of the specimen PW

b Comparison of full stress-strain curves of different pre-made crack angles

Figure 5   Full stress-strain curves of the specimen

Figure 5b shows the comparison of full stress-strain curves of different pre-made crack angles. It can be seen that the main differences between these curves are in their peak stress, σ_0, and the strain at peak stress, ε_0. The initial parts of the curves are nearly the same. The peak stresses, σ_0, of specimens with pre-made cracks are lower than that of PW sample as well as the strain at peak stress, ε_0. And finally, all curves tend to a similar residual strength σ_u. The variations of ε_0, σ_0 and σ_u with α were shown in Figure 3.

MODELLING OF THE σ–ε CURVES OF CONCRETES WITH CRACKS

According to the continuum damage theory, a scalar degradation damage variable $0 \leq D < 1$ is generally used to represent the isotropic material damage. In one dimension case, the damage variable D can be expressed as a function of strain ε [4]:

$$D(\varepsilon) = 1 - \frac{\sigma(\varepsilon)}{E_n \varepsilon} \tag{1}$$

where E_n is the modules of non-damaged material. From Figure 4, it can be seen that there is no significant variation in the initial elastic module E_o with α, which indicates that the pre-made cracks do not have a distinct influence on the stress-strain curves at the initial part. Therefore, initial damage is assumed to be neglected, and E_o takes the identical value of E_n.

Then the damage degree can be evaluated by the D value. $D = 0$ means no damage in the initial state of the concrete and $D = 1$ means the concrete is totally damaged. D value evolutions with the strain and pre-made crack angles are shown in Figures 6. It can be seen that generally the damage evolution is slow in the initial stage ($\varepsilon < 0.001$, before the elastic limit **a** in Figure 5).   A rapid damage evolution occurs with an increase in strain from $\varepsilon = 0.002$ to 0.006. Then the damage evolution tends to slow again with a further increase in strain. It can also be seen that the D value reaches the maximum when α is about 30^0 at the same strain level when the ε is between 0.003 and 0.006. However, no significant differences between the D values for different α at the same ε were found when $\varepsilon < 0.002$ and $\varepsilon > 0.008$. This means that those pre-made cracks mainly affect the damage evolution in the crack developing stage. They have a very small effect on the damage evolution in the initial damage stage or in the stage of concrete near to totally damaged.

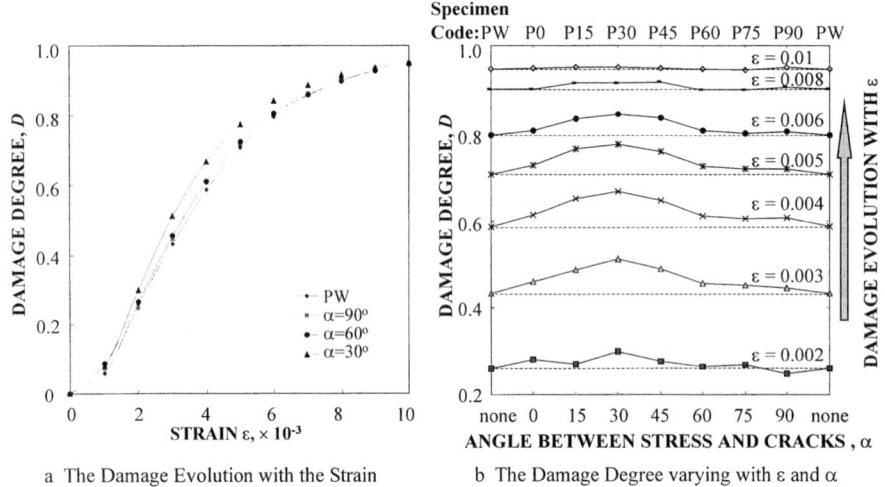

a The Damage Evolution with the Strain　　　b The Damage Degree varying with ε and α

Figure 6　The Damage Evolution

For modelling stress-strain curves of concretes with cracks, it was considered that the total damage degree, D, of the concrete can be divided into two part: damage from the non-linear elastic deformation, D_1, and the damage from the further accelerated cracking in the descending stage, D_2. Therefore:

$$D(\varepsilon) = D_1(\varepsilon) + D_2(\varepsilon) = 1 - \frac{\sigma(\varepsilon)}{E_o \varepsilon} = 1 - \frac{\sigma_{el}(\varepsilon)}{E_o \varepsilon} + \frac{\Delta \sigma_{so}(\varepsilon)}{E_o \varepsilon} \qquad (2)$$

Where, σ_{el} is the calculated stress assuming that all deformation, ε, is the non-linear elastic deformation; $\Delta\sigma_{so}$ is the further stress softening with the deformation increase because of the cracking; $D_1(\varepsilon) = 1 - \sigma_{el}/E_o\varepsilon$, and $D_2(\varepsilon) = \Delta\sigma_{so}/E_o\varepsilon$,

The equation proposed by Seanz [8] has been initially selected for calculating the non-linear elastic deformation, D_1, because this equation has been widely applied in the commercial finite element analysis packages, such as NONSAP and ADINA. The Seanz two-parameter equation is in quadratic form and can be written as

$$\sigma_{el}(\varepsilon) = \frac{E_o \varepsilon}{1 + (\frac{E_o}{E_c} - 2) \bullet \frac{\varepsilon}{\varepsilon_o} + (\frac{\varepsilon}{\varepsilon_o})^2} \tag{3}$$

So the damage by the non-linear elastic deformation, D_1, equal:

$$D_1(\varepsilon) = 1 - \frac{1}{1 + (\frac{E_o}{E_c} - 2) \bullet \frac{\varepsilon}{\varepsilon_o} + (\frac{\varepsilon}{\varepsilon_o})^2} \tag{4}$$

The damage from the further accelerated cracking, $D_2(\varepsilon)$, is determined by the further stress softening, $\Delta\sigma_{so}$. When $\varepsilon \leq \varepsilon_o$, $\Delta\sigma_{so} = 0$ and when $\varepsilon > \varepsilon_o$, $\Delta\sigma_{so}$ will increase quickly with an increase in ε and then tend to a constant when the σ-ε curve enter the converging branch. These characters can be well described by the following function:

$$\Delta\sigma_{so} = \frac{C}{A + e^{-B(\frac{\varepsilon}{\varepsilon_o})}} + G \tag{5}$$

The parameters C and G can be determined according to the conditions $\varepsilon = \varepsilon_o$, $\Delta\sigma_{so} = 0$ and $\varepsilon = \varepsilon_u$, $\Delta\sigma_{so} = \sigma_{el}(\varepsilon_u) - \sigma_u$. Therefore, the shape of the descending- branch of the σ-ε curve will be depend on parameters A and B, which can be obtained from the tested curves.

Comparisons between calculated stress-strain curves with measured data from the samples of different pre-made crack angles are shown in Figure 7. The points represent the tested data whilst the curves are calculated from the models. It can be seen that the calculated results from the proposed modified Seanz model much improved the results from the Seanz σ-ε equation (Figure 7a), and agree with the experimental data well.

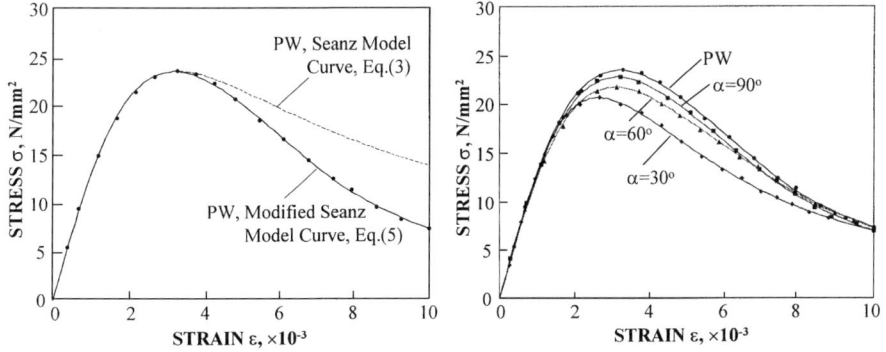

a Comparison between Seanz and modified Seanz models.

b Comparison between calculated and measured stress-strain relationships

Figure 7   Comparisons between calculated stress-strain curves with measured data

CONCLUSIONS

From the study reported, the following conclusions can be obtained for C25 plain concrete with pre-made cracks in different angles under unilateral loading condition:

i)   The failure patterns of the specimens can be divided into three typical types, which are stud compressive failure, shearing failure and crack-through failure pattern. In different angles between stress and pre-made cracks, the main failure patterns are different.

ii) The angles of the pre-made cracks have a significant effect on the macro mechanical properties of compressive strength (ultimate stress) and ultimate strain. The crack $\alpha=30°$ is the most unfavourable angle and it reduces the ultimate stress and strain significantly.

iii) The initial modulus of elasticity and the residual strength of the concrete seem not affected by the pre-made cracks. Therefore, these pre-made cracks have a little effect on the initial part and the final part of the stress-strain curves.

iv) The damage evolution is also not affected by the pre-made cracks on the initial part and the final part of the strain increment. However, the pre-made cracks have a significant effect on the damage evolution with the increase of strain from $\varepsilon=0.003$ to $\varepsilon=0.008$. The crack with α about $30°$ is the most unfavourable one in concrete damage evolution.

v)   The Seanz equation, as modified by considering the additional damage from further accelerated cracking starting after the peak stress, gives good estimations of the stress-strain curves of the concrete with pre-made cracks.

ACKNOWLEDGEMENT

The research reported in this paper was financially supported by Research Grant from Natural Science Grant Committee of Zhejiang Province, P. R. China. The author gratefully acknowledges this generous support.

REFERENCES

1. HSU T. T. C., SLATE F. O., STURMAN G. M., WINTER G., Microcracking of plain concrete and the shape of the stress-strain curve. ACI Journal, Vol.60, 1963, pp 209-224

2. NEVILLE A. M., Properties of Concrete, 4th ed, Pitman Publishing Limited, London, 1995

3. LEMAITRE J., CHABOCHE J. L., Mechanics of Solid Materials, Cambridge University Press, New York, 1990

4. KACHANOV L. M., Introduction to Continuum Damage Mechanics, Martinus Nijhoff Publishers, Dordrecht, 1986

5. YU T., Mechanical Damage Theory and Energy Damage Theory, Proceedings of 4th Symposium of Engineering Materials of Rock and Concrete, Beijing, 1989

6. LEE J., FENVES G. L., Plastic Damage Model for Cyclic Loading of Concrete Structures. Journal of Engineering Mechanics, ASCE, Vol.124, No.8, 1998, pp 892-900

7. JIANG J., FENG N., Concrete Mechanics, China Railway Publishing House, Beijing. 1991

8. SAENZ L. P., Discussion of equation for the stress-strain curve of concrete by Desayi and Krishnan, ACI Journal, Vol.61, 1964, pp 1229-35

IN SITU MECHANICAL DETERMINATION OF ELASTIC MODULUS OF CONCRETE

G Bocca

M Crotti

Turin Polytechnic

Italy

ABSTRACT. This article describes a new experimental-numerical method based on the pull-out test, for the on site determination of the elastic modulus, E, of concrete structures. The method consists of working out the Load - Displacement $P(\eta)$ curve by pulling out a steel bolt introduced into the concrete structure through a hole produced with a drill. The pulling force is applied with the aid of a hydraulic jack and a steel ring interposed between the jack and the concrete structure. The state of stress and strain in the concrete around the hole and under the steel ring is also analysed numerically. To ensure the applicability of the method, the values of the elastic moduli obtained on prism shaped specimens were correlated to the values obtained from the pull-out tests. All tests were conducted on concrete specimens produced with Portland cement and belonging to three different strength classes: poor (mix "A"), average (mix "B"), good (mix "C"). Furthermore, a few corrective factors, which are a function of the testing method and the material, have been developed in order to widen and perfect the field of application of the method.

The article describes the testing apparatus used for the tests, the laboratory tests conducted on concrete specimens, the results and the relative analysis, the concrete possibilities offered by this method for the characterisation of existing structures directly on site and at affordable costs.

Keywords: Pull-out, Concrete, Elastic modulus, On site, Testing material.

Professor P G Bocca, is an Ordinary Professor at Turin Polytechnic, Department of Structural Engineering and Geotechnics, Italy.

Mr M Crotti, is a PhD Student in Laboratory of Non Destructive Testing at the Department of Structural Engineering and Geotechnics, Turin Polytechnic, Italy.

INTRODUCTION

The advent of new checking and design methods in the field of reinforced concrete and prestressed concrete structures, the ever greater attention devoted to serviceability conditions (and hence to maximum deflection), and, above all, the widespread use of structural design computation codes, have brought to the fore the importance of a viable method for the determination of the elastic modulus to be used both in the formulation of design provisions and in the assessment of existing structures, so to be able to define (for instance through finite element analyses) the utilisation and safety conditions of such structures in seismic areas.

For new structures, in fact, the elastic modulus of concrete can be determined effectively by means of compressive tests on prism shaped specimens, while for existing structures the method adopted to estimate the Young modulus of concrete is a more complex process, which, in most cases, does not yield sufficiently accurate results.

On account of the variability of the material's properties, the current methods for on-site determination often prove unsatisfactory.

The aim of this investigation was to evaluate the possibility of determining the elastic modulus of ordinary concrete through a newly conceived method - the pull-out test - fine-tuned at the Non-Destructive Test Laboratory of the Polytechnic of Turin. After the execution of a number of preliminary tests, that supplied indications on the behaviour of the block subjected to the pull-out tests, a simple enough, accurate and repeatable testing procedure was developed to work out the stiffness value of the pull-out/concrete system. With this method, a suitable number of tests was performed on the concrete mixes –25 tests each concrete mix-, to be able to analyse the test results by making effective use of statistical tools.

The pull-out test was also analysed through a numerical simulation. The chosen method was a FEM code which made it possible to produce a model closely in keeping with actual testing conditions. A mathematical correlation linking model stiffness to the Young modulus of concrete was worked out on the basis of this simulation.

The curve obtained with the FEM was calibrated against the experimental results in our possession, with the aim of developing a measuring method for use on existing structures.

MATERIALS TESTED AND DESCRIPTION OF THE TESTS

Compressive Tests and Determination of the Elastic Modulus

This section illustrates the different types of concrete examined in the course of the laboratory testing campaign.

The following test pieces were produced for each type of mix:

- No. 3 cubes, sized 16x16x16 cm (for the compressive tests).
- No. 3 prisms, sized 16x16x50 cm (for the elastic modulus tests).
- No. 1 slab, sized 100x100x16 cm (for the pull-out tests).

Table 1   Mix characteristics

CHARACTERISTICS / MIXES	MIX "A"	MIX "B"	MIX "C"
Proportions (only variable characteristic)	200 kg/m^3	250 kg/m^3	300 kg/m^3
Type of cement	425	425	425
Cement density	3200 kg/m^3	3200 kg/m^3	3200 kg/m^3
Gravel density (0 - 7)	2800 kg/m^3	2800 kg/m^3	2800 kg/m^3
Sand density (7 – 15)	2700 kg/m^3	2700 kg/m^3	2700 kg/m^3
Quantity of gravel used (by weight)	45%	45%	45%
Quantity of sand used (by weight)	55%	55%	55%
Max. aggregate size	15 mm	15 mm	15 mm
Water cement ratio	0.7	0.7	0.7

The three cubes for each type of mix were subjected to a simple loading test carried on until specimen failure. The prisms were used to determine the elastic modulus of the relative concrete. Strains were detected by means of two strain gauges with measuring base extending over 30 mm and by means of two extensometers with 100 mm measuring base placed on the centre of the four specimen faces. Knowing, from the failure tests conducted previously, the maximum load of each mix, three loading levels were defined to be used in the tests for the determination of the elastic modulus. The loading range consisted of a maximum threshold corresponding to 1/3 the failure load, and a minimum value, corresponding to 1/10.

For each loading level, 3 loading - unloading cycles were performed to achieve optimal stabilisation of the mechanical behaviour of the test pieces.

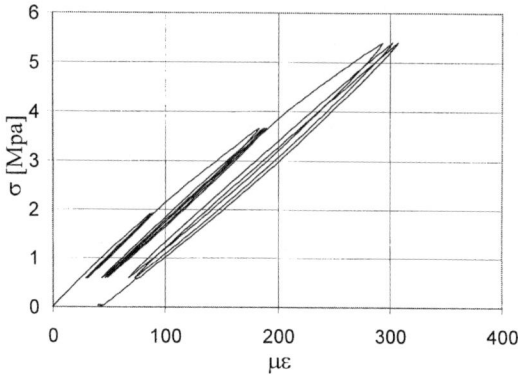

Figure 1 $\sigma(\varepsilon)$ curve obtained from the tests for the determination of the elastic modulus, E

As can be seen, differences in compressive strength were observed for each of the three mixes, resulting in a strength spectrum that is fairly representative of the typical strength values of existing structures.

Table 2   Values of failure strength and elastic modulus

MIX	TEST PIECE NO	SECTION SIDES (cm x cm)	AREA (cm²)	LOAD (daN)	σ (daN/cm²)	AVG. σ (daN/cm)	E MODULUS (MPa)
	1	16.1x15.9	255.99	42680	166.73		
A	2	16.0x16.1	257.60	42520	165.06	165.42	
200	3	16.0x15.9	254.40	41840	164.47		
	Prism	16.0x16.0x50					21.306
	1	16.0x16.1	257.60	74580	289.52		
B	2	15.9x16.1	255.99	72160	281.89	286.83	
250	3	16.0x16.0	256.00	74005	289.08		
	Prism	16.0x16.0x50					25.298
	1	15.6x16.0	249.60	93440	374.36		
C	2	15.8x16.0	252.80	91740	362.90	367.31	
300	3	16.1x16.0	257.60	93940	364.67		
	Prism	16.0x16.0x50					28.334

Testing Set-Up And Procedures

The testing apparatus consisted of an electric-hydraulic system for the gradual pull-out of a metal insert, embedded in the concrete, and for the continuous measurement of both the load applied and the displacement of the extractor. The electrical signal given out by the measuring instruments is conditioned, amplified and eventually acquired by a digital system. The operation of the testing apparatus is fairly simple: by means of a manual pump, the oil in the hydraulic circuit is pressurised and the hydraulic jack is activated; the data acquisition system reads and saves the electrical signals received from the transducers and the pressure cell via a signal amplification and processing unit.

The insert consists of a mechanical block that is embedded in the concrete through geometric type expansion.
Following a careful study of the various types of inserts commercially available, we chose a block that is widely used.

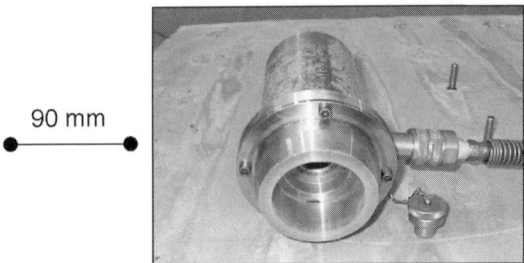

90 mm

Figure 2   Extractor with steel ring

Figure 3   Testing apparatus complete with extractor and displacement transducers

In order to obtain a load - displacement curve suitable for the determination of the actual stiffness of the structure, it is necessary to perform a certain number of loading - unloading cycles during each test, so as to eliminate all phenomena of mutual sliding between the various element of the structure.

Accordingly, the choice of the loading range within which to perform the cycles (at least four for each test) is of decisive importance. The loading range was defined by conducting a series of tests with different loading cycles and plotting the stiffness value (whose meaning will be explained later on) as a function of the ratio between loading level (F) and failure strength (Fr).   An acceptable zone for the execution of the loading cycles was defined, ranging from 66% to 82% of the failure value. The maximum value of the loading range was fixed at 2/3 of the failure value, to make sure it would fall within the material's linear-elastic response interval. The minimum value identified, of ca 500 daN, corresponds to the loading level necessary to eliminate initial settlement phenomena.

Having plotted the complete load - displacement curve for the four loading/unloading cycles (see Figure 4), we can calculate the values (K) of the slopes of the straight portions of the curves at the unloading stage; during the latter stage, in fact, we can achieve optimal control of the load reduction rate, something that cannot be done equally well at the reloading stage.

After calculating the four K values for the structure (one for each cycle) according to the simple expression of the determination of stiffness, it was decided to assume as the representative value the one obtained for the fourth cycle, by analogy with the procedure adopted in traditional elastic modulus tests.

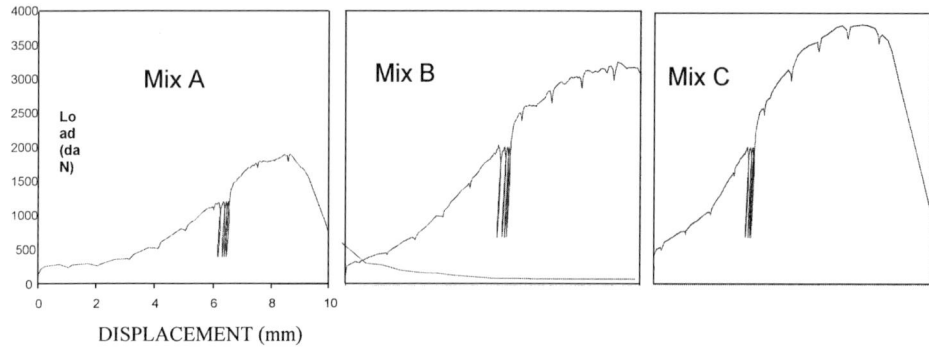

DISPLACEMENT (mm)

Figure 4   Load - displacement P(η) curve plotted experimentally for the determination of stiffness, K (in unloading conditions during the 4th cycle)

The stiffness value obtained from the tests, hereinafter denoted with Knet, includes the elongation of the extractor stem connecting the insert to the hydraulic jack. Accordingly, it proves necessary to correct the stiffness value obtained to eliminate the stiffness associated with the connecting stem, Kstem, through the Formula:

$$Knet = (Ktest \times Kstem) / (Kstem - Ktest)$$

(whose simple demonstration is omitted).

The stiffness of the other components of the testing apparatus can be assumed to be infinite.

The testing procedure includes the following steps:

1. Determining the pull-out load through a preliminary test.

2. Completing four loading - unloading cycles within the loading range defined according to the criteria described in this paragraph.

3. Increasing the load gradually until the concrete cone is pulled out (Figure 5).

4. Acquiring the data and constructing the complete load - displacement curve for the test.

5. Determining the 4 net values of stiffness, which must always be read on the unloading curve of the testing cycles described above.

6. Compiling the summary Table for each mix tested; in addition to listing the test results, these Tables are also used for a complete and significant analysis of the data.

ANALYSIS OF THE RESULTS

The Table below lists the average stiffness values (average Knet) obtained.

The statistical considerations on the test results are illustrated in graphical form in Figure 6.

Table 3   Values of Knet for the mixes considered

MIX	Knet (N/mm)	s.q.m.	v.c. [%]
A – 200	125354	19933.9	15.9
B – 250	151873	8860.7	9.51
C – 300	154197	14671.5	8.93
D - Fibre reinforced concrete	178053	3775.4	2.12

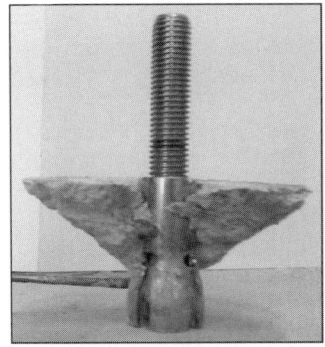

Figure 5   Concrete cone pulled out during the test

The chart in Figure 6 correlates the results obtained with the strength of the three mixes (A, B, C).

The results obtained on fibre reinforced concrete are also given, as a further confirmation of the validity of the method.

The values of the variation coefficient are seen to be high for low quality concrete, smallest for better quality concrete. The v.c. determined for fibre reinforced concrete appears to be very low, owing to the higher ductility afforded by the fibres.

It was decided to illustrate the test results in graphic form, without showing the interpolation curve between the average stiffness points (Kaverage) obtained, which were but 3 (one per mix).

Through a finite element simulation it proved possible to obtain a greater number of points, by changing the elastic modulus of the material at will.

The correlation curve with the FEM is approximated by the quadratic Equation:

$$E=3x10^{-6}K^2-0.5345K+C$$

where:

E = elastic modulus of the material [MPa]
K = stiffness of the pull-out apparatus [N/mm^2]
C = known term

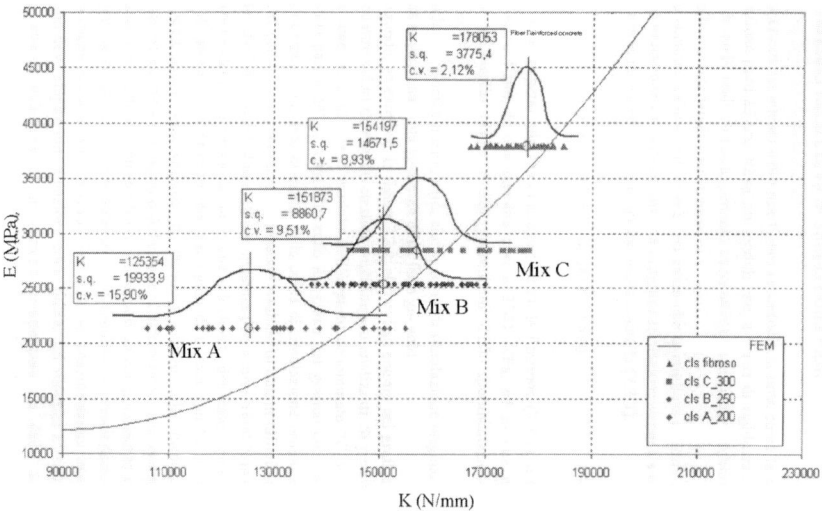

Figure 6   Correlation between the test results and the FEM simulation curve

CONCLUSIONS

The primary goal of this investigation was to work out a method for the determination of the elastic modulus of concrete through pull-out tests with post-inserted blocks, to be able to propose the pull-out test as a non-destructive procedure suitable for the estimate of the mechanical deformability of concrete in existing structures of any kind.

As for the classification of pull-out tests as "non-destructive" it can be easily seen that the damage caused to a structure is decidedly limited, can be repaired readily with a small quantity of mortar, and will not give rise to a weak point in the structure.

The major advantage of the pull-out test lies in its ease of execution: the entire testing apparatus can be moved without difficulties even to places that are hard to reach, and is sturdy enough for use at building sites, even underground.
The other primary types of non-destructive tests developed for on-site use are: the pulse wave test, the rebound test, and core drilling.

The first two are influenced by the heterogeneous nature of the material, humidity and testing conditions in general, as well as the presence of surface discontinuities or reinforcing steel.

Core tests should preferably be performed in the laboratory, since the test pieces drilled out have to be tested in compression with the aid of a laboratory testing machine. Furthermore, the results of core tests exhibit considerable variability due to the disturbance introduced into the material by the hollow punch, which means that sample size has to be increased to a considerable extent.

The variation coefficient obtained in the pull-out tests performed in the course of this investigation ranged from 15.9% for low quality concrete to 8.9% for higher quality concrete. This indication is confirmed by the coefficient obtained for the fibre reinforced concrete specimens used as a further term of comparison, which turned out to be as low as 2.1%.

Since the tests are performed in controlled displacement conditions, it proves possible to formulate energy related considerations on ductility/brittleness by analysing the post-failure stage.

The method should be further refined to pass from the stiffness value, Kcorr, to the Young modulus, E, especially where low values of E are concerned.

In a future testing campaign, the pull-out test will be performed on concrete types having different elastic moduli so as to obtain a greater number of experimental points (Kaverage, Ecls) and be able to plot the interpolation curve.

Finally, it should be noted that the results obtained are correlated to the type of block employed in the tests and, if the insert to be used is of a different sort, its behaviour should be carefully be evaluated before starting the tests.

REFERENCES

1. OTTOSEN, N S, Nonlinear Finite Element Analisys of Pull-out Test, Journal of the Structural Division, ASCE, Vol 107, No ST4, April 1981, p 591-603.

2. BICKLEY, J A, The Variability of the Pullout Tests and In-Place Concrete Strength, Concrete International: Design & Construction, Vol 4, No 4, 1982, p 44-51.

3. YENER, M, CHEN, W F, On In-Place strength of concrete and pullout tests, Cement, Concrete and Aggregates, Vol 6, No 2, 1984, p 96-99.

4. STONE, W C, GIZA, B J, Statistical Methods for In-Place Strength Predictions by the Pullout Test, ACI Journal, Proceedings, p 745-756, September-October 1986.

5. BOCCA, P, CARPINTERI, A, VALENTE, S, Evaluation of concrete fracture energy trough a pull-out testing procedure, in Proceedings of the International Conference on Recent Development on the Fracture of Concrete and Rock, (Eds. S P Shah, S E Swartz, B Barr), p 347-356, Elsevier Applied Science, Amsterdam, 1989.

6. BOCCA, P, CADONI, E, VALENTE, S, On concrete fatigue fracture in pull-out tests, in Fracture and Damage of Concrete and Rock, p 637-646, Boston, Chapman & Hall, 1992.

7. COSSU, P, POZZO, E, Correlation between pull-out force and compression resistance in concrete for different curing age, L'Industria Italiana del Cemento, No 7-8, 1995, p 434-439.

8. FOGLIATO, D, REL BOCCA, P, BALLATORE, E, BELLINO, F, Pull-out tests for the determination of the mechanical charateristics of fibre reinforced concrete, Polytechnic of Turin, 1999.

9. BOCCA, P, CROTTI, M, Method for the determination of the elastic modulus through pull-out tests, Proceedings of 7th International Conference on Inspection, Appraisal, Repairs & Maintenance of Buildings and Structures, Notthingham Trent University, 11-13 September 2001.

MICROWAVE NON-DESTRUCTIVE TESTING OF FIBRE CONCRETE USING FREE SPACE MICROWAVE MEASUREMENTS

H M A Al-Mattarneh

D K Ghodgaonkar

W M bin W A Majid

MARA University of Technology

Malaysia

ABSTRACT. Microwave nondestructive testing (MNDT) techniques such as reflection, transmission and dielectric measurements have advantages over other NDT methods regarding low cost, good penetration in nonmetallic materials, good resolution and contactless feature of microwave sensor (antenna). In this paper, reflection coefficients, transmission coefficients, dielectric properties (ε' = dielectric constant and ε'' = loss Factor) were measured to detect fibre distribution and concentration in fibre-reinforced concrete (FRC). For FRC specimens, reflection (S_{11}) and transmission (S_{21}) coefficients were measured in the frequency range 8 – 12.5 GHz by using free-space microwave measurement (FSMM) system. FSMM system consists of a pair of spot-focusing horn lens antenna, mode transitions, coaxial cables and vector network analyzer. ε' and ε'' values can be evaluated from the measured S_{11} and S_{21}. FRC specimens were manufactured with polypropylene (PP) fibre (Fibermesh® fibres, length = 19 mm) and steel fibres (Convotex fibres, length = 30 mm). FRC specimens were cast with fibre concentration of 0.0, 0.6, 0.9 and 1.2 kg /m^3 for PP fibres and 0, 10, 20, and 30 kg/m^3 for steel fibres. The results show decreasing reflection coefficients, transmission coefficients, and dielectric properties with increasing curing age. These properties also show the potential to be used to determine the percentage of fibre content in concrete.

Keywords: Microwave, Fibre concrete, Nondestructive, Free-space, Dielectric properties.

Dr H M A Al-Mattarneh, is a Research/Teaching Fellow at the Faculty of Civil Engineering, MARA University of Technology, Shah Alam, Malaysia. His main research interests include the nondestructive testing of construction materials, high performance concrete, the statistical modeling and evaluation of the actions and materials properties.

Professor D K Ghodgaonkar, is an Associate Professor in the Faculty of Electrical Engineering, MARA University of Technology, Shah Alam, Malaysia. His research interest is in the area of microwave nondestructive testing of composite materials.

Professor W M B W A Majid, is a Professor in the Faculty of Civil Engineering, MARA University of Technology, Shah Alam, Malaysia. His research interest is in the area of nondestructive testing of construction materials.

INTRODUCTION

Conventional concrete has many advantages over other structural materials such as; it is the lowest cost structural material, it can be produced from materials found in abundance all over the world, it is resistant to water and it provides good protection for steel. Unfortunately, conventional concrete also has limitations and disadvantages such as; its hydration reaction produces a decrease in volume that result in shrinkage cracking; it is brittle and must be reinforced to improve mechanical strength and toughness. As each of these problems has been identified, research and development in concrete industry has proposed new additives to solve it. Several studies have been conducted to investigate use of different types of fibres to solve some of these problems [1]. The fibres can be broadly classified as; metallic, polymetric, mineral and natural. Metallic fibres are made of either steel or stainless steel. The polymeric fibres in use include acrylic, aramid, carbon, nylon, polyester, polyethylene, and polypropylene fibres. Glass fibre is the predominantly used mineral fibre. Various types of natural fibre such as cellulose are also being used [2].

Due to the use of fibres in concrete, major improvement occur in the area of ductility and fracture toughness, but flexural strength increases were also reported. Other function of fibres is to lock the coarse aggregate together and prevent micro cracks [3]. The major application of fibre concrete are: shotcrete for slope stabilization; tunnel/mine linings; slap on grade; and precast element [4]. Type of fibre, physical properties, aspect ratio and shape are the major factors in determining the performance of each fibre [5].

In the last twenty years, the use of fibre concrete in the construction industry is increasing. However, the need for quality test is ever increasing. The available test methods are destructive, costly, limited to laboratory and time consuming. A nondestructive test could save time and save cost on some of the measured properties such as moisture content, fibre characterization, fibre orientation, voids, defects and fibre volume fraction. Microwave nondestructive testing seems promising; since it is easy to adopt to use in site, it is contactless, and it is continuous and intensive measurement method.

Microwave nondestructive testing (MNDT) techniques using free-space methods, are mainly used for nonmetallic materials. These techniques have advantages over other NDE methods (such as radiography, ultrasonic, eddy current) regarding low cost, good penetration in nonmetallic materials, good resolution and contactless feature of the microwave sensor (antenna) [6]. R. M. Redheffer was the first researcher to suggest a simple free-space method for measurement of dielectric constant from the measured phase of transmission coefficients [7]. Bassett was the first researcher to measure complex permittivity in free-space using spot-focusing antennas at a frequency of 9.4 GHz [8]. In the last twelve years, a number of free-space methods were developed for measurement of dielectric properties [9]. Previous research work shows that there is potential in the use of electromagnetic properties of concrete to determine w/c ratio, strength and moisture content [10-11].

In this paper, the feasibility of nondestructive testing of fibre concrete using free-space microwave technique will be investigated. Two type of fibre concrete will be used namely: steel fibre concrete and polypropylene fibre concrete manufactured using different fibre contents. Microwave nondestructive testing (MNDT) will be carried out for characterization of fibre volume fraction. The MNDT system consist of transmit and receive horn lens antennas, a vector network analyzer, mode transitions, and a printer. The horn lens antennas are used for minimizing diffraction effects due to the edges of the sample.

DIELECTRIC CONSTANT AND LOSS FACTOR

Most of the construction materials such as concrete are dielectric materials. A dielectric material can be characterized by two independent electromagnetic properties namely, complex permittivity ε^* and complex permeability μ^*. However, most dielectric materials including wood and concrete are nonmagnetic, making the permeability very close to the permeability of free space. So the discussion is limited to the complex permittivity ε^* which defined as[12]:

$$\varepsilon^* = \varepsilon' - j\varepsilon''$$

(1)

Where ε' is the real part of the complex permittivity, ε'' is the imaginary part of the complex permittivity and $j = \sqrt{-1}$. Dividing equation 1 by the permittivity of free space ε_0 the property becomes dimensionless and relative to free space;

$$\varepsilon_r^* = \frac{\varepsilon^*}{\varepsilon_0} = \varepsilon_r' - j\varepsilon_r''$$

(2)

Where ε_r' is the real part of the relative permittivity called dielectric constant and ε_r'' is the imaginary part of the relative permittivity called loss factor. The dielectric constant is a measure of how much energy from external electric field is stored in a material. The loss factor is a measure of how dissipative or lossy a material is to an external electric field due to current conduction [12].

The ratio of the energy lost to the energy stored in a material is given as loss tangent;

$$\tan\delta = \frac{\varepsilon_r''}{\varepsilon_r'}$$

(3)

where $\tan\delta$ is the loss tangent. These electromagnetic properties are not constant. They change with frequency, temperature, moisture, and mixture of the material. Figure 1 shows a planer sample of thickness d placed in free-space. Reflection and transmission coefficient were measured using the free-space microwave system to determine the dielectric constant and loss factors of fibre concrete [9].

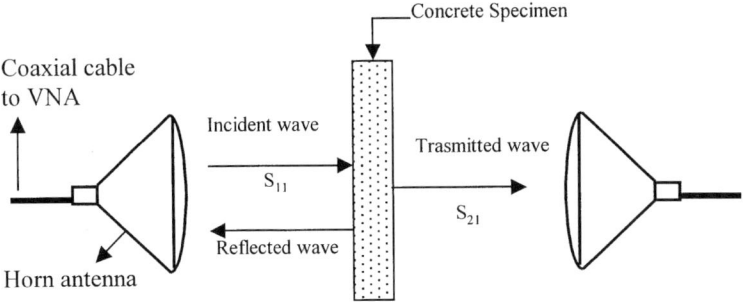

Figure 1   Schematic diagram of planar sample

MEASUREMENT SYSTEM SETUP

Figure 2 gives a schematic diagram of the FSMM system, which was used for microwave nondestructive evaluation of composite materials [9].   A pair of spot-focusing horn lens antennas (model no. 857012X-950/C) were manufactured by Alpha Industries, Woburn, MA (USA).   These antennas have two equal plano-convex lenses mounted back to back in a conical horn antenna.   One plano-convex lens gives an electromagnetic plane wave and the other plano-convex lens focuses the electromagnetic radiation at the focus.   This measurement set up covers a frequency range of 8 - 12.50 GHz.   But, the same setup can be used in the frequency range of 7.5 -40 GHz by appropriate change of mode transitions.

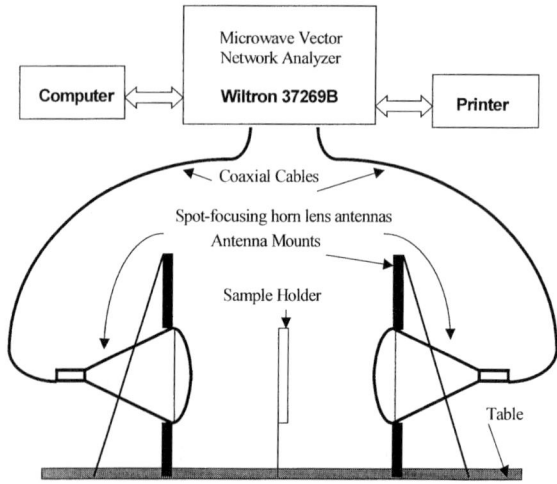

Figure 2   Free-space microwave measurement system for NDT of fibre concrete

The focused antennas are connected to the two ports of the Wiltron 37269B vector network analyzer by using precision coaxial cables, rectangular-to-circular waveguide adapters and coaxial-to-rectangular waveguide adapters. This network analyzer is used to make accurate S-parameter (reflection and transmission) measurements in free-space using line-reflect line calibration model. A personal computer along with appropriate measurement software can be used for overall automation of the free-space measurement system.

MATERIALS AND SPECIMENS

Concrete specimens were made using Ordinary Portland Cement, fine aggregate (natural sand), coarse aggregate (granite), water and fibre. All mixes prepared with water to cement ratio 0.5, cement to aggregate ratio 1:5 and fine to coarse aggregate 2:3. Four mixes were prepared using different percentage of steel fibre (SF); 0, 10, 20 and 30 kg per cubic meter and another 4 mixes were prepared using different percentage of polypropylene fibre, 0, 0.6, 0.9 and 1.2 kg per cubic meter of concrete. The result of quality control test (28 days compressive and flexural strength test in N/mm^2) are shown in Table 1.

The steel fibre used in this study was Novotex IV manufactured by Novocon steel fibre, USA. The properties of novotex steel fibre are: length = 30 mm, the diameter = 0.7 mm, aspect ratio 43, tensile strength 1150 N/mm^2, flattened ends with round shaft, bright and clean wire. The polypropylene fibre was fibremesh manufactured by Synthetic Industries, Inc, USA. The properties of polypropylene fibre are: specific gravity 0.91, modulus of elasticity 3.5 KN/mm^2, fibre length 13-19mm, diameter 18 microns, specific surface area 225 m^2/kg and melting point 160-170 C.

Table 1   Properties of steel and polypropylene fibre concrete

FIBRE CONCRETE STRENGTH (N/mm^2)	STEEL FIBRE CONTENT			
	0 kg/m^3	10 kg/m^3	20 kg/m^3	30 kg/m^3
28 days compressive strength	35.6	37.0	36.0	33.9
28 days Flexural strength	3.18	3.49	3.55	3.87
FIBRE CONCRETE PROPERTIES (N/mm^2)	POLYPROPYLENE FIBRE CONTENT			
	0 kg/m^3	0.6 kg/m^3	0.9 kg/m^3	1.2 kg/m^3
28 days compressive strength	36.3	37.0	36.5	34.9.
28 days Flexural strength	3.18	3.45	3.94	3.29

EXPERIMENTAL RESULTS

Steel Fibre Concrete

The measurements of steel fibre concrete were conducted in the frequency rage of 7.5-12.5 GHz. The effects of frequency on the reflection and transmission coefficients of steel fibre are shown in Figure 3 and 4 respectively. These results show that reflection coefficients increase with increasing frequency while Transmission coefficients decrease with increasing frequency.

The reflection and transmission coefficients of steel fibre concrete were evaluated during the 28 days of curing. The relationships between the reflection and transmission coefficients are shown in Figures 5 and 6 respectively. From the result in Figure 5, it is clear that the reflection coefficients of all fibre concrete samples decrease with increasing curing age which can be attributed to the loss of free water during curing time since part of the free water is lost by evaporation and the rest is bound in the cement paste.

The result of the transmission coefficients in Figure 6 increases with increasing curing time because with increasing curing time the free water decreases by evaporation process, this will increase the penetration of the microwaves in steel fibre concrete which increases the transmission coefficients. Also the Figure shows that the higher fibre contents in the concrete indicate the lower transmission coefficients.

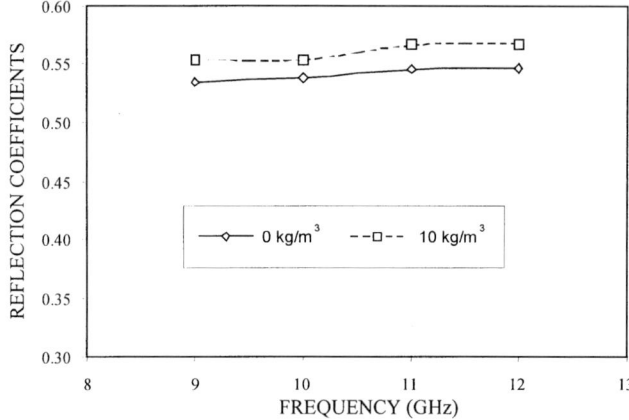

Figure 3   Effect of frequency on the reflection coefficients of SF concrete

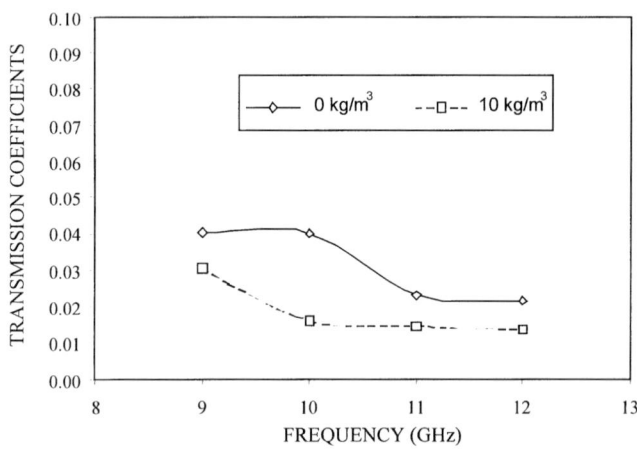

Figure 4   Effect of frequency on the transmission coefficients of SF concrete

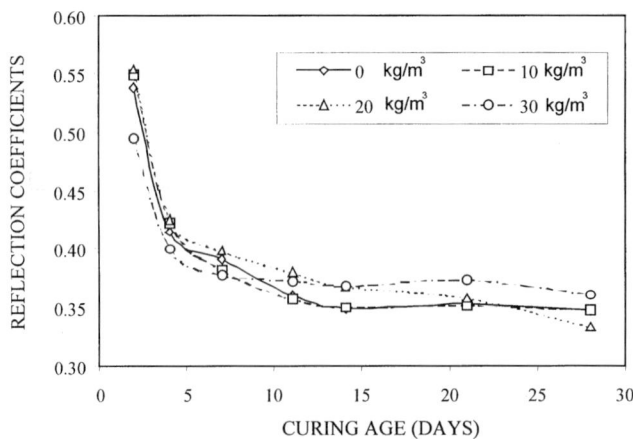

Figure 5   Effect of curing age on the reflection coefficients of SF concrete at 10 GHz

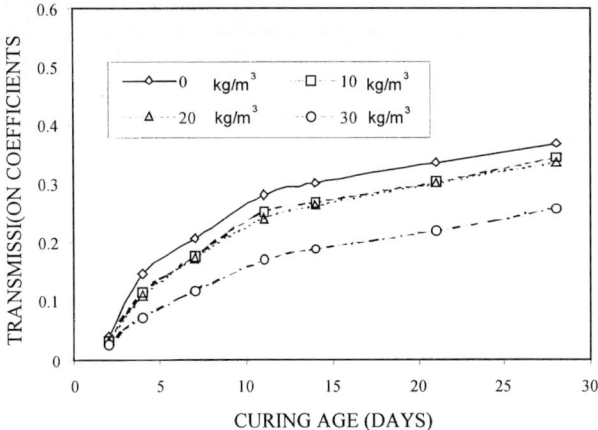

Figure 6   Effect of curing age on the transmission coefficients of SF concrete at 10 GHz

The relationship between the steel fibre content in concrete and the transmission coefficients are shown in Figure 7. The results show that with increasing the fibre content in concrete, the transmission coefficients were decreasing. The relationship between the fibre content and the transmission coefficients were established at 3, 7, and 28 days of curing. The relationship was linear with high correlation coefficients R square.

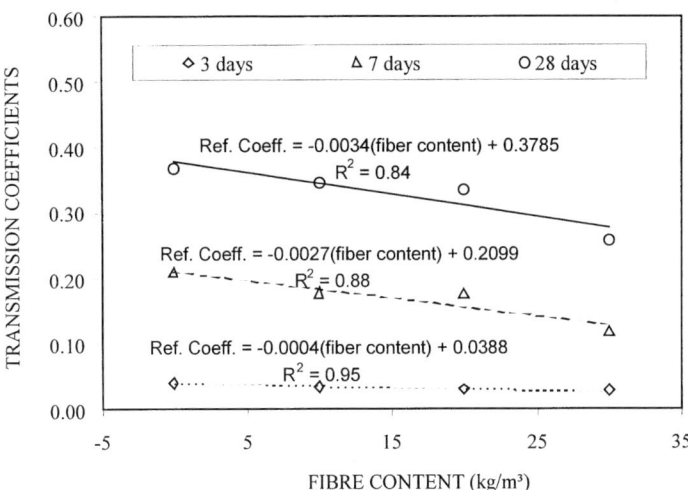

Figure 7   The relationship between the transmission coefficients and SF content at 10 GHz

The dielectric constant and loss tangent for all steel fibre content were calculated using the formulation given in [9]. Results of dielectric constant and loss factor are shown in Table 2.

Table 2   Dielectric constant and loss factors of steel fibre concrete at 10 GHz

SAMPLE	FIBRE CONTENT	CURING AGE = 3 DAYS		CURING AGE = 28 DAYS	
		ε'	ε''	ε'	ε''
ST0	0 kg/m^3	5.103	-0.949	3.748	-0.241
ST1	10 kg/m^3	4.813	-0.473	4.397	-0.064
ST2	20 kg/m^3	4.711	-0.322	4.076	-0.307
ST3	30 kg/m^3	4.646	-1.057	4.356	-0.975

Polypropylene Fibre Concrete (PPFC)

The measurements of PP fibre concrete were conducted in the frequency rage of 7.5-12.5 GHz. The effects of frequency on the reflection and transmission coefficients of PP fibre concrete are shown in Figures 8 and 9 respectively. The results of reflection coefficients shown in Figure 8 do not show any trends. The reflection coefficients of PP fobre concrete takes its minimum values at 10GHz and maximum values at 11 GHz.

Concrete containing high fibre content show low minimum reflection coefficients and high maximum values compare to concrete containing low fibre content.   The results of transmission coefficients of PP fibre concrete shown in Figure 9 decrease with increasing frequency for all PP fibre content. Also the transmission coefficients concrete with high PP fibre content are higher than the transmission coefficient of concrete with low PP fibre content. This trends is correct over all frequencies.

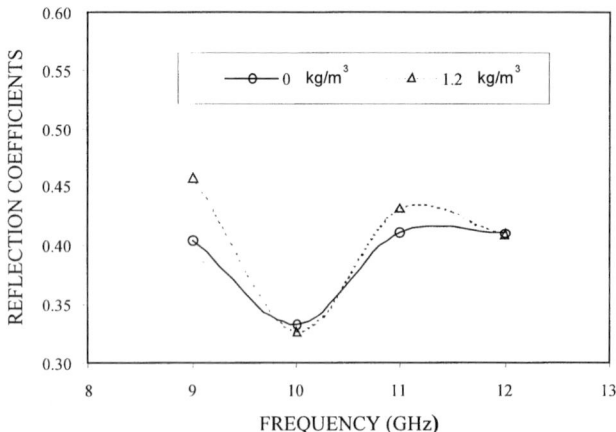

Figure 8   Effect of frequency on the reflection coefficients of PP fibre concrete

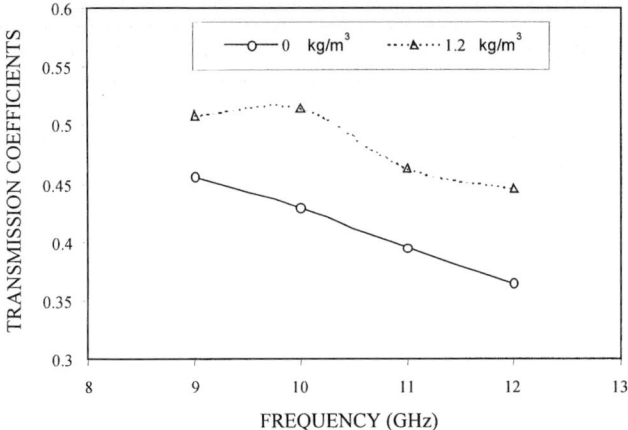

Figure 9   Effect of frequency on the transmission coefficients of PP fibre concrete

From the result in Figure 10, it is clear that the reflection coefficients of all fibre concrete samples decrease with increasing curing age this can be attributed to the loss of free water during curing time since part of the free water is lost by evaporation and the rest is bound in the cement paste. The result of the transmission coefficients in Figure 11 increasing with increasing curing time because with increasing curing time the free water decreases by evaporation process, this will increase the penetration of the microwaves in PP fibre concrete which increase the transmission coefficients. Figures 10 and 11 show that the reflection coefficients decrease with increasing PP fibre content except for 0 fibre (concrete without fibre) while the transmission coefficients increase with increasing PP fibre contents. These results indicate that reflection and transmission coefficients in the microwave frequency can be used to determine the PP fibre content in concrete. So, the variation of these measurements from point to point can be used as quality test for homogeneity of fibre concrete (uniform distribution of the fibre in concrete).

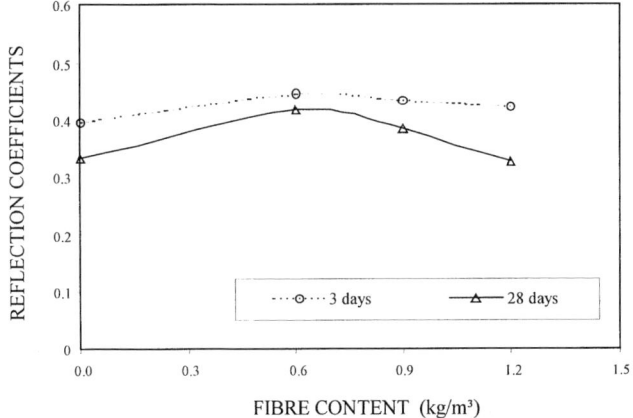

Figure 10   The relationship between the PP fibre content and reflection coefficients at different curing age and frequency of 10 GHz

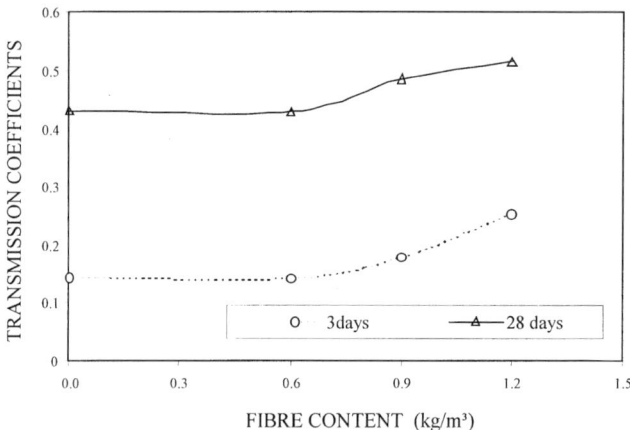

FIBRE CONTENT (kg/m³)

Figure 11   The relationship between the PP fibre content and transmission coefficients at different curing age and frequency of 10 GHz

The results of dielectric constant and loss factor of PP fibre concrete at 10 GHz are shown in Table 2. From these results, it is observed that the dielectric constant of PP fibre increase with increasing fibre content which can be used to estimate the fibre content in concrete.

Table 3   Dielectric constants and loss factors of PP fibre concrete at 10 GHz

SAMPLE	FIBRE CONTENT	CURING AGE = 3 DAYS		CURING AGE = 28 DAYS	
		ε'	ε''	ε'	ε''
PPC0	0 kg/m³	5.109	-0.984	4.029	-0.148
PPC1	0.6 kg/m³	5.811	-0.221	5.554	-0.168
PPC2	0.9 kg/m³	5.866	-0.335	5.629	-0.156
PPC3	1.2 kg/m³	6.974	-0.659	6.497	-0.382

CONCLUSIONS

The conclusions of this study were made as:

1. The reflection coefficient of steel fibre concrete increase with increasing frequency while the transmission coefficients of steel fibre concrete decrease with increasing frequency. Reflection coefficients of SF concrete decrease with increasing curing age while the transmission coefficients of SF concrete increase with increasing curing age. The transmission coefficients of SF concrete decrease with increasing the SF content in concrete. The dielectric constants of SF Concrete decrease with increasing SF content.
2. The transmission coefficient of PP fibre concrete decrease with increasing frequency. The reflection coefficients of PP fibre concrete decrease with increasing curing age and decrease with increasing fibre content. The transmission coefficients of PP fibre concrete increase with increasing curing age and increase with increasing fibre content. The Dielectric constants of PP fibre concrete increase with increasing PP fibre content.

3. The free space microwave system using reflection coefficients, transmission coefficients and dielectric properties of fibre concrete can be used to estimate the fibre content in concrete and to test the dispersion of the fibre in the concrete matrix.

REFERENCES

1. NATIONAL MATERIALS ADVISORY BOARD, 1997, Nonconventional concrete technology, Renewal of the Highway Infrastructure, NMAB, National Academy Press, Washington, D.C.

2. BALAGURU, P N, SHAH, S P, Fiber reinforced cement composites, McGraw-Hill Book Co, Singapore, 1992.

3. ROBERT NICHOLIS, Composite construction materials handbook, Prentice-Hall, New Jersey, 1976, p 373-387.

4. TAYLOR, G D, Construction materials, Longman Scientific & Technical, Singapore, 1991.

5. GOPALARATNAM, V S, SHAH, S P, Fracture Toughness of fiber reinforced concrete, A report of concrete materials research council, ACI Materials Journal, July – August, 1991.

6. BOTSCO, R J, CRIBBS, R W, KING, R J, MCMASTER, R C, Microwave methods and applications in nondestructive testing, In Nondestructive Testing Handbook, Paul McIntire, Vol 4, Section 18, 1986, American Society of Nondestructive Testing.

7. REDHEFFER, R M, The measurement of dielectric constant, In Techniques of Microwave Measurements, (Ed. C G Montgomery), Vol. 2, Dover, 1966, p 591-657.

8. BASSETT, H L, A free-space focused microwave system to determine the complex permittivity of materials to temperature exceeding 2000°C, The Review of Scientific Instruments, Vol 42, 1971, p 200-204.

9. GHODGAONKAR, D K, VARADAN, V V, VARADAN, V K, Free-space method for measurement of dielectric constants and loss tangents at microwave frequencies, IEEE Transactions on Instrumentation and measurement, Vol 38, 1989, p 789-793.

10. AL-MATTARNEH, H M A, GHODGAONKAR, D K, MAJID, W M B W A, Determination of compressive strength of concrete using free-space reflection measurements in the frequency range of 8-12.5 GHz, Asia Pacific Microwave Conference, December 3-6, 2001, National Taiwan University, Taipei, Taiwan, R.O.C.

11. AL-MATTARNEH, H, ZAIN, M F M, GHODGAONKAR, D K, MAJID, W M B W A, Senseng of Moisture content of PCC using microwave nondestructive testing in the frequency range of 7-13 GHz, 7th International Conference on Concrete Engineering and Development, 5-7 June, 2001, Shah Alam, Malaysia.

12. ARTHUR VON HIPPEL, 1995, Dielectric Materials and Applications, Artech House.

DIAGNOSTIC, ASSESSMENT AND REPAIR OF HONEYCOMB CONCRETE IN ALGIERS NEW AIRPORT REINFORCED CONCRETE TERMINAL BUILDING

S Kenai

University of Blida

R Bahar

University of Tizi-Ouzou

Algeria

ABSTRACT. This paper reports on the diagnostic and repair of honeycomb concrete in the new Algiers airport building. The building is a reinforced concrete building of more than 80 thousand square meters of floor slabs supported on more than 1700 foundation piles. The main cause of honeycomb concrete in the beams and the walls was severely congested reinforcing steel and inadequate spacing between parallel layers. Inappropriate mix design with respect to coarse aggregate size of concrete, construction errors such as lack of cover and the use of low slump concrete without plasticiser admixtures, poor placement and inadequate vibration also contributed to honeycombing and segregation. The repair work proposed, involved the application of sprayed polymer modified mortar with and without fibres on more than ten thousand square meters of honeycomb concrete.

Keywords: Concrete, Low strength, Honeycomb, Segregation, Non-destructive testing, Cores, Repair, Sprayed mortar.

Professor S Kenai, is an Associate Professor and Chairman of the Scientific Committee of the Civil Engineering Department of the University of Blida. He obtained his Ph.D from Leeds University in 1988. His main interests include concrete technology, cement replacement materials, durability of concrete structures, building failures and construction repair.

Professor R Bahar, is an Associate Professor at the University of Tizi-Ouzou and Chairman of the Civil Engineering Department Scientific Committee. He obtained his Doctorate degree from Ecole Centrale of Lyon (France) in 1992. His main interests are building failures, non-destructive testing of materials, foundations and soil mechanics.

INTRODUCTION

The new Algiers airport terminal building is a reinforced concrete building of more than 80 thousand square meters of floor slabs supported on more than 1700 foundation piles designed to handle about six million passengers a year. It is composed of two symmetrical modules in the form of a circular arc of an internal radius of 132 m and external radius of 175 m and is located near the actual Algiers international airport. The project involved the casting of more than 100 thousand cubic meters of concrete. The construction work was stopped for financial difficulties as the structural work was almost complete. The new owner felt it necessary to evaluate the structure before completing it as many shortcomings were observed. The assessment study was done in 2000 by a team led by the first author to evaluate the structure and propose remedial work for the repair of honeycomb concrete

The evaluation approach included visual inspection of the concrete, seismic parallel method for piles testing, non-destructive testing of concrete with Schmidt hammer, ultrasonic-pulse velocity measurements, laboratory testing of concrete cores to determine compressive strength, carbonation tests and also ambient vibrations of the structure. This paper reports on the diagnostic and repair of honeycomb concrete observed mainly in beams and walls. More details on the assessment study and the repair work are given elsewhere [1].

ASSESSMENT OF HONEYCOMB CONCRETE

The main deficiency which worried the owner and was spread on most elements was segregation and honeycombing. Hence, a visual inspection of all reinforced concrete elements were conducted by the team. The main affected elements were all the thirty four long beams in the passenger hall. These beams are about 42 m long , 0.5 m wide and with a web depth of 2 m and about 7.4 m long 55° sloped part. The central span of the beam of over 22 m is at 11.5 m from the floor level with cantilevers on both sides. Figure 1 showed a section of this beam with views of the reinforcement on the bottom side whereas Figure 2 shows a general view of these beams and the passenger hall. Typical honeycombs in these beams and at other structural elements are shown in Figure 3.

Factors Causing Segregations and Honeycombs

The honeycombs are due to many reasons. The main reason could be attributed to the contractor who was a local firm without any previous proven experience in this type of buildings and lacking concrete technology skills. The contractor was chosen to do the construction work, after an international bid.

Detailed inspection of design details, concrete testing reports, technical site meeting reports and other construction records revealed that the main contributing factors to segregations and honeycombs observed are inappropriate mix design with respect to coarse aggregate size, construction errors such as lack of cover and the use of low slump concrete (80 to 100 mm), the non-use of plasticizer admixtures, poor placement and inadequate compaction. Severely congested reinforcing steel and inadequate spacing between parallel layers not conforming to standards requirements also contributed to the appearance of honeycombs.

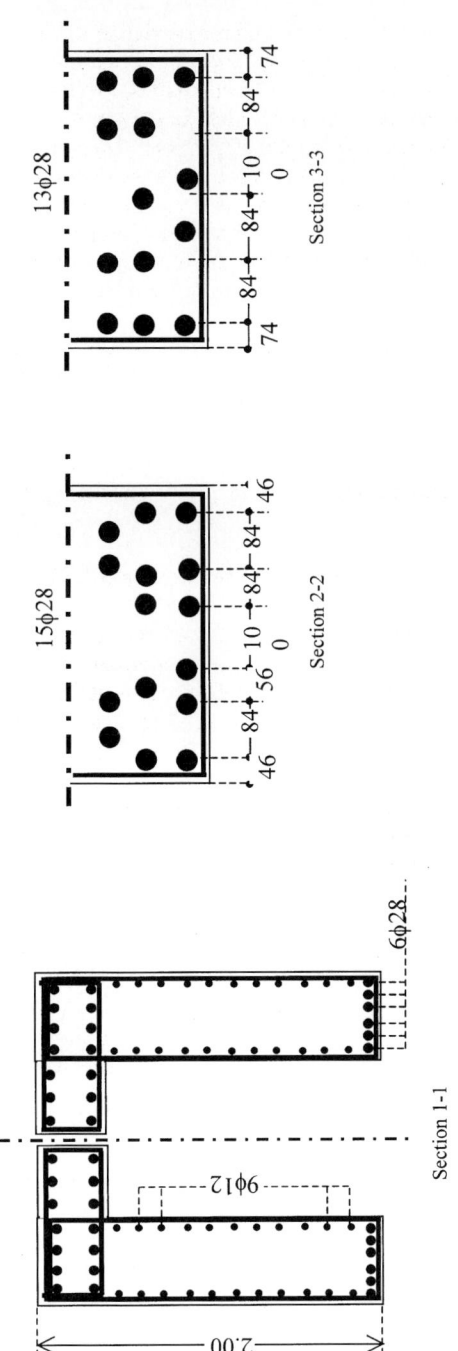

Figure 1   A view of beam and reinforcement details

Figure 2   A general view of the passenger hall and the main beams

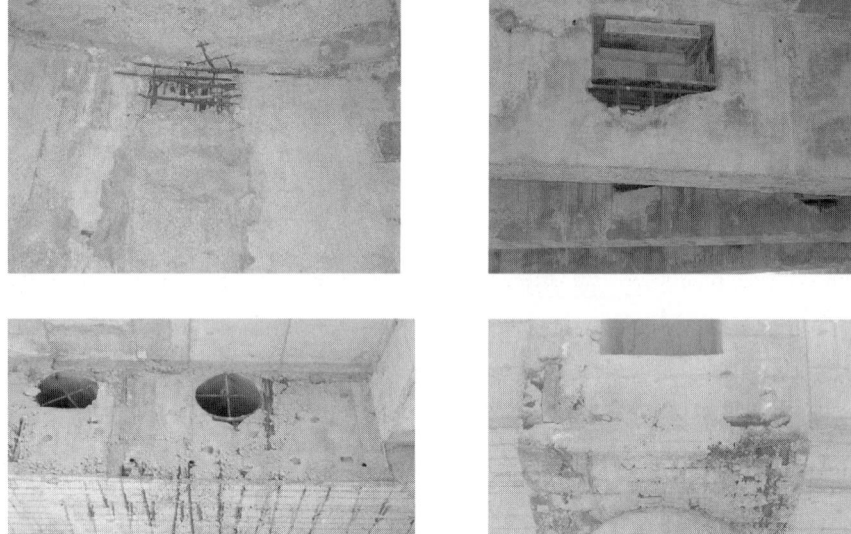

Figure 3   Honeycombs near openings, slab-wall junction, and beam-column junction

Concrete Mix Design

The concrete mix was designed to get a cube characteristic strength at 28 days of age of 25 MPa according to the design requirements. Maximum aggregates size specified by the site engineer was 25 mm but inspection of some segregated areas revealed higher sizes in some zones proving that the aggregates were not well graded. Many quarries were used for aggregates supply as Algiers area is lacking natural aggregates but no action was taken to modify the mix when aggregates source is changed. Site testing reports revealed that the sand used had excess fines and sometimes a low sand equivalent. Fine aggregates contained sometimes higher level of impurities. Beach sand, which is usually not well garded and very fine, was sometimes used to replace river sand without any size correction or blending with another sand. Typical laboratory tested mixes proposed for one cubic metre of concrete are given in Table 1.

Table 1   Typical concrete mixes used

	MIX 1	MIX 2	MIX 3
Cement Content (kg/m^3)	350	350	400
Sand (kg)	650	577	577
Gravel (3/8)	295	190	345
Gravel (8/15)	180	260	262
Gravel (15/25)	685	635	672
Water (l)	210	210	210
Plasticiser (kg/m^3)	--	---	---
Slump (mm)	50	80	60
28 days cylinder compressive strength (MPa)	24.8	28.8	25.6

It can be seen that although cement content was high, low strength concrete was obtained. It seems that this is due to the low strength cement used which was a CEM II/A-32.5 type with about 15% limestone additions and to the high water cement ratio. During the investigation, trial mixes (Table 2) were prepared using aggregates left by the contractor on site to prove to the owner that good mixes with higher workability and higher compressive strength could be obtained if basic precautions are taken, superplasticisers are used, and careful mix design is realised.

Table 2   Trial mixes prepared during the assessment study

	MIX 1	MIX 2	MIX 3
Cement Content (kg/m^3)	350	400	400
Sand (kg)	779	720	810
Gravel (5/20)	624	620	852
Gravel (12.5/25)	362	350	---
Water (l)	210	180	200
Plasticiser (kg/m^3)	--	1.6	1.6
Slump (mm)	70	80	160
28 days cylinder compressive strength (MPa)	25.6	34.6	35.0

Reinforcement Congestion

Reinforcement congestion was observed on many structural elements. However the congestion in the long beams of the passenger hall was more apparent. The beam drawings (Figure 1) showed two to three rows of 28 mm bars both top and bottom and 10 mm stirrups spaced at 150 mm. The bar spacing at the bottom was between 56 to 84 mm with a cover of

46 mm. The clear spacing between individual bars at the bottom layer was according to the drawings as low as 28 mm. The design consultant used the German standards DIN and hence the 28 mm diameter bars proposed which are not available in the local market which used the French standards AFNOR. For this reason, 20 and 25 mm bar diameters were used to substitute the 28 mm diameter ones resulting in even smaller bar spacing as shown in Figure 4 where some bars are in contact. The mix was not designed so that the largest aggregate size can pass between adjacent bars and between form and the reinforcement. ACI 318 [2] requires that the nominal maximum size of coarse aggregate shall not be larger than 3/4 the minimum clear spacing between individual reinforcing bars and that clear distance between parallel bars in a layer shall be not less than the nominal diameter of the bar nor 25 mm. Similar conditions are required by the French standards for reinforced concrete BAEL 91 [3] which requires that the the clear spacing between individual bars should be greater than the nominal bar diameter and also greater than 1.5 x maximum aggregates size. The concrete mix design contained 30 mm maximum aggregate size and a theoretical clear spacing of 28 mm for 28 mm diameter bars which has to be reduced to about 20 mm when 25 mm diameter bars are used. Consequently, both ACI and BAEL conditions were violated. In addition to that, workability of the mix was low and did not permit the concrete to be worked around reinforcement causing segregation. The analysis of testing report data gave an average slump test results of 70 to 100 mm. Segregation was very important in particularly in the bottom of the sloped parts of the beams which seem not to be well vibrated by external form vibrators as clusters of coarse aggregates with little mortar were found. However, it should be noted that even the use of a high workability mix and good vibration could not have remedy this situation as coarse aggregates cannot pass between the reinforcing bars. In some instances, only a small layer of laitance was covering the bottom reinforcing layer which could be removed by hand.

The honeycombs could have been avoided using appropriate mix design with lower aggregate size (15 or 20 mm maximum aggregates size), appropriate higher workability (100 to 150 mm slump) with the use of superplasticisers and thorough consolidation

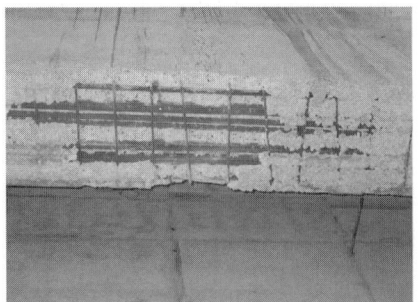 

Figure 4   Reinforcement congestion in beams

Poor Concreting Practice

Lack of compaction and also poor workmanship were evident. Quality control on site and supervision during construction were not done accurately. No precautions were taken to take into account the effect of hot environment on the mix design and concreting work.

Concrete Compressive Strength

The specified concrete strength was either of a characteristic 200 mm cube strength (F'_c) of 25 MPa for most of the elements. More than 3300 cube test results of grade C25 concrete were analyzed and gave a satisfactory average compressive strength at 28 days of age of 30 MPa but a high standard deviation of 3.4 MPa showing a large variability of strength. In order to ascertain whether the in-situ strength of concrete is acceptable for the designed loading system, 349 structural elements (beams, columns, walls or slabs) suspected, as low strength elements after the visual inspection, were tested by a combined method Schmidt hammer and ultrasonic pulse velocity. The results confirmed that the concrete strength of the elements is satisfactory.   The estimated in-situ strength based on both methods was comparable.   Pulse velocity measurements varied from 3300 to 4600 m/s indicating an irregular concrete production. More than forty elements showed an estimated in-situ strength less than the characteristic strength and were checked by drilling cores on them. 126 cores were drilled from 42 elements suspected for low strength.   The cores testing results confirmed the low strength but most of the elements satisfied the ACI conditions [1] of structurally adequate concrete. A good correlation was found between core tests and the non-destructive tests.

Cover Achieved On Site

Reinforcement cover was measured by an electromagnetic cover-meter on more than 80 elements. Carbonation depth was measured on all cores by spraying with phenolphthalein solution on the freshly taken drilled cores and compared to the steel cover. An average carbonation depth of 10 to 25 mm was found compared to an average cover of 5 to 30 mm which showed clearly that most of the concrete cover is carbonated and cover specified was not achieved. It should be noted that there were many cases where reinforcement was seen at the bottom of the slabs and beams with no cover at all. In addition, even where cover was satisfactory, low quality concrete spacers with high porosity and water absorption higher than 10% were used. Inadequate compaction and improper curing seems to be the cause of the carbonation and carbonation induced corrosion problems which started to occur in this relatively short time (5 to 10 years) in some elements.

REPAIR METHOD AND MATERIALS

The repair work proposed involved the application of sprayed mortar on more than ten thousand square meters of honeycomb concrete. Other deficiencies such as plastic shrinkage cracking in beams, expansion joints full of mortar and debris and corrosion were also assessed and repaired. The selection of the appropriate repair material was based on its intrinsic properties as well as its compatibility between the repair material and the existing concrete substrate. Hence, cement based repair materials were chosen for their low cost and compatibility.

The first step of repairing of honeycombs and segregations was surface preparation by removing laitance and loose concrete by either pneumatically driven or hand-operated lightweight jack hammers and rust from the reinforcement bars by sand blasting and final cleaning by water jet. Simple geometrical prepared surface shapes were used to avoid differential drying shrinkage. Following the cleaning of the reinforcement, the bars were treated with a cement based coating containing a corrosion inhibiting admixture. No bonding agent was applied. Final substrate cleaning was carried out immediately before repair work to prevent contamination of the prepared surfaces and to get a saturated surface-dry surface. An imported commercial prepackaged cement based polymer modified mortar was mixed and cast according to the manufacturer's instructions and used for the repair of honeycombs. For deep honeycombs, a similar mortar was used but incorporating 7% by weight of cement of silica fume and also polypropylene fibres in order to achieve higher thickness in one layer of sprayed mortar. Curing was started immediately by sprinkling water for a minimum of three days. It was difficult to predict the amount of work required to complete the repair work in a satisfactory manner. However, the estimation given to the client was satisfactory after the repair work was finished. The repair of the honeycombs and segregation constituted about 2/3 the cost of all the repairs realised at a cost of about three million US dollars.

CONCLUSIONS

The assessment study of this building showed that thee main deficiencies observed were honeycomb and segregation in beams. The main reason for the honeycombing is poor concrete mix design and non-conformity to standards recommendations concerning reinforcement details and in particular aggregate size, and steel spacing. The absence of the designer on site, the use of non-familiar standards and the lack of good supervision and quality control on site contributed to the deficiencies observed. Violating basic standard rules has led to costly repairs and delay. The repair work was conducted at a cost of more than three millions US dollars.

REFERENCES

1. KENAI, S, BAHAR, R, Evaluation and Repair of Algiers New Airport Building, International Symposium on Materials and Infrastructure and Development, ASME/ASCE Summer Meeting, San Diego, June 27 – 29, 2001.

2. ACI 318M-83, Building Code Requirements for Reinforced Concrete, Second edition, March, 1985.

3. BAEL 91, Technical Rules, Design and Calculation of Reinforced Concrete Constructions to the Limit State, Eyrolles Edition, Paris, 1992.

NON-DESTRUCTIVE TESTING OF THE MICROSTRUCTURAL DEVELOPMENT IN HARDENING CEMENT-BASED MATERIALS BY ULTRASONIC PULSE VELOCITY MEASUREMENTS

G Ye   K van Breugel   A L A Fraaij

Delft University of Technology

Netherlands

ABSTRACT. In this contribution, a four-phase model for the hydration of cement paste is described, based on the numerical simulation model HYMOSTRUC. Ultrasonic Pulse Velocity (UPV) measurements were utilized with the particular aim to trace the development of the microstructure in cement-based materials. Experimental tests were performed both on cement paste and on concrete mixtures with water/cement ratio 0.40, 0.45 and 0.55. The samples were cured isothermally at 20, 30 and $40^{°}$ C. The measurements were done until an age of 72 hours for the cement paste and of 160 hours for the concrete. Based on the experimental data, a mathematical model for the evolution of the UPV for cement paste is presented and "extended" to a two-phase concrete model. The correlation between the measurements and the simulated values for cement paste and concrete was very good.

Keywords: Cement paste, Concrete, Early age, Microstructure, Bridge volume, Ultrasonic pulse velocity, Modelling.

G Ye, is a Researcher at the Delft University of Technology. His research topic is numerical simulation of the microstructure, porosity and permeability of cement-based materials, in particular non-destructive test at early age.

K van Breugel is a Professor at Delft University of Technology. His research topics are early age concrete, design for imposed deformations and concrete structures for environmental protection.

A L A Fraaij is an Associate Professor at the Delft University of Technology, his research interest is fly ash in concrete, using of recycle material as an aggregate in concrete and durability of concrete.

INTRODUCTION

The evolution of the microstructure in cement-based materials during hydration involves a great number of variables. The kinetics of the chemical reactions between cement and water on the one hand and the complex physical transformations on the other hand, are subject of intensive research. The material properties can be predicted once a clear profile of the rule of these variables is known. Experimental methods, such as scanning electron microscopy, small angle X-ray scattering and mercury intrusion porosimetry [1-3] are widely used for this purpose. However, all these methods have their specific limitations. Several attempts based on a non-destructive method, the ultrasonic pulse velocity measurement, have been reported in the literature [4-6]. This technique was specially used in the monitoring of the cement hydration processes at the early age, eg from a few minutes up to one week after mixing.

Based on a basic principle of physics, the motion of any wave will be affected by the medium through which it travels [7]. The wave velocity depends on the stiffness and density of the material. The stiffness follows the development of the microstructural formation during cement hydration. Both the characteristics of the acoustic wave propagation and the hydration of the cement-based material permit to correlate the change in UPV and the development of the microstructure at early age.

The aim of this research was based on the experimental investigation and modelling study, to explore a new non-destructive method for monitoring the development of the microstructure of cement-based materials at early age. In particular, to investigate the relationship between microstructural parameters and the evolution of the UPV in hardening cement paste. The "bridging volume" concept [8, 9] was used to model the development of the microstructure. The influence of different isothermal curing temperatures and water/cement ratios was studied. From the experimental data, a model that describes the evolution of the UPV for cement paste was deduced. This model was extended to concrete, considering it as a two-phase system.

MODELLING OF HYDRATION AND MICROSTRUCTURAL DEVELOPMENT

According to the HYMOSTRUC model [10, 11] for cement hydration, the connectivity between hydrates has been labeled as "bridge volume". In this model, the cement particles are randomly distributed in the paste and the hydrating cement grains are simulated as gradually expanding spheres. As hydration takes place, the cement grains gradually dissolve and a porous shell of hydration products is formed around the grains. The hydrates around the cement grains first cause the formation of small isolated clusters. Bigger clusters are formed when small cement particles become embedded in the outer shell of other particles, which results in a growth of these particles. The contacts between clusters are formed by cement particles that are not embedded completely in a cluster. These particles are considered as "bridging" particles. As the hydration progresses, the growing particles become more and more connected because of the "bridging" process. Thus, at a certain degree of hydration α, the hydrating cement paste consists of an anhydrous cement phase ($p_{(\alpha)}$), cluster volume ($q_{(\alpha)}$), bridge volume ($r_{(\alpha)}$) and free capillary pore water and air ($s_{(\alpha)}$). These volume phases are illustrated in Figure 1 and are calculated as follows:

$$V_{0,paste} = V_{0,cem} + V_{0,wat} + V_{0,air} \qquad (1)$$

$$p_{(a)} = \frac{(1-a)V_{0,cem}}{V_{0,paste}} \tag{2}$$

$$q_{(\alpha)} = \frac{V_{clus(\alpha)}}{V_{0,paste}} \tag{3}$$

$$r_{(\alpha)} = \frac{V_{br(\alpha)}}{V_{0,paste}} \tag{4}$$

$$s_{(\alpha)} = \frac{(1-a)V_{0,(wat+air)}}{V_{0,paste}} \tag{5}$$

where, $V_{0,paste}$, $V_{0,cem}$, $V_{0,wat}$ and $V_{0,air}$ are the volume fractions of the cement paste, the initial volume of the anhydrous cement, of the capillary pore water and air respectively.

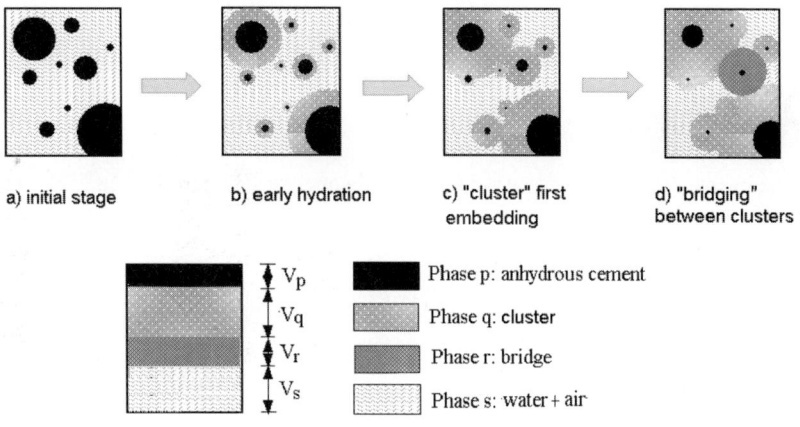

a) initial stage   b) early hydration   c) "cluster" first embedding   d) "bridging" between clusters

V_p Phase p: anhydrous cement
V_q Phase q: cluster
V_r Phase r: bridge
V_s Phase s: water + air

four phases cement paste component

Figure 1   Formation of the microstructure in hardening cement paste (after [9])

The volume of the clusters V_{clus} and of the bridging particles V_{br} can be determined directly with the HYMOSTRUC model. The total cluster volume (V_{clus}) must be equal to the total volume of the remaining cores of anhydrous cement grains plus the volume of the cement gel:

$$V_{clus(\alpha)} = (1-\alpha)V_{0,cem} + \alpha \cdot v \cdot V_{0,cem} \tag{6}$$

where α: the degree of hydration.
v: the ratio of the volume of the cement gel and the anhydrous cement from which the hydration products are formed.

As already mentioned, the bridge volume V_{br} is the volume of the particles that are able to act as a bridge to neighboring clusters. In order to act as a bridge, the bridge particles should exceed a certain minimum size that is related to the thickness of the outer shell of the free particle. The minimum particle size can be expressed in a bridging criterion. This criterion states that the diameter D of the bridging particle must be b times the thickness S of the outer shell of a free particle (Figure 2). The factor b is called the bridging-length factor. Details about how this bridging length factor affects the bridge volume were discussed in [9].

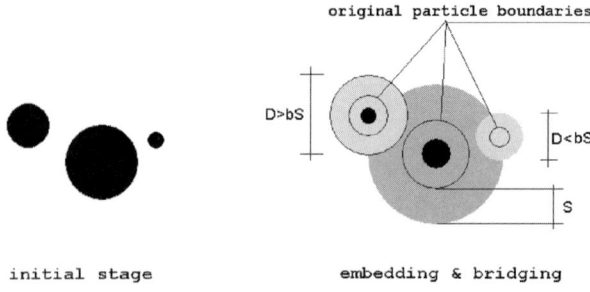

Figure 2   The criterion of the bridge particles [9]

MATERIALS AND METHODS

The experimental setup both for cement paste and concrete was described in [12]. The main features of this setup are:

1.   A 54 kHz transducer and a $150 \times 150 \times 200$ mm^3 steel mould were used according to the sample of concrete and size of particles to avoid the attenuation of the ultrasonic pulse signal.

2.   The ultrasonic transducers are coupled directly to the specimen through a plastic membrane, (Figure 3).

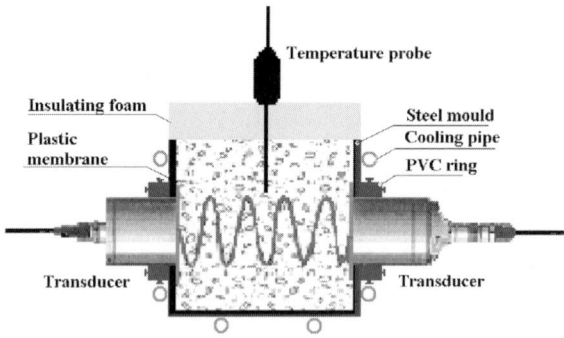

Figure 3   Experimental set-up for monitoring UPV in young concrete (schematic)

3.   The contact pressure of the transducer is adjusted by four springs to guarantee a good contact with the specimen.

4.   The whole system is temperature controlled by a cooling/heating system controlled by a cryostat.

5.   Computer controls the experiments and automatically records the hydration time and ultrasonic pulse transition time.

A series of experiments was performed both on cement paste and concrete. Table 1 contains the main data of the tested specimens. The main variables considered are water/cement ratio (w/c), aggregate content and different isothermal curing temperatures. The maximum aggregate size is 16 mm. No special treatment was imposed to remove the extra air bubbles from the specimens except for vibration. The measurements started about 15 minutes after mixing.

Table 1 Mix proportions

SPECIMEN NO.		CEMENT TYPE	CEMENT CONTENT [kg/m³]	AGGREGATE CONTENT [kg/m³]	W/C [-]	TEMPERATURE [°C]	CURING AGE [days]
Concrete	PCA16350-40	CEM I/32.5	350	1942	0.40	20, 30,40	7
	PCA16350-45	CEM I/32.5	350	1884	0.45	20, 30,40	7
	PCA16350-55	CEM I/32.5	350	1792	0.55	20, 30,40	7
Paste	PCP-40	CEM I/32.5			0.40	20, 30,40	3
	PCP-45	CEM I/32.5			0.45	20, 30,40	3
	PCP-55	CEM I/32.5			0.55	20, 30,40	3

RESULTS AND DISCUSSIONS

From the comparison of the cement pastes and the concretes, a clear pattern of the evolution of the UPV can be found for the influence of curing temperature, water/cement ratio and age.

The influence of curing temperature during the first 24 hours is shown in the case of w/c ratio 0.4 in Figure 4a for cement paste and in Figure 5a for concrete.

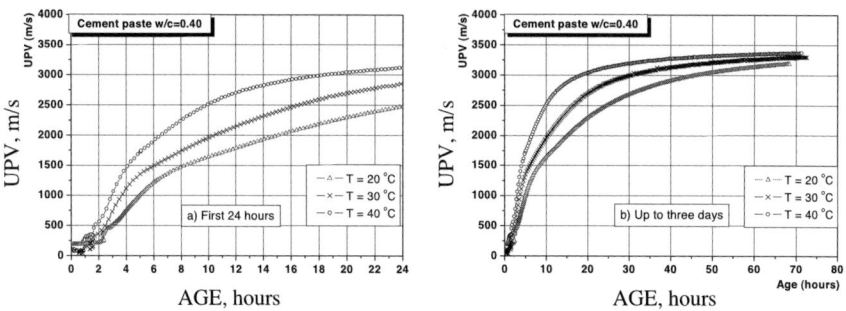

Figure 4 Influence of curing temperature on UPV of cement paste

In the first 2 to 4 hours relatively low values of the pulse velocity were found for all specimens. As already explained in [13], the ultrasonic signals in the fresh mixtures and strongly attenuated by the reflection of huge number of tiny air bubbles.

The time from casting to the point when the velocity increases depends on the temperature and the water/cement ratio. This period is generally referred to as the "dormant stage". In image-based simulations of the hydrating cement paste, this point was found to correspond to the percolation threshold of the solid phased.

The influence of curing temperature on UPV is shown in Figure 4b for cement paste and in Figure 5b for concrete. Three principal features were individuated. Firstly, during the first 40 hours, the pulse velocity rapidly increases in all specimens. Secondly, after 40 hours curing, the pulse velocity reaches a certain point and increases only slightly thereafter. Moreover, the higher the isothermal curing temperature, the quicker the increase in pulse velocity during the first 24 hours. In other words, the UPV takes a shorter time to reach a plateau value for higher temperatures. After the plateau is reached, the temperature shows less influence on the pulse velocity.

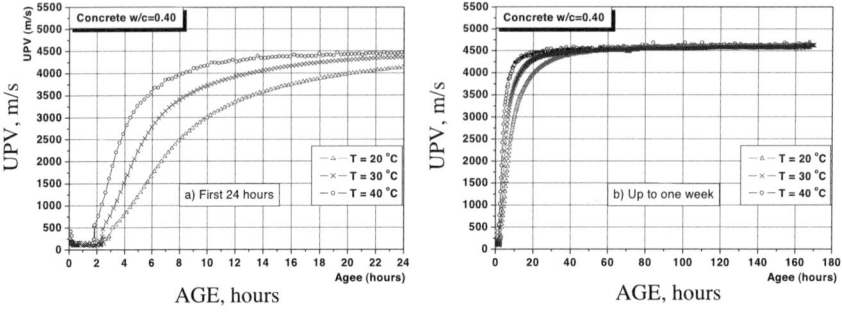

Figure 5   Influence of curing temperature on UPV of concrete

The effect of water/cement ratio on the ultrasonic properties was examined for mixtures with a water/cement ratio of 0.4, 0.45 and 0.55. Results are shown in Figure 6. The same shape was found for both cement paste and concrete. Mixtures with a higher water/cement ratio show lower values of the pulse velocity. A quicker increase in pulse velocity is displayed in concrete than in cement paste. he pulse velocity was almost 8~10% higher in concrete with w/c ratio of 0.45 than in w/c ratio 0.55. These can be related to the effect of the microstructure of cement paste and to the different amount of aggregate in each concrete mixture.

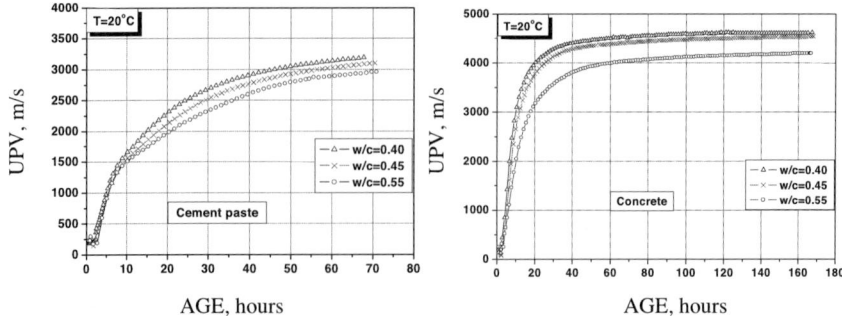

Figure 6   Influence of the water/cement ratio on UPV of cement paste and concrete

MODELLING THE UPV IN CONCRETE

Correlating the UPV and the Increase of Bridge Volume During Cement Hydration

In cement paste, the degree of hydration α and the bridge volume $r_{(\alpha)}$ were calculated with the HYMOSTRUC model. Figure 7 shows the calculated degree of hydration as a function of time whereas Figure 8 presents the bridge volume as a function of the degree of hydration.

It is noticed that the simulation model makes allowance for the fact that at higher curing temperature a dense hydration product is formed [8]. This phenomenon is modeled with a temperature dependent expansion factor v (see Eq 6).   This results in less interparticle contacts and higher capillary porosity and, hence, in a lower bridge volume.

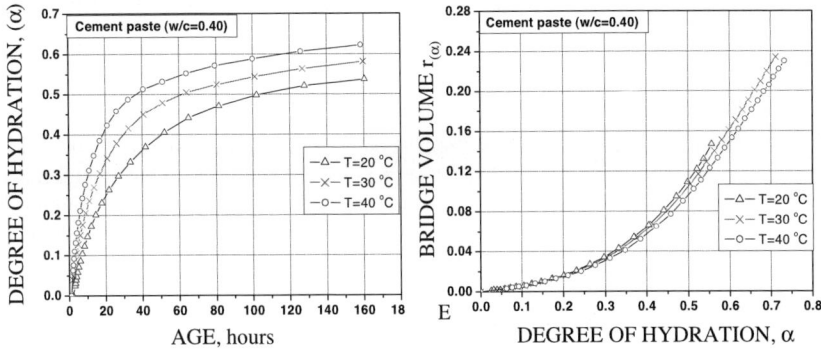

Figure 7   Evolution of degree of hydration

Figure 8   Degree of hydration vs. bridge volume

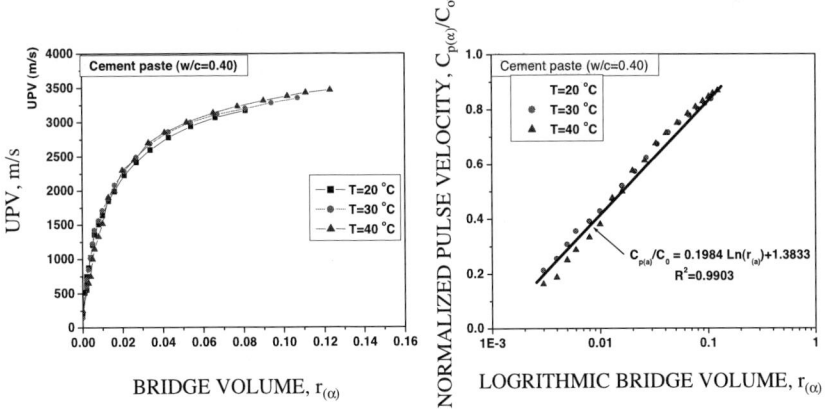

Figure 9 UPV as a function of the bridging volume in the cement paste for three different temperatures sample with w/c = 0.4

It must be noticed that the bridge volume is the volume of the particles that are able to act as bridge to neighboring clusters.   Hence this microstructural parameter indicates the connectivity of solid phases. In Figure 9 the normalized pulse velocity $C_{p(\alpha)}/C_o$ is plotted against the bridge volume parameter $r_{(\alpha)}$ at different temperatures for the 0.4 w/c ratio sample. $C_o = 36\ 00$ m/s is the longitudinal ultrasonic velocity in the dense cement paste calculated according to [7].

A logarithmic relationship between $C_{p(\alpha)}/C_o$ and $r_{(\alpha)}$ was found, as shown in Figure 9. The coefficient of determination R^2 is equal to 0.990. From this relationship, it results that the $C_{p(\alpha)}/C_o$ is independent of the curing temperature. Obviously, the connectivity of the solid phase is a dominating parameter that determines the UPV in the cement paste. Compared to Figure 8, higher curing temperature lead to a lower bridge volume, hence, a more porous microstructure. From acoustic test, a lower bridge volume resulted in a lower UPV. Thus, the UPV in the cement paste, $C_{p(\alpha)}$, can be derived mathematically according to the increase of the bridge volume.

Pulse Wave in A Homogenous Multiphase body

In order to extend the model of the UPV from cement paste to the concrete, Jones's [14] multiphase theory was utilized. Once consider a macroscopically homogeneous multiphase body subjected to a pulse wave, the transit time of the front of the pulse should be approximately equal to the sum of the transit time in each phase. Thus, the pulse velocity C through this multiphase body can be described by:

$$\frac{1}{C} = \sum_{i=1}^{n} \frac{V_i}{V \times C_i} = \sum_{i=1}^{n} \frac{N_i}{C_i} \qquad (8)$$

In this expression, C_i is the longitudinal pulse velocity through each phase and N_i is the volume percentage of each phase. Concrete can be considered as a two-phase medium consisting of a cement paste phase and an aggregate phase [Figure 10].

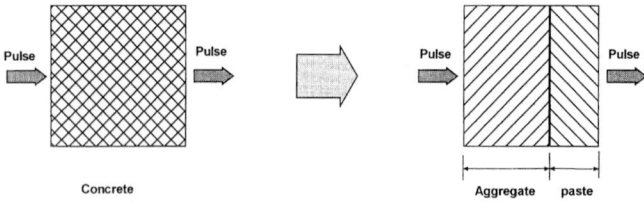

Figure 10   Concrete considered as a two-phase model

Consequently, according to Eq.8, the UPV of concrete $C_{con(\alpha)}$ as a function of the evolution of pulse velocity $C_{p(\alpha)}$ in cement paste can be derived:

$$\frac{1}{C_{Con(a)}} = \frac{V_p}{C_{p(\alpha)}} + \frac{V_{agg}}{C_{agg}} \qquad (9)$$

where, V_p and V_{agg} are the volume percentages of cement paste and aggregate in the concrete. $C_{agg} = 5000$ m/s is the longitudinal ultrasonic velocity in the dense aggregate; calculated from the mechanical properties of rock [7].

The simulation results of the evolution of UPV in concrete are shown in Figure 11 together with the experimental test. The simulation results are in excellent agreement with the experimental results.

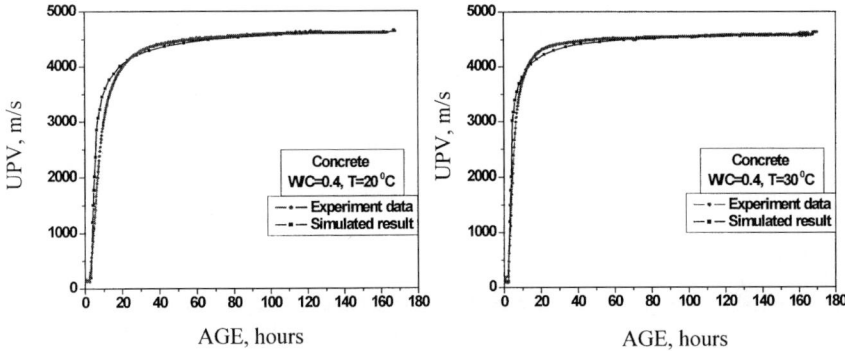

Figure 11 Simulated and measured pulse velocity as function of time

CONCLUSIONS

Based on the numerical simulation model HYMOSTRUC, a four-phase model for the hydration of cement paste was described, in which the cement paste consisted of anhydrous cement, cluster volume, bridge volume and free capillary pore water. The bridge volume was modeled as the volume of the particles that are able to act as a bridge to neighboring clusters. It is also a parameter that indicates the connectivity of the solid phase. From the simulation results, lower bridge volume is found if curing take place at higher temperature. Hence, associated with this, a more porous microstructure is obtained. This has been demonstrated by the measuring of the evolution of UPV. Both experimental and simulation results showed that a lower bridge volume results in a low pulse velocity. A logarithmic relationship between bridge volume and UPV was found, which is independent from curing temperature. This indicated that the connectivity of the solid phase dominates the speed of the ultrasonic pulse propagating in hardening cement paste. Based on this relationship, a two-phase model for concrete was proposed. The simulation results were in good agreement with the measurement. This result indicates that the UPV measurements can successfully be used as a non-destructive technique for monitoring the evolution of the microstructure in hardening cement based materials.

ACKNOWLEDGEMENTS

The authors wish to thank Mr E M Horeweg for his expert assistance in conducting the experiments and solving numerous technical problems. Useful discussions with Mr P Lura are gratefully acknowledged. The research was financially supported by the Dutch Technology Foundation (STW), which is gratefully acknowledged.

REFERENCES

1.   WILLIS, K.L., ABELL, A.B., LANGE, D.A., Image-based characterization of cement pore structure using wood's metal intrusion, Cement and Concrete Research, Vol 28, No 12, 1998, pp 1695-1705.

2.   KRIECHBAUM, M., DEGOVICS, G., LAGGNER, P., TRITHART, J., Investigation on cement pastes by small-angle X-ray scattering and BET: the relevance of fractal geometry, Advances in Cement Research, Vol 6, No 23, July 1994, pp 93-101.

3.   JI, X., CHAN, S.Y.N., FENG, N., Fractal model for simulating the space-filling process of cement hydrates and fractal dimensions of pore structure of cement-based materials, Cement and Concrete Research, Vol 27, No 11, 1997, pp1691-1699.

4.   KEATING, J.,HANNANT, D.J., Correlation between cube strength, ultrasonic pulse velocity and volume change for oil well cement slurries, Cement and Concrete Research, Vol 19, 1989, pp 715-726.

5.   SAYER, C.M., DAHLIN, A,. Propagation of ultrasound through hydrating cement parts at early times, Advance Cement Based Materials, Vol 1, 1993, pp 12-21.

6.   BOUMIZ, B.,VERNET, C., COHEN, T. F., Mechanical Properties of Cement Pastes and Mortars, Advance Cement Based Materials, Vol 3, 1996, pp 94-106.

7.   KRAUKRAMER, J., KRAUKRAMER, H., Ultrasonic testing of materials. 4th, Springer-Verlag, 1990.

8.   BREUGEL, K. van., ARK, I.A., Predication of the evolution of strength of hardening cement-based systems as a function of microstructural development. 2nd International Symposium on Cement and Concrete Technology in the 2000s, 6-10 September, Istanbul, Turkey. Vol 1, 2000, pp 208-217.

9.   LOKHORST, S.J., "Deformational Behavior of Concrete Influenced by Hydration-related Changes of the Microstructures". Research Report, Delft University of Technology, 1998.

10. BREUGEL, K. van., Simulation of hydration and formation of structure in hardening cement-based materials, Dissertation, Delft University of Technology, The Netherlands, 1991.

11. KOENDERS, E.A.B., Simulation of volume changes in hardening cement-based materials, Dissertation, Delft University of Technology, The Netherlands, 1997.

12. YE, G., BREUGEL, K. van., FRAAIJ, A.L.A., Experimental study on ultrasonic pulse velocity evaluation of the microstructure of cementitious material at early age, 2001, submitted to HERON.

13. POVEY, M.J.M., Ultrasonic techniques for fluids characterization. Academic Press. 1997.

14. JONES, R., Non-Destructive Testing of Concrete. Cambridge Press, 1962.

ELECTROCHEMICAL SYSTEMS FOR REPAIR OF REINFORCED CONCRETE STRUCTURES

N Davison

A C Roberts

J M Taylor

Fosroc International Ltd

United Kingdom

ABSTRACT. The corrosion of reinforcing steel in concrete structures is a global problem. A number of highly technical repair materials are now available to repair concrete structures suffering steel corrosion damage. Often, however, an incipient corrosion problem remains in untreated contaminated areas that may lead to subsequent failure. This paper describes a number of electrochemical systems, which can be applied to corrosion susceptible structures, which focus on addressing the fundamental electrochemical nature of corrosion. A system for enhancing longevity of patch repairs to reinforced concrete is discussed with particular reference to its ability to overcome incipient anode formation. A cost effective extension of this technology to globally treat corroding reinforced concrete structures will also be discussed. An example of the application of galvanic corrosion protection technology to reinforced concrete piles in a marine environment will be presented. Finally the use of an Impressed Current Cathodic Protection (ICCP) system using discrete anodes will be discussed.

Keywords: Electrochemical, Chloride, Incipient anode, Sacrificial zinc, Discrete anode, Galvanic, ICCP, Marine environment.

Dr Nigel Davison is a technology manager within the corrosion control section of Fosroc Group Development. Gained a Ph.D. in inorganic/physical chemistry from Aston University. He joined Fosroc in 1996, having previously worked in product development for Kodak Ltd, Pozament Ltd and 3M(UK) Ltd.

Mr Adrian C Roberts is a chartered chemist focusing on the product development of electrochemical solutions for the repair of reinforced concrete at Fosroc International Ltd where he has worked since 1991.

Mr John M Taylor is a development chemist within the corrosion control section of Fosroc International ltd focusing on electrochemical solutions for the repair of reinforced concrete. He has worked for Fosroc since 1991

INTRODUCTION

The majority of reinforced concrete structures do not suffer from rebar corrosion, despite the fact that the concrete/steel composite is inherently inhomogeneous. The highly alkaline nature of concrete leads to the formation of a passive oxide layer on the surface of the reinforcing steel that reduces corrosion to negligible levels. However, the finite permeability of concrete can allow the ingress of chemical agents that lead to a breakdown in the protective passive layer and subsequent corrosion of reinforcing steel [1]. The two most commonly encountered processes leading to rebar corrosion are (a) carbonation in which a pH reduction in the concrete pore solution is induced by the action of carbon dioxide and (b) chloride attack . These processes lead to breakdown of the passive oxide layer and subsequent formation of expansive corrosion products, which can lead to cracking and spalling of the concrete surface.

Chloride attack on reinforcing steel can be as a result of chloride salts which have diffused in from the external surface, for example, de-icing salts, or from cast-in chloride salts used to promote set-time. Above a certain threshold chloride level [2], steel corrosion is initiated. A variety of solutions to the problem of reinforcement corrosion exist which can be used to treat structures in the initiation/propagation phases or following failure [3]. In recent years, improved understanding of the corrosion process has allowed development of electrochemical techniques, which attempt to modify the chemistry involved. This paper details some technical advances in the use of galvanic/ICCP technology for protecting reinforced concrete structures, with examples of actual installations and supporting data obtained.

DISCRETE GALVANIC ANODES IN CONCRETE PATCH REPAIR

A very common method of repairing concrete spalling on structures due to chloride induced rebar corrosion, is reinstatement with a low permeability repair mortar. This involves removal of loose concrete and further breakout to clean steel, prior to mortar application. As the corrosion previously occurring in the repair patch has been eliminated, its influence in effectively cathodically protecting the surrounding steel is also lost. However, this repair technique does not guarantee removal of chloride bearing concrete, which may remain in areas adjacent to the repair. Thus, new electrochemical corrosion cells may be set up between steel in the fresh repair (0% chloride) and the adjacent chloride contaminated concrete ,which can ultimately lead to failure at the periphery of a repair. This is commonly referred to as the 'incipient anode effect' [4](or 'ring anode') - an example of this problem encountered on bridge columns in the UK is shown in Figure 1.

In order to avoid triggering this problem of 'incipient anode' formation around the repair zones, it is desirable to incorporate some form of intentional 'cathodic prevention', which can be accomplished by incorporating sacrificial anodes at the periphery of repair patches. One sacrificial anode design comprises a sacrificial zinc alloy, surrounded by a specifically formulated mortar to optimize lifetime [5]. The mortar facilitates zinc dissolution by preventing the formation of an interfering passive layer, which allows the less noble zinc to corrode and sacrificially protect reinforcing steel to which it is attached, thus countering the formation of anodic sites outside the periphery of the patch repair.

Figure 1   Corrosion around repair patch resulting from formation of incipient anodes

Site Trial of discrete sacrificial anodes

A patch repair site trial using sacrificial anodes was carried out on a reinforced concrete structural crossbeam on a road bridge in the UK. The bridge was suffering from considerable reinforcement corrosion caused by chloride contamination from de-icing salts leaking through the bridge deck joints. The area selected for repair had experienced concrete cracking and spalling, chloride levels were assessed and found to be in the range of 0.8 to 2.2% w/cement.

All cracked and spalled concrete was initially removed from the area and additional breakout was carried out until non-corroding steel was found, as per standard concrete repair practice. Twelve sacrificial anode units were installed into the trial area, distributed around the perimeter of the patch to maximize their effectiveness in preventing incipient anode formation and each unit was monitored using an externally sited system. Significant levels of current output have been generated (Table 1) for >750 days indicating a sustained level of current has been passed to the steel.

Table 1   Current output results from UK bridge trial

NO. OF DAYS	A1 μA	A2 μA	A3 μA	A4 μA	A5 μA	A6 μA	A7 μA	A8 μA	A9 μA	A10 μA	A11 μA	A12 μA
41	266	261	405	145	225	195	220	186	179	325	-	-
50	261	240	395	137	312	186	216	183	183	302	-	-
112	352	285	345	230	322	279	257	295	337	236	-	-
218	245	158	200	115	177	139	151	150	231	116	-	-
323	145	95	130	70	121	88	115	77	158	72	-	-
497	333	231	308	173	270	243	275	242	345	198	292	177
616	222	202	199	129	162	203	206	149	212	153	198	153
785	318	295	422	186	228	351	325	227	296	387	279	289

Current densities calculated by assuming an extension of the protected area to 400 mm beyond the edge of the repair, indicate values of between 0.5-2 mA/sqm steel . This level of current density compares favorably with that proposed by Pedeferri [6] for cathodic prevention of steel in reinforced concrete. It is also interesting to note that a current density of 0.4 mA/sqm has been shown to counter the initiation of steel corrosion in reinforced concrete samples containing 2% chloride w/c [7].

DISCRETE GALVANIC ANODES IN GLOBAL CORROSION PROTECTION

As an extension of the discrete anode system discussed in section (A), further developments based on similar galvanic technology allow protection of structures prior to the requirement for patch repairs. In this case, the enhanced sacrificial anode core/active mortar composite is formed into a geometry, which facilitates installation onto a reinforced steel concrete structure, to allow galvanic protection outside the region of conventional patch repairs. This system has been installed on a number of structures, suffering from chloride-induced reinforced concrete corrosion, and a typical example is discussed.

Evaluation of in-situ performance of discrete sacrificial anodes outside patch repairs

A trial of 'global' sacrificial anode protection was instigated on a reinforced concrete multi story car park, exhibiting clear signs of chloride-induced corrosion resulting in a significant degree of spalling. However, due to the global nature of the chlorides located throughout the deck, the structure required treatment additional to patch repair to extend its working life.

For the purposes of this evaluation an area of the deck was selected that had not shown any signs of spalling but had significant levels of chloride contamination (between 1.7 and 2.8% chloride w/c) and half-cell potentials indicating steel corrosion. Twenty sacrificial units were installed in a square grid pattern along with six embedded Ag/AgCl reference electrodes.

Table 2 details the current outputs from each chain of sacrificial anode units, over a period of 250 days. Although current outputs appear to have stabilized, they are clearly responding to environmental conditions, indicating a degree of 'intelligent behaviour'.

Ignoring the beneficial effects, which may be achieved from the high early current outputs [8], the calculated current densities (Figure2) appear to be comparable to values that have been effective in impressed current cathodic protection installations.

Table 2   Individual current results obtained from car park

AGE/ DAYS	CHAIN 1 CURRENT/ µA	CHAIN 2 CURRENT/ µA	CHAIN 3 CURRENT/ µA	CHAIN 4 CURRENT/ µA
1	1445	1270	1683	1427
6	4800	3300	3300	4000
14	2660	2070	1950	2170
26	2825	2605	2345	2474
56	1630	1620	1396	1155
84	1755	1692	1453	1209
141	1245	931	720	655
251	1645	1125	1078	915

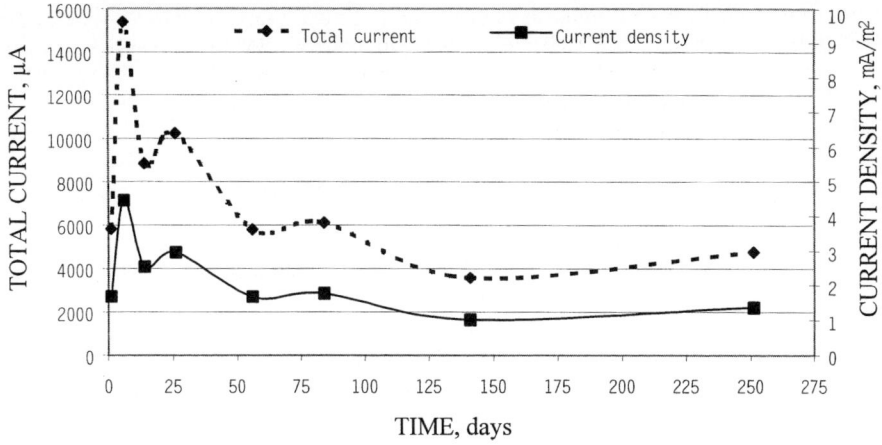

Figure 2   Total current and Current density results from Car park

Four-hour depolarisation results (Figure 3) are consistently positive indicating that the sacrificial anodes are polarising the steel to a significant level and thus offering cathodic protection. Depolarised steel potentials are steadily becoming noble with time, indicating that the applied galvanic current is facilitating the movement of the reinforcing steel to a more passive state.

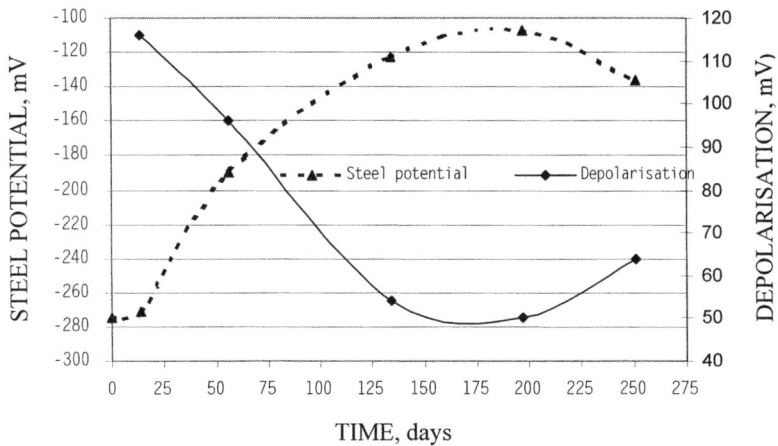

Figure 3   four-hour depolarisation and depolarised steel potential results vs time

GALVANIC PROTECTION IN THE MARINE ENVIRONMENT

Marine environments are aggressive to steel in reinforced concrete owing to the high concentration of chloride ions in seawater. However, not only does the chloride ion catalyze corrosion, but also its presence, amongst others, ensures that sea water is a good electrolyte. In addition, wet and dry tidal cycling is particularly aggressive and many structural problems occur in this zone.

One repair option for marine structures suffering rebar attack is to cathodically protect the steel by application of a DC current, through use of an external power supply (ICCP). However, cathodic protection using a galvanic system is also feasible and has indeed been effectively used on bridges in Florida for over 10 years [9].

In addition to the advantage of requiring no additional monitoring or control, the galvanic technique is compatible with pre-stressed steel commonly used in bridge piles, which may otherwise be susceptible to hydrogen embrittlement.

Site trial of galvanic system for use in marine environments

A trial of a galvanic pile jacket system was installed onto reinforced concrete columns on a Quay in the Channel Islands, suffering from chloride-induced corrosion. The system comprised a glass reinforced plastic (grp) jacket containing a high surface area zinc mesh, which was placed around two test piles.

The jackets were then in filled with a pumpable cement repair material which performed two tasks; repaired the broken out or spalled concrete and also electrolytically connected the zinc to the steel reinforcement in the host concrete. Electrical connections were made between the wire mesh, a bulk anode that was positioned such that it was immersed on every high tide, and the steel reinforcement to complete the galvanic cell. Note on only one of the piles was a bulk anode used (column two), column one relied solely upon the jacket.

The steel to zinc connections were made through a data-logging device to continually record current output from the zinc. In addition, Ag/AgCl reference electrodes were embedded in the host concrete and placed close to the steel inside the jacket for half-cell potential measurements. Further reference cells were installed in the adjacent untreated column for control data.

The reference electrode half-cell measurements from the control column showed large changes in steel potential through tidal wetting, indicating shifts in the corrosion activity of the reinforcing steel as the environment changes.

As the protection system is galvanic in nature, then the level of protection offered varies concomitantly with tidal movements. This effect is demonstrated in Figure 4, where the average currents for column 1 were compared against the total monthly time that the jacket was at least partially submerged.

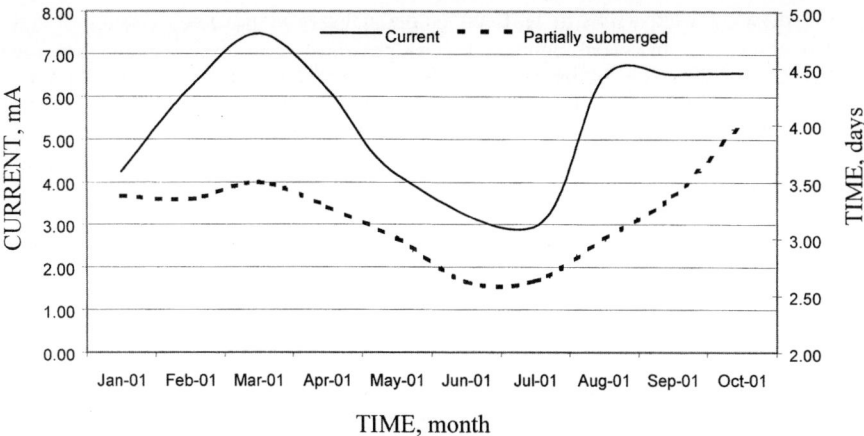

TIME, month

Figure 4   Average monthly current output Vs total time the jacket was at least partially submerged for column 1

Table 3 summarizes the mean current output from column 2 by month over a period of nineteen months. The sacrificial zinc consistently produces a protective current of ~10 mA/sqm steel within the pile jacket, which compares favorably with values associated with ICCP systems [6, 8], but with the additional advantage of maintaining highest current output when steel corrosion conditions are worst case.

Table 3   Mean current output by month for column 2

MONTH	MEAN CURRENT / MA
1	30.49
2	27.87
3	26.15
4	24.09
5	25.00
6	32.49
7	33.24
8	26.87
9	24.62
10	22.57
11	24.36
12	28.88
13	25.91
14	25.08
15	25.16
16	24.05
17	26.72
18	27.1
19	33.2

Figure 5 shows a comparison of half-cell potential between protected column 2 and the untreated column, it indicates significant steel polarisation of the treated column when compared against the control column.

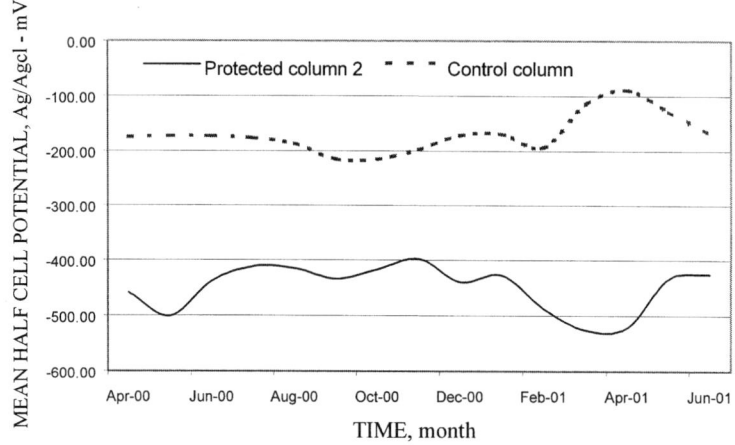

Figure 5   Mean half-cell potential of protected column vs control

Further evidence of the effectiveness of the galvanic cathodic protection system was gained from observation of depolarisation results taken from column two 24 hours after disconnecting the zinc from the steel. The polarisation and depolarisation data are very similar, both being between 200 and 250 mV throughout a seven month period.

DISCRETE ICCP ANODES

This technique involves the application of a low, direct current from a permanent anode to the reinforcing steel. Sufficient current is used (ranging from 0.25 to 25 mA/m^2 of steel) to maintain the reinforcing steel in a cathodic state. One of the features of the system is the permanent requirement for a DC power source and the requirement for regular monitoring and maintenance. Anodes for use in ICCP applications can take a variety of forms dependent on the type of structure to be protected and the practicalities of installation. Types of ICCP anodes include MMO coated titanium wire/ribbon , conductive paints/mortars and discrete anodes as discussed in this paper. These discrete anodes are formed using electrically conductive ceramic tubes of varying sizes (7-28 mm diameter x 75-600 mm length) depending upon application. The ceramic material is highly corrosion resistant to enable high current densities to be passed to protect reinforcing steel.

Discrete anodes are installed by encapsulating into pre-drilled holes using an acid resistant grout, which aids longevity. In general multiple anodes are electrically connected via titanium wire to a rectifier so that current densities can be adjusted to suit structure requirements. Each discrete anode unit is fitted with a tube to allow anodic gases to be vented to the surface of the concrete. The design and variable sizes of the discrete anodes enable protection of deeply buried steel or multiple layers where surface mounted anode systems may not provide sufficient protection. By reducing spacing between units it is also possible to provide full protection to localized areas with heavy reinforcement.

Site evaluation of discrete ICCP anodes

A trial application was undertaken on a deep basement in UAE suffering from heavy chloride contamination as a result of ingress of highly saline ground water through poor quality waterproofing. Discrete ICCP anodes were installed on a grid pattern at locations between two layers of reinforcing steel at spacings of 600 mm and 1000 mm. The extent of protection was then measured at various times using a copper/copper sulphate reference electrode to obtain surface half-cell potential measurements on an extensive grid. Table 4 records the steel depolarization observed at various distances from the discrete anodes, 100 mV depolarization criteria detailed in NACE 0290-90 is achieved at a distance of over 50 cm suggesting cathodic protection is obtained.

Table 4   24 hour depolarization results from UAE discrete anode trial

DISCRETE ANODE REFERENCE	DISTANCE FROM REFERENCE ELECTRODE TO ANODE IN CM	24 HOUR DEPOLARISATION RESULT IN MV
A1	40	172
	50	163
	60	69
	70	89
A2	20	195
	50	119
	60	80
	65	79
A3	25	193
	45	102
	60	136
A4	10	158
	45	123
	50	111

SUMMARY

A number of systems for the treatment of corroding reinforced concrete structures have been demonstrated. Technical evaluation of actual site installations indicates that these can be utilized in a variety of guises to meet various technical requirements.

Data from pile jacket, discrete galvanic anodes and discrete ICCP anodes indicates the capability to meet the widely accepted 100 mV steel depolarisation criterion identified in the draft European standard for cathodic protection of steel in concrete [10]. The level of current outputs measured, 0.5-10 mA/sqm for the galvanic systems, are in the ranges that would be expected for cathodic prevention/protection [6, 10] of reinforced concrete.

The evidence from the site installations discussed in this paper demonstrates the practicalities of applying galvanic protection and discrete ICCP anodes to reinforced concrete structures in a range of environments and the simplicity of monitoring the effectiveness of these systems.

REFERENCES

1. BAMFORTH, P.B (1996), Definition of exposure classes and concrete mix requirements for chloride contaminated environments, 4th Int. Symp. Corrosion of Reinforcement in concrete construction, Cambridge, UK, 176-188.

2. BUILDING RESEARCH ESTABLISHMENT DIGEST 264, The durability of steel in concrete: Part 2, Diagnosis and comment of corrosion-cracked concrete, Building Research Station, Garston, UK, 1982.

3. BROOMFIELD, J. P., Corrosion of steel in concrete, E and FN Spon, London, 1997.

4. EMMONS, P.H.; VAYSBURD, A.M. (1997), Corrosion protection in concrete repair: Myth and reality, Conc. Int., March, 47-56.

5. SERGI, G.; PAGE, C.L.(1999):Sacrificial anodes for cathodic protection or reinforcing steel around patch repairs applied to chloride contaminated concrete, Eurocorr99, Aachen, topic 10, paper no.12.

6. PEDEFERRI, P. (1996), Cathodic protection and cathodic prevention, Constr. and Build. Mat., Vol 10, No.5, 391-402.

7. BERTOLINI ET AL. (1998), Cathodic protection and cathodic prevention in concrete : principles and applications, J. App. Electrochemistry, 28, 1321-1331.

8. GLASS, G.K. ET AL (2001), Cathodic protection afforded by an intermittent current applied to reinforced concrete, Corr. Sci., 43, 6, 111-1131.

9. LENG, D.L. (2000),Zinc mesh cathodic protection systems, Mat. Perf., 39, 8, 28-33.

10. CEN STANDARD, PR EN 12696-1, Cathodic protection of steel in concrete, Part 1:Atmospherically exposed concrete."

TEES VIADUCT CHLORIDE EXTRACTION TRIAL

D A Kimberley
Scott Wilson
United Kingdom

ABSTRACT. As part of the A19 DBFO scheme Autolink Concessionaires (A19) Ltd, a consortium of Amey and Sir Robert McAlpine are responsible for the operation and maintenance of the A19 Trunk Road from Dishforth to Tyne Tunnel. Scott Wilson as their appointed designers have trialed Chloride Extraction as a possible means of repair to the Tees Viaduct Substructures. The substructure of the bridge exhibits deterioration resulting from leakage of the deck joints and subsequent chloride ingress through the cover concrete. Repairs have been ongoing for over a decade and have included the complete demolition and replacement of some pier bents. The trial was carried out to prove the effectiveness of the technique to remove chloride ion to a target threshold level and to examine the risk associated with the potential side effect of generating Alkali Silica Reaction.

Keywords: Chloride extraction, Concrete repair, Deterioration, Trial, Alkali silica reaction.

Mr D Kimberley, is a Chartered Engineer and a Principal Engineer within the bridge design and maintenance division of Scott Wilson in Chesterfield. In addition to being design team leader for refurbishment works on the A19(T) DBFO he is responsible for project management of bridge inspection and assessment projects and also remains a key member of the new build design team.

INTRODUCTION

Like many reinforced concrete structures on the UK's Highway Network the Tees Viaduct substructures exhibit considerable premature deterioration as a result of chloride ion ingress from the seasonal application of de-icing salts on the road above. Refurbishment of the 1.8km long viaduct, which on average carries in excess of 40,000 vehicles in each direction each day along the A19 Trunk Road over the River Tees, is a major part of the A19 Design Build Finance and Operate (DBFO) Contract. Since February 1997 operation and maintenance of the A19(T) from Dishforth to Tyne Tunnel has been managed on behalf of the Highways Agency (HA) by Autolink Concessionaires (A19) Ltd, a consortium of Amey and Sir Robert McAlpine. Appointed as principal designer to the concessionaire Scott Wilson has been responsible for the design of the refurbishment works, which includes demolition of certain piers and repair of others.

The process of deterioration of reinforced concrete due to corrosion of the reinforcement through chloride ion ingress has been well documented over the years by others [1], [2]. Repair has traditionally involved removal of contaminated concrete and replacement with proprietary repair concrete. Whilst this method has proved successful over the years, it carries a high cost premium and can in many cases require temporary propping to the structure and extensive traffic management with disruption to road users. It is therefore not surprising that in recent years structure owners and contractors have been looking for alternative, less disruptive repair methods. A commonly used alternative method is Cathodic Protection (CP) which in particular has been used extensively over the last 10 to 15 years at Midland Links [3]. Despite the considerable cost benefit of using CP instead of concrete removal the method has one main disadvantage in that it is only effective as long as the system is functioning correctly. This leads to long term maintenance cost associated with monitoring the system and replacement of deteriorated items which may even include the anode itself at some time.

A second alternative method of treatment is Chloride Extraction (CE) often referred to as Desalination. This is also an electrochemical treatment but one which is applied over a short period of time, typically six to eight weeks. This method does not have a significant track record of use in the UK other than on multi storey car parks but it has been used more extensively in the United States. In the UK the method has mainly been the subject of trials and it has been used more recently for the removal of chloride ion from the cover concrete of piers to bridges around Junctions 38 and 39 of the M6 [4], [5]. Although in general the method has been proven to successfully reduce the chloride ion content in concrete it does theoretically have a number of potential side effects. Adhering to maximum voltage recommendations, which have been developed, can control most of the side effects relating to changes in the mechanical properties of the concrete [6]. Although, the same research concluded there is a risk of a reduction in bond strength due to dissolution of corrosion product. This in particular gives cause for concern where the reinforcement comprises plain round bars. As a consequence use of CE on structures with plain round bars is generally restricted to non-bond sensitive areas. A further potentially serious problem with using CE, which is more difficult to deal with, is the risk of generating Alkali Silica Reaction (ASR) when the treatment is applied to concrete, containing reactive aggregates. It has been shown in the laboratory [7] that the increased alkalinity generated locally at the reinforcement can be sufficient to trigger the reaction. Petrographic analysis carried out previously on core samples indicates that the concrete at Tees Viaduct has a significant content of reactive aggregates in particular Chert and Siliceous Limestone.

To investigate whether it is possible to carry out CE effectively, in terms of chloride ion content reduction and also without generating ASR a further trial of the technique was initiated.

TRIAL SPECIFICATION

When any product or system undergoes a trial it is essential to first establish the aims and objectives. In this case they were as follows:

1. To reduce the chloride ion content at the depth of the reinforcement to an acceptable threshold level. A chloride ion threshold level of 0.3% was chosen for the trial, a typical figure deemed to give a low risk of initiating further corrosion.
2. To replicate as close as possible the methodology for the treatment application and monitoring of results which would be employed in any subsequent full scale application to ensure that the trial is truly representative.
3. To carry out the treatment effectively and in a controlled manner to ensure ASR is not generated unnecessarily.
4. To ensure that the trial would be comprehensive, sufficient for evidence to be gained to enable un-qualified conclusions to be made.
5. To carry out the trial within an allocated budget and timescale.

Bearing in mind the aim and objectives listed above the trial specification was written to include the following four stages:

Stage 1 Testing in Advance of the Treatment

a. Petrographic analysis.
b. Chloride ion content.
c. Half-Cell potential.
d. Cover survey.

Stage 2 Application of the Treatment and Regular Monitoring

e. Chloride ion content, measured at approximately 2 week intervals during the treatment.
f. The current output to each trial area was continually monitored by the computerised monitoring system.

Stage 3 Laboratory Analysis of Concrete Samples to Assess the Degree of ASR Development

g. Cores were taken from each trial area at 16 weeks after treatment and 1 year after treatment. At both stages the cores were initially stored for 3 months in a laboratory, in conditions, which would accelerate any potential ASR reaction, i.e. high temperature (38°C) and high humidity. All cores were then thinly sliced and subjected to a Petrographic Analysis.

TRIAL ARRANGEMENT AND SPECIFICATION

The trial was carried out to the west column of Pier N14, with an arrangement of four separate trial areas as shown in Figure 1. This site was chosen for the following reasons:

1. The level of chloride contamination on the column was typical of the contaminated concrete on the viaduct in general i.e. in excess of 1% measured by mass of cement.
2. The extent of delaminated and spalled concrete on the column needing to be repaired in advance of the treatment was small, approximately $2m^2$.
3. Pier N 14 is a pier programmed for demolition and replacement in the future therefore the consequences of ASR if generated would not be as significant.
4. Accessibility to the pier was generally good.

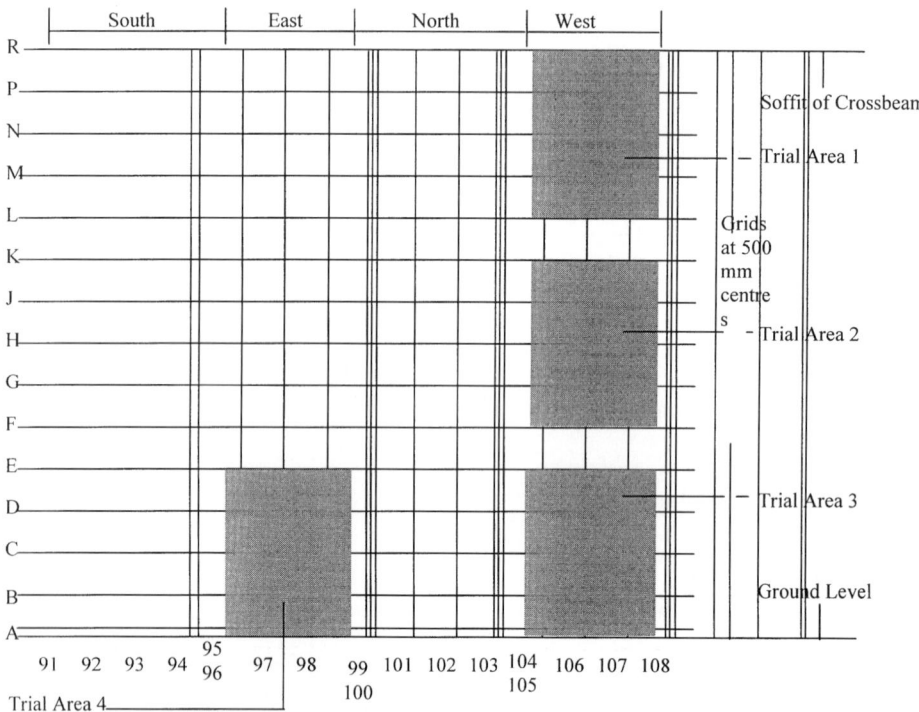

DEVELOPED ELEVATION OF PIER N14 WEST COLUMN

Figure 1   Arrangement of the trial areas

Trial Area 1 received chloride extraction treatment with a calcium hydroxide electrolyte in accordance with the Norcure™ Specification.

Trial Area 2 received the same treatment as Area 1 but upon completion was impregnated with silane. This replicated the proposed coating treatment should the trial have been successful.

Trial Area 3 received chloride extraction treatment with a lithium borate electrolyte. In this case lithium borate was used in the electrolyte because of its known properties in restricting the development of ASR.

Trial Area 4 was treated using a corrosion inhibitor but this is not the subject of this paper.

Stage 1 Testing in Advance of Treatment

Table 1 indicates the chloride ion content recorded at each node position from dust samples taken from within the 25mm to 50mm depth range.

Table 1   Pre Treatment chloride ion content test results at 25mm to 50mm depth range (measured as percentage chloride ion by mass of cement)

TRIAL AREA	GRID REF	GRID REF			
		105	106	107	108
1 & 2	R	1.34	0.88	1.26	0.94
	P	1.04	1.44	0.88	1.56
	N	0.3	0.99	1.2	0.74
	M	1.19	1.45	0.69	1.26
	L	1.61	0.69	1.08	1.13
	K	1.01	0.88	1.28	0.81
	J	1.04	1.4	1.82	0.89
	H	1.56	1.38	1.87	1.25
	G	1.01	1.51	1.71	1.21
3	E	1.43	1.63	1.53	0.83
	D	0.86	1.53	1.56	0.58
	C	0.97	0.8	1.55	0.77
	B	1.48	1.69	1.32	0.49
	A	1.27	1.58	0.72	0.33
Average		1.16			

The initial chloride content measurements were reasonably consistent with most being in the range 0.8% to 1.6%. This high level of contamination combined with the visual evidence, which included some areas of spalled and delaminated concrete, suggested that ongoing corrosion of the reinforcement was likely. This level of contamination was typical of other contaminated areas of concrete at the viaduct in general, which was an important consideration in ensuring that the trial would be representative.

Not surprisingly, the Half-Cell Potential readings also indicated a high risk of corrosion activity in certain areas. Generally the readings fell within the range –100mV to –350mV with isolated readings at around –450mV. Unlike many structures suffering premature deterioration the concrete cover measurements were generally found to be satisfactory with almost all measurements being within a range of 40mm to 65mm.

However a petrographic analysis of thin sections taken from 70mm diameter cores cut from each trial area concluded that the cement paste exhibited a moderate to high microporisity with well dispersed microcracks indicative of drying shrinkage effects. It was also known from previous testing work that the gravel aggregate used in the concrete mix was generally of high permeability. All of this evidence goes some way to explaining the rapid ingress of chloride ion into the concrete irrespective of the generally good cover levels since the bridge was opened to traffic in 1974. The petrographic analysis also confirmed the presence of reactive aggregates although no evidence of reaction was observed in any of the samples tested.

Stage 2 Application of Treatment and Regular Monitoring

Following installation of the temporary anode and the cassette shutters, shown in Figure 2 the system was activated on the 17th February 2000.

Figure 2   Cassette shutters which contain titanium mesh anode and electrolyte

In the early stages of treatment the first thing that became apparent was the difference in current being applied to the three trial areas. After only a few weeks of treatment Area 3 had received almost twice the number of amp-hours per square metre of reinforcement when compared to Areas 1 or 2.

The initial target of 1000Ahrs for Area 3 was achieved in a little over four weeks, at the same time Areas 1 and 2 were at approximately 450Ahrs of treatment. The rate of treatment to the three areas is shown in Figure 3.

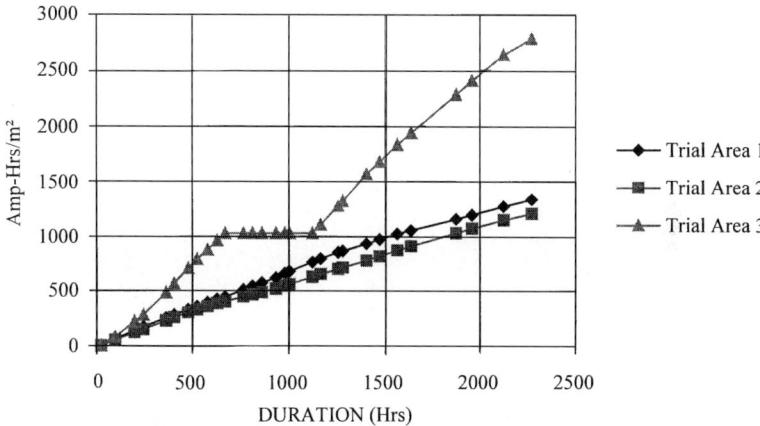

Figure 3   Summary Ahr chart

The contractors own electrochemical experts confirmed that this could have been expected as the use of a lithium borate electrolyte helps to maintain low resistivity. In comparison the resistivity in the areas treated with calcium hydroxide tends to increase during treatment, although Figure 3 indicates that over the trial period the rate of treatment was reasonably constant throughout.

Treatment to Trial Area 3 was shut down once the initial target of 1000Ahrs had been achieved. However analysis of chloride ion content indicated that the threshold of 0.3% had not been achieved therefore the system was re-activated approximately 3 weeks later. This pause period appeared to have no significant effect on the subsequent rate of treatment.

The collection of dust samples for chloride ion content analysis was found to be quite a disruptive operation as each time it involved shutting down the treatment and temporarily removing the cassette shutters to gain access to the concrete surface. In an actual treatment, as opposed to a trial, such regular testing would not be practical.

Nevertheless persisting with testing at approximately two week intervals during the trial provided valuable information about the rate of chloride ion reduction and the effectiveness of the treatment with time. The results of the testing are shown in Tables 2 and 3, and Figure 4.

Table 2 Chloride ion content testing at depth range 25mm to 50mm
(Measured as percentage chloride ion by mass of cement)

		105		106		107		108	
		Start	Finish	Start	Finish	Start	Finish	Start	Finish
Trial Areas	R	1.34	/	0.88	/	1.26	/	0.94	/
1 & 2	P	1.04	0.69	1.44	0.99	0.88	1.02	1.56	0.58
	N	0.3	0.33	0.99	0.78	1.2	0.17	0.74	0.33
	M	1.19	0.59	1.45	0.84	0.69	0.76	1.26	0.83
	L	1.61	0.64	0.69	0.81	1.08	0.73	1.13	0.96
	K	1.01	0.4	0.88	0.7	1.28	0.61	0.81	0.61
	J	1.04	0.16	1.4	0.4	1.82	0.83	0.89	0.14
	H	1.56	0.2	1.38	0.43	1.87	0.51	1.25	0.37
	G	1.01	0.52	1.51	0.68	1.71	0.74	1.21	0.66
	Average	1.122	0.441	1.180	0.704	1.310	0.671	1.088	0.560
Trial Area 3	E	1.43	1.19	1.63	0.45	1.53	0.13	0.83	0.46
	D	0.86	0.68	1.53	0.26	1.56	0.93	0.58	0.21
	C	0.97	0.02	0.8	0.8	1.55	0.4	0.77	0.36
	B	1.48	0.54	1.69	0.17	1.32	0.46	0.49	0.14
	A	1.27	0.61	1.58	0.89	0.72	0.7	0.33	0.26
	Average	1.20	0.61	1.45	0.51	1.34	0.52	0.60	0.29

Table 3 Summary of chloride ion content testing at depth range 25mm to 50mm
(Measured as percentage chloride ion by mass of cement)

AVERAGES	AREAS 1 & 2	AREAS 3	TOTAL
Start	1.18	1.15	1.16
Finish	0.59	0.48	0.54
% age reduction	45	53	49

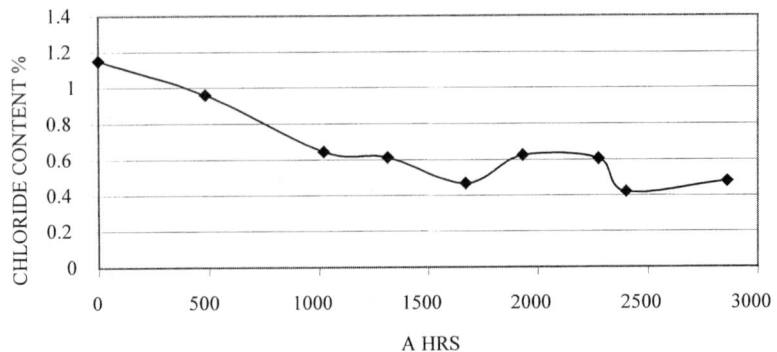

Figure 4 Rate of chloride reduction at Trial Area 3

Table 3 shows that on average, over the three trial areas, the treatment successfully reduced the chloride ion content to approximately 50% of the initial test results. However, a closer look at Table 2 indicates considerable variation in the results to the extent that at a number of grid locations the reduction was negligible, whereas at others the reduction was over 90%. At only 9 out of 52 locations was the chloride ion content below the 0.3% threshold. Figure 4 shows that the rate of chloride ion reduction up to approximately 1000Ahr of treatment was reasonably constant however there was no significant further reduction beyond that point. This is consistent with the initial target of 1000Ahrs of treatment as recommended by the Norcure[TM] specification. Chloride reduction in Trial Areas 1 and 2 followed a similar profile.

Stage 3 Laboratory Analysis of Concrete Samples to Assess the Degree of ASR Development

Sixteen weeks and then one year after completion of the treatment one core was taken from each trial area. The cores were typically 70mm diameter and 90mm long. The samples were taken to a laboratory and were stored for three months in a controlled water saturated environment at 38°C as recommended in BRE Digest 330 [8]. Three sets of Demec points were positioned on each core at 120° apart and measurement of expansion was carried out at approximately seven-day intervals.

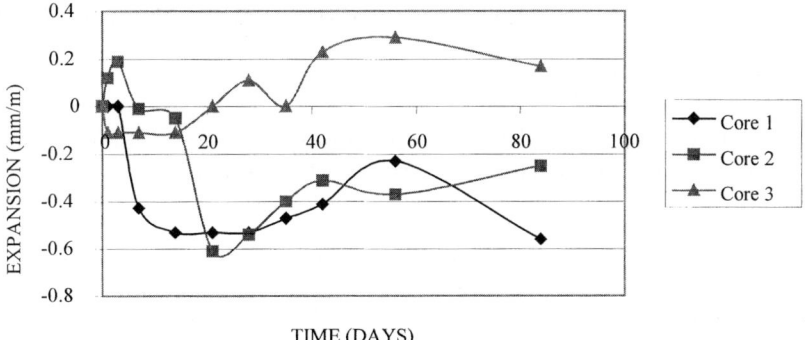

Figure 5   Summary of expansion measurements from the 16 week cores
(Core 1 was taken from trial area 1, core 2 from trial area 2 etc.)

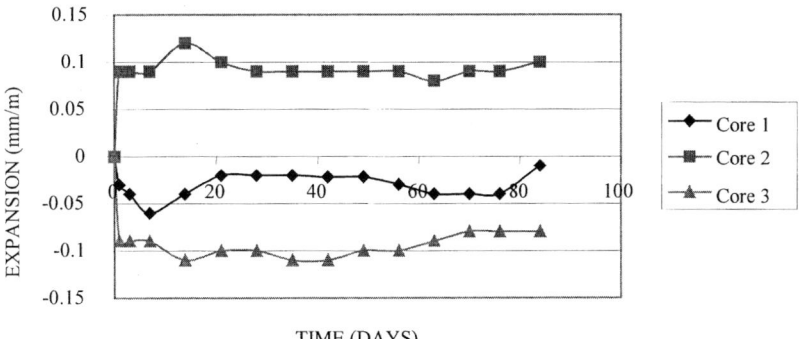

Figure 6   Summary of expansion measurements from the 1 year cores
(Core 1 was taken from trial area 1, core 2 from trial area 2 etc.)

In comparison with the expansion curve shown in BRE Digest 330 for concrete seriously affected by ASR, the expansion curves shown in Figures 5 and 6 suggest that the expansion recorded on the Tees Viaduct samples was not significant. The one possible exception is core 3 of the 16 week samples which exhibited expansion, which when compared to the Digest, could be considered as consistent with that of a sample showing slight ASR. However, visually none of the samples showed any signs of cracking or the presence of silica gel on the surface of the concrete. Four out of the six samples actually exhibited shrinkage; it is possible that this was due to further hydration of the sample resulting from the elevated temperature and the increased moisture content.

At the end of the monitoring period for both the 16 week and 1 year samples the cores were sliced in half longitudinally and thin sections were prepared from within three depths, namely 0-30mm, 30-60mm and 60-90mm. The thin sections were subject to examination by transmitted light microscopy using both normal and polarised light. Of the 16 week cores the three slices from core 1 showed no sign of ASR however one sample from the three slices taken from cores two and three had what appeared to be traces of crystalline gel within one crack and the samples were deemed to be suffering from slight ASR. Examination of the one year cores, found traces of gel within one slice taken from core No 2. No evidence of ASR was found in the slices taken from cores 1 or 3.

The petrographic analysis of samples taken before the treatment had been applied showed no sign of ASR and it was considered that the minimal evidence of reaction found in the post treatment samples need not be attributable to the application of the CE treatment. From the number of thin slices examined post treatment, 9 from each of the 16 week and one year sets of core samples, it was not surprising to find one or two with slight signs of ASR when examined microscopically. In addition based on the limited amount of expansion recorded during the monitoring period and also the lack of visual evidence of ASR in the form of cracking and presence of gel at the core surface it was considered that the treated concrete was not suffering from disruptive ASR.

CONCLUSIONS

1. Chloride Extraction successfully reduced the chloride ion content of the concrete at the depth of the reinforcement to approximately 50% of the initial level of contamination. However the acceptance criteria of 0.3% chloride ion by mass of cement was not achieved at the majority of test locations.

2. The treatment would be better suited to situations where the initial contamination is less than approximately 0.6%-0.8% chloride ion by mass of cement and also where the contamination is not deeper than the cover concrete.

3. Replacement of calcium hydroxide electrolyte solution with one containing lithium borate considerably reduces the period of treatment, a finding that has been taken forward by the contractor for use in other projects. No conclusion could be drawn as to whether using Lithium Borate prevented ASR development.

4. There appears to be little benefit in continuing CE treatment much beyond a 1000Ahrs. This is consistent with the Norcure™ Specification, which recommends an initial 1000Ahrs of treatment plus an additional 500Ahrs if chloride threshold level has not been achieved.

5. Whilst it is theoretically possible to generate ASR when Chloride Extraction is used on concrete containing reactive aggregates, little evidence of this was found during the trial. As the aggregate type and quantities vary and also the original alkali content is unlikely to be known it is felt that proposals to use this treatment on structures containing reactive aggregates should be assessed on a case-by-case basis by carrying out an initial trial application.

6. With regard to the way forward for repairs to the Tees Viaduct substructures the lack of success in the level of chloride reduction has resulted in a search for an alternative repair method. Consequently Scott Wilson and Autolink are planning trials of two alternative electrochemical treatments scheduled to commence in the spring of 2002.

REFERENCES

1. MALLETT, G. P., TRL, Repair of concrete bridges, Thomas Telford, Published 1994, 194 pages.

2. BYARS, E. A., Carbonation, chloride, sulphate and acid attack, Repair of Concrete Structures Conference, Centre for Cement and Concrete, University of Sheffield, May 1997, p 28-41.

3. GOWER, M., BEAMISH, S., Cathodic protection on the Midland Links Viaduct, Construction Repair, July/August 1995, p 10-13.

4. TOTTON, B., Electric potential, Surveyor, July 2001, p 11–13.

5. JACKSON, D., HA Press Release Ref HA/NW/298/01, November 2001.

6. BROOMFIELD, J. P., BUENFELD, N R, Effect of electrochemical chloride extraction on concrete properties, Transportation Research Record 1597, p 77–81.

7. PAGE, C. L., YU, S. W., Potential effects of electrochemical desalination of concrete on alkali-silica reaction, Magazine of Concrete Research, 1995, p 23–31.

8. BRE DIGEST 330, Alkali-silica reaction in concrete, 1999.

PROPERTIES OF PRE-TENSIONED PRESTRESSED CONCRETE MEMBERS SUPPLIED WITH CATHODIC PROTECTION FOR TEN YEARS

T Aoyama   **K Igawa**   **H Seki**

PS Comp Limited   Nakabou Tech Comp Ltd   Waseda University

M Abe

Port and Harbor Research Institution

Japan

ABSTRACT. This paper evaluates the effect of cathodic protection applied to prestressed concrete members. These members were T-type pretensioned PC beams, two of which had no-cathodic protection and three that were supplied with various levels of cathodic protection. NaCl was added at mixing stage to accelerate corrosion of the PC strands. The beams were exposed to severe environmental conditions for ten years. Measurements were carried out by way of electrical-chemical tests on the embedded PC strands, load-carrying tests on the beams, and after the beams were broken, visual observation of corrosion and the tensile tests on the PC strands. These test results through and after ten years indicated the followings: 1) beams protected with some type of cathodic protection had the same loading properties as beams tested at the initial stage (age zero years), 2) PC strands embedded in concrete had no corrosion for two types of anode materials and maintained suitable mechanical properties.

Keywords: Pretensioned PC members, PC strands, Corrosion, Cathodic protection, Anode materials, Sea environments

T Aoyama is a civil engineer of PS Comp. Ltd. , Tokyo, Japan. He is engaged in the design of PC members and the maintenance of structures.

H Seki is belonging to Dept. of Civil Eng. at Waseda Univ. ,Tokyo, Japan and have been doing research on durability and life cycle cost of concrete members.

M Abe is Chief Researcher of Port and Harbor Research Institution belonging to Ministry of Public Work and Transportation, Yokosuka, Japan. He has been undertaking into corrosion of metal and concrete structures.

K Igawa is Chief Engineer of Nakabou Tech. Comp. Ltd. , Tokyo, Japan. His main research work relates with cathodic protection and maintenance of public structures.

INTRODUCTION

Reinforcing steel for RC and PC members can be fundamentally protected from corrosion by cathodic protection. However, there are some negative effects: i) alkaline ions concentrated around reinforcing steel softens concrete after an electric current has been applied from anode materials to cathode for a long time [1], ii) softening of concrete reduces the bond strength between concrete and reinforcing steel [1][2], iii) reinforcing steel is damaged by hydrogen embrittlement after an electric current has been applied for a long time [1][2]. This paper clarifies these disadvantages and the effect of cathodic protection by a mainly mechanical approach. Specimens were T-type pretensioned PC beams, and they were exposed to marine environments for ten years. The experimental work consisted of electrical-chemical tests on embedded PC strands and load carrying tests on beams. After breaking the concrete of the beams, the corrosion of the PC tendons were visually observed and the tensile properties of the PC strands were tested.

EXPERIMENTAL PROCEDURES

Outlines Of Test Beams

Figure 1 shows the dimensions of the pretensioned PC test beams, which were 325mm deep and 4 m long. The PC tendons consisted of 7 twisted strands 9.3 mm in diameter, 5 tendons in the lower flange and 2 in the upper flange. The tensile stress in the PC strands at initial tension was 1,170 N/mm^2. The reinforcing stirrups were 6 mm in diameter. The concrete had a design strength of 50 N/mm^2 and was mixed with NaCl of 15 kg/m^3 in order to accelerate corrosion. Two reference electrodes were buried in each beam, as shown in Figure 1, to measure the electrical potential of the PC tendons during the exposure period under the marine environment.

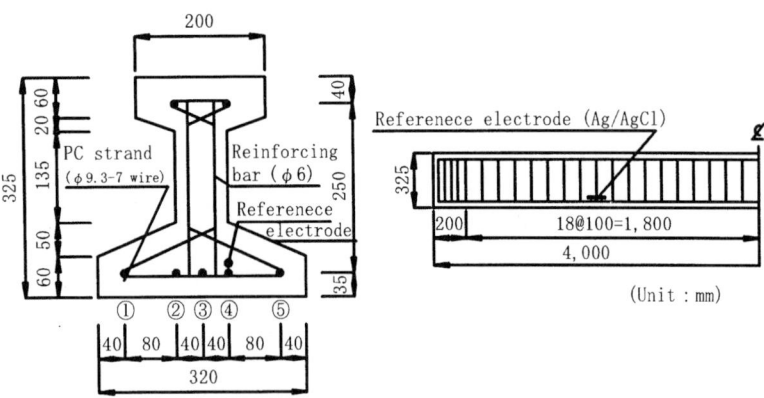

Figure1   Dimensions of PC Beams

Cathodic Protection System

Figure 2 and Table 1 indicates the cathodic systems for the test beams. A "dummy beam" was used for the load application test immediately after the beam was fabricated.

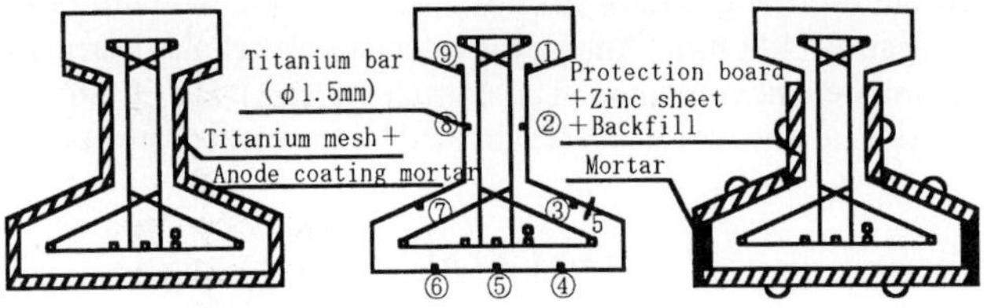

Figure 2 Type of cathodic system

Table 1 Type of test PC Beams

SPECIMEN		NOTE
Initial Stage		Load carrying test performed at zero years
No-cathodic Protection		-
Titanium mesh	Impressed anode	Current density (10mA/m² to 5mA/m²)
Titanium bar		Current density (10mA/m²)
Zinc sheet	Sacrificial anode	-

Two cathodic protection systems were applied: one was the impressed current system and the other was the scarified anode system. The anode materials of the impressed current system were titanium mesh attached to the beam surface (denoted by "titanium mesh" in this paper), and titanium wire anode buried 5 mm from concrete surface(denoting "titanium wire" in this paper). The scarified anode system consisted of a protection board, a zinc sheet and a back-fill material(denoted by "zinc sheet" in this paper). The applied current density was basically 10 mA/m^2.

Environment Exposure

The test beams were placed outdoors facing to the seashore. A sea water shower was automatically sprayed onto the beams twice a day (three hours per spraying).

Specimens

The testing comprised electrical-chemical tests, visual inspection of beams and PC tendons, load application tests, measurement of corroded areas and tensile strength tests on PC cables. In the electrical-chemical tests, the potentials in the PC tendons were measured both just after cutting-off of the electric current and 24 hours later. This was done by way of reference electrodes embedded in the beams. Depolarization potential was defined as the discrepancy between the two potentials.

After ten years exposure, the beams were visually inspected for items such as cracks, delamination of concrete cover and rust stains on the beam surface. After the loading test, cover concrete was broken out and corrosion of PC tendons was inspected. The load carrying tests were carried out on beams with a span of 3,200 mm and the pure flexural-moment test on beams with a span of 400 mm. Analytical calculation was also carried out by way of a two-dimensional finite element method. The corroded area of the PC strands was expressed as a corroded area rate, ie, the corroded area divided by the surface area of the PC tendons.

TEST RESULTS

Depolarization of PC Strands

Figure3 indicates the relationship between beam exposure time and depolarization. Beams equipped with titanium mesh maintained a depolarization of 100 mV to 300 mV, and the PC strands were definitely protected against corrosion. Beams equipped with titanium wire did not satisfy the standard depolarization value of 100 mV, and were considered to be ineffective for corrosion protection. It might be reasoned that impressed current did not evenly flow into all the PC strands buried in the beam and concrete surrounding the anode bars deteriorated.

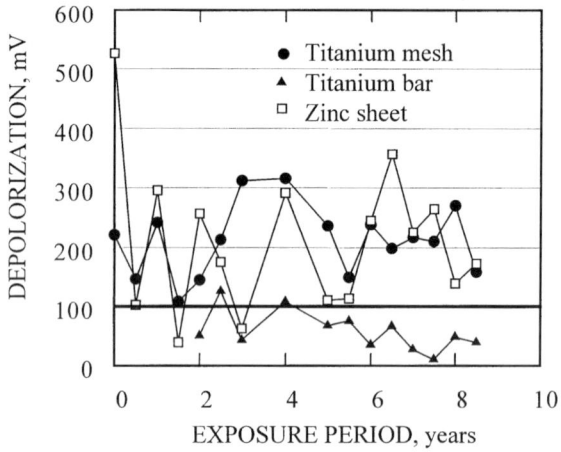

Figure 3   Relationship between exposure period and depolarization

For the beam equipped with zinc sheet, most of the data satisfied the depolarization of over 100mv, although depolarization depended on temperature and rain. For depolarization of less than 100mV, that depolarization speed was low because of the damp condition of the concrete. During the wet season, the instant-off potential was nearly –1,000mV, which was approximately the half-cell potential of zinc. It is therefore assumed that the PC tendons were fully protected against corrosion.

Visual Inspection and Corrosion of PC Tendons

The beams with no-cathodic protection and the beams equipped with titanium wire were visibly deteriorated. Figure 4 indicated cracking, delamination of cover concrete and rust stains on the beam surface. Some cracks were up to 1mm wide. Cracking observed on the surface of the beams equipped with titanium wire coincided with the position of the PC strands. Figure 4 also shows corrosion of the PC strands.

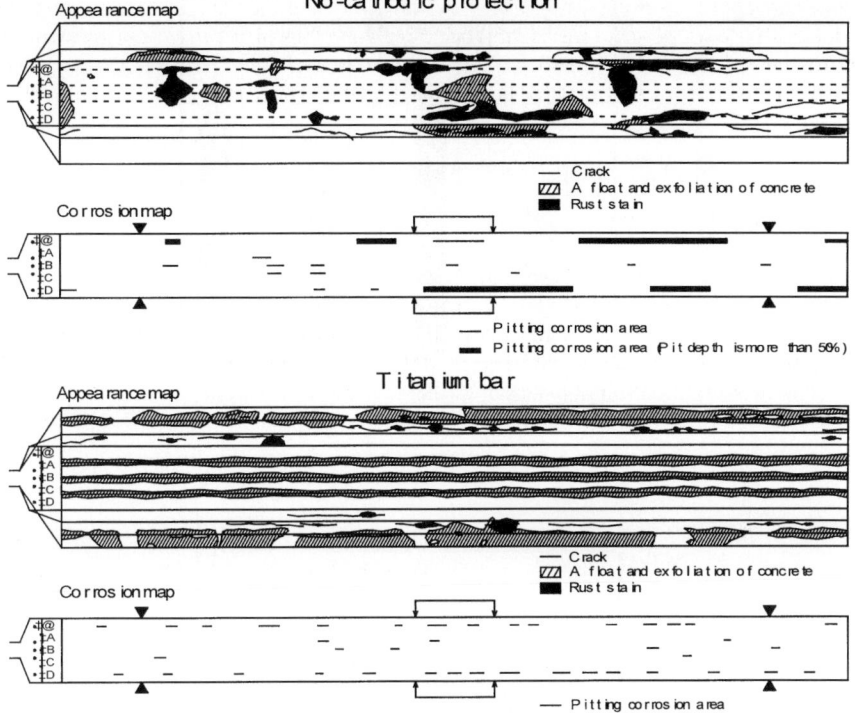

Figure 4   Visual observation of deteriorated test beams

Load Application Test

Cracking and Ultimate Load

Figure5 indicates the ratio of cracking load to ultimate load. This ratio means the load value divided by that at initial loading (zero years test). The initial cracking load and ultimate load were 78.5 kN and 205.0kN, respectively. As shown in this figure, neither beams with titanium mesh and zinc sheet had decreased flexural characteristics after 10 years of cathodic protection. However, the ultimate loads for both greatly decreased, and the ratio decreased to approximately 75 percent. The PC strands in these beams were greatly corroded, and two tendons positioned in the outer area were fractured before or during loading.

Deflection

The load/deflection relationship is shown in Figure 6. Deflections of non-deteriorated beams closely coincide with the calculation results, in which the initial concrete strength was introduced and the tensile strength of the PC strands was tested after the beam loading test. Another calculation was carried out assuming that two of the five PC strands were lost and the prestress in the PC strands was lowered to 70 percent of the initial prestress. Calculation results based on this assumption closely coincided with those for the beams with no-cathodic protection.

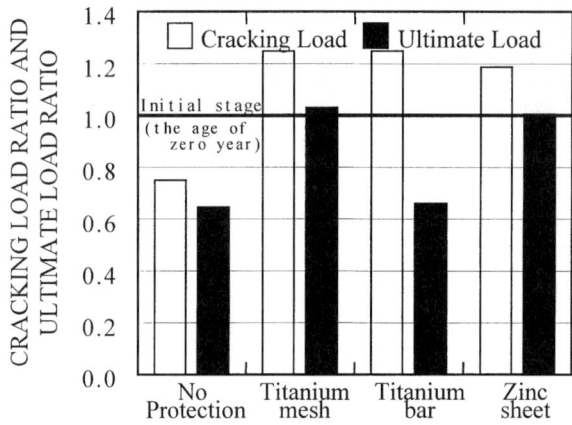

Figure 5   Cracking load and ultimate load at flexural loading test

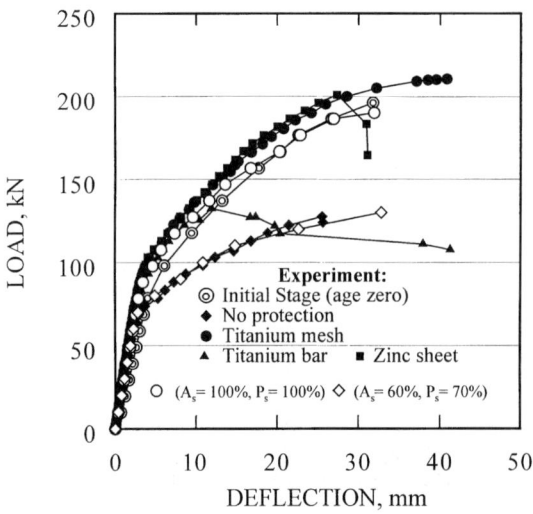

Figure 6   Deflection of test beams

Cracking Pattern

Figure 7 shows the cracking patterns of beams after the loading tests. The beams with titanium mesh anode and those with zinc sheet had nearly the same pattern as that of the beam at the initial stage, and are considered to maintain good bond properties.

However, greatly deteriorated beams showed fewer cracks. This was mainly caused with the decrease in cross-section of the PC tendons and longitudinal cracking.

Area of Corrosion for PC Strands

The test results for the corroded area are shown in Figure 8, in which 1~5 in the graph correspond to the positions 1~5 of the PC tendons of Figure 1. The figure indicates that the beams with titanium mesh and those with zinc sheet had little corrosion, coinciding with the test results for depolarization. For the beams of no-cathodic protection and those with wire bars, the PC strands buried in the outer area of the flange were greatly corroded. This is probably because the ingress of chloride ion into the concrete was severe and chloride ion easily accumulated at this area.

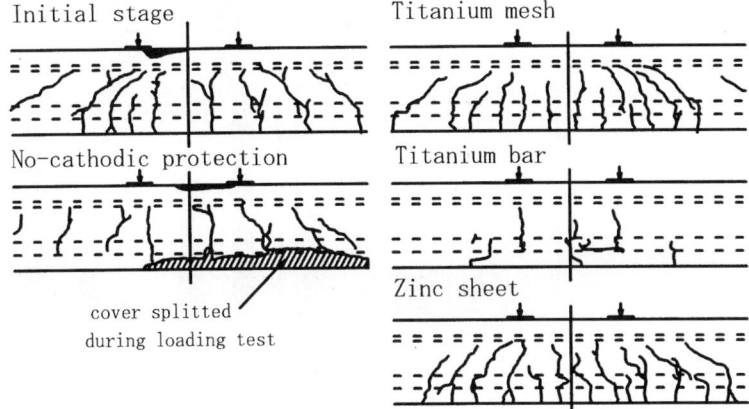

Figure 7   Cracking patterns after loading

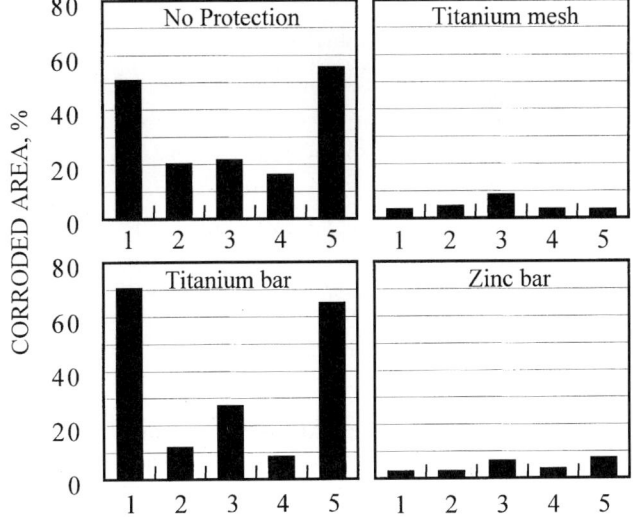

Figure 8   Corroded area of PC strand

Tensile Properties of PC Strands

Figure 9 shows the tensile test results. The ratio (%) shown at the axis of the ordinates is the value in which test data are divided by the standard value [4] (ultimate tensile load: 88.8 kN, elongation: 3.5%). Most of the PC strands in both the beams with titanium mesh and those with zinc sheet had mechanical properties that exceeded the standard value. The strand 4 in the beam with titanium mesh and the strand 3 in the beam with zinc sheet were damaged when the concrete of beams were broken.

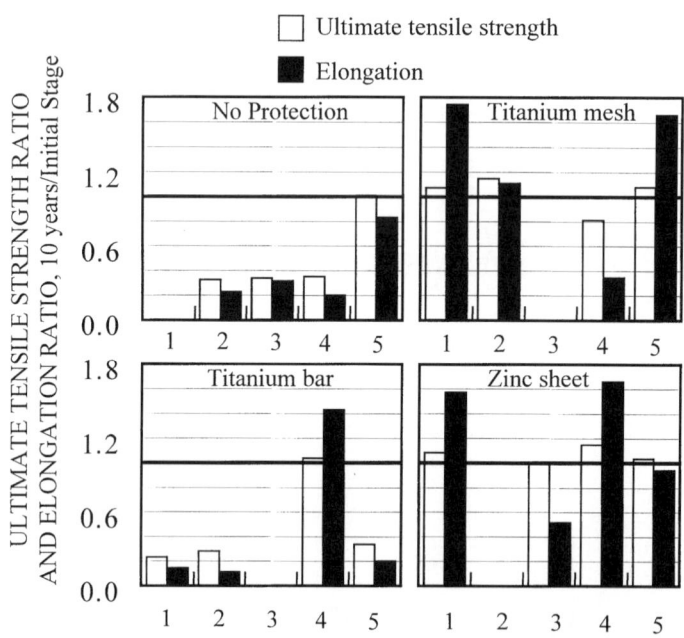

Figure 9   Mechanical properties of PC strands before and after exposure tests

Therefore, it was made clear that cathodic protection operated for a longer time did not damage the mechanical properties of the PC tendons. However, most of the PC strands in the beams with no-cathodic protection and those with titanium wire failed at the corroded section. The tensile load was approximately 30% of the standard value. The elongation properties showed a similar tendency at the ultimate tensile load.

CONCLUSIONS

Ten-year tests were carried out on T-type pretensioned PC beams exposed to severe environmental conditions. PC strands buried in some beams were protected from corrosion by a cathodic protection system.

The test results obtained are summarized as follows.

1. Beams with titanium mesh had the same flexural properties as beams tested at the initial stage (age of zero years). Beams with zinc sheet had the same inclination as beams with titanium mesh. However, beams with no-cathodic protection had considerably decreased ultimate flexural capacity at the age of ten years.

2. PC strands embedded in the concrete had no corrosion for beams of either mesh anode or galvanic system, and maintained suitable mechanical properties.

ACKNOWLEDGMENTS

The authors wish to express their appreciation and thanks to Dr.Tsutomu Fukude, Dr.Hidenori Hamada, Dr. Kouji Ishii, Mr.Takehiko Sako and Mr.Kiyoshi Kirikawa for their constructive comments and assistance in the laboratory work.

REFERENCES

1. ISHII, K, Application of Cathodic Protection to Pretensioned Prestressed Concrete Members, Doctoral Thesis (at Waseda University in Tokyo), 1996.10

2. TAKAO, U, Application of Desalination Method for Concrete Members Deteriorated with Salt Attack, Doctoral Thesis(at Kyoto University in Kyoto), 1996.6

3. Research Activities on Corrosion of Reinforcing Bars and Repairing for Reinforced Concrete Members, Japan Society of Civil Engineers, Concrete Technical Series, No.40, 325pp., 2000.12 (published by JSCE in Tokyo)

4. Standard Code of Concrete - Volume of criterions -, Japan Society of Civil Engineers, 611pp., 1999 (published by JSCE in Tokyo)

THE INFLUENCE OF REBAR ORIENTATION ON ELECTROCHEMICAL CHLORIDE EXTRACTION

D W Law

Heriot Watt University

A N Fried

Kingston University

United Kingdom

ABSTRACT. The use of electrochemical chloride extraction (ECE) has in recent years become an accepted technique for the rehabilitation of reinforced concrete structures subject to chloride attack. The technique has primarily been applied to structures where the chloride ions have ingressed from the external environment and have not progressed much beyond the first layer of reinforcement. However, the technique also has the potential to be applied to structures containing cast-in chlorides or structures where the chloride ions have progressed significantly beyond the first layer of reinforcement. To study the effect of applying desalination to this type of contaminated structure specimens have been cast with reinforcing steel at 50mm and 100mm centres and chloride levels at 3% by weight of sample. The percentage chloride extracted and the extraction profiles are presented.

Keywords: Concrete, Chlorides, Desalination, Reinforcement, Electrochemical, Rehabilitation.

Dr D W Law, is a Research Associate at Heriot Watt University, Edinburgh. His principal areas of research are electrochemical treatment of reinforced concrete, non-destructive testing of reinforced concrete structures using electrochemical techniques and the development of life management strategies for reinforced concrete structures.

Dr A N Fried, is a Reader at Kingston University, Surrey. His principal areas of interest are the properties of concrete and masonry.

INTRODUCTION

One of the major problems that confronts the construction industry is that of chloride induced corrosion of reinforcing steel. To counter this problem corrosion repair techniques have evolved from the use of patch repair to the use of electrochemical techniques such as cathodic protection and more recently electrochemical chloride extraction (ECE).

The first trial of the electrochemical desalination of concrete was conducted in the US by Slater [1], [2] in the mid 1970's. The technique has been developed over the course of the last twenty five years and is now an established treatment for the rehabilitation of reinforced concrete structures subject to chloride attack.

The technique uses an external electrode contained within an electrolyte held in a reservoir which is in contact with the concrete surface. The reinforcing steel acts as the cathode and the external electrode as an anode. Typical current densities applied to the concrete surface are of the order $1\text{-}2\text{A/m}^2$ of concrete surface and the treatment is generally operated over a four to ten week period. During the process the chloride ions are removed from the concrete and hydroxyl ions are generated at the steel with the result that corrosion is halted and a passive film is restored to the steel.

Since the technique was first used a number of concerns have been raised about the process including the possible reduction in bond strength [3], [4], microcracking [4], the possible instigation of Alkali Silica Reaction (ASR) [5], [6] and hydrogen embrittlement [4]. Research has been conducted on all of these areas as well as on the system efficiency [7-11], the composition of the pore solution [10], [11] and more recently modelling of the extraction process [12], [13] as well as practical applications on structures.

Chloride can be present in concrete either in the bound or free state. In cast-in chlorides a higher percentage of the chlorides will be in a chemically bound state compared to concrete with environmentally ingressed chlorides. Previous research has shown that free and loosely bound chlorides may be removed under the influence of electrochemical treatment [10], [11]. However, recent work has reported that bound chloride presents some risk of corrosion as these can disassociate to become free chlorides. This process is brought about to maintain equilibrium as the free chlorides are removed from the concrete. Hence, chloride threshold levels are best judged on total chloride content [13], [14].

Thus, ECE can be regarded as a two stage process. In the first stage free chloride ions are removed, while in the second stage the bound chlorides will be released and then removed. The initial stage is fast while the second stage is comparatively slow. This is reflected by a sharp fall in the efficiency of the ECE process at around the four to six week point in the application of ECE.

A number of factors effect the desalination process, these include the amount of reinforcement, spacing and orientation with regard to the face at which extraction is to take place [15], [16]. Other factors include the type of contamination, the amount of chloride, the water/cement ratio, temperature and cement content.

To date the majority of applications and laboratory research have concentrated on the application of the technique to structures where ingressed chloride has yet to reach or has only just passed beyond the steel reinforcement. This is due to the potential difficulty in extracting chlorides from behind the reinforcing steel.

In structures where the chloride ions are beyond the first layer of reinforcement the orientation, location and number of reinforcing bars will be crucial factors in determining the route by which the chlorides can be removed from the concrete.

EXPERIMENTAL

Sample Preparation

The experimental programme was devised to investigate the effect of applying electrochemical chloride extraction to concrete containing cast-in chlorides and the effect of rebar spacing on the level of chlorides that could be removed. Two sets of test specimens, 100mm x 350mm x 200mm, were cast. Each specimen contained two 12mm mild steel bars at either 50mm or 100mm centres, Figure 1. Two of each type of specimen were manufactured, to give four specimens in total.

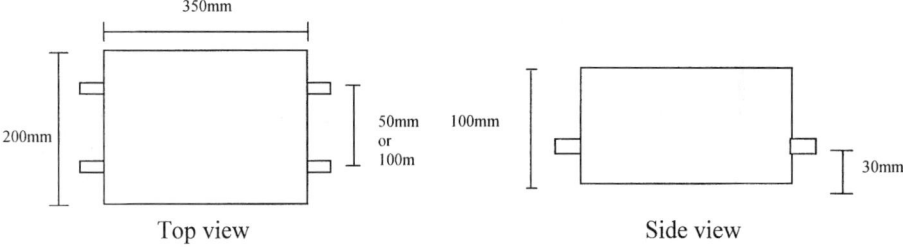

Figure 1   Test specimen

The concrete was a standard $40Nmm^{-2}$ mix, Table 1. An artificially high chloride content of approximating to 3% by weight of sample was selected to enable the extraction pattern within the concrete to be clearly determined. The samples were cured for 28 days in water, allowed to dry for 24 hours and then had all faces painted with epoxy, other than that to which the electrochemical chloride extraction would be applied. The epoxy coating was applied to prevent evaporation through any other face affecting the extraction process. The actual strength was $43.5Nmm^{-2}$ as determined from three 100mm cubes tested in accordance with BS 1881: Part 116.

Table 1   Mix design

CONSTITUENT	QUANTITY kg/m^3
Cement	12.75
Aggregate Fine	27.8
Aggregate Coarse	47.2
Water	6
Sodium Chloride	1.15

Test Procedure

The experiment used an external anode of platinised titanium mesh and an electrolyte of saturated calcium hydroxide. The prisms were placed on thin wooden supports in plastic containers with the electrolyte covering the exposed face. A direct current of $1A/m^2$ of concrete surface was applied to all of the test specimens. All specimens were treated for a total duration of five weeks.

Chloride concentration profiles for all specimens were determined on dust drillings taken using a 10 mm drill bit. All chloride analyses were conducted in accordance with BS:1881 Part 124, using acid extraction and electrochemical analysis. Drillings were taken at five separate locations, A-E, perpendicular to the steel reinforcing bars. Locations A and E were equidistant between the steel reinforcing bar and the edge of the specimen, locations B and D were at the steel reinforcing bars and location C was at the equidistant point between the steel reinforcing bars.

For specimens with bars at 100mm centres locations A and E were 25mm from the bar and the edge of the specimen and location C was 50mm from each bar. For specimens with bars at 50mm centres locations A and E were 37.5mm from the bar and the edge specimen and location C was 25mm from each bar.

Duplicate drillings were taken at two randomly selected points at each of the five locations, A-E. The two dust samples taken from each location were combined for analysis. Samples were analysed in 10mm segments over the complete, 100mm, depth of the specimen. The surface of the specimen exposed to the ECE was taken as 0mm, the steel bars were located at 30mm depth.

RESULTS

The chloride profiles following extraction are presented in Figures 2 to 7 . The results are presented as the average value of the two specimens at 50mm centres and the two specimens at 100mm centres.

The initial chloride content was determined by taking two set of drillings at randomly selected points on all four specimens. The dust samples were analysed in 10mm segments and gave an overall chloride concentration of 2.93% by weight of sample. This value was taken as the base level when determining the percentage chloride removed. The amount of chloride removed in the 0-40mm depth range are presented in Tables 2-3.

Table 2   % chloride removed in cover zone 0-40mm, 50mm centre specimen

LOCATION	DISTANCE FROM BAR (mm)	% CHLORIDE REMOVED
A(50)	37.5	21
B(50)	0	55
C(50)	25	45
D(50)	0	56
E(50)	37.5	27

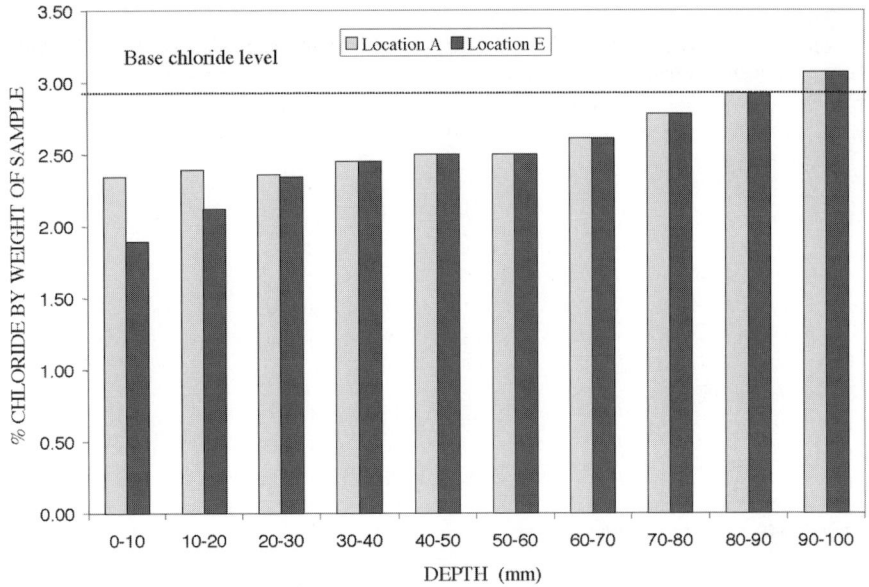

Figure 2   Chloride profile, location A and E, 50mm centre

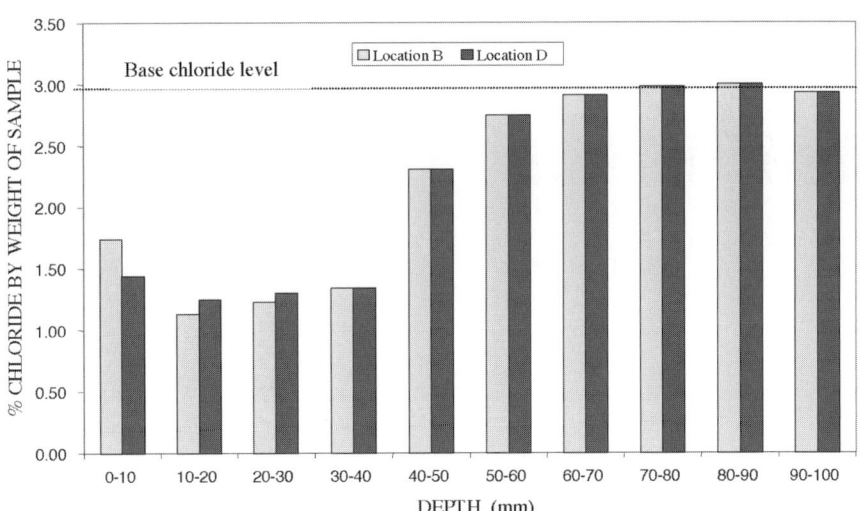

Figure 3   Chloride profile, locations B and D, 50mm centre

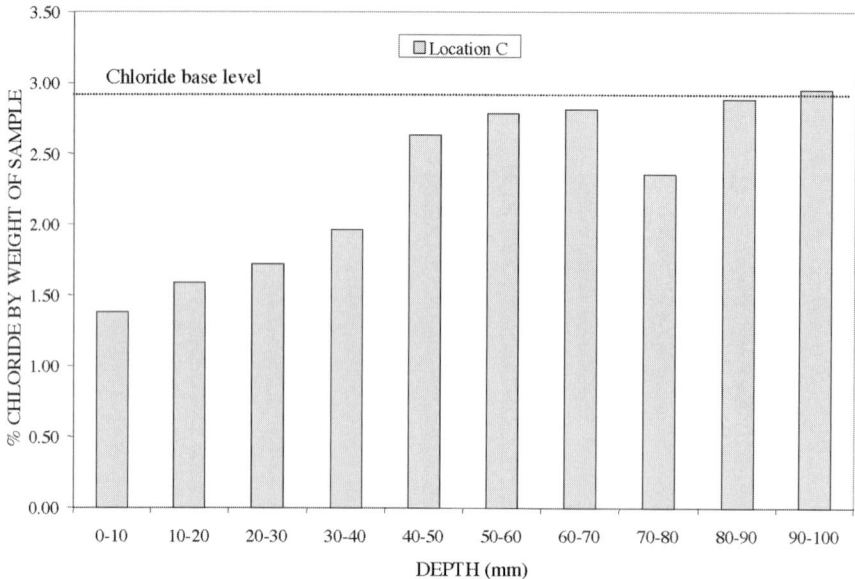

Figure 4   Location C, 50mm centre

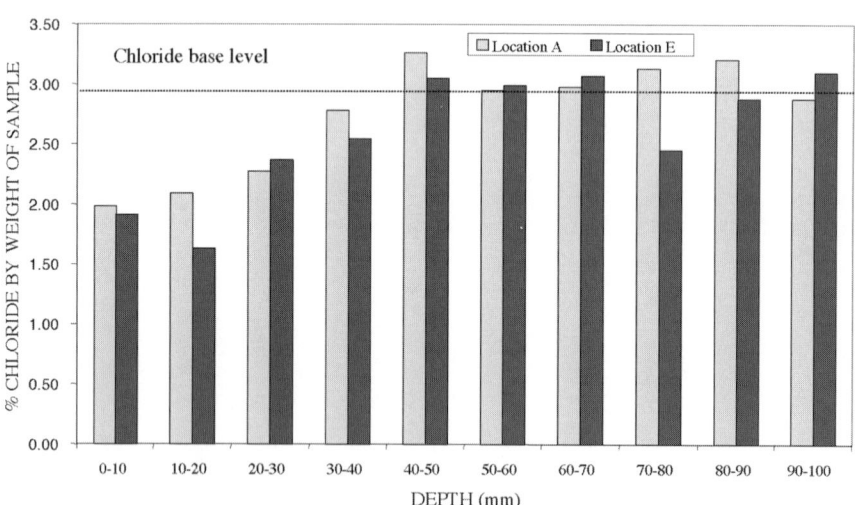

Figure 5   Chloride profile, locations A and E, 1000mm centre

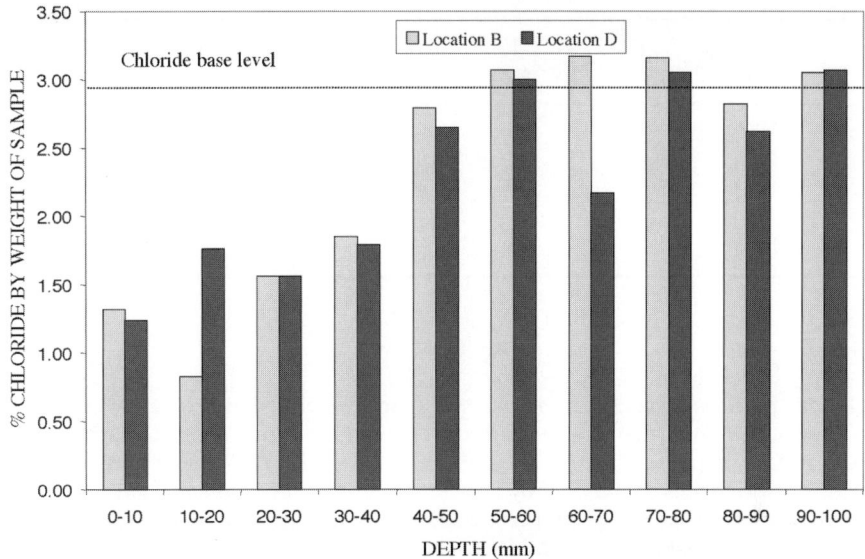

Figure 6   Chloride profile, locations B and D, 100mm centre

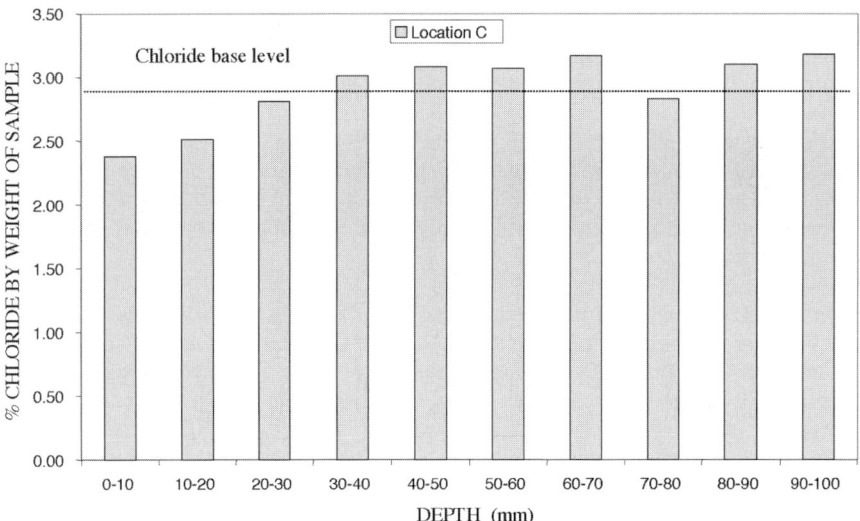

Figure 7   Chloride profile, location C, 100mm centre

Table 3   % chloride removed in cover zone 0-40mm, 100mm centre specimen

LOCATION	DISTANCE FROM BAR (mm)	% CHLORIDE REMOVED
A(100)	25	24
B(100)	0	54
C(100)	50	11
D(100)	0	47
E(100)	25	30

DISCUSSION

The chloride profiles indicate that, for both sets of specimens, the highest levels of chloride ions are removed from the concrete directly between the steel reinforcing bar and the concrete surface, depth 0-40mm. Taking the base level of 2.93% chloride by weight of sample the chloride ions are reduced by an average of 55% in the 50mm centre specimens, and an average of 50% in the 100mm centre specimens, locations B(50), B(100), D(50) and D(100). This is as expected as the highest current density will be centred in this zone during the application of the ECE.

As the distance from the bar increases the level of chloride ions removed falls, dropping from 43% at C(50), a distance of 25mm, to 9% at C(100), a distance of 50mm. It was also noted that the combination of the extraction field produced by two bars gave significantly higher chloride reduction than was achieved by a single bar. Location C(50) gave a 43% reduction, being the mid-point between two bars while locations A(100) and E(100), being between the bar and the edge of the specimen, gave reductions of 22% and 28% respectively. All three locations were 25mm from the closest bar. The indication of these findings is that the application of ECE to structures with bar spacings at or in excess of 100mm will result in a low level of chloride being removed from the concrete between the bars. This may result in the chloride ions remaining diffusing, under the concentration gradient present in the concrete, to the vicinity of the reinforcing bars and re-initiating corrosion.

Further analysis of the chloride profiles indicates that chloride levels behind the bars are not reduced by a significant factor by the extraction process. A small reduction is observed for the 50mm centre specimens, in the depth range 40-60mm. No measurable reduction is achieved in the 100mm centre specimens. Beyond 60mm depth the chloride concentrations at all locations are consistent with the base level in the original mix. These data would indicate that the extraction process has a limited influence beyond the reinforcement and the extraction of chlorides from behind the bar, either due to ingressed or cast-in chlorides must be doubtful. However, within actual structures more than a single layer of reinforcement is likely to be present and the subsequent layers may exert an influence over the extraction process which could result in chloride ions being extracted from behind the bars.

CONCLUSIONS

i   Electrochemical chloride extraction can remove significant quantities of chloride ions from contaminated concrete when the chloride ions are between the rebar and the concrete surface, the greatest amount of chloride is removed is the region directly between the bar and the concrete surface.

ii   The application of electrochemical chloride extraction can remove high levels of chloride at distances of 25mm adjacent to the bar, but this falls significantly at distances of 50mm.

iii   The interaction of two bars raises the level of chloride that can be extracted compared to a single bar

iv   The extraction process can remove chlorides at distances of 10mm to 20mm behind the bar, at distances greater than 20mm little extraction is achieved.

REFERENCES

1.   LANKARD, D R, SLATER, J E, HEDDEN, W A, NIESZ, D E, Report No. FHWA-RD-76-90, 1975.

2.   SLATER, J E, LANKARD, D R, MORELAND P J, Report No. DOT-FH-11-9026, 1976.

3.   BUENFELD, N R, BROOMFIELD, J P, Effect of chloride removal on rebar bond strength and concrete properties, Proc. Corrosion and Corrosion Protection of Steel in Concrete, Sheffield, Vol 2, 1994, p1438-1450.

4.   MILLER, J B, Structural aspects of high powered electro-chemical treatment of reinforced concrete, Proc. Corrosion and Corrosion Protection of Steel in Concrete, Sheffield, 1994, p1499-1514.

5.   SERGI, G, PAGE, C L, Advances in electrochemical rehabilitation techniques for reinforced concrete, Proc. UK Corrosion 95, Harrogate, November 1995.

6.   PAGE, C L, YU, S W, Potential effects of electrochemical desalination of concrete on alkali-silica reaction, Magazine of Concrete Research, Vol 47, 1995, p 23-31.

7.   POLDER, R P, WALKER, R J, PAGE, C L, Electrochemical desalination of cores from a reinforced concrete coastal structure, Magazine of Concrete Research, Vol 47, 1995, p 321-327.

8.   ARYA, C, SA'ID-SHAWQI, Q, VASSIE, P R W, Factors influencing electrochemical removal of chloride from concrete, Cement and Concrete Research, Vol 26, 1996, p 851-860.

9.   TRITTHART, J, PETTERSSON, K, SORENSEN, B, Electrochemical removal of chloride from hardened cement paste, Cement and Concrete Research, Vol 23, 1993, p 1095-1104.

10.   BERTOLINI, L, YU, S W, PAGE, C L, Effects of electrochemical chloride extraction on chemical and mechanical properties of hydrated cement paste, Advances Cement Research, Vol 8, 1996, p 93-100.

11.   TRITTHART, J, Electrochemical chloride removal – a case study and laboratory tests, Proc. Corrosion of Reinforcement in Concrete Construction, Cambridge, July 1996, p 433-447.

12.   ANDRADE, C, DIEZ, J M, ALAMAN, A, ALONSO, C, Mathematical modelling of electrochemical chloride extraction from concrete, Cement and Concrete Research, Vol 25, 1995, p 727-740.

13.   HASSANEIN, A M, GLASS, G K, BUENFELD, N R, A mathematical model for electrochemical removal of chloride from concrete structures, Corrosion, Vol 54, 1998, p 323-332.

14.   GLASS, G K, BUENFELD, N R, The presentation of chloride threshold level for corrosion of steel in concrete, Corrosion Science, Vol 39, 1997, p 1001-1037.

15.   IHEKWABA, N M, HOPE, B B, HANSSON, C M, Carbonation and electrochemical chloride extraction from concrete, Cement and Concrete Research, Vol 28, 1996, p 267-282.

16.   LAW, D W, FRIED, A N, The Electrochemical Chloride Extraction of Concrete Containing Cast In Chlorides, Proc. BCA Education Conference, Cardiff, 1999, p 65-74.

THEME THREE:
ENHANCEMENT OF EXISTING STRUCTURES

STRUCTURAL CHALLENGE OF HISTORIC STRUCTURES A CASE STUDY ON RENEWING THE REICHSTAGS BUILDING FOR GERMAN PARLIAMENT IN BERLIN

M Maier

Leonhardt, Andrä und Partner

Germany

ABSTRACT The following paper covers the special structural design features of structural alterations and methods used to safeguard, improve and evaluate the fitness for use and stability of existing historic structures and parts of historic buildings. Special reference is made to the need for the responsible engineers to have an understanding of the craftsman principles and an engineering sensitivity for the work beyond the standard mathematical theory taught them. A pure application of codes and design regulations and an uncritical, blind believe in scientific finite element analysis will definitely lead to an uneconomic and unsatisfactory result. As a case study on the latest structural alterations of the Reichstag Building in Berlin, which was totally renewed for German Parliament to the design of Architect Sir Norman Foster, London between 1995 to 1999 this paper points out the special challenge of historic structures for designers and engineers and the necessity of a sensitive designers approach with respect to the former state of art of building and the historic materials. This paper illustrates the sequence of important design procedures, beginning with the thorough analysis of existing, often incomplete building documents, the surveying and mapping of the complete building, the evaluation and modelling of the existing structural systems and material strengths and their condition or grade of deterioration. Once the condition of structural members and materials are assessed, the most suitable out of the full range of strengthening methods have to be selected. For historic structures this requires an experienced designer's knowledge. As shown in this paper load testing of complete structural members can be a helpful tool to find out structural redundancy in static systems and to calibrate the static analysis by incorporating actual test results.

Keywords: Historic structures, Structural alteration, Condition assessment, Load tests.

Mr M Maier, received his structural engineering degree from University of Stuttgart in 1993 and has been working since then with Leonhardt, Andrä und Partner on the design of bridges and buildings. He was in charge of the "Reichstag-Project" from the very beginning and responsible for the structural design and alteration works of the historic wings. Since 1999 he has been Head of R&D at Leonhardt, Andrä und Partner and mainly involved in the development of the Leoba prestressing system for CFRP tendons. Since 2000 he has been external Research Assistant at the Institute of Structural Concrete and Building Materials, University of Leipzig and is currently finishing his PhD thesis.

INTRODUCTION

The Reichstag Building in Berlin, Germany has figured prominently in Germany`s history over the last hundred years and is regarded today as a memorial, a symbol and a monument. In the conversion of the Reichstag Building to the seat of the German Bundestag (Parliament) between 1995 to 1999, the design of the brand-new plenary room and the domed shaped roof are particularly evident to the outside world. An unnoticeable, but in engineering terms at least comparable if not even more demanding task was that of the structural alterations and in the change of use and reuse of existing historic structural elements and parts of the historic building, especially the foundations, the crown structures and the arched vaults and the connection between the existing and new constructions. The Architects Sir Norman Foster and Partners, London have attempted to sympathetically integrate the existing structure into the overall design of a new, expanded facility in order to reflect the complex symbolic aspects of the building. The large dimensions and the new open spaces have called for a high technology/low energy approach that has been a challenge for the structural engineers. Thoughtless application of existing standards would have necessarily led to complete demolition. The classification of these structures in a level of stability and fitness for use complying with that given in the standards demanded not only the application of modern calculation techniques from the structural design and checking engineers, but a real sensitivity for their work and traditional master builder virtues. The intention is to describe these in the following examples.

Some "milestones" in the history of the Reichstag Building are shown in Figures 1 to 4.

Figure 1   Finished Reichstag Building to the design of Paul Wallot 1880

Figure 2   The ruin of the Reichstag Building after Second World War

Figure 3 "Wrapped Reichstag" artist-project by Christo & Jeanne-Claude, 1995

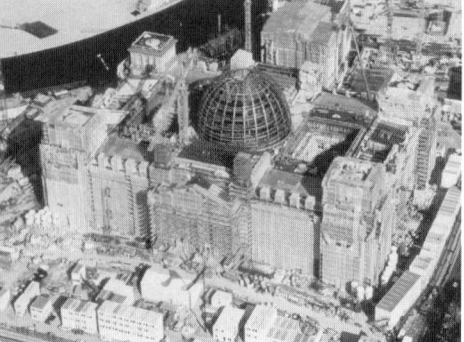

Figure 4   Reichstag under reconstruction again, design Sir Norman Foster, 1995-1999

HISTORICAL BACKGROUND

The Reichstag building was built in the years 1884 to 1894 to designs by Paul Wallot. With a built area of 11,200 m² and a useful area over four storeys of around 12,600 m² the building costs totalled 24 million gold marks. The main building material used was masonry clay brick. They were used for the foundations, pillars, walls and vaulted ceilings. If the 500,000 m³ building volume, around was 140,000 m³ solid masonry and vault mass. The very massive effect of the masonry walls and pillars with walls up to 2·m thick were interspersed with a highly branched pipe system for the then very modern air heating and ventilation installations of the firm David Grove from London, in parts with shafts with a cross-section of several square metres in size and bricked in clay and cast iron groups of pipes. The building was seriously damaged by the Reichstag fire in February 1933 and the shooting during the Soviet storming of the building during the final days of the Second World War effectively left behind a ruin. In the years thereafter, following initial ideas of total demolition the collapsed dome was blasted (1954). The clearance period began (approx. 15,000 m³ rubble) and initial, isolated securing and substance maintenance measures were carried out on the load-bearing structure which was also steadily worsening due to the effects of weathering (1957). Individual areas and rooms were converted for all kinds of different uses without the existence of an overall structural and architectural concept. The decision was then made to reconstruct the complete building to plans by the architect Paul Baumgarten in the years 1961-1972.

The originally monolithic masonry structure was replaced in the central area by a steel and prestressed concrete load-bearing structure, which was structurally separated from the remaining building. In the outer wing areas, additional stiffening horizontal supports and column wraps made of reinforced concrete were constructed. Some of the existing masonry shafts mentioned above were backfilled for stabilising purposes. Destroyed Wallot vaults and Prussian cap vault ceilings were mostly replaced by reinforced concrete waffle slabs. The alteration works reflected the state of architectural sensitivities of the sixties i.e., it evidenced little respect for the existing structure. Many areas were straightened and the historical wing areas were mostly covered with wire netting plaster layers (Rabitz plaster construction). Little consideration was given to the existing historical building substance on the inside. Following reunification and the decision of the German Bundestag made in June 1991, to relocate parliament in the capital Berlin, the architect Sir Norman Foster and Partners was commissioned to design the conversion of the Reichstag building for future use by the German Bundestag. The engineers Leonhardt, Andrä and Partner were commissioned to carry out the structural design for the alteration, conversion and new designs.

ALTERATION AND PARTIAL REMOVAL

The design produced by Sir Norman Foster and Partners primarily involved knocking back the structures and installations to the retained building substance from Wallot's time. This meant completely removing the installations in the central plenary hall area and the upper storeys of the wings carried out under Paul Baumgarten in the sixties, insofar as economically and structurally feasible in the historic basements. The development of the structural concept for the removal works was preceded by intensive examination of the available and examinable documents since Wallot. The allocation of individual items in the sparse structural documents to the structural elements in the building proved to be particularly difficult. Due to the "concept-less" individual conversion measures carried out after the war and the fact that the structural design changed during the Baumgarten conversion and some completed areas had already been reworked or reinforced, this allocation was extremely difficult. Descriptions of items such as the "Ceiling above the reading room" and changing axis systems and names made the research a real puzzle task. The knowledge gleaned was supplemented by on-going surveys and investigations on site, to verify the structural systems on which the alterations would be based. The proximity of those involved in the planning and design to the Reichstag building was a great advantage. The structural elements identified and their structural position including the load assumptions made were mapped in specially produced drawings, which were to prove to be the most important aid during the whole construction phase. During the "Back (wards) construction" (alterations – i.e. partial removal) the work was comparable to a Mikado game i.e. to find a sequence, which guaranteed that none of the load-bearing parts made unplanned movements during the alterations phase. The sequence of alterations was consequentially planned before the start of the works, both with respect to the structural aspects and with respect to the construction programme. Under structural design aspects, additional dependencies of the transport paths within the building had to be taken into consideration. Low vibration methods and techniques were particularly important for the alteration works, in particular to safeguard the historic areas with a building conservation order.

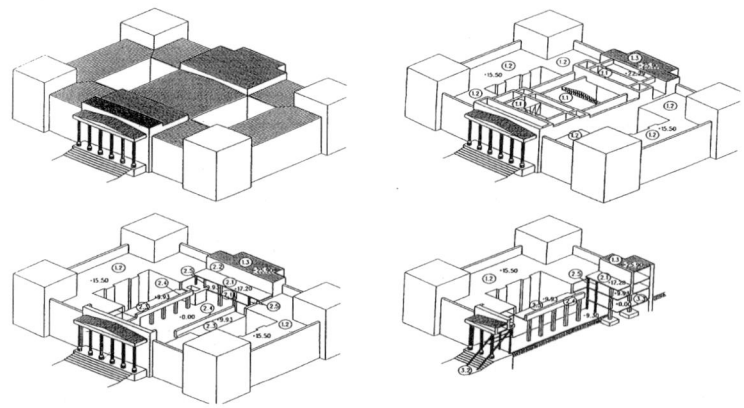

Figure 5   Schematic of the alteration phases

The result was a construction programme with over 200 individual steps and detailed demolition instructions which were recorded in the form of drawings and descriptions. Supplementary to this, a working model was produced for the alteration works, which was used to explain the procedures and interdependencies to those involved in the construction works. This plastic representation enabled above all the demolition form to be quickly and efficiently introduced to the alterations planning.

In order to avoid temporary stiffening and securing measures of intermediate building conditions as much as possible and minimise their costs, the attempt was made to initially leave parts of building substance to be broken off wherever possible and purposeful to be used as stiffening elements, and thus to maintain the stability in the existing structure, especially the horizontal stiffening during the conversion period. This principle of interaction between the alterations and new build works for stiffening purposes was applied above all in the area of the facades by leaving ceiling strips including their supports. Following the reconstruction of adjacent and stiffening parts of the building and their coupling with the existing structures, these initially left parts could then be removed.

Figure 6   Cutting back the shear prestressed concrete walls with an oxygen lance and rope saw

Figure 7   Demolition at the East Gate using a concrete cracker fitted onto a 40m extension arm and video controlled stirring

CONDITION ASSESSMENT

The structural design required a thorough analysis of all existing building documents since the time of Wallot. The building was completely surveyed and mapped, and local samples were taken to assess and model existing structural systems and material strengths. Even though the material properties could partly be ascertained from the original building documents, these properties had to be verified since they might have changed in the course of time from fire and weather and the Baumgarten modifications. Also, the quality of structural elements added in the 1960s, such as waffle slabs, only partially comply with present building codes. Numerous pits were dug to explore the geometry and material quality of the foundation masonry. Samples from the masonry, steel girders and reinforced concrete elements were tested to identify strengths, damage and weldability.

The masonry design strength was evaluated from test results according to Eurocode 6 regulations, where slenderness effects of walls and columns and the quality of workmanship were considered. The results were statistically evaluated and the allowable stresses were partly reduced for practical purposes to avoid the explicit consideration of slots and recesses in the stress analysis.

VAULTS

In the historical side wings primarily barrel vaults with flat arch rises from the time of Wallot were found in the cellar and still in a very good condition, while the ground floor vaults – mainly built as cross vaults with semi-circular shaped interior curves – had suffered badly during the course of time due to fire, shooting and weathering. This was reflected in partially widespread-flaked areas and shell cracks running parallel to the surface. The appearance of the vault interconnections themselves appeared to be crack free to the eye. Special attention was paid to the vault geometry and the retained condition or degree of damage to the ground floor vault undersides. The geometric course of the undersides of the vault arches and their interior curve lines were recorded by photogammetric levelling. Extensive core drills during the course of the material investigations carried out throughout the whole building enabled both the thicknesses of the caps and the type of arch infill as well as the properties of the materials such as the apparent density (self weight) mortar, stone and resulting vault brickworks strength to be determined. Additional comparisons by recordings taken on site of parallel, scheduled vault demolition works were able to support the geometry assumptions made. This resulted in an adequate picture of the type and geometry of the historic vault. The assessment of the degree of damage and its effect on the load bearing capacity proved to be far more difficult. The over 40 different types of vaults with their varying geometric parameters (spans, arch pitch, bearing joint direction, cap thickness, arch shape) and material parameters (type of stone, stone strength, mortar strength, stone density, density of the backfill) were also divided into classes of damage (lack of cracks, flaking, shell cracks,).

Calculated verification of adequate stability of the vaults using the thrust line theory (plane consideration of the load bearing performance assuming the formation of three hinges) was not possible, especially for a half-sided live load across the layout diagonals – even with adequately precise knowledge of the aforementioned vault parameters. Even the simulation of a spatial, shell-like load bearing performance using FE analysis and a non-linear material performance with unsuppressed tensile stresses gave impermissibly high eccentricities of the resulting pressure so that according to the calculations, the gaping joints appeared in the typical third points, which led to a reduction of the pressure area and thus to the exceeding of the permitted compressive stresses.

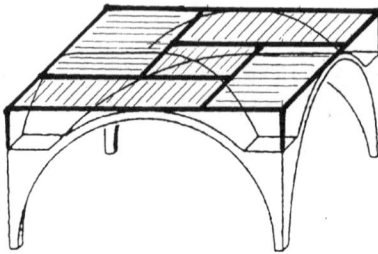

Figure 8   Cross vault exposed from above   Figure 9   Initially planned composite steel
with clear thickening of the crossribs   casement of a cross vault

In accordance with Sir Norman Foster's design, the floor height of the ground floor was to be lowered to the historical measure of +/- 0.00 m, which if safeguarding correct building physics floor construction heights would lead to a reduction in the unfinished floor height and thus a reduction in the vault crown thickness by up to 11 cm. Due to a lack of available useful height, the previously planned self-supporting vault casement as a composite steel floor structure had to be dropped. The peak bricks of the vault crown, which is under a preservation order, were dry milled, vibration free, using a multiple disc diamond cutter and then relined with 20 mm plastic hardened composite mortar.

This weakening of the caps with subsequent partial replacement was not able to be even approximately modelled with the available calculation methods.

For this reason, tests were carried out on structural elements with in-situ loading tests, to analyse the influence of the real boundary conditions on the load bearing capacity of the vaults and to be able to indicate any system reserves not taken into account in the calculations. At first an, under all aspects optimally classified, cross vault was load tested to assess the prospects for the successful analysis of the loading safety. After positive results were obtained, a further 3 vaults were selected as being representative for their vault class, including the worst under all the assessment criteria. Under γ-fold loads with up to 47 t per vault field – including for the most unfavourable half-sided live load case for the curve support line – the deformation and expansion measurements resulted in a completely elastic behaviour. The results of these influences needed to be analysed locally and globally, since safety and serviceability cannot be assessed on the basis of isolated load tests, but need to be accompanied by structural analyses. Calculating back from the expansions measured resulted in values for the compressive stresses much less than the values permitted from the material investigations.

Thus, the joint load-bearing effect of the vault filling of ventilated bricks was thus able to be verified.

Figure 10   Loading and measuring device for a loading test on the arched vault damaged by fire

Figure 11   Transferring the loads via tension rods from above onto the respective quarters of the arched vault

Structural vault strengthening therefore only had to be carried out in areas in which the existing curved support system of the vault, for example by upper tie beams for vaults with lost adjacent vaults for taking up the vault thrust.

FOUNDATIONS

The soil conditions, types of foundations and their conditions were greatly varied throughout the Reichstag building. There were no as built drawings for the existing foundations of the Reichstag building, merely fragmented sketches of the principles from specialist magazines of the time were available (Figure 12). The foundation geometry, type and property, their materials and strength values were therefore investigated by a large number of trial pits and core drills of the foundations. Here mostly masonry (both limestone rubble and clay brick masonry) but also concrete foundations made of clay brick grit concrete with a low strength were found. In the central area and beneath the northern towers, soil improvement measures had been carried out using rammed timber piles due to the lack of load-bearing soil there. The timber piles with an average diameter of 25·cm and a length of 5·m were driven in at an axis spacing of approx. 1.0·m. The timber piles are bonded in a 15·cm depth of the concrete foundation above them. Aware that the groundwater had been lowered during the 1930's it was not possible to be sure that the pinewood piles had not been damaged. To clarify to what extent the timber piles would contribute to the load bearing for the remaining period of building use loading tests were planned. They should provide data to verify the reliability for the limiting pile load on the one hand (peak pressure and possible loss of the sleeve friction due to rotting) and on the other hand to provide information about the load bearing capacity of the soil between the displacement piles. As a result of bacterial rotting of the outer timber layer the sleeve friction was classified as being very low both in the plate loading test as well as tin tensile tests. By loading the samples, the primarily displacement effect of the timber piles was able to be verified. The long-term retention of this displacement effect even by the timber piles with rot was able to be confirmed by a timber expert.

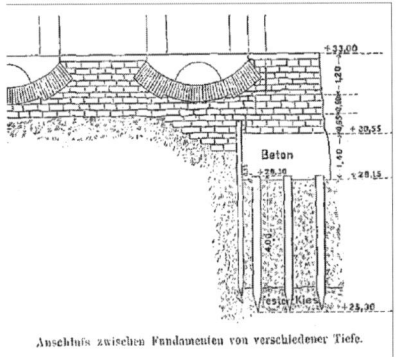

Figure 12   Detail of the foundation principle published in the Central Paper of the Building Administration of 1884

Figure 13   Trial pit showing an reversed arch pier foundation

Figure 14   Exposed wooden pile heads in the central area

Figure 15   Testing set up for the wooden pile loading test

Based on this, extensive structural calculations were carried using a finite element model to verify that these areas could be left without any replacement or strengthening measures for the "pile-work concrete". Only the new stiffening cores in the central area were given a deep foundation penetrating the wooden pile layer to avoid long-term settlement.

A further special feature of the existing foundations was presented by the "reversed dry arches" (Figure 13), which as a brickwork connection between individual pier foundations enable loads to be transferred to adjacent foundations. Thus a load-bearing effect similar to that of a grillage or even a foundation slab is produced by the narrow spacing between the foundations.

When calculating the overall foundations, in agreement with the soils engineer, a maximum settlement difference of l/500 between adjacent foundations had to be verified by calculation. The stability and verification of suitability for use was produced with the verification of the safety against ground seepage and of compliance with the prescribed settlement values and soil pressures.

The settlement values were determined for various building conditions with the aid of an interactive load-bearing structure – foundation – soil model. The modelling was carried out as a finite element model of the load-bearing structure in the lower storeys and of the body of the foundations and characterisation of the soil using the stiffness module method. The areas with timber piles in the soil were idealised as a compacted soil layer. The reduction in strength of the timber piles was incorporated in the investigations as a special load case. The model imaging included half the layout of the Reichstag building. The calculations were carried out on quarter of the system for capacity reasons.

Apart from determining the settlement values, it was possible to use the model to make a statement concerning the loading of the foundations and the load-bearing structure.

EXAMPLES OF MODIFICATION AND STRENGTHENING

Loads from the new building on existing structural elements, especially existing foundations and columns, are generally significantly higher than the loads from the previous building. This is due to one additional storey, new high quality interior cladding, sound insulation and the generally high specifications of installations.

Due to the differentiated considerations, both with respect to modelling as well as the investigation of the building condition, the strengthening of the load-bearing structure required according to a load balance could often be avoided.

As noted, great efforts were made to realistically assess the actual loads from the new building and the strength of the existing structural elements in order to try to avoid strengthening and modification works. Some strengthening measures, however, proved to be necessary:

- Lesene-like installation of steel columns in masonry wall recesses. This enabled fixing the cantilever loads of the gallery installation in the existing masonry to be achieved using the steel frame effect.

- Separate transfer of concentrated new structure loads from areas above the existing storeys through the building detached from the existing load bearing structure with independent foundations on small bore injected piles.
- Strengthening of reinforced slabs and girders by externally bonded steel plates and CFRP strips: higher dead loads of the flooring materials required partly additional reinforcement of the reinforced concrete members.
- The reopening of the original arches in the entrance lobby is a typical modification example. These arches were badly damaged during and after the war and were totally closed in the 1960s, leaving only small central door openings. The reopened arches now have to carry concentrated loads from additional storeys. In order to avoid overloading of the reopened arch spans, the flow of forces was influenced by an additional head beam such that vertical loads act upon the columns rather than on the arch apex.
- Strengthening of the Wallot Prussian cap vault ceilings by a jointly carrying, plastic hardened screed as an additional "pressure plate". Tests on the structures, including on the Prussian cap vault ceilings displayed this co-bearing effect of the arched cap between the steel girders, which without additional composite securing can only be explained by mechanical frictional bonding as a result of the cap impost forces. This proven raising of the pitch line was incorporated in the structural calculations and with the aid of the stiffening of the pressure zone by the jointly load bearing screed resulted in an otherwise reinforcement free construction.

NEW CENTRAL AREA INSIDE HISTORIC FACADE

The structural design of the new central building inside the historic facade was rather straightforward. The structure is dominated by twelve circularly arranged columns which directly support the dome ribs. The columns themselves are individually supported by large diameter bore piles. The 20 m long columns are spincast precast elements erected from the central crane.

The roof above the main assembly room is an axisymmetrical ring plate cantilevering out from the circle of the columns. The installation storey behind the column circle had to be kept completely free from structural obstructions such as walls or truss diagonals.

The cantilever could thus not be balanced by continuous girders in the back span. Equilibrium for the cantilever was achieved by coupling the interior compression ring 2 m above the level of the column heads with the exterior tension ring over the columns. The contractor proposed to use partially precast elements so that a perfect concrete surface could be achieved (Figures 16 and 17).

The roof is capped by a 23 m tall glass and steel dome that rises to a height of 47.3 m above ground and has the same diameter as the column circle, 38 m (Figure 18). The viewing platform in the dome is 16.7 m above the floor. Its main structural elements include:

- steel ribs and ring beams
- two ramps
- a viewing platform
- a light-gathering cone structure with mirror elements
- cladding and sun protection.

The main structure is a space frame shell with vertical ribs and horizontal ring beams. The ramps are also needed to stabilise the main structure and the viewing platform.

The ramps are supported along their periphery only by tension and compression rods, permitting an unobstructed view through the dome and easy access to the viewing platform. Result from wind tunnel tests were used to evaluate the wind pressure distribution on and in the dome and the flow characteristics of the waste air exhaust. Installations inside the light cone include smoke and waste-air exhaust fans with sound absorbers and heat exchangers. Pipes and ducts for service are hidden in the ramps.

The erection began with setting the ramps on a temporary assembly stand. The main steel ribs and ring beams were subsequently added. The viewing deck was assembled at the base of the dome and lifted into position. After the tie rods were connected, the ramps, platform and the space frame structure was stable and the assembly stand could be dismantled.

Figure 16   Erection of assembly room ceiling

Figure 18   Dome during erection

Figure 17   Ceiling of assembly room from inside during construction:

SUMMARY

The conversion of the Reichstag building to the seat of the German Bundestag in Berlin was completed on time with the first sitting of the Bundestag on 19th April 1999. Renovation and enhancement of existing structures requires a tight and strictly disciplined interaction of all members of the planning team and innovative engineering is necessary, to structurally combine a historic structure with a modern architects design to a running parliament building according latest standards. In case of the Reichstag building it is clearly visible, that the involved planning team succeeded in doing this.

REFERENCES

1. WALLOT, P, Das Reichstagsgebäude in Berlin 1897/1913, Cosmos Verlag für Kunst and Wissen.

2. LEONHARDT, ANDRÄ, PARTNER, Beratende Ingenieure VBI GmbH, Dez., 1995, Gutachten zum baulichen Zustand des Reichstagsgebäudes in Berlin.

3. ANDRÄ, H P, FINK, R, Untersuchungen an der historischen holzpfahlgründung and am historischen mauerwerk beim umbau des reichstagsgebäude zum sitz des deutschen bundestages in Berlin, 4, Internationales Kolloquium Werkstoffwissenschaften und Bauinstandsetzung, Esslingen, Fraunhofer IRB Verlag, 1996, R.v.Halasz and Czempin Gutachten zum baulichen Zustand des Südflügels am ehemaligen Reichstagsgebäude in Berlin, 12.03.1968.

4. CULLEN, M S, Der Reichstag, Parlament, Denkmal, Symbol, Berlin be.bra verlag, 1995.

PRODUCTS AND SYSTEMS FOR THE PROTECTION AND REPAIR OF CONCRETE STRUCTURES: THE CURRENT POSITION OF EUROPEAN STANDARDS

G C Mays

Cranfield University

United Kingdom

ABSTRACT: This paper provides an update on the position within Europe regarding the standardisation of products and systems for the protection and repair of concrete. Included are surface protection products, concrete repair materials, structural bonding agents, crack injection systems, and products for the anchoring of reinforcement and reinforcement corrosion prevention. The general principles for the use of these products and systems were laid down in a prestandard (ENV 1504-9) in 1997. Within each product group, the performance characteristics associated with a number of repair principles and methods have been identified. A separate harmonised standard then specifies the identification and performance requirements for a range of material properties. These specification standards are, in turn, supplemented by a range of test method standards. In some cases these have been based on proven existing techniques; in other cases further research has been necessary before they can be recommended for standardisation purposes. The work is being co-ordinated within CEN TC 104 SC8, of which the author is European Convenor, the Secretariat being provided by AFNOR. The paper will concentrate on presenting the significant progress made since the last Dundee Conference in 1999.

Keywords: Concrete, Protection, Repair, Standardisation, Europe

Geoff Mays is Professor of Civil Engineering and Deputy Principal for Cranfield University at the Royal Military College of Science, Shrivenham, UK. He specialises in the repair and strengthening of concrete structures, with particular reference to the use of structural adhesives in bonded external reinforcement. He has been European Convenor of CEN/TC 104/SC8 – Protection and Repair of Concrete - since 1996 and of its Working Group on Structural Bonding Products since 1990.

INTRODUCTION

At either the design or assessment stages, new or existing concrete structures may be deemed to need protection from potential future damage or deterioration. Alternatively, for various reasons, such structures may suffer damage or deterioration during their working lives and as a result may subsequently require repair. The design of an appropriate protection or repair scheme is a complex process involving:

- Assessment of the condition of the structure;
- Identification of the causes of deterioration;

- Deciding the objectives of protection and repair;
- Selection of the appropriate principles for protection and repair;
- Selection of methods;
- Definition of properties of products and systems;
- Specification of maintenance requirements following protection and repair.

The selection of products and systems for the protection and repair of concrete structures requires consideration of all of the above factors in order to ensure that the materials are appropriate for the intended use. New packages of European Standards are currently in an advanced stage of development to provide guidance in this selection process.

EUROPEAN CONCRETE REPAIR STANDARDS

European Standards for the protection and repair of concrete are being drafted by Sub-Committee 8 (SC8) of CEN Technical Committee 104, the secretariat being provided by the Association Française de Normalisation (AFNOR). The mirror organisation in the UK is BSI Committee B/517/8 and its members are playing a key role in developing the standards. The basic structure and responsibilities within SC8 for preparing the Standard pr EN 1504 – Products and systems for the protection and repair of concrete structures – is summarised in Table 1.

Each of the key product performance requirements of the relevant materials are being embodied within the "Specification Standards" pr EN 1504 Parts 2 to 7, respectively. These standards are intended to be "harmonised" standards in support of the essential requirements of the Construction Products Directive of the European Commission which will allow "CE" marking of products. They define the main performance characteristics of the various types of protection and repair products and specify both the "identification" and "performance" requirements associated with each characteristic. The Working Groups are also developing "Test Method Standards" to evaluate material characteristics against these requirements. Over 60 new test method standards are included within the work programme of SC8. prEN 1504 Part 8 specifies procedures for quality control and evaluation of conformity, including initial type testing, factory production control and the marking and labelling of products. It is currently on circulation for public comment.

ENV 1504 Part 9, which defines the general principles for the use of products and systems for the protection of repair of concrete was published as a "Draft for Development" in 1997. Its sequel, pr EN 1504 Part 10 dealing with site application and quality control, is at an advanced stage of preparation and is expected to be published shortly.

Table 1.   Structure of CEN/TC104/SC8 and pr EN 1504

STANDARD	ACTIVITY	STATUS	GROUP	CHAIRMAN
EN 1504 Part 1[1]	General scope and definitions	Published 1998	SC8	Prof GC Mays (UK)
pr EN 1504 Part 2	Surface protection	CEN enquiry	WG1[1]	Dr R Stenner (Germany)
pr EN 1504 Part 3	Structural and non-structural repair	CEN enquiry	WG2	Mr JDN Shaw (UK)
pr EN 1504 Part 4	Structural bonding	CEN enquiry	WG3	Prof GC Mays (UK)
pr EN 1504 Part 5	Concrete injection	CEN enquiry	WG4	Mr J Wiertz (Belgium)
pr EN 1504 Part 6	Grouting to anchor reinforcement or to fill external voids	In preparation	TG5[2] within WG2	Mr JDN Shaw (UK)
pr EN 1504 Part 7	Reinforcement corrosion prevention	In preparation	WG7	Prof M Raupach (Germany)
pr EN 1504 Part 8	Quality control and evaluation of conformity	CEN enquiry	Ad-hoc Group	Dr H Davies (UK)
ENV 1504 Part 9[2]	General principles for use of products and systems	Published 1997	WG7	Prof HR Sasse (Germany)
pr EN 1504 Part 10	Site application of products and systems and quality control of the works	Formal vote	WG9	Mr F Dyton (UK)

Notes
(1)   WG = Working Group
(2)   TG = Task Group

EC MANDATE

The European Commission (EC) has issued the "Mandate" M/128 to CEN for standardisation work on harmonised standards for "Products related to concrete, mortar and grout". A mandate is a political request from the EC, as agreed upon by the Member States, in support of legislative work or an industrial policy action from the EC.

Product mandates lead to the development of "Harmonised Standards" in support of "Essential Requirements" which allow the CE marking of the products.   For concrete protection and repair products the harmonised standards will be the EN 1504 series.   The essential requirements of the EC are laid down in the "Construction Products Directive" (CPD) [3] and are summarised in Table 2.

Table 2   Essential requirements of Annex I of the Construction Products Directive (CPD)

CODE	REQUIREMENT
1	Mechanical resistance and stability
2	Safety in case of fire
3	Hygiene, health and the environment
4	Safety in use
5	Protection against noise
6	Energy economy and heat retention

The mandate also directs CEN as to which system of attestation of conformity shall be specified in the harmonised standards. The systems of attestation of conformity under the CPD are summarised in Table 3. For concrete protection and repair products the following systems will be used:

- For uses with low performance requirements in buildings and civil engineering works – System 4;
- For other uses in buildings and civil engineering works – System 2+.

SC8 advises that System 2+ be reserved for use in the repair and protection of reinforced or prestressed concrete designed to Eurocode 2 (EC2) and use of repair materials associated with the essential requirements 1 to 4 of the CPD.

Table 3   Systems of Attestation of Conformity under the CPD

ANNEX III OF CPD	2(i)		2(ii) First		2(ii) Second	2(ii) Third
Conformity attestation: CEC numbering system	1+	1	2+	2	3	4
Tasks for the manufacturer						
1 Factory production control	✓	✓	✓	✓	✓	✓
2 Further testing of samples taken at factory according to prescribed test plan	✓	✓	✓	-	-	-
3 Initial type testing	-	-	✓	✓	-	✓
Tasks for the approved body						
Initial type testing	✓	✓	-	-	✓	-
4 Certification of FPC	✓	✓	✓	✓	-	-
5 Surveillance of FPC	✓	✓	✓	-	-	-
6 Audit testing of samples taken from 7 the factory on the market or on construction sites	✓	-	-	-	-	-

2(i)	=	certification of conformity of the product by an approved certification body
2(ii)	=	declaration of conformity by the manufacturer
FPC	=	factory production control

GENERAL PRINCIPLES FOR USE AND SITE APPLICATION

The scope of ENV 1504 Part 9 includes:

- The need for inspection, testing and assessment before, during and after repair;
- Protection from and repair of defects caused by the influence of certain environments and chemical substances;
- The repair of defects from such causes as mechanical damage, differential settlement, loading, biological attack, inadequate construction or the use of unsuitable construction materials;
- Protection and repair in order to decrease the progress of alkali-silica reaction;
- Meeting the required structural capacity in repair by;
    - replacement or addition of embedded or external reinforcement;
    - filling of external voids between elements to ensure structural continuity;
- Meeting the required structural capacity by replacement or addition of concrete;
- Waterproofing as an integral part of protection and repair;
- Protection and repair of pavements, runways, hard standings and floors, as an integral part of protection and repair;
- Methods of protection and repair including:
    - treating cracks;
    - restoring passivity to reinforcement;
    - reducing the rate of corrosion of reinforcement by limiting moisture content;
    - reducing the rate of corrosion of reinforcement by electrochemical methods;
    - controlling corrosion of reinforcement with coatings.

The basis for the choice of products and systems is founded on 11 principles of protection and repair. These are based on the chemical and physical laws which allow prevention or stabilisation of the chemical or physical deterioration processes in the concrete or the electrochemical corrosion processes on the steel surface. These 11 principles, and associated methods of protection and repair covered by pr EN 1504, are summarised in Table 4 for defects in concrete and for reinforcement corrosion, respectively.

Whereas Part 9 is concerned with principles of use of products and systems, Part 10 deals with site application and quality control. As such its scope may be defined as:

- The preparation of the concrete or reinforcement before application of products and systems
- The minimum requirements as to environmental conditions for storage and application of products and systems
- Controlling the quality of the repair work

Table 4   Principles and Methods related to defects in concrete
and reinforcement corrosion

	PRINCIPLE	DEFINITION	METHODS OF PROTECTION AND REPAIR
1	Protection against ingress	Reducing or preventing the ingress of adverse agents	Impregnation Surface coating Filling cracks
2	Moisture control	Adjusting and maintaining the moisture content in the concrete within a specified range of values	Hydrophobic impregnation Surface coating
3	Concrete restoration	Restoring the original concrete of an element of the structure to the originally specified shape and function	Applying mortar by hand Recasting with concrete Spraying concrete or mortar
4	Structural strengthening	Increasing or restoring the structural load bearing capacity of an element of the concrete structure	Installing bonded rebars Plate bonding Adding mortar or concrete Injecting cracks, voids or interstices Filling cracks, voids or interstices
5	Physical resistance	Increasing resistance to physical or mechanical attack	Overlays or coatings Impregnation
6	Resistance to chemicals	Increasing resistance of the concrete surface to deterioration by chemical attack	Overlays or coatings Impregnation
7	Preserving or restoring passivity	Creating chemical conditions in which the surface of the reinforcement is maintained in or is returned to a passive condition	Increasing cover to reinforcement Replacing contaminated or carbonated concrete Realkalisation of carbonated concrete by diffusion
8	Increasing resistivity	Increasing the electrical resistivity of the concrete	Limiting moisture content by surface treatments, coating or sheltering
9	Cathodic control	Creating conditions in which potentially cathodic areas of reinforcement are unable to drive an anodic reaction	Limiting oxygen content by saturation or surface coating
10	Cathodic protection [1]		
11	Control of anodic areas	Creating conditions in which potentially anodic areas of reinforcement are unable to take part in the corrosion reaction	Painting reinforcement with coatings containing active pigments Painting reinforcement with barrier coatings Applying inhibitors to the concrete

[1]   Covered by EN 12696-1

HARMONISED SPECIFICATION STANDARDS

The Specification Standards pr EN 1504 Parts 2 to 7 will identify the performance characteristics of the particular family of products and systems, which are associated with the principles and methods of protection and repair described in Part 9.

For example, Table 5, reproduced from pr EN 1504 Part 4, lists the performance characteristics for structural bonding agents associated with the principle "Structural Strengthening" (see Principle 4, Table 4) for each of the repair methods "Bonded plate reinforcement" and "Bonded mortar or concrete". Further tables in the Specification Standards then define the "Identification" and "Performance" requirements and the associated "Test Method" Standards.

Table 5. Performance characteristics for structural bonding agents

MAIN PERFORMANCE CHARACTERISTICS	PRINCIPLE OF REPAIR STRUCTURAL STRENGTHENING REPAIR METHOD	
	Bonded Plate Reinforcement	Bonded Mortar or Concrete
1 Suitability for application: a) to vertical surfaces & soffits b) to top horizontal surfaces c) by injection	 □ □ □	 □ □ □
2 Suitability for application and curing under the following special environmental conditions: a) low or high temperature b) wet substrate	 □ 	 □ ■
3 Adhesion: a) plate to plate b) plate to concrete c) corrosion protected steel to corrosion protected steel d) corrosion protected steel to concrete e) hardened concrete to hardened concrete f) fresh concrete to hardened concrete	 ■ ■ □ □ 	 ■ ■
4 Durability of composite system: a) thermal cycling b) moisture cycling	 ■ ■	 ■ ■
5 Material characteristics for the designer: a) open time b) workable life c) modulus of elasticity in compression d) modulus of elasticity in flexure e) compressive strength f) shear strength g) glass transition temperature h) coefficient of thermal expansion i) shrinkage	 ■ ■ ■ □ ■ ■ ■ ■	 ■ ■ ■ □ ■ ■ ■ ■ ■

■ = a material characteristic which shall be considered for all intended uses
□ = a material characteristic which shall be considered for certain intended uses

An identification test is carried out to verify a required property of the product or system in terms of consistency of production. The requirements are usually stated in terms of the property falling within a specified percentage of a declared value provided by the manufacturer. A performance test is undertaken to verify directly a required property of the product or system in terms of its specified performance during application and use.

The requirements may be stated in terms of a threshold value, a declared value or as a pass/fail criteria. Examples of how the identification requirement for "pot life" and the performance requirement for "adhesion" of structural bonding agents are specified are given in Table 6.

Table 6. Examples of Identification and Performance Requirements

IDENTIFICATION REQUIREMENTS

Property	Test Method	Requirements
Pot life	EN ISO 9514	Declared value ±20%

PERFORMANCE REQUIREMENTS

Principle:	Structural Strengthening		
Repair Method:	Bonded Plate Reinforcement		
Property	Reference concrete/mortar	Test Method	Requirements
Adhesion	pr EN 1766 MC (0.40)	pr EN 12188	The tensile stress carried by the bonded joint in a pull off test shall not be less than 15 MPa.
			The slant shear strength of scarf-jointed prisms tested in compression at various angles θ shall not be less than the values x MPa tabulated below.

θ	x(MPa)
50°	50
60°	60
50°	70

Each Specification Standard includes an "Annex ZA" which contains those clauses that specifically address the provisions of the CPD. It identifies those performance characteristics which the mandate M/128 specifies shall be covered by the harmonised standard, together with the relevant requirements clauses.

For structural bonding products these essential characteristics are:

- Bond/adhesion strength
- Shear strength
- Compressive strength
- Shrinkage/expansion
- Workability
- Sensitivity to water
- Modulus of elasticity
- Coefficient of thermal expansion
- Glass transition temperature
- Durability

The Annex ZA also contains information on the relevant systems of attestation of conformity, including the assignation of tasks for the manufacture with respect to initial type testing and factory production control, and for the notified body with respect to certification of factory production control. Finally, guidance is provided on CE marking.

TEST METHOD STANDARDS

pr EN 1504 Parts 2 to 5 make reference to 65 test method standards, the majority of which are new methods being drafted by SC8. In most cases these are based upon proven existing techniques; in other cases further research has been, or will be, necessary before test methods for standardisation purposes can be recommended.

At the time of writing, 14 of the new test methods have been published, 8 are at the stage of formal vote and 29 have been submitted to the CEN enquiry. However, it is not appropriate in a paper of this nature to enter into details of the proposed test methods.

CONCLUDING REMARKS

1. Significant progress has been made within Europe towards the standardisation of products and systems for the protection and repair of concrete since the last Dundee International Conference [4].

2. The development of a coherent set of specification and test method standards within the CEN framework and applicable across the European Union will potentially benefit all branches of the concrete repair industry. The whole industry will be working from a common set of tests and hence will be armed with comparable data. This should result in economic benefits because of the reduced need for multiple-testing and certification procedures.

3. Concrete protection and repair products are included within the EC mandate for products related to concrete. This will lead to the development of harmonised standards in support of the essential requirements of the Construction Products Directive and allow the CE marking of these products.

REFERENCES

1.   BRITISH STANDARDS INSTITUTION. BS EN 1504-1: 1998. Products and systems for the protection and repair of concrete structures – Definitions, requirements, quality control and evaluation of conformity – Part 1: Definitions.

2.   BRITISH STANDARDS INSTITUTION DD ENV 1504-9: 1997. Products and systems for the protection and repair of concrete structures – Definitions, requirements, quality control and evaluation of conformity – Part 9: General principles for the use of products and systems.

3.   COUNCIL OF THE EUROPEAN COMMUNITY. The approximation of laws, regulations and administrative provisions of Member States relating to construction products. Council Directive 89/106 EEC, 21 December 1988.

4.   MAYS, G.C. Materials for the protection and repair of concrete: progress towards European Standardisation. Concrete Durability and Repair Technology (Ed R.K. Dhir & M. J. McCarthy). Proceedings of the International Conference, University of Dundee, 8-10 September 199, Thomas Telford, pp 481-491.

STRENGTHENING OF CONCRETE STRUCTURES WITH EXTERNALLY BONDED REINFORCEMENT: PRACTICAL APPLICATIONS IN BELGIUM

S Ignoul   K Brosens

Triconsult N.V.

D Van Gemert

Catholic University of Leuven

Belgium

ABSTRACT. The strengthening of reinforced concrete structures with externally bonded reinforcement has become common practice. Both steel plates and carbon fibre sheets can be used for strengthening in bending and in shear. It is also possible to combine the two materials. In that way the specific properties and benefits of steel and CFRP are used. Steel plates are rather cheap and have good tensile and stiffness properties. They are most effective when used in bending to reduce or limit deflections and deformations. CFRP laminates are lightweight and have a high tensile strength. They are very appropriate to carry tensile loads in bending and in shear. Rather complex shapes can be applied with the flexible CFRP sheets. Over the last few years, an increasing number of projects concerning concrete strengthening with externally bonded CFRP reinforcement was executed. The paper discusses two important projects executed in Belgium. Each project will be discussed from the preliminary material investigation and the design to the supervision and inspection on site. The used techniques are clarified. Special attention is paid to the anchoring problems.

Keywords: Concrete strengthening, Externally bonded reinforcement, CFRP bonding, Steel bonding, Hybrid solutions, Steel/CFRP.

Mr S Ignoul, is Project Engineer at Triconsult N.V., a spin-off company of K.U.Leuven (B). Triconsult's major activity is the investigation, testing and development of materials, and design of components and structures. Triconsult carries out design and stability analyses as well as supervision and coordination of strengthening and consolidation projects.

Dr K Brosens, obtained his PhD degree at the Department of Civil Engineering, K.U.Leuven (B) in 2001. His research concerns the strengthening and retrofit of concrete structures with externally bonded steel plates and fibre reinforced materials. At present he is working as a Project Engineer at Triconsult N.V.

Professor D Van Gemert, is Professor of Building Materials Science and Renovation at the Department of Civil Engineering, K.U.Leuven (B). He is Head of the Reyntjens Laboratory for materials testing. He is actively involved with research and projects on repair and strengthening of constructions, deterioration and protection of building materials and concrete polymer composites.

INTRODUCTION

From the end of the 1970's, the first practical applications of repairing or retrofitting of existing structures with externally bonded reinforcement took place in Belgium. At the beginning, only steel plates were used. Both bending and shear strengthening works were dealt with. The first years, very few applications were realized but the number of applications increased year after year. At the end of the 1980's the technique of externally bonded steel reinforcement had become common practice.

From the middle 1990's, new materials like CFRP became valuable alternatives for steel. Nowadays, both materials, steel and CFRP, are frequently used for repairing and strengthening of structures.

The preliminary material investigation and the structural analysis for numerous applications in Belgium was done by the Reyntjens Laboratory of the Department of Civil Engineering of K.U.Leuven. The design of the external reinforcement and the supervision of the repairing works on site was done by Triconsult N.V., a spin-off design office of the Department of Civil Engineering of K.U.Leuven. This combination of theoretical research and participation in practical applications, allowed to build up a great experience in the field. In this paper, two recently executed and representative case studies are presented. The first case study deals with a hybrid strengthening case. Both steel plates and CFRP laminates are used for strengthening the ribbed floor slab of a future library in bending and in shear. This hybrid strengthening technique combines the benefits of both materials. The second case study discusses the strengthening of a roof structure made with fibre-reinforced cement wallboard panels. Due to an overload on the roof, several roof panels were cracked. They were strengthened with externally bonded CFRP-laminates.

CITY LIBRARY AT LEUVEN

Introduction

In 1998, a former school building in Leuven, Belgium, was transformed into a city library with a considerable increase of load as a consequence. The floor slabs had to be strengthened to increase the bearing capacity from 3 kN/m^2 to 6 kN/m^2. These floor slabs consist of ribs spaced every 55 cm. The thickness of the floor slab is 50 mm.

An extensive material investigation was done to determine the material properties and the condition of the construction [1]. Six concrete cores (Ø113 mm) were drilled to determine the concrete compressive strength, resulting in a characteristic value of 22.1 N/mm^2. The concrete tensile strength at the surface was measured by a pull-off test, giving 2.96 N/mm^2. The location and the dimensions of the internal steel reinforcement were found using electro-magnetic waves. The longitudinal reinforcement in the ribs consists of two rebars Ø16 mm. No internal steel stirrups were found.

The concrete was not affected chemically. No steel corrosion could be observed. The chloride and sulphate content were far below the maximum allowable values, whereas the carbonation depth was restricted to a few millimeters.

Both additional bending reinforcement and shear reinforcement were required, since there were no internal stirrups present in the ribs of the floor slab. The idea arose to use a hybrid strengthening method. Steel plates could be used as bending reinforcement to increase the bearing capacity and to limit the additional deflections, whereas CFRP laminates could be applied as shear reinforcement. A very cost effective way would be the application of the CFRP laminates at only one side of the beam, as a strengthening method against shear forces. Since such a hybrid strengthening solution was not yet done in Belgium before and no examples from the literature were available, a small preliminary test program was executed to check the feasibility of the proposed hybrid system. This test program is discussed in [2,3].

On Site Realization

The strengthening procedure of the ribbed floor slab was twofold, Figure 1 [4]. Firstly, externally bonded shear reinforcement had to be provided since no internal steel stirrups were present. The decision was taken to use CFRP sheets. The CFRP sheets used are Forca Tow Sheets FTS-C1-30. The carbon fibres have a tensile strength of 2450 MPa and the equivalent fibre cross section of one layer CFRP is 167 mm^2/m width. The Young's modulus ECFRP is 235000 MPa [7].

The experimental program learned that externally bonded CFRP sheets at one side of a beam as shear reinforcement are almost as effective as CFRP sheets bonded at both sides of a beam. For that reason, two layers of CFRP sheets were applied at only one side of the ribs in order to increase the shear capacity of the floor slab. For the first layer the carbon fibres are oriented vertically while for the second layer, the carbon fibres are oriented horizontally. Before bonding the CFRP sheets, the concrete surface was roughened slightly by sandblasting. Thereafter the surface is cleaned carefully and an epoxy primer is used to guarantee good bonding. Then the first layer of CFRP is applied. A roller is used to give a good penetration of the resin through the laminate. It is very important that every fibre is surrounded by epoxy resin to guarantee full composite action. After four to five hours, the second layer is applied.

Secondly, the flexural rigidity of the ribs had to be increased in order to carry higher bending loads. Therefore an externally bonded steel plate (70 x 14 mm^2) was applied at the bottom side of each rib. The anchorage of this steel plate is done by two bolts Ø 16 mm. Before gluing the steel plate, the concrete surface is roughened by sandblasting and cleaned carefully. A filled epoxy glue is used to bond the steel plate to the concrete surface. The plate end shear crack [6] is prevented by a CFRP stirrup with a width of 150 mm, Figure 2. Before applying this stirrup, all cavities have to be filled, the corners have to be rounded and the surface has to be smoothened with an epoxy repair mortar.

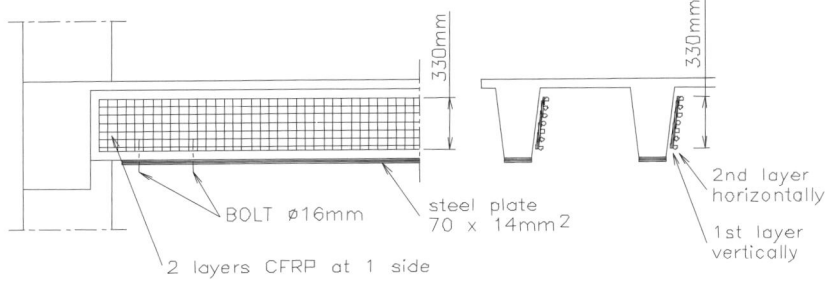

Figure 1 Hybrid strengthening of the ribbed floor slab

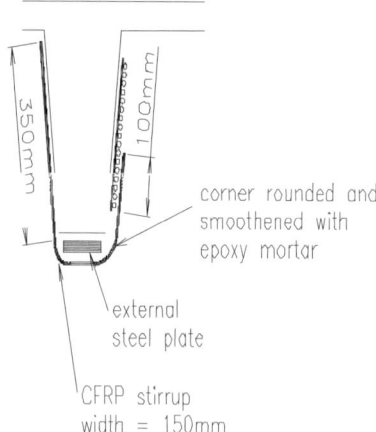

corner rounded and
smoothened with
epoxy mortar

external
steel plate

CFRP stirrup
width = 150mm

Figure 2   CFRP stirrup to prevent plate end shear crack

Figures 3 and 4 show the repair works and the final result. The application of the CFRP sheets was very easy. Especially when there is a high degree of repetition, labour costs can be kept very low. One skilled worker can easily bond the CFRP sheets to one side of the ribs one by one. When he has finished the first layer of CFRP on the last rib of the floor slab, the CFRP sheet on the first rib has already hardened enough and the second layer of CFRP can be applied. In this way, he can complete the whole floor slab without any loss of time. The alternative, bonding steel stirrups, requires much more working hours and is therefore less economical.

Figure 3   Application of the CFRP sheets

Figure 4   Hybrid strengthening of ribbed floor slab

SWIMMING POOL AT TERVUREN

Introduction

The roof structure of the swimming pool at Tervuren, Belgium, consists of fibre-reinforced cement panels. These panels were originally manufactured to be used as wall panels, but were also often used as roof panels for small swimming pools. The great benefit of these kind of panels is the absence of corrosion sensitive interior steel reinforcements. During the renovation of the swimming pool, the false ceiling was removed and large deformations of some roof panels could be observed. One panel was even cracked. In each panel several holes has been drilled for the fixation of the false ceiling and the electrical wiring.

Due to the presence of asbestos in the roof panels, the removal of these panels would be very expensive and a strengthening solution with externally bonded CFRP-sheets was proposed.

Repair and strengthening strategy

The main concern for these kind of panels is that they are very brittle, which means that no warning is given before collapsing. In the past, a sudden roof collapse due to an overload on such roof panels already happened [7]. These sudden collapses can be initiated around small holes in the roof panel. The tensile stress around a hole in the roof panel can mount up to three times the average stresses in the panel. All the holes, already present in the roof panels of the swimming pool of Tervuren, are potentially dangerous and needed to be strengthened.

The second problem is the presence of asbestos. For health reasons, it was not possible to roughen the surface of the roof panels.

Figure 5   Test cyclinders

Therefore, it was necessary to test the bonding strength of a smooth roof panel. For this reason, 10 aluminium cylinders were glued to the panels to test the bonding capacity. Two epoxy resins were used with different viscosities. Before gluing the cylinders, the surface was cleaned with a solution of methylene chloride. Some of the test cylinders, glued to the roof panel, are shown in Figure 5.

The first epoxy resin (more viscous) gave very poor test results and was therefore rejected for further use. The second resin provided a minimum bonding strength of 0,5 MPa. This value is only a fraction of the normal bonding strength for roughened concrete surfaces, so special care had to be taken to limit the shear stresses between the external reinforcement and the smooth roof panels. For the design of the anchorage length of the externally bonded CFRP-sheets a safety factor of 2 is used for the bonding strength, resulting in a design bonding strength of 0,25 MPa.

On-Site Realization

The total length of one roof panel is 4m. The width of the roof panel is 600 mm. The thickness of the panels is 100 mm. Each roof panel was strengthened with 2 layers of flexible CFRP-sheets with a width of 300 mm [8]. To limit the high shear stress peaks at the end of the reinforced section, the total length of the CFRP-sheets was 3.7 m, almost equal to the complete length of the panel. By doing so, the shear stresses at the end of the sheets are limited to 0,17 MPa < 0,25 MPa.

The application of the CFRP-sheets to the roof panels can be seen in Figures 6 and 7.

Figure 6   Application of CFRP-sheets

Figure 7   Impregnation of CFRP-sheets

The numerous bore drilling holes in the panels are protected with 2 layers of CFRP-sheet with a length of 500 mm and a width of 150 mm. The application of these sheets prevents the initiation of cracks caused by the stress concentrations around the hole. These reinforcements are shown in Figure 8.

Figure 8   Reinforcement of drilling holes

CONCLUSIONS

Externally bonded reinforcement is very effective for strengthening of reinforced concrete structures. A lack of bearing capacity in shear and in bending can be solved by adding additional reinforcement.

The first case study deals with a hybrid strengthening technique. Hybrid CFRP/steel solution benefits from both materials. CFRP laminates are most effective for strengthening in shear by bonding an orthogonal net of carbon fibres at one or two sides of a beam. An increase in shear capacity of about 50 % can be obtained. Research pointed out that CFRP laminates bonded at only one side are almost as effective as those bonded at both sides. Special attention must be given at the anchorage in the end zones of steel plates for bending strengthening. It can be done by anchor bolts. Plate end shear cracks can be prevented by applying a CFRP stirrup at the end of the steel plate. On the basis of this research, a former school building was renovated and transformed into a library. A hybrid solution - CFRP laminates for shear strengthening and steel plates for bending strengthening - was carried out. The ribbed floor slabs were strengthened in a very effective and economical way. The research and experience in hybrid CFRP/steel reinforcements resulted in a lot of similar applications in Belgium, as there are the strengthening of the supporting beams of a floor structure in a warehouse in Denderleeuw [9] or the shear strengthening of a pre-stressed I-shaped concrete beam in Oostende [10].

The second case study deals with the strengthening of brittle fibre cement roof panels. Due to the presence of asbestos it was impossible to adequately prepare the surface of the panels.

So special attention was given to limit the shear stresses in the interface between the externally bonded CFRP-sheets and the roof panels. A special reinforcement was designed to prevent crack growth around bore holes in the panels. The use of CFRP-sheets to reinforce the roof panels avoided a very expensive replacement of the roof.

REFERENCES

1. LADAN,G C, VAN GEMERT, D, Rito gebouw Leuven, structureel onderzoek, Internal Report 28174A, K.U. Leuven Research & Development, 1995.

2. BROSENS, K, AHMED, O, VAN GEMERT, D, IGNOUL, S, Strengthening of R.C. Beams - Hybrid steel/CFRP solutions, Structural Faults & Repair 99, 8th International Conference, 13-15 July 1999, London, United Kindom.

3. BROSENS, K, AHMED, O, VAN GEMERT, D, IGNOUL, S, ULRIX, E, Performance of hybrid CFRP/steel strengthening of RC constructions, Damstruc 2000, 2nd International Conference on the Behaviour of Damaged Structures, 1-3 June 2000, Rio de Janeiro, Brasil.

4. BROSENS, K, VAN GEMERT, D, Rito gebouw Leuven - Dwarskrachtversterkingen met uitwendig gelijmde CFRP laminaten, Internal Report 114, Triconsult N.V., 1998.

5. TONEN FORCA TOWSHEET, Technical Memo, Tonen Corporation, 1997, Japan.

6. JANSZE, W, Uitwendig gelijmde wapening; einde-plaat afschuifmodel voor dwarskracht en verankering, Cement, Vol 50, No 5, May 1998, p 22-26.

7. VAN GEMERT, D, Swimming pool Celestijntje at Heverlee, Internal Report 82-004, 1982, (in Dutch).

8. IGNOUL, S, VAN GEMERT, D, Zwembad Tervuren - gelijmde wapening, Internal report 163, Triconsult N.V., 1999.

9. IGNOUL, S, VAN GEMERT, D, (2000a), Delhaize Denderleeuw - gelijmde wapening, Internal Report 204, Triconsult N.V., 2000.

10. IGNOUL, S, VAN GEMERT, D, (2000b), Kinkhoorn Oostende - dwarskrachtversterking voorgespannen I-ligger, Internal Report 206, Triconsult N.V., 2000.

FRP FOR BRIDGE DECK STRENGTHENING

A F Daly
Transport Research Laboratory
B Sadka
Highways Agency
United Kingdom

ABSTRACT. This paper describes the development of design requirements and guidelines for the use of bond FRP for strengthening concrete bridges, and research into a number of key areas that affect the performance of this technique.

Keywords: Fibre reinforced plastic, FRP, Bridges, Bridge strengthening, Bond, Pull-off tests, Standards.

Albert F Daly, is a Senior Research Fellow at the Structures Department, Transport Research Laboratory, UK, where he is carrying out a programme of research relating to the behaviour of bridges. Research interests include assessment of load capacity, effects of deterioration in concrete, fibre reinforced plastic deck systems and bridge strengthening.

Ben Sadka, is a technical advisor at the Safety, Standard and Research Directorate of the Highways Agency, where he is responsible for design and technical approval issues relating to concrete bridges and the use of new technologies and materials.

INTRODUCTION

The Highways Agency manages around 10,000 trunk road bridges in England. As a result of the deterioration of structural components and increasing volumes and weights of Heavy Goods Vehicles, it is inevitable that a number of these bridges will require strengthening in order to maintain an acceptable margin of safety against collapse.

Bridge strengthening schemes should be economic, minimise disruption to traffic and be aesthetically in sympathy with the structure. Use of Fibre Reinforced Polymer (FRP) has great potential for strengthening because it is strong, light and can be installed quickly using adhesive bonding. Trial structures and laboratory tests in many parts of the world over the past 20 years have provided confidence, and FRPs are now being used where traditional materials do not provide economic long-term solutions.

Technical guides, codes and standards are being developed: these will help ensure safety, efficiency and durability as well as promoting education and training into FRP use. As a major client, the Highways Agency has taken a leading role in UK research and development, and strengthening existing bridges with FRP is a significant step towards full structural use.

BONDED FRP FOR BRIDGE STRENGTHENING

Flexural Strengthening

Because of the ease of handling and speed of application, FRP has potential benefits in the rehabilitation of bridge decks. The strengthening of bridge decks in flexure using bonded FRP laminates has been developing since the late 1980s and has been successfully applied to a number of bridges in the UK[1] and around the world[2]. FRP can be applied as an alternative to the more conventional steel plate bonding. The laminate can consist of a preformed FRP plate which is glued in the same way as a bonded steel plate. Alternatively, the FRP can be supplied in the form of a sheet of fabric to which the resin is applied on site. Multiple layers of the material can be installed to provide the correct amount of strengthening using a process called hand lay-up.

Hand lay-up FRP systems are especially suitable for use in confined spaces, such as the underside of concrete slabs, or in situations where the likelihood of corrosion makes the use of steel plates unsuitable. The flexibility of the FRP laminates makes them much easier to install, particularly where the shape or profile of the concrete surface makes the bonding of rigid steel plates difficult. The complex support system required to lift and hold steel plates in position is not necessary. The use of bolts, which is normally required for steel plate bonding, is not necessary for FRP laminates.

Indeed the use of bolts should generally be avoided as they may result in local damage to the FRP material. They may be used to provide additional anchorage at the ends of the laminates where sufficient anchorage length cannot be provided or where the FRP is to be stressed. This requires a carefully designed anchorage detail. As for steel plates, the bonding operation can take place with no (or minimal) disruption to traffic.

FRP laminate bonding has been used in preference to steel plate bonding for deck strengthening on a number of bridges including the A34 Barnes Bridge. This structure carries the main A34 trunk road into Manchester over the M60 Outer Ring Road, one of the most heavily trafficked and congested sections on the network. In order to ensure safe and consistent introduction of this technique, the HA has recently issued a design standard, BD 85[3] for the use of bonded FRP in bridge strengthening applications.

Shear Strengthening

Bonding FRPs to the webs of beams can be used to increase both flexural and shear strength of a section. For shear strengthening, the bonded material acts as external reinforcement to prevent or limit cracking, and can be designed as external stirrups in the same way as steel links. However, it is more difficult to add shear reinforcement to concrete beams because of the problem of providing sufficient anchorage. This is especially true when the beams are not easily accessible, such as when the beams are closely spaced or contiguous. It is best carried out by completely encasing the web of the beam, for which a flexible material such as FRP sheet or fabric is most suited. To date, the use of FRP for strengthening in shear has been limited to laboratory investigations[4] and limited site trials[5].

As with bonded steel plates, the preparation carried out to the concrete surface is critical to the effective performance of the strengthening system. It is important that all defects such as wide cracks, spalling, etc, are repaired prior to installation. It is likely that strengthening schemes will be carried out on deteriorated structures so that in some cases, the condition of the concrete will not be ideal. All concrete surfaces must be carefully cleaned, be as dry as possible and all loose material removed if an effective durable system is to be achieved.

Strengthening Columns Against Vehicle Impact

In 1997 the Highways Agency embarked on a research programme to investigate the feasibility of using FRP to strengthen sub-standard (in terms of sustaining vehicle impact) bridge supports. On the basis of this research HA has produced a Standard, BD 84[6], for the use of FRP in "column wrapping" applications.

Strengthening circular columns for vehicle impact utilises confinement by hoop fibres and the resulting enhancement in the concrete strength/strain and well as tensile (and compressive when prevented from buckling by an overlay of hoop fibres) reinforcement provided by axial fibre. The technique of wrapping concrete columns with FRP is well-established, particularly in the USA and Japan, where it is used to protect columns against collapse during earthquakes. HA investigated whether the same method could be used to strengthen bridge supports in the UK that were at risk from vehicle impact. The advantages of FRP wrapping as against conventional column strengthening methods are:

- Strengthening can be carried out quickly with minimal need for lane closure.

- Improved aesthetics with little change to the structural form.

- Reduced strengthening and traffic management cost.

On the basis of this research HA developed design rules and simple-to-apply design charts which utilise:

1. Concrete strength enhancement due to confinement by hoop fibres.

2. Additional tensile and compressive (when prevented from buckling by an overlay of hoop fibres) reinforcement provided by axial fibres.

To date a number of bridge columns have been strengthened with FRP. They include the Coopersale Lane Bridge, a three-span reinforced concrete bridge over the very busy M11 artery into London. FRP resulted in savings of about 50% as against conventional strengthening methods and the work was carried out with only minimal need for lane closures of the motorway. In addition, the solution was aesthetically more desirable than the alternative strengthening option of linking columns together with infill concrete.

Development of BD 85

The recently developed Highways Agency Standard[3] is based on laboratory tests and design practice from the UK and abroad. An important issue being addressed that of interface shear and failure. The Standard does not cover the strengthening of metallic structures or the use of prestressed plates or other systems in which FRP is subjected to sustained long-term loading. However, some general guidance on such techniques is provided. Only bridge decks which can be shown to be at least capable of supporting nominal dead plus superimposed dead plus assessment live load with all partial safety factors, including those applied to material strengths, set to unity, are considered suitable for strengthening using this technique. This requirement acknowledges the fact that FRP technology is developing and that the strengthening is dependent the efficacy of the interlaminar bond.

DESCRIPTION OF PROJECT

Objectives

The objective of the research described here is to examine aspects of BD 85 which may be too conservative or unrefined, or where there is some doubt as to the efficacy of their basis, and to produce requirements and guidance, which more accurately represent the performance of bonded FRP.

Methodology

The research project includes a review of literature and current practice to determine the current state of the art. Recent research in this area will be studied and incorporated in the standard. Existing tests and material standards are being reviewed and will be adapted where appropriate. Test methods will be devised for the relevant mechanical properties and durability of FRPs and components. Experimental work will be needed to develop and calibrate certification tests.

The safety and reliability of FRP systems, including strengthening systems, are being investigated through a testing programme. The focus of the work is on bonded systems for particular application to concrete bridge strengthening. A concrete test specimen was devised to study the parameters affecting the strength of the bond between the concrete and a bonded FRP laminate. The following issues are being addressed:

Bond strength: FRP is commonly applied to concrete surfaces, eg, to strengthen a concrete beam, slab or column. All of the force carried by the FRP has to be transferred from the adjacent concrete. The transfer depends on the strength of:

• the near surface concrete.

• the bond between the concrete and resin.

• the resin.

• the bond between the resin and the fibres.

The weakest link in this chain is likely to be the strength of the near-surface concrete.

Transmission length: This is the length over which a given force can be transferred from concrete to FRP. It is related to bond strength and peeling, but is another area where evidence is needed on which to base a design method. Previous testing has indicated that the longitudinal shear stress that can be transferred between the FRP and the concrete is not independent of the bonded length, as is typically assumed in the design of imbedded steel reinforcement. Thus the ultimate anchorage capacity can be much less than the tensile capacity of the FRP.

Surface condition: The condition of the concrete surface has a significant effect on bond strength, but it is difficult to obtain a quantitative measure of condition to enable bond strength to be predicted. Often, there is considerable pressure to carry out work in less than ideal conditions. Loose material and contaminants can be removed by grit blasting, but it may be difficult to dry a large concrete member. The effect of various moisture levels on the performance of the concrete/FRP bond is being studied.

Lack of bond: Defects are inevitable in any material or process. The type most likely to be significant for site applied FRP, especially where the wet lay-up process is used, is incomplete bond between the FRP and the concrete. Small, isolated unbonded areas could initiate failure. It is necessary to quantify the relationship between unbonded area and loss of strength, specify acceptable levels and allow for the effects in design. A related defect, the presence of voids in the FRP, needs to be controlled to an acceptable level. However, this may be specified by reference to existing rules for other industries, so tests may not be needed.

Minimum bend radius: The fibres likely to be used on bridges have very high tensile strength, but some are easily damaged by transverse loading and rely on protection by the resin. Situations where fibres may be wrapped around sharp corners include non-circular columns, shear reinforcement applied to the sides/soffit of beams, and beam-column connections. The cost of providing large radius may be significant so it is necessary to establish the minimum radius needed to for safety for each fibre type (glass, aramid, carbon).

Peeling failure: Peeling failure often occurs before the full strength of the FRP can be achieved. In addition it is a sudden, brittle mode of failure. Peeling is prevented by continuing the FRP beyond the point it is required to a region of low or zero stress. However, evidence is required to support a method of determining the point at which the FRP can be safely terminated.

Concrete cracks: Cracks on the concrete surface will cause local peak strains at the concrete/FRP interface. These peaks may be sufficient to initiate a progressive debonding failure. In some ways, the failure mechanism is similar to peeling. It may be necessary to require limits on crack widths.

Shear strengthening: The actual (as opposed to design) shear capacity of reinforced concrete members is difficult to predict, being a function of concrete strength, area of main reinforcement, aggregate interlock, dowel action, and shear span. The addition of external FRP is an added complication. Although many small studies have been reported in the literature and many strengthened beams have been tested, a design method suitable for general use is not yet available.

The additional shear capacity provided by the FRP depends on several parameters. It is beyond the scope of the present project to solve the question of shear in sufficient detail to produce clauses for a design standard. Instead existing data will be collated and used to see if simplified rules can be drawn up as an interim measure and to specify the requirements for a design method suitable for a design standard.

Test Specimens

The proposed experimental work utilises a range of tests. The preliminary tests are being carried out on small scale concrete prisms, while further testing will comprise a series of 3m beams loaded using single or two point loading. The proposed test programme focuses on the parameters outlined in the previous section.

The prism specimens were devised with the objective of maximising the control and accuracy of the test procedure. The prisms were designed to be loaded in tension in a testing machine, which has the advantage of offering greater precision and control than a test rig. The machine used is an Instron 150kN two-column computer controlled screw-jack machine. It can be programmed to carry out test procedures at pre-determined strain rates.

The specimens were produced as shown in Figure 1, using a standard 100mmx100mmx 500mm cast iron mould. This produces prisms with the accuracy and repeatability necessary for a comparative study. The moulds are set up with perforated end plates and plastic tube to enable threaded sockets to be positioned through the centre of the prism. A void is incorporated in the centre of the prism to facilitate the insertion of bolts onto the sockets. Steel anchor plates are cast into the prism at the ends of the void to provide bearing surfaces. After removing the prism from the mould, it is clamped using a threaded bar to prevent the unreinforced concrete from cracking.

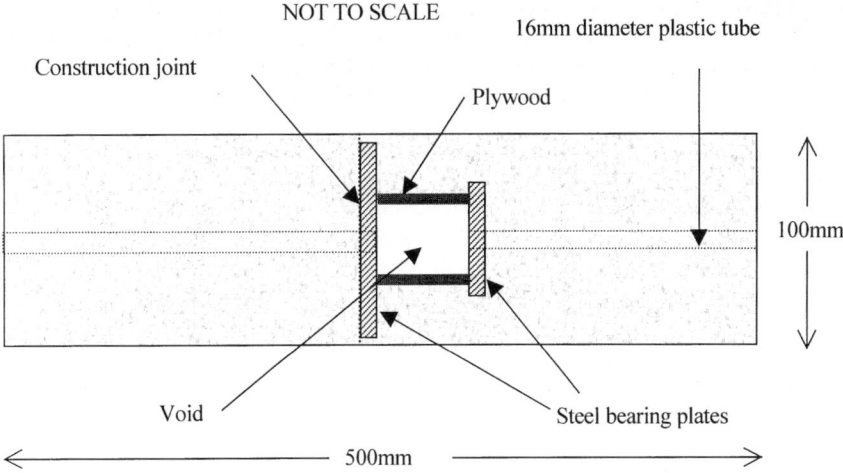

Figure 1   Prism specimen

A debonding joint was incorporated in the specimen by placing a greased plastic sheet against one of the steel plates. This predetermines the bond length being tested. The FRP was bonded onto the sides of the specimen at least 28 days after casting, when a concrete strength of about $50N/mm^2$ was achieved. The recommendations of the FRP supplier was followed in terms of the concrete surface preparation and the application of primers, adhesives and overcoats. After the specimens were cured, the threaded sockets were put in place, with nuts provided to bear on the cast-in plates. These sockets were then fitted to the testing machine cross-heads and actuator. Figure 2 shows a typical specimen after failure.

Figure 2   Photograph of a failed specimen in test machine

Results to Date

In all, 45 tests have been completed to date. The following paragraphs summarise the conclusions derived to date based on a preliminary examination of the test data.

Figure 3 shows the pull-off strength of carbon and aramid laminates as a function of bonded length. For very short bonded lengths, the shear capacity is clearly a function of bonded length, but this quickly reaches a maximum value at a bond length of about 75mm for both carbon and aramid FRP. In all cases, the failure was instigated by rupture of the near surface concrete. The pull-off strength was only about 20% of the ultimate strength of the carbon FRP, with slightly higher values recorded for aramid. Values for glass FRP are not yet available.

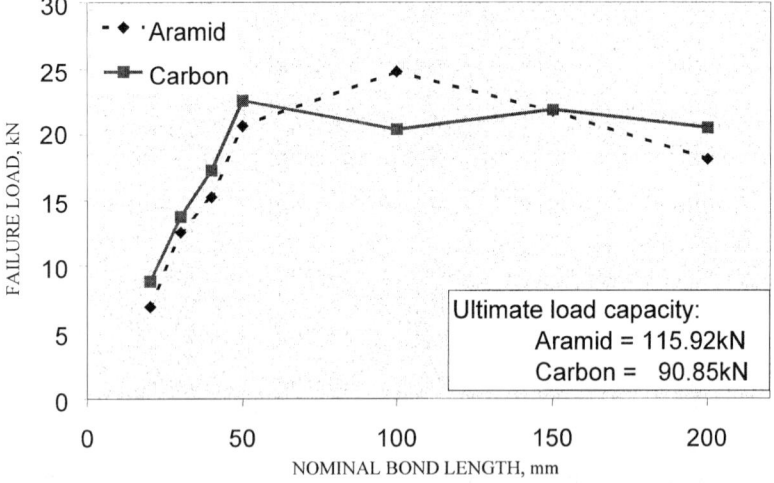

Figure 3   Failure load as a function of bond length

Figure 4 shows the effect of using multiple layers of carbon FRP for a bond length of 50mm. This shows that the increased strength from using multiple layers is not in proportion to the additional FRP. In fact, expressed as a ratio of the FRP strength, the failure load falls from 25% to less than 10%. The performance of bonded FRP systems depend to a great extent on the preparation of the bonded surfaces. To quantify this effect, a number of tests were carried out with defects built into the bond to simulate less than ideal surface preparation. Figure 5 illustrates the change in pull-off capacity as a function of the following defects:

- Central transverse de-bonded strip
- Central longitudinal de-bonded strip
- Rear transverse de-bonded stripB
- Moist concrete surface: water retained on paper
- Moist concrete surface: Water not retained on paper
- Bonding carried out in 100% relative humidity.

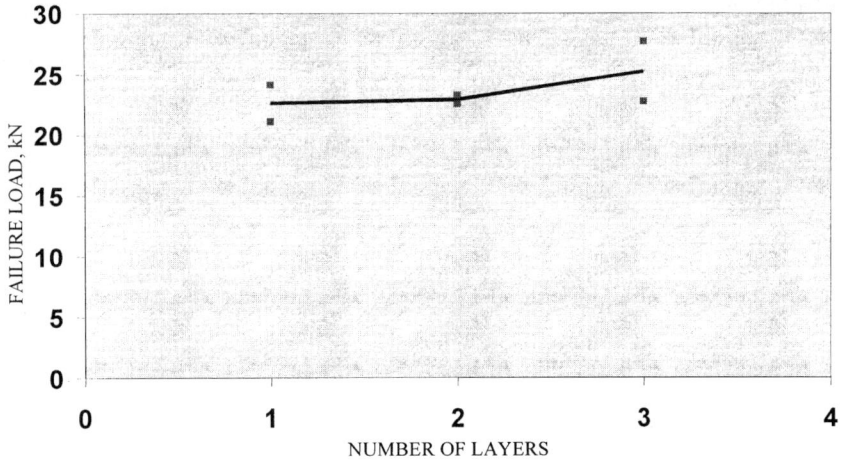

Figure 4   Failure load as a function of number of layers of carbon

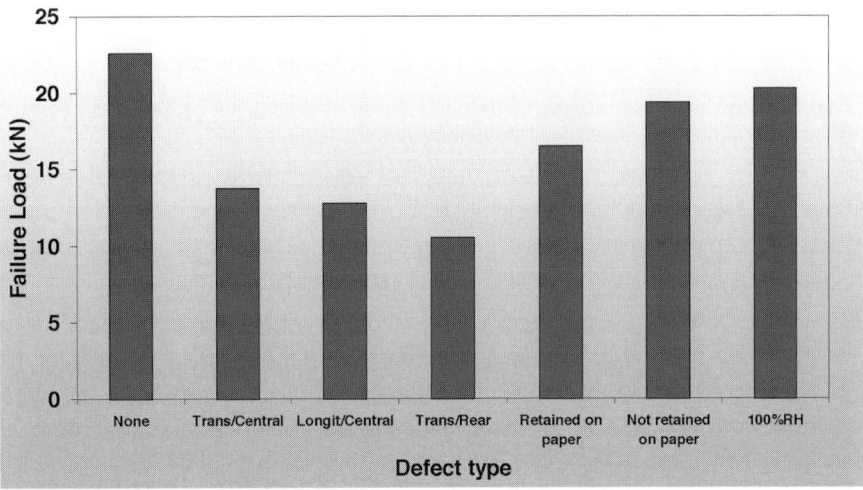

Figure 5   Failure load as a function of defect

The figure indicates that the bond strength is very dependent on surface preparation, particularly where de-bonded areas are present. These pull-off tests are continuing and will be reported fully in due course. To date, no beam tests have yet been carried out.

CONCLUSIONS

The paper describes an on-going research project into the performance of bonded FRP systems and has highlighted the differences between the behaviour of FRP and conventional structural materials. The project focuses on the behaviour of the bond between concrete and FRP and how it is affected by a number of parameters.

Preliminary results confirm that increasing the length of the bond does not increase the anchorage bond strength and that careful attention to detail is required if the possibility of brittle failure is to be avoided. The importance of surface preparation is also being investigated and the initial testing has shown that the presence of defects can have a considerable detrimental effect on the bond between FRP and concrete.

FRPs provide economic and rapidly installed bridge strengthening options and there is little doubt that they will play an ever-increasing role in the future. However, engineers should be aware that in FRPs, they are using hi-tech materials that require a level of care in specification and application, which is not usually demanded in conventional construction. There is no doubt that, in the same way, our knowledge of the behaviour and performance of FRPs will constantly be updated and their use refined. The Highways Agency is keen to see a wider use of FRPs in bridge construction and strengthening applications, as research and good practice permit.

REFERENCES

1. HOLLAWAY, L., LEEMING.   Structural strengthening with bonded fibre-reinforced polymer composites. Woodhead Publishing, 1998.

2. MEIER, U.   Carbon fibre-reinforced polymers: modern materials in bridge engineering. Structural Engineering International, IABSE, Zurich, Switzerland, 1992.

3. HIGHWAYS AGENCY.   BD 85: Strengthening of concrete highway structures using externally bonded fibre reinforced polymers. HMSO, 2002.

4. AL-SULAIMANI, G. J., A. SHARIF, I. A. BASUNBUL, M. H. BALUCH, B. N. GHALEB.   Shear repair for reinforced concrete by fibreglass plate bonding.   ACI Structural Journal, Title No 91-S45, July-August, 1994.

5. DRIMOUSSIS, E., J. J. R. CHENG.   Shear strengthening of concrete bridge girders using carbon fiber-reinforced plastic sheets.   Conference Proceedings 7: Fourth International Bridge Engineering Conference, Transportation Research Board, USA, 1995.

6. HIGHWAYS AGENCY.   BD 84: Strengthening of concrete bridge supports for vehicle impact using fibre reinforced polymers. HMSO, 2002.

FIBRE REINFORCED MORTAR FOR THE REHABILITATION OF HISTORIC RC BUILDINGS – THE LIRICK THEATRE IN ASSISI

M Mezzi   **R Radicchia**

University of Perugia   Freelance Engineer

G Mantegazza   **A Gatti**

Ruredil Chemicals

Italy

ABSTRACT. The "Montedison Factory" in Assisi, an existing reinforced concrete industrial building over fifty years old, has been retrofitted to transform it in the "Lirick Theatre". Due to the age of the construction, the concrete was dramatically decayed. Structural works has been carried out in order to rehabilitate the r/c decayed elements and to strengthen the building to resist the actions prescribed by present codes, including earthquake. Traditional steel and r/c elements, respectful of the architectural and historical worth, have been inserted to strengthen the structure. Innovative materials have been widely used: CFRP strips have been used to improve tensile strength; "Exocem PVA", a Synthetic Fibre Reinforced Mortar (SFRM), have been laid on the shell surfaces both to rehabilitate and to strengthen the thin shells of the construction. Both the design and the realization of the building are illustrated. The design strategies and the strengthening techniques used in the rehabilitation are reported, making special reference to the specific new materials and adequate application techniques adopted for the repair, rejuvenation and enhancement of existing r/c structures.

Keywords: Fibre reinforced mortar, Carbon fibre reinforced plastic, Rehabilitation, Retrofitting, Strengthening, Historical buildings, Reinforced concrete.

M Mezzi is researcher at the Department of Civil and Environmental Engineering and Professor in charge of Testing, Inspection and Control of Construction at the University of Perugia (Italy). His researches focus mainly on seismic design and structure retrofitting.

R Radicchia is freelance Civil Engineer operating in Perugia (Italy). He has carried out the structural design of a number of new and retrofitted buildings.

G Mantegazza is the Technical Manager of Ruredil Chemicals for buildings, located in San Donato Milanese, Milano (Italy).

A Gatti is Ruredil Technical Manager's assistant. She is in charge of New Composites Materials Laboratory.

INTRODUCTION

The preservation of the existing constructions representing a historical witness of their period, independently from their specific aesthetic value, is a requirement of the modern architectural thought. Therefore, the concept of cultural heritage is spreading out and, for example, a careful attention of architectural culture is now turned to the re-evaluation and re-use - as social, public or service building - of the existing industrial settlements, dismantled in the last years, due to the obsolescence of traditional production techniques, the growth of the new economy alternative to the industrial production, the urban development of the cities which have encompassed old industrial suburbs pushing out factories. The rehabilitation of historical buildings always involves structural aspects because it usually requires both retrofitting and strengthening works, depending on the decay of materials associated to age and neglect, and the need for adjustment of the strength to modified load levels and safety factors. In this paper, the project of the Montedison factory in Assisi, an existing r/c industrial building over fifty years old, retrofitted to be transformed into the "Lirick Theatre", is illustrated. Structural aspects connected to the use and application techniques of specific new materials - Carbon Fibre Reinforced Plastic (CFRP) and Fibre Reinforced Mortars (FRM) - adopted in retrofitting, are discussed. The design strategies, the project planning, and the strengthening techniques used in the rehabilitation are described. Authors thank Dr Arch Vincenzo Maia who has drawn up the architectural design, "La Fenice" who was the construction firm, and Ruredil Spa who furnished the FRM and CFRP.

EXISTING BUILDING

The building to be transformed into the "Lirick Theatre" is a part of an industrial complex located in a suburb of the city of Assisi, it has a significant architectural value, being attributed to the school of the Italian architect Pier Luigi Nervi. It was built over fifty years ago and it hosted Montedison chemical working. The activity was interrupted many years ago and the building was abandoned. The building consists of a 30 m wide and 16 m high longitudinal shed, having regular rectangular plan and two transversal foreparts at its centre: the general plan is thus cross-shaped. The longitudinal shed has a parabolic cross section with 8 m long external penthouses. The central foreparts have two stories with a "pilotis" ground level and depressed barrel r/c vaults as roof. Over these foreparts the parabolic body has a multi-storey superstructure rising up to 22 m. Figure 1 shows an image of the building before the works and drawings of the plan and the longitudinal and transversal sections.

The project regards only the central part and the eastern parabolic shed of the entire complex: the western shed was already retrofitted and it hosts a county sporting centre. The building is divided into three sections by structural gaps. The first two sections are made up of four transversal parabolic arches, located at a distance of about 10 m, at 5 m from the ground level, lateral cantilevers jut out from the arches and support the cylindrical shell of the penthouses; above the cantilever level the shed is covered by a thin shells, having a 70 mm medium thickness, spanning among the arches. The central body has a framed superstructure based on the current parabolic arches having reduced distance in this section, and an independent framed structure is present at the interior. The building stands on non-connected shallow footings posed at a 2.5-3.0 m depth on alluvial of medium density. The most important characteristic of the building is the complex equilibrium, which controls the stability of the single curvature very thin shells and required a careful definition of the critical rehabilitating and strengthening works.

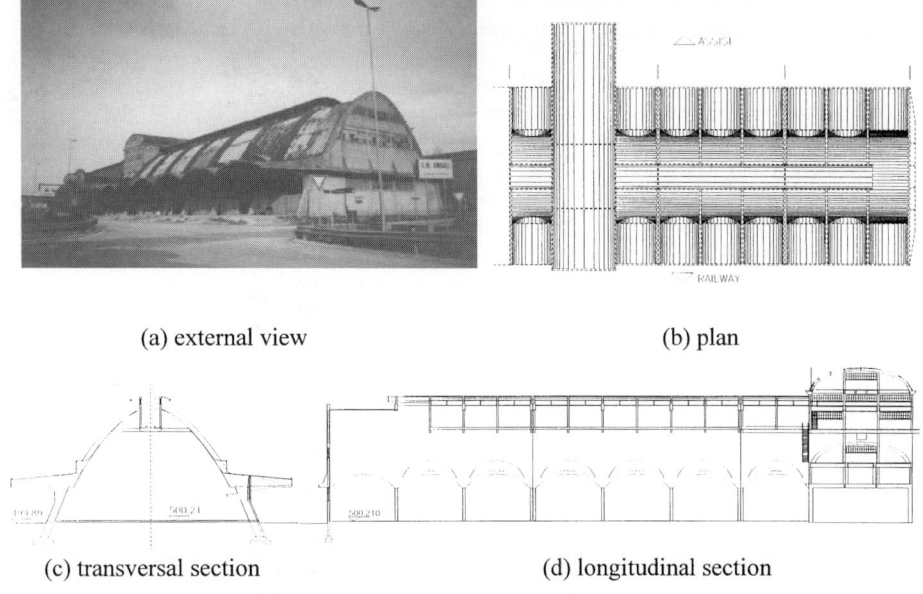

(a) external view   (b) plan

(c) transversal section   (d) longitudinal section

Figure 1   Existing building

PRELIMINARY CONTROLS AND DAMAGE SURVEY

Before drawing up the design, investigations have been carried out to evaluate the existing structures, as summarised in the following list.

• Collection of documentation on the structural works already made on the section previously retrofitted.

• Detailed survey, using optic-electronic instruments and CAD restitution, of the configuration of the main structural elements.

• Excavations to assay and survey the actual typical configuration of foundations.

• Boring and penetrometer tests to establish the geological and geotechnical data of foundation soil.

• Non-destructive tests (drilling, pull-out and rebound tests on concrete and sampling on steel) to identify the mechanical characteristics of materials. The potential cube strength of concrete was estimated in 15.1 N/mm^2 for columns, 24.4 N/mm^2 for arches, and 31.7 N/mm^2 for beams: the last value was attributed to the shells too, where tests were not allowed. The yield strength of the reinforcing steel resulted 368.4 N/mm^2.

• Electro-magnetic survey and direct sampling to locate the reinforcement in typical structural elements and in critical areas.

- Chemical tests to define the chemical decay of materials in terms of carbonation and sulphation of concrete. An up to 45 mm generalised carbonation depth was pointed out.

- Mapping of structural, mechanical and chemical decay. Thin shells of the roofing had a slight mono-layered reinforcement, often uncovered and corroded, and the concrete presented a deep decay and crumbling caused by both cracking, due to overstress, and environmental attacks (icing/de-icing cycles, washing away, corroded reinforcement expansion) associated to the loss of the waterproof protection. Extended areas of shells were partially collapsed (Figure 2) or had such large deformations as to appear near to collapse. Linear elements presented wide decayed surfaces characterised by pull-out of concrete cover and deep corrosion of steel reinforcement.

 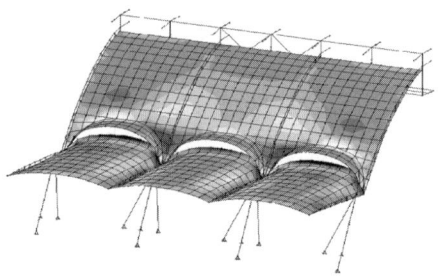

Figure 2   Partially collapsed shells   Figure 3   Stress diagrams on thin shells

NUMERICAL ANALYSIS ON FEM MODEL

Numerical analyses were carried out on detailed finite elements models reproducing both linear (beams, columns, arches) and bi-dimensional elements (thin shells). Both the existing and the retrofitted structure configurations were analysed. An increased stiffness was considered for the shells retrofitted with FRM: an equivalent thickness has been computed assuming the actual Young modulus of the existing and strengthening layers.

The internal forces due to service loads - dead and permanent loads, snow, wind and live loads - were calculated. A modal analysis with response spectrum was carried out, with reference to the longitudinal and transverse direction, to account for seismic forces. The Italian spectrum was used, which provides a response acceleration of 0.105 g for zones of medium seismicity, like the building site.

The 3D models were useful to identify some particular global structural behaviours and to point out the most stressed zones: in Figure 3 is illustrated the diagram of the longitudinal stress on the main shell, which evidence its behaviour as a global membrane. The results obtained were a guide in deciding the strengthening work, especially the position of CFR strips inserted to carry the tensile stresses. Also the lateral seismic behaviour was clearly shown by the analyses, evidencing a relevant longitudinal deformability.

WORKS ON EXISTING STRUCTURES

The areas to be treated were defined, basing it on the preliminary investigations and the results of the numerical calculations. The works were pointed out and differentiated in: (1) collapsed structural elements to be rebuilt; (2) structural elements to be demolished and rebuilt; (3) areas to be strengthened; (4) areas to be restored; (5) new elements to be inserted and connected. The strengthening aim was pursued by different tools: strengthening of existing elements by plating and covering, preserving aesthetics by maintaining the original shape; demolition and reconstruction using materials having higher performances; inserting new structural elements explicitly "evidenced" with respect to the original building. The works had a number of limitations in order to safeguard existing structures, to operate in the time limit imposed by the customer, 12 months, and to grant the safety standard to workmanship operating in a construction near to collapse.

The existing building had shallow foundations not tied together. The soil resistance did not allow the increase of forces due to the updated service loads and to seismic actions. Therefore, the existing foundations have been strengthened with "collars" cast around the existing footings and supported on micro-piles transferring the loads to the deeper soil thus limiting the disturb to the stress status below the shallow foundation. The "collars" are connected with r/c beams which, together with a slab at ground level, make up a rigid floor at the base of the building which distributes the seismic shear and co-operates to transfer it to the soil.

The main works carried out, located as shown in Figure 4, are listed below.

a) Strengthening of the lateral cantilever beams using CFRP strip glued at the top edge.

b) Strengthening of penthouse and shed shells (detailed in Figure 5a), carried out in a succession of steps: (1) removing of decayed concrete; (2) hydro-demolition of the exterior surfaces and application of inhibitors to prevent corrosion; (3) gluing of CFRP strips (Figure 5b), as defined by the numerical analysis results; (4) laying down PVA-FRM characterised by high adherence and compatibility with the existing concrete and high mechanical characteristics (compression strength $\geq$ 75 N/mm^2, Young modulus = 35000 N/mm^2, adhesion to concrete $\geq$ 3 N/mm^2); (5) surface finishing.

c) Demolition of single fields of the parabolic shells which appeared deeply decayed and their rebuilding using admixtured concrete with balanced shrinkage. Particular curing was carried out and scaffolds were maintained in place till complete curing.

d) Rebuilding of the collapsed r/c roof shell of the transverse forepart in front of the railway.

e) Longitudinal strengthening of the r/c roof shell of the transverse forepart in front of Assisi using steel plates glued at the exterior.

Generally the following works were decided when important structural decay required the rehabilitation of the structural performances of shells and r/c members and their protection from future decay: (1) removing of decayed concrete (from carbonation, erosion, cracking, icing); (2) protection of the reinforcement; (3) restoration and local strengthening of the sections; (4) application of protective covering. In some cases it was necessary to rebuild parts of the structures partially collapsed or so damaged as to be dangerous or impossible to be repaired. Demolitions were decided on the base of the decay mapping and direct control of the concrete consistency and reinforcement status. Reconstruction of missing and demolished parts was preceded by surface treatment: cleaning, washing, laying down adhesive paste and epoxidic resins, chemical anchoring of steel bars.

A number of strengthening work have been provided to improve the structure capacity against lateral seismic actions.

- Longitudinal bracing of the first level of the building. Inclined steel tubes are bolted at the top of the free span of the arches pillars and are connected to the foundation "collars" anchored with inclined bulb-shaped micro-piles. The existing structure showed a lateral behaviour characterised by a longitudinal flexibility (1st mode, period of 1.3 s) associated with the deformation of the first part of the pillars of the arches reaching a 34 mm drift, not compatible with the non structural elements of the perimeter facades, and inducing a "pilotis" effect reducing the ductile performance of the structure under severe earthquakes. The stiffened structure has a 0.55 s period longitudinal mode with displacement of 6 mm.
- Strengthening of the existing sections of the pillars at the interior side. Usually this is obtained with a r/c element cast against the existing pillar and connected with anchored reinforcement. Where the dimensions had to be reduced steel plates, glued with resins, were used in place of the r/c elements.
- Strengthening of the exterior side of the pillars with one CFRP strip.
- Bracing of superstructure frames and internal suspended footbridge with crossing steel bars.
- Connection of the gaps of the internal framed structures of the central section.
- Casting of shear walls infilled in the existing frame of the central section to obtain a seismic-resistant system having the suitable lateral and torsional stiffness and strength.

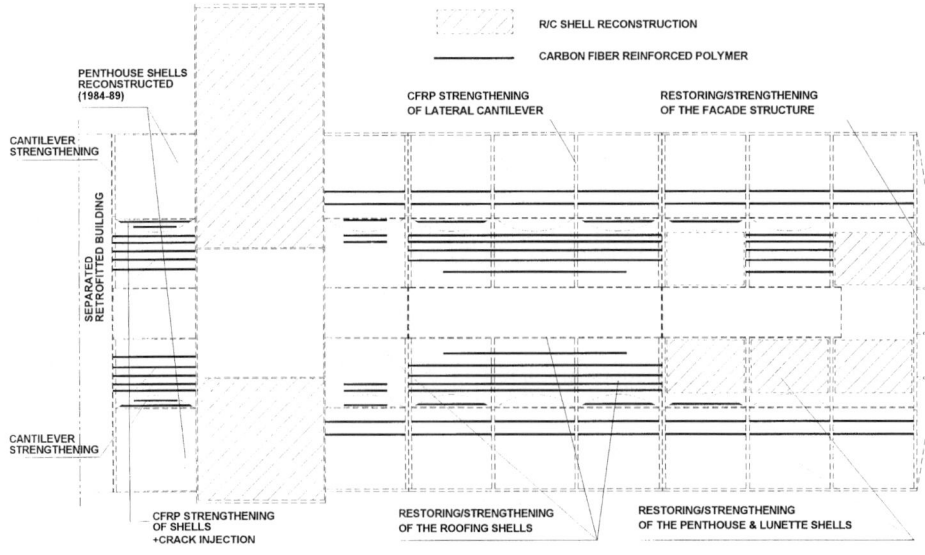

Figure 4 Plan of the main works

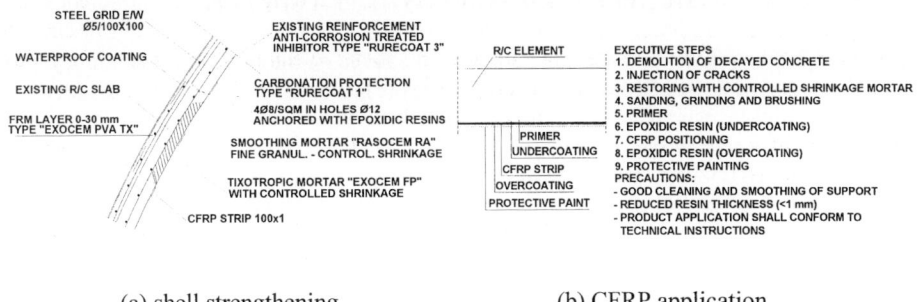

(a) shell strengthening   (b) CFRP application

Figure 5 Working specifications

CONSTRUCTION STEPS

The structural works had to follow a rigorous order with the aim of both obtaining the optimum behaviour of the rehabilitated structures, under the forces assumed at the design stage, and respecting the limitation imposed by the constructive requirements.

The succession of the performed phases was strictly defined in the design drawings, in compliance with the following priority order: (1) foundation strengthening; (2) longitudinal bracing with inclined tubes; (3) column strengthening; (4) cantilever strengthening; (5) penthouse shell strengthening; (6) placing of internal CFRP strips to parabolic shells; (7) external strengthening of parabolic shells with FRM; (8) application of waterproof covering; (9) bracing of the footbridge; (10) connection of floor gaps; (11) casting of seismic-resistant r/c walls; (12) bracing of tower frames. In each phase is provided to carry out the restoration of the concerned concrete.

(a) application of FRM   (b) hydrodemolition

Figure 6 Works on penthouse shells

MULTI-LAYER STRENGTHENED SHELLS

In this project the rehabilitation of the r/c thin shells, based on the application of thin layers of "Exocem PVA", a PVA-FRM, has been widely used (Figure 6a) and the actual strength of the composite multi-layer structures has had to be defined as a function of the thickness of the strengthening mortar aiming at minimising it. The parameters which influence and control the structural behaviour of the composite rehabilitated elements are those characterising the mortar (fibre type and contents, cement type and contents, mix design), the existing r/c elements (concrete strength, carbonation, reinforcement, etc.), and the interface between the existing r/c structure and the new FRM layer (surface roughness, chemical and physic-mechanic characteristics of the two materials).

FRM are a composite material made up of a cement mortar matrix including a volume of fibres dispersed in it to improve the mechanical characteristics, in particular the tensile strength and the deformation energy. Different kinds of fibres can be used: steel, glass, polypropylene, polyvinyl-alcohol are the most usual materials, and the actual behaviour depends on many parameters: the mechanical characteristics of the matrix and fibres, the fibre configuration, the volume of fibres. From a practical point of view the FRM can be treated as an isotropic material in the elastic range with post-elastic behaviour characterised by a plastic threshold in compression and by a cracking peak followed by a reduced post-cracking threshold in tension.

Due to the lack of experimental tests carried out on mortars, mechanical and physical-chemical characteristics of the FRM are derived from the experimental tests carried out on FRC (Fibre Reinforced Concrete) and reported in literature. Tests [1, 2] showed that tensile strength rises from 10% to 30%, increasing the fibre volume. The ultimate deformation goes up from 50% (using PVA fibres) to 100% (using steel fibres). Ductility is always improved with softening effect given by all kind of fibres, but steel fibres, presenting a hardening effect. Tests, simulating environmental attacks, on the durability of FRC [3] showed that if steel fibres are used, both strength (-10%) and toughness (-25%) are reduced; reductions are even more relevant when glass fibres are used. On the contrary, using polypropylene fibres the reductions are limited and, for low volume percent, practically null.

Different mechanisms can provide the stress exchange through the interface, ensuring the monolithic behaviour of the composite element: mechanical connection by anchored pins, "gluing" of the two layers by adhesives, bonding induced by special surface working of the substratum. Shear stresses varying from 3.25 to 2.17 N/mm^2 were measured [4] in case of bare connections: values decrease while increasing the specimen area. The deformation at peak resulted about 0.1 mm.

Interface behaviour is governed by the following parameters: roughness of the substratum surface; physical-mechanical characteristics of substratum and mortar; chemical characteristics; surface working. Hydro-demolition (Figure 6b) has appeared as the most effective technique of surface working because it dramatically reduces the substratum micro-cracking and increases the percent of micro-pores, then the effective area of the interface, and allows, changing the water pressure, to easily modify the depth of the removed material. The humidity of the materials slightly influences the tensile strength but modifies the location of the rupture (in concrete, for air surface wet, or at interface, for air surface dry). In polymer-FRM the influence of humidity is controlled by the polymer content which constitutes an obstacle to the water transfer from mortar matrix to the substratum.

The parameters which influence and control the structural behaviour of the composite rehabilitated elements composed by the existing concrete substratum and the reinforcing layer are the main characteristics of the three components: substratum, interface, mortar. Advanced numerical non linear models, developed using the computer code LUSAS [5], have been used to reproduce the behaviour of the composite elements and parametric analyses have been carried out to investigate the influence of the main parameters: the mesh of the model; the stress-strain behaviour of the interface connection; the non-linear behaviour of the support structure (r/c existing structure); the stress-strain relationship characterising the FRM of the strengthening layer. Detailed results are reported in [6].

3D and 2D non-linear models, straight and curved, have been investigated. Both the concrete (70 mm thick) and mortar (30 mm thick) layer is modelled with two layers of elements, a steel bar reproduces the steel reinforcement of the existing concrete. The main mechanical parameters of the materials are resumed in Table 1. FRM behaviour is taken according to RILEM [7], concrete cracks in tension and is indefinitely elastic in compression. The model is simply supported at its ends and is loaded with an increasing vertical load having reference (starting) value equivalent to a uniform load of 2.5 kN/m^2.

Table 1   Mechanical parameters of the materials

MATERIAL	E [N/mm^2]	v	STATUS	σ_y [N/mm^2]	ε_y [1/1000]	σ_s [N/mm^2]	ε_s [1/1000]	σ_u [N/mm^2]	ε_u [1/1000]
FRM	35000	0.15	Compr	24	0.68	---	2.0	60	3.5
			Tens	12	0.34	4.5	0.44	1.5	10.0
CONCRETE	30000	0.15	Compr	∞	---	---	---	---	∞
			Tens	0.5	0.016	---	---	---	3.0
STEEL	206000	0.3	Compr	260	1.26	---	---	360	10.0
			Tens	260	1.26	---	---	---	10.0

The interface element here used is characterised by a stress-displacement behaviour having a first elastic branch followed by a softening plastic branch: the diagram is identified by the stress, τ_e, and the deformation, δ_e, at the elastic limit (peak of the curve) and by the total deformation energy that defines the maximum deformation, δ_u. Similar diagrams are valid for all the three possible collapse cases: shear in two orthogonal directions and tension. The non-linear constitutive relationship is elastic-perfectly plastic and uses a limited tension criterion normal to the plane of the interface and a Mohr-Coulomb criterion tangential to the plane of the connection.

Coupled delamination has been assumed. Interface characteristics are changed assuming strength values $\tau_e = 1.5, 3.0, 6.0$ N/mm^2 and corresponding elastic deformation $\delta_e = 0.001$, 0.1,   0.3   mm.   A   single   value   of   the   ultimate   deformation, $\delta_u = 1.0$ mm, is used. No practical influence can be observed, on the structure response, varying the parameters: stress variations are less than 2%.

The load-displacement behaviour of the straight model is always similar, with yielding of steel preceding FRM crushing: the two crises always happen at the same load level. The effect of the substratum stiffness is more evident at the first levels of load, before concrete cracking: a decrease of 15% of the tensile stress in concrete and an increase of 12% of the compression in mortar come out when passing from E=30 GPa to E=1.8 GPa. At the lower load levels, when the interface is compressed and the substratum is not cracked, there are differences between 60% and 90%, but, when the neutral axis is included in the FRM layer and the concrete cracking is advanced, the differences are negligible. The ultimate limit load is independent from the considered variations of the interface and materials parameters: the steel yielding always corresponds to a load factor of about 11 (27.5 kN/m^2). The FRM cracking appears successively, for a load factor greater then 13. In all the cases these critical events precede the interface crisis and delamination.

The curved 2D model has the geometry and the mechanical characteristics of the actual thin shell of the lateral penthouse of the "Lirick Theatre" in Assisi. The mesh layering is maintained, but the values of some mechanical parameters differ: σ_y = 215 N/mm^2 and ε_y = 1.04 $^0/_{00}$ for the steel, E=25000 N/mm^2 and ν = 0.2 for the concrete. The starting value of the normal load is a uniform vertical load of 1.2 kN/m^2. The stress status is dominated by the compression. The stress distribution, here not reported, evidences a plastic redistribution with a tendency to assume a uniform distribution across the section when the load increases. The ultimate limit load is independent of the considered variations of the interface and materials parameters in the fields here investigated. The concrete crushing (σ = 16 N/mm^2) corresponds to a load factor of about 7 (8.4 kN/m^2). The FRM reaches its compression limit successively. The interface crisis and delamination never take place before the crisis of the component materials, therefore the ultimate strength of the elements can be well estimated by simplified procedures based on the current methods of structural design.

(a) view of the interior of the theatre   (b) exterior view of the retrofitted building

Figure 8   The final configuration of the building

CONCLUSIONS

The project of the "Lirick Theatre" in Assisi, which occupies an old industrial building, has been illustrated from the preliminary survey to the construction phases, with special regard to the use of innovative materials, requiring adequate application techniques, as CFRP strips used to improve the tensile strength and FRM layered on the thin shells. The parametric analyses carried out show that a suitable choice of the FRM mechanical characteristics, with respect to those of substratum, can improve the composite performance and so do the interface characteristics. The complex parameters, characterising the interaction between the substratum and the strengthening layer do not influence the ultimate load which can be adequately estimated by simplified procedures based on the current methods of the structural design taking into account the actual material behaviour. The stage and parterre of the theatre have been designed as completely independent new structures and are not described in this paper. The theatre was inaugurated on May 28, 2000. Figure 8 shows the internal and external view of the final configuration of the building.

REFERENCES

1.   LI, Z., LI, F., CHANG, T.P., MAI, Y.W., Uniaxial tensile behavior of concrete reinforced with randomly distributed short fibres. ACI Journal Mat. Vol 95. 1998.

2.   XU, G., MAGNANI, S., HANNANT, D.J., Tensile behaviour of fibre-cement hybrid composites containing PolyVinyl Alcohol fibres yarns. ACI Journal Mat. Vol 95. 1998.

3.   KOSA, K., NAAMAN, A.E., HANSEN, W., Durability of fibre reinforced mortar. ACI Journal Mat. Vol 88. 1991.

4.   CHOI, D.U., JIRSA, J.O., FOWLER, D.W., Shear transfer across interface between new and existing concrete using large powder-driven nails. ACI Journal Str. Vol 96. 1999.

5.   FEA, LUSAS Modeller. Kingston Upon Thames, UK. 1999.

6.   MEZZI, M., PARDUCCI, A., Numerical modelling of strengthened r/c thin shells. 7th Intern. Conf. Modern Building Materials, Structures and Techniques. Vilnius, Lithuania. May 2001.

7.   RILEM TC 162-TDF, Test and design methods for steel fibre reinforced concrete. Materials and Structures. Vol 33. 2000.

ACCIDENTAL IMPACT LOADING OF CONCRETE STRUCTURES IN THE MARINE ENVIRONMENT

T Browne

P Watry

Collins Engineers Incorporated

United States of America

ABSTRACT. The most frequent accidental damage to structures in the marine environment arises from vessel impact. While available maneuvering area; existing environmental conditions; and the capabilities of the vessel and its crew play a vital role in preventing accidental impact damage to concrete structures, the designers and managers of these marine structures also have an important role in minimizing the damage (type, development, and extent), as well as the resulting consequences. Depending on the type and primary function of the structure, design requirements are different. This paper will discuss both concrete berthing structures as well as non-berthing structures. In general, the primary function of berthing structures is to support frequently applied lateral forces of a vessel, while a non-berthing structure would only be subjected to vessel forces during an accident. Bridges are the most common type of non-berthing structure that frequently must withstand accidental impact loadings from vessels. However, accidental impact loading can also occur on berthing structures by overstressing protection fendering or by applying the forces at unsuspected locations, directions, or magnitudes. Following an illustration of typical vessel forces that may be possible during an accidental impact loading, the factors of the vessel's size, geometry and velocity are discussed. Furthermore, the response actions including repair design philosphy after vessel impact damage has occurred at a facility are discussed.

Keywords: Vessel / Ship collisions, Accidental impact, Overload, Assessment, Marine environment, Safety, Concrete repair.

T M Browne, P.E., is a Project Manager with Collins Engineers, Inc., which is headquartered in Chicago and has its international operations office in Dublin. Mr. Browne is an active member of several technical societies including ACI, ISMA, PIANC, SNAME and ASNE.

P M Watry, E.I.T., is a Project Engineer with Collins Engineers, Inc.

INTRODUCTION

Concrete structures in the marine environment are typically protected by designed fenders as well as independent other structures, such as mooring cells, individual piles, or pile clusters (often referred to as dolphins). However, accidental impact loads may overstress these protection systems, or occur at a vulnerable direction or location and contact directly on a concrete structure. Therefore, this paper will not specifically discuss the design of various types of protection systems, but the events surrounding when these systems are absent or ineffective.

Accidental impact loading can be defined as an applied load causing an untended contact that may or may not have been accounted for during the design of the structure. Accidental impact loading occurs on berthing structures as well as non-berthing structures. However, the reported cases are usually greater for non-berthing structures (such as bridges, breakwaters, etc.) generally because these incidents typically attract more attention. Although berthing structures are designed to support "reasonable" lateral impact loads, these structures also suffer from accidental "extreme" impact loads, as well as forces applied to unprotected vulnerable members.

INTERNATIONAL DATA

The Permanent International Association of Navigational Congresses (PIANC) indicated the following points based on a study of 151 bridge collisions in 13 countries (Belgium, Germany, France, Japan, Netherlands, Denmark, Finland, Iceland, Norway, Sweden, Spain, United Kingdom, and the United States of America) from 1960 to 1998 in a PIANC - INCOM (Inland Navigation Commission) Working Group 19 Report published in 2001 [1]:

- Types of vessels in the accidental impacts were: a Tugboat With Barge on 46% of occurrences, General Cargo Ship on 31% of occurrences, Oil Tanker on 16% of occurrences, and undocumented vessel types on 7% of occurrences.

- Bridge damage was categorized as: collapsed at 30% of occurrences, moderate at 27% of occurrences, minor at 21% of occurrences, and undocumented at 22% of occurrences.

- Vessel damage was categorized as: moderate at 26% of occurrences, minor at 23% of occurrences, negligible at 11% of occurrences, and undocumented at 40% of occurrences.

- The majority of accidents occurred when the wind speed was less than 5 meters per second, and current velocity was less than 2 meters per second (4 knots).

- Most accidents occurred when the general weather conditions were fair, during daytime and in good visibility.

- 53% of all collision accidents were due to human error, 19% due to technical failure and 23% due to extreme environmental circumstances.

- This statistical information in the PIANC 2001 Report corresponded well with a 1983 IABSE (International Association for Bridge and Structural Engineering) Investigation Report, and it seems that these trends have not changed in the past 15 years.

United States of America

In the United States, the U.S. Coast Guard (a division of the United States Department of Transportation (USDOT)) investigates vessel impacts. Furthermore, the National Transportation Safety Board (NTSB) also investigates these incidents if significant. The NTSB is an independent federal agency that conducts special safety investigations in every mode of transportation in the United States. A federal study in the United States covering the period between 1970 and 1974 reported 811 towboat and barge collisions with bridges on the inland waterway system in America. In 1980, the collapse of the Sunshine Skyway Bridge, which crosses the Tampa Bay in Florida, was a major turning point in the awareness and increased concern for accidental impact loadings on bridges in America.

However, the Marine Safety Evaluation Branch of the U.S. Coast Guard still documented 222 bridge-vessel collisions occurrences between 1981 and 1986. In 1988, the Federal Highway Administration of the USDOT along with various local DOT's undertook a research project to develop vessel collision design provisions for bridges. The research project's final report was adopted by the American Association of State Highway Transportation Officials (AASHTO) as a Guide Specification in 1991 [2].

However, not all bridges over navigable waterways in the United States have been designed or rehabilitated since 1991 thus not truly benefiting from the published AASHTO Guide Specification. And, a significant number of bridge-vessel collisions have still occurred between 1991 and 2000 in the United States. It is estimated that approximately 35 accidental impact load collisions on average are reported everyday to the U.S. Coast Guard Headquarters. Most of these impacts are from commercial shipping activity along the Mississippi River, according to the USDOT.

United Kingdom

In the U.K., the Marine Accident Investigation Branch (MAIB, a division of the Department of Transport Local Government Regions) investigates vessel impacts. The MAIB receives between 30 and 50 reports each year involving vessel impacts on structures. Although the structure material type is not recorded in their database, it is estimated that a large portion of these incidents occurred while berthing or unberthing vessels at typical concrete port structures. According to the MAIB, the cause of the accidents is commonly human error - a misjudgment of prevailing weather or tidal conditions, or more generally a lack of attention.

Japan

In Japan, the Japanese Coast Guard (JCG, a division of the Ministry of Land Infrastructure and Transport) investigates vessel impacts. The JCG estimates that approximately eight vessel impacts occur everyday in Japan. However, only 69 vessels were involved in accidental impacts on concrete structures during the 2000 calendar year. During the 2000 calendar year, the impacts involved 3 bridges, 22 berths, 22 breakwaters, 4 tetrapod structures, and 7 other structures totaling 58 of the 69 incidents occurring in port.

DAMAGE EXTENT

Accidental impact loading incidents represent a serious threat to public safety and environmental protection in many locations in the world. Although this paper discusses the related damage to concrete structures, impacts can also significantly damage the vessels. Refer to Figures 1 and 2 for views of damaged bows. The primary types of concrete structures discussed herein are bridge substructures for non-berthing structures, and pile-supported piers or concrete bulkheads for berthing structures. Each type of concrete structure has inherent capabilities to resist accidental loads, as well as vulnerable locations to such loads. Figure 3 illustrates an unprotected two-column bridge substructure due to high water. Figure 4 illustrates a multiple-column substructure with integral collision wall protection. Although the integral solid collision wall extends sufficiently above the water for protection from barge impacts, vessels with taller freeboards or projecting hull geometry could still significantly damage the columns during an impact. Figure 5 illustrates an impacted berthing marine structure due to lack of attention on the Captain's part. Figure 6 documents the vessel *Bright Field's* impact on a waterfront facility in 1999 due to mechanical failure.

Figure 1   Overall view of vessel with damaged bow

Figure 2   Close-up view of a vessel with damaged bow

Figure 3   Two-column pier with ineffective collision wall at waterline due to high water conditions of waterway

Figure 4   Multiple-column pier with effective concrete collision wall for barges

Figure 5   Concrete bulkhead dock wall impacted by a vessel

Figure 6   Vessel *Bright Field* impact at Poydras St. Wharf

The extent of damage and resulting consequences induced from an accidental impact load depends on the impact orientation, magnitude, damage type, contact location, as well as the capabilities of the physical structure and involved individuals to deal with such incidents. While all structures do not need to be designed for the absolute worst-case accidental loading possibility, all designers do need to evaluate the risk and probability of vessel impact during their design. Risk is generally defined as the exposure to the chance of loss, which is the combination of the probability of an event occurring and the significance of the consequence of the event occurring.

Impact Orientation

Orientation of an impact is typically measured from a perpendicular line off the structure's fascia and can vary from "90 degree angle (purely abrasion)" to "zero angle straight on" impact. Impact energy increases with the reduction of the impact angle, as well as the reduction of contact/center of gravity distance [3]. Therefore, more kinetic energy must be transferred and absorbed by a structure when the impact orientation angle is low. Also, the related impact damage is typically greater when the point of contact is closest to the vessel's center of gravity. When a ship contacts a structure at an angle, the ship rotates around its own center of gravity. The radius of gyration is calculated based on the distance between the point of contact and the vessel's center of gravity.

Damage Type

Frequently, all damage is wrongly referred to as impact related after a vessel collision. Accidental loading incident damage can result as either Impact-Related or Overload-Related. Both impact- and overload- damage can range from superficial surface damage to complete failure of the concrete member.

Generally, the consequences of Impact-Related damage are localized structural defects ranging from abrasion, cracks, voids, chipped corners, and local spalling to major structural distress. Impact damage can cause complete failure of a structural element or can accelerate future corrosion of steel reinforcement by reducing the concrete cover depth. Impact-Related damage is usually located near the contact area, and can frequently be seen on inadequately protected concrete elements in the splash zone of a marine structure.

Overload-Related damage is caused when loads are applied to a structure exceeding its capacity. When loads cause stresses in excess of the tensile stress capacity of the concrete or the yield stress of the steel reinforcement, overload-related damage will occur.

Impact or overload may cause localized overstressing of concrete members and can cause severe cracking with potential for additional damage from corrosion and contamination. Overstressing is caused by external loads, which cause high internal stresses that exceed the strength of the concrete member. The result is an overstress crack, which is characterized by its sharp edges and small wedges of missing concrete along the length of the crack.

Overstressing damage does not necessarily occur at the point of the accidental impact, but at locations where the stress exceeds the concrete's structural capacity. Often this location may occur below the waterline and therefore only be detected during an underwater inspection performed by qualified structural inspection divers. Overstress cracks should not be confused with temperature and shrinkage cracks or general cracking associated with steel corrosion. All types of cracks are different and must be evaluated appropriately to determine the correct repair.

Loading Magnitudes

Loading magnitudes primarily vary on the size and velocity of a vessel. The impact energy is proportional to the displacement tonnage of the vessel and to the second power of the velocity. Refer to Table 1 for a brief sample of loading magnitudes based on a variety of vessels [4]. The loadings presented in Table 1 were calculated in this example with a variety of approach velocities in order to illustrate the significant relative magnitude change related to the size and velocity of a vessel. A velocity of 0.2 m/s is equal to approximately 0.5 knots and would be an example of a vessel under control while berthing. A velocity of 4.0 m/s is equal to approximately 8 knots and would be an example of a vessel traveling under way at a relatively slow speed.

As shown in Table 1, the minimum and maximum impact energy increase varies from a factor of 280 to 400 times the original magnitude respectively based on the approach velocity increase from 0.5 knots to 8 knots. It should be noted that for the example data in Table 1, the orientation angle was assumed close to purely abrasion, which resulted in a radius of gyration assumption of approximately 0.5. If the radius of gyration was assumed larger to illustrate more of a straight on collision, the impact energy values would be extremely higher that shown in Table 1.

Loading magnitudes are presented in Table 1 in terms of impact energy (tonnes-m). The actual collision impact force (tonnes) calculation is extremely complex and depends on a variety of additional variables from the geometric shape of both the vessel and structure, to the material strength of both the vessel and structure.

The vessel collision force equation utilized in the United States is based on the research conducted by Woisin in Hamburg, Germany between 1967 and 1976. Woisin's results have been found to be in good agreement with other research conducted by other vessel collision investigators worldwide, according to U.S. DOT publications.

Table 1   Loading magnitudes based on size and approach velocity of a vessel

SAMPLE VESSEL TYPE	DISPLACEMENT (TONNES)	OVERALL LENGTH (M)	MASS FACTOR	IMPACT ENERGY (TONNES – M)		
				At Velocity = 0.2 m/s (0.5 knots)	At Velocity = 1.0 m/s (2 knots)	At Velocity = 4.0 m/s (8 knots)
PASSENGER						
Large	75,000	315	1.65	126	3,150	50,395
Small	30,000	230	1.71	53	1,311	20,970
FREIGHTER						
Large	580,000	435	1.73	1,025	25,607	409,700
Small	20,000	165	1.88	39	961	15,362
CONTAINER						
Large	73,500	290	1.80	136	3,377	54,019
Small	9,600	143	1.68	17	413	6,593
MIXED CARGO						
Large	20,000	165	1.88	39	961	15,362
Small	1,000	60	1.82	2	47	744
MILITARY						
Carrier	90,950	332	1.57	146	3,646	58,329
Destroyer	7,810	172	1.76	15	351	5,611
Tug Boat	80	20	1.69	0.2	3.5	56

Contact Location

The three possible contact locations for impact loads can generally be described as the deck/superstructure, substructure (piers, piles, etc.), and the associated appurtenances (fenders, curb, railings, fascia panels, mooring devices, utilities, and other facilities). Direct impact contact to the substructure elements typically results in the highest potential for significant damage. However, the substructure elements can also be significantly affected even without direct contact due to overload as previously discussed. Frequently, deck/superstructure and associated appurtenances experience accidental impacts because the geometry of an impacting vessel differs from that anticipated in the protection design for the structure. A setback distance may be required if the vessel geometry or stopping distance may affect other land based facilities. Refer to Figure 7 for a view of the *Bright Field* impact causing additional damage to land-based facilities.

Figure 7   Vessel *Bright Field* damages land-based mall

ASSESSMENT AND MANAGEMENT

All managers of waterfront facilities should have an Incident Management Plan (IMP). Three items typically found in an IMP include: Procedures for Immediate Incident Notification, Procedures for the Assessment of the Structure and Vessel; and Required Action Plans. There are three phases for the assessment of a concrete structure after impact damage has occurred: 1) Rapid Damage Assessment (RDA), 2) Detailed Damage Assessment (DDA), and 3) Formal Engineering Evaluation. At least one Damage Survey Team with 24-hour emergency contact information should be cited in the IMP so that immediate notification can result in a Rapid Damage Assessment. Depending on the findings of the RDA, the area may be restricted until a DDA can be performed.

A RDA typically involves individuals familiar with the structure looking for related, changed- conditions on the structural members. A RDA should be performed immediately after impact. While it is desirable, it is not always necessary that these individuals be structural engineers since the RDA is more of a cursory overview of the damage, rather than a scientific assessment.

A Detailed Damage Assessment (DDA) should always be performed after significant abnormal loading incidents have occurred as a follow-up to a RDA. A DDA typically involves structural engineering experts specializing in waterfront assessments to perform a complete evaluation of the impacted area and related underwater elements. The DDA team typically includes engineer-divers for the underwater elements, as well as the difficult to access underside portions of waterfront facilities. A DDA should be typically performed within a few days of the impact incident. Formal documentation of findings and any recommendations are typically provided during a DDA.

Depending on the findings in the DDA, a formal engineering evaluation may be warranted. A Formal Engineering Evaluation typically involves additional testing or analysis, and provides additional documentation for the most cost-effective, durable, and feasible recommendation. DDA's are also often necessary to evaluate areas with a history of impacts or impacted structures with obvious reduction in structural integrity. Refer to Figure 8 for a graphical display of impacts at the Port of New Orleans wharves. A formal engineering evaluation frequently also includes a structural analysis of remaining capacity after obvious integrity reduction has occurred.

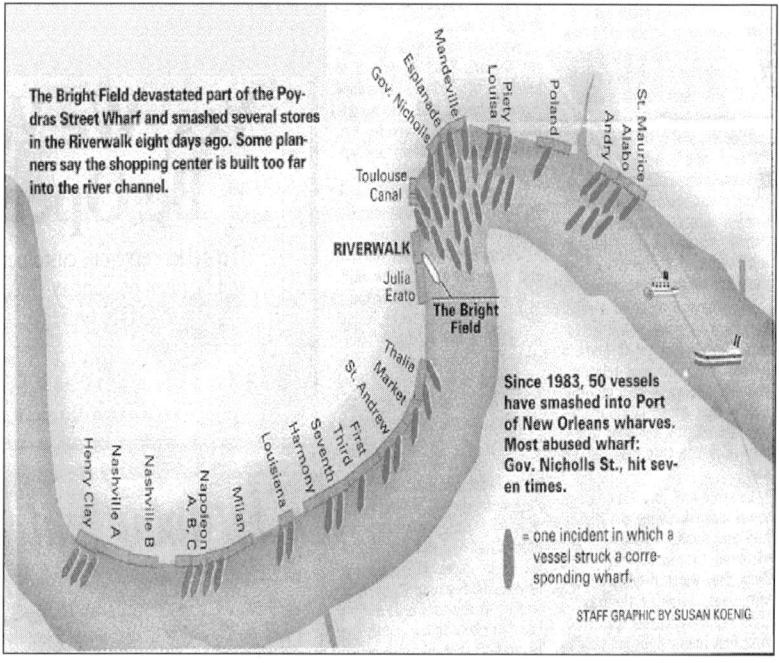

Figure 8   Graphical display of vessel impacts along waterway

EMERGING TECHNOLOGIES AND PRACTICES

State-of-the-Art design practices have utilized impact computer simulation models, as well as even bridge superstructures capable of carrying primary loads even after one pier has failed. In Australia, the superstructure of a large bridge was recently designed to transfer load paths in an event that one entire pier failed due to vessel impact. Although this increases the cost of the superstructure, it decreases the cost of impact protection, as well as increases overall safety.

State-of-the-Art maintenance practices utilize real-time tracking of vessels and computerize databases to document history at a location. Computer maintenance programs also are frequently being used to document impact collisions, repairs, and condition of concrete members. These maintenance programs may also be used to track the successfulness of repair alternatives. Refer to Figures 9 and 10 for emerging technologies and design practices.

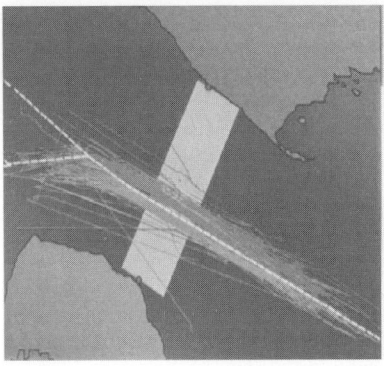

Figure 9   Real-time computer tracking

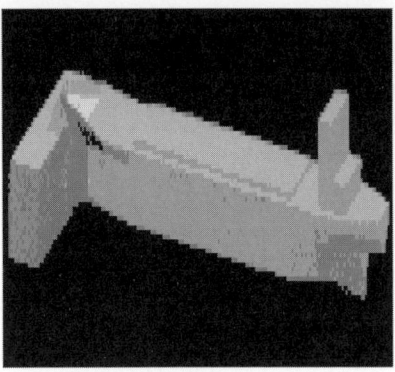

Figure 10   Computer modeling of impact

CONCLUSION AND RECOMMENDATIONS

- Designers should always list the categorical and geometrical type and size of vessel that a structure was designed for, as well as designed loading forces and any assumptions on the structure plans.

- Designers should always evaluate impact risks, probabilities, and needed setback variances in case of accidental impacts.

- Owners should evaluate structures continuously for any vessels different from those listed on the design plans.

- Owners should always keep original design plans and an IMP available, as well as information on previous damage and repair documentation on file. A record should be kept for all incidents in order to evaluate trends and improve future conditions.

- Vessel Pilots and Captains should always be knowledgeable and familiar with the waterfront facilities and the local environmental conditions.

- Vessel Pilots and Captains should inform structure owners of any hazardous conditions, as well as make suggestions to improve the safety of a structure.

ACKNOWLEDGMENTS

The authors would like to acknowledge the statistical support provided for this paper by the U.K. Marine Accident Investigation Branch, the U.S. Department of Transportation, and the Japanese Ministry of Land, Infrastructure and Transport. Furthermore, appreciation is acknowledged for the utilization of many figures in this paper from a variety of different sources, including news media and websites.

REFERENCES

1. SHIP COLLISIONS DUE TO THE PRESENCE OF BRDIGES, PIANC INCOM REPORT OF WG 19, 2001.

2. VESSEL COLLISION DESIGN OF HIGHWAY BRIDGES.   FHWA Publication No HI-92-050, July 1992.

3. AGERSCHOU H., et al, Planning and Design of Ports and Marine Terminals, 1983.

4. RECOMMENDATIONS OF THE COMMITTEE FOR WATERFRONT STRUCTURES, GERMAN SOCIETY FOR SOIL MECHANICS AND FOUNDATION ENGINEERING, EAU 1985.

THE PRINCIPLE OF INFORMATION ENTROPY IN FRACTURE MECHANICS OF BUILDINGS - ELEMENTS OF THEORY OF CATASTROPHES

S M Skorobogatov

The State Academy of Architecture and Building Sciences

Russia

ABSTRACT. The paper is devoted to elaboration of the elements of technical and natural catastrophes. The paper describes a generation of results of researches in patterns of forming cracks in massive concrete and reinforced concrete structures. In the stage of pre-failure the massive oversize solids (concrete, rock) transmute into the hierarchy of unitized, enclosed each into other blocks divided by long and broken cracks. Disclosing indeterminacy in a pattern of cracks, from the beginning and up to the end of loading, is for the first time described with the help of the information entropy under multi-step independent communication from many sources (signals of ruptured links between blocks). The conception of obtaining a stage of a structure under every degree of loading includes the parameter of reserve of serviceability for a structure. The curve of the parameter of serviceability coincides with the curve of the information entropy by character of change. This permits to use the scale coefficient in design. Using the criterion of serviceability means a design on a new state, i.e. the longitudinal crack resistance. This determines an assessment of expedience rehabilitation of a structure damaged with unknown load.

Keywords: Catastrophe, Oversize, Information entropy, Serviceability, Crack hierarchy, Longitudinal cracks.

Doctor, Professor S M Skorobogatov in 1978 organized a new section of building structures for training civil engineers at the Urals State University of Railway Transport (Russia, Ekaterinburg). Previously he was a professor at The Urals Polytechnical University and took charge of the investigation on the fundamentals of the endurance theory for deformed bar reinforcement and the principles of its design. He took part in inspection, enhancement and rejuvenation of many buildings including ones damaged in the well-known catastrophe at the railway station of Sverdlovsk-Sortirovochniy. During the past decade he has been creating the elements of the theory of technical and natural catastrophes. The theory was described in his book (monographe): – "The principle of information entropy in fracture mechanics of buildings and rock seams", Yekaterinburg, 2000, pp 420. Prof. S.M.Skorobogatov has published widely and served on many technical committees. Now he is Corresponding Member of The State Academy of Architecture and Buildings Sciences and Honoured Science Worker of Russia.

INTRODUCTION

Chernobyl, Arzamas, Sverdlovsk-Sortirovochniy, Ufa, Spitak, Leninakan, North Sakhalin and a framework of the energy project in Moscow suburb have become the dismal symbols of natural and technical catastrophes in the past twentieth century.

In spite of poorness of publication and lack of publicity, the history of civil engineering along with scientific and technical essays, special and popular-scientific literature affirm that during last centuries there were great wrecks and catastrophes of massive over-size engineering structures.   Logic from the civil engineering history shows that the catastrophes of engineering structures and buildings are possible in the future.   Moreover, there is an escalation of large sizes in unique complicated structures and buildings including ones made with reinforced concrete: more than 500m TV Towers in Moscow and Toronto, a bridge with 330m span in Panama, hundreds of high multistorey houses in New York and more than hundred storey building in Malaysia and so on.

Statistics show that year after year the number of industrial breakdowns increases.   It is assumed that the concentration of potential energy in numerous engineering buildings (reservoir for liquefied natural gas, containment vessels for nuclear reactors) is so high that the civil-engineering means are insufficient to create protective buildings which could keep the breakdown products in closed space.   For example, the potential energy of the liquefied natural gas of the reservoir of 50,000 cubic meters capacity is 10 times more than the energy of the atom bomb thrown down in Nagasaki.

VALIDITY OF A NONTRADITIONAL APPROACH
TO ELABORATION OF A RELIABILITY THEORY

Due to the development of new, experimental reinforced concrete massive structures, there appears to be much more complicated static, design diagrams and hence necessity of new approach to designing structures.   There appears to be new, unknown in practice, shortcomings in designed preconditions, whereupon there are miscalculations and errors in design and thereof they result in breakdowns and catastrophes in construction.

The case is that the catastrophe is a spasmodic change (leap) arising in the form of a sudden reaction of the structure system to smooth change under external condition.   The main property of the catastrophe being considered is hypothetical character of its manifestation. The hypothetical character of the catastrophe theory excludes the possibility of using only determinational theories because great dispersion of concrete properties exists inside of a large structure.

At present, for the purpose of designing a massive reinforced concrete, some scientists are developing theories of reliability based on statistical laws, arithmetic mean value and standard deviation obtained from test results of mass units.   As a rule, these statistics cannot be obtained for a separate (sole) or for the unique structures.   To create a reliability (serviceability) theory for a separate structure the researcher should resort to physical peculiarity of texture of concrete that is fracture accumulation and crack hierarchy inside of a structure.   Then the researcher discloses indetermining of crack hierarchy with the help of the information entropy.

NEW ENTROPY PARADIGM IN CATASTROPHE SCIENCE
FOR MASS CONCRETE

The author makes an attempt to combine an entropy probabilistic approach to fracture accumulation causing development of catastrophe and a determinational approach to calculation of serviceability on the basis of refined knowledge of stress-strain state of a structure. Attractors of catastrophe are weak places in concrete massif due to great dispersion of strength in a separate structure. The intermittent process of crack accumulation and crack pattern can be a more statistically reliable forerunner of catastrophe than magnitudes of strength and deformation of concrete.

The presence of such two operated factors for concrete as loading and cracking makes it possible to choose such a well known kind of elementary catastrophe as a "gather". The "gather" can cause bifurcation in the process of deformation of concrete as a consequence of heterogeneity in its texture. But tear-off phenomenon and intermittent, discontinuous crack hierarchy is not described by traditional mathematical classification that is why the author resorted to the method of information entropy [1,2,3].

Unexpected dispersion in physical and mechanical properties spreads over concrete and is enhancing with increasing a structure size. At present, here are sufficient experimental data concerning crack hierarchy in heterogeneous, poros solids in general (with type of concrete, granite, limestone, sandstone and so on). Exposure of properties of these materials has been described in a such quickly developing branch of science as structural geophysics. From the beginning of loading, the process of cracking from microcracks up to the main long cracks is connected with the irreparable loss of energy earlier spent for the manufacture of a structure. The irreparable loss of energy results from the well-known process of entropy in work of a structure (under loading) in accordance with universal second law of thermodynamics. The accumulation of fracture or the growth of the irreparable loss of energy is assumed to be a change of energy with positive sign.

When a structure is at the beginning of testing or loading, much information is required to describe the space position of every point of a structure. Here, indeterminacy of the fact of failure is great because the fact of failure is so far from the beginning of loading. When a structure is under the end of testing or loading, it is turned gradually to a debris cluster or to chaos. The state of chaos means absence of a necessity of strict fixing the space position of every point and of their description. This means a full loss of the information concerning the solid state. Indeterminacy of the fact of failure is becoming equal to zero.

As the test process goes on, the information entropy goes on in the opposite direction to positive entropy (loss of energy) that is from the largest value to zero. This is known as negentropy. From the point of view of negentropy "a having received telegram but not read it" presents the largest value of indeterminacy or scientific information. As soon as the telegram has been read, the value of its indeterminacy (or scientific information) reaches zero, whereupon the telegram may be thrown out. The information entropy follows from the universal second law of thermodynamics. If the number of brittle fractured ties between grains or blocks is treated as signals of their fracture, then there appears a possibility to use information entropy.

TURN-ROUND, TEAR-OFF FRACTURE MODE OF CONCRETE

The initial flaws or microcracks as boundary cracks nucleate grains or grain groups for the first level of concrete texture. The porosity and its heterogeneity more or less distributed throughout the concrete massif predetermine the disposition of weak places. It makes possible for microcracks to develop into mezocracks, macrocracks and long cracks for the higher levels of concrete texture for a oversize structure. Real material divided by broken cracks as boundaries, in dependence on the size of a structure can be simulated by an initial pseudo-grain mode as an weak-packed assemblage of strong grains, on the one hand, and with thin weakened ties between those grains. The role of the weakened ties plays small isthmuses, interspaces of random size between grains or grain groups. Because of small transverse cross, section the strength of material of isthmuses is of not great practical importance. Thus, in the state of pre-failure the strength mode of concrete presents as pseudo-grainy skeleton consisting of strong grains or blocks combined with weakened links. One of the main reasons of forming initial cracks is initial processes of volumetrical changes in matrix materials (paste) at the interfaces (contacts).

Independent of kind of bonding, no matter whether it is adhesion or cohesion, the volumetrical changes produces radial tensile forces relative to the centre of aggregate grain. Through the middle of contact zone, interfaces between aggregate grain and mortar one can draw conditional perpendicular lines. These lines cross each other almost in the centre of the aggregate (Figure 1). There appears to be a changeable cinematics diagram (Figure 2). During turning an aggregate grain under loading, this statically indeterminate system easily breaks the links around the grain or block.

The weakened ties or small isthmuses between grains and blocks are usually subjected to brittle fracture. This "turn-round, tear-off" fracture mode makes it possible to account for the increasing of the absolute size of a structure with the increasing concrete blocks. Naturally, the strength mode excludes similarity theory and the well-known scale principle of such a theory. The turn-round tear-off fracture mode of concrete results in the new following interesting main physical and mechanical properties [1]. Among them there exists one fundamental property of porous heterogenous solid bodies (concrete, rock and so on) that is the low relationship between tensile strength and compessive strength $\left(R_{bt}/R_b \cong 0.10...0.05\right)$. The morphological peculiarities of binder components at atomistic or molecular levels cannot influence on this low relationship.

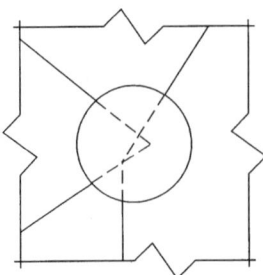

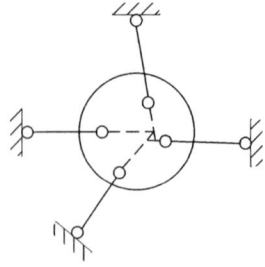

Figure 1 Cracks in cement paste around an aggregate

Figure 2 Turning an aggregate grain because of changeable cinematics system

CONVENTIONAL NORMALIZED CRACK HIERARCHY AND UNITIZED BLOCK ENCLOSED SYSTEM IN OVER-SIZE STRUCTURES

Reinforced concrete structures with size of 0.3...0.6 m contain specific cracks of three orders: millimeters, centimeters and tens of centimeters and consequently three levels of crack hierarchy and block enclosed system in a structure. In their investigations M.M.Kholmyansky (1981) shown that there existed polimodal shape of the distribution of crack lengths in concrete during the process of fracture accumulation and formation of cracks. The availability of three or more modes in the distribution confirms three or more steps mechanism of formation of cracks in structures with such a size.

The microtexture and accordingly its grains at the beginning of loading have the greatest quantity of cracks and the greatest magnitude of probability P_1 in its existence. As the load under a structure is increased, the role of the mezocracks P_2 and then the macrotexture P_3 increases too. Immediately ahead of the rupture of a structure, the role of the macrotexture P_3 is becoming prevalent. At that instant of time the rupturing structure is resulting in maximum ordinary information about the impending rupture. But indeterminacy (scientific information) of the fact of the rupture is approaching zero.

As the case stands, we consider that a concrete massif converts into hierarchy system of enclosed each into other grains, grain groups and blocks. The hierarchy of enclosed block system is also observed in granite rock massifs. From the data by E.N.Peresypkin (1989), worth attention is given to block size as 8 cm. It corresponds to the size of mezocracks after which the intensity coefficient K_{1C} (from fracture mechanics) becomes a constant value. Researches in structural geophysics by M.A.Sadovskiy, V.A.Petrov, A.V.Drumya and N.V.Shebalin have indicated that sizes of rock blocks between cracks vary within wide range but in the average by 3.5 times. Taking into account the magnitude of the "jump" of 3.5 we shall have the following nominal size of concrete block and levels of the hierarchy: 30 cm for the third level, one meter for the fourth level, 3.5 m for the fifth level, 12.0 m for the sixth level, 42.0 m for the seventh level and so on. Crack patterns of the 4-th, 5-th and 6-th levels can be confirmed by many investigations on reinforced concrete structures published in scientific and technical literature. Relying on such analyses of many investigations we propose the conventional normalized hierarchy, scheme with micro-, mezo-, macro- and long cracks.

THE PRINCIPLE OF INFORMATION ENTROPY FOR DISCLOSING INDETERMINACY IN CRACK PATTERN

As mentioned above, the local character of fracture of ties between grains is the first concept of the physical basis of the theory. The second concept of the theory is the admission of a multi step developing crack pattern and accordingly multilevel cracks under loading. These two physical concepts are extremely important for choosing the kind and the length of a formula by well-known american scientist C.E.Shannon [1,2,3].

Such an abstraction as applied to strength texture of concrete made it acceptable because the quantity of information is measured with a number (bit) not depending on absolute magnitude and kind of information just as the volume of a body does not depend on its shape.

Omitting the bulky deduction of the equation, we give the final formula by C.E.Shannon

$$H_i = -p_1 \log_2 p_1 - p_2 \log_2 p_2 - p_3 \log_3 p_3 \text{ at } \sum_1^3 p_i = 1 \qquad (1)$$

Where p_1, p_2, p_3 – probabilities of fractured ties according to micro-, mezo- and macrolevel of crack hierarchy.   At equal probability of all the components of $p_1=p_2=p_3=0.333$ and at $\sum p_i = 1$, we have the largest magnitude of information entropy $H_i = 1.585$ bits.   The information entropy determines the process in the average. At the three-step information, the largest magnitude $H_i = 1.585$ bits, means the greatest indeterminacy of the fact of rupture. When rupture, $H_i = 0$, the indeterminacy of rupture is equal to zero. This means that there is no doubt about the fact of the rupture.

If we analyse the data by E.N.Peresypkin (1989), K.A.Maltsev (1957) and A.V.Karavaev (1976) [1], we can conclude that within the limit of depth of specimen up to 200cm the curve of the informative entropy and the curve of the relative tensile strength of concrete coincides in their character. The fact of coincidence makes it possible to built up a scale safety factor. If we accept three step hierarchy as the basis, then ratio $\gamma_m = H_i / H_3$ can be a partial safety factor for the size of an over-size structure. For example, designing an over-size reinforced hydro-technical dam with the depth h=3.5m of the cross section we have five-step hierarchy in concrete. At that the maximum magnitude of $H_i = H_s$ can be calculated with such formula as

$$H_s = -\sum_1^5 (p_i \log_2 p_i) = 2.322 . \qquad (2)$$

This results in the largest magnitude for the five-step crack hierarchy as $H_s = 2.322$. The assumed magnitude of three step crack hierarchy is the basis for calculation: $H_3 = 1.585$. As a consequence, the design tensile strength R_{bt} should be lowered by 2.322/1.585=1.465 times. Such a value will greatly affect the results of designing large size structures. Using the new lowered design tensile strength R_{bt} in traditional calculation is one of the main method for preventing catastrophe.

REFINED METHOD OF CALCULATING STRESS-STRAIN

To stop the dangerous development of fracture accumulation it is necessary to modify exactly the stress-strain state corresponding serviceability of a structure. Based on this conception, serviceability can be considered as a define residual resource (in bits) in work of a structure due to the fracture accumulation.   The complexity of solving the problem of defined calculation of stress-strain state is compensated with introducing the plane section hypothesis and accordingly an equation to the well-known equilibrium of $\sum X = 0$ and $\sum M = 0$ (the sum of projections on the longitudinal axis and the sum of moments of external and internal forces). According to the aim of calculation (for research or for designing) one can use either arithmetical mean stress-strain curve or the design and the characteristic stress-strain curves of reinforced steel and concrete. As a preliminary it is accepted that the stress-strain curve of concrete is of the shape of FIP-ECB recommendation.

The minimum value of the ultimate bending moment corresponds to the case therein the equilibrium equation $\sum X = 0$ does not have its solution because the resultant of the compression zone does not equalise the result of the main longitudinal tensile reinforcement. The construction of curves of bending moment relative to steel elogation makes it possible to discover new earlier unknown properties of a reinforced concrete bending elements.

ASSOCIATION BETWEEN SERVICEABILITY AND INFORMATION ENTROPY

On the basis of analysis of elements under bending the stress-strain state of a structure can be determined by the special-purpose parameter, i.e. reserve of serviceability H_{ser}

$$H_{ser} = \frac{\sigma_{bc}\xi_{bt}}{R_{bt}\xi_{bc}}, \text{ bit} \qquad (3)$$

where σ_{bc} – stress in the extreme fibre of compression zone of a bending elements;

R_{bt} – tensile strength according to the size of a structure;

ξ_{bc}, ξ_{bt} – relative depths of compression and tensile zones of an element under bending.

H_{ser} accounts for all changes of stress-strain according to the forming of cracks and up to the failure of elements.

As for dimension properties and character of the curve H_{ser} and the curve of information entropy H_i, they coincide (Figure 3). Then it should be recognised that serviceability H_{ser} implies fracture accumulation and consequently levels of development of cracks. Here first of all it should be noted that at the beginning of loading there is a suprising coincidence of $H_i=1.585$ bits and $H_{ser}=1.585$ bits for some beams with average reinforcement ratio and concrete strength. At the end of failure we have $H_{ser}=0$ and accordingly $H_i=0$. This means that the fact of failure has been known and there is no inderterminacy of information concerning the failure. As the beam is approaching failure, the difficulty of prognosis the problem in assessment of the measure of scientific information concerning the failure is gradually disappearing.

For the bending moment of serviceability there is a criterion of serviceability $\lim H_{ser} \geq 1.376 \pm 0.015$. Sometimes the corresponding value of the bending moment is less than the design moment calculated with Building Code (1985). There are two reasons for that. The first reason is the early formation of microcracks in concrete of low and middle concrete grades. This fact was discovered by O.Ya.Berg (1964). The second reason is the large exceeding actual stress σ_{bc} against the value σ_{bc} from the rectangular stress compression block (or alike) accepted in the Building Code. To stop fracture accumulation means to limit the magnitude of H_{ser}. In work [1] there is a specification of the creterion $\lim H_{ser}$ of serviceability and recommendation on its use in dependence of concrete strength, reinforcement ratio and value of its prestressing. Use of the serviceability criterion $\lim H_{ser}$ means a design on a new state in the normal method of the limit states i.e. on the longitudinal crack resistance. Such a design comes into the second group of the limit state and provides durability of an over-size reinforced concrete structure. Use of the new criterion of serviceability promotes determining a stage of work and assessment of expedience for rehabilitation of a structure damaged by unknown load.

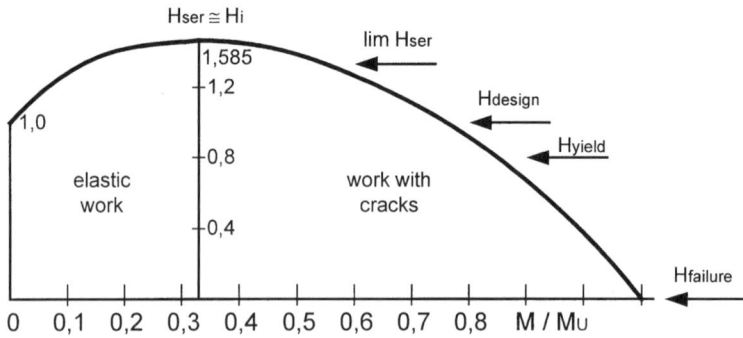

Figure 3   Changes of reserve of serviceability H_{ser} under loading M/M_u

CONCLUSIONS

1. The elements of the proposed theory of catastrophes include the following concepts:Physical basis of the theory consists in ascertainment of turn-round, tear-off fracture modeinducing crack accumulation, crack hierarchy, and utilized block enclosed system in anover-size structure.

2. Probabilistic basis of the theory consists in the disclosing of indeterminacy in a crackpattern with the information entropy.

3. The methodical basis of the theory consists in creating the special-purpose parameter, i.e reserve of serviceability (longitudinal crack resistance).

REFERENCES

1.   SKOROBOGATOV, S.M., The Principle of information entropy in fracture mechanics of buildings and rock seams. Ekaterinburg, URGUPS, 2000, 420pp (Summary and contents in English).

2.   SKOROBOGATOV, S.M., Design of structures using crack indeterminacy and information entropy. Proceedings of International Conference "Concrete 2000". Scotland, University of Dundee, 7-8 September 1993, 10 pp.

3.   SKOROBOGATOV, S.M., Residual service life of reinforced concrete structures. Proceedings of International Congress, Scotland, University of Dundee, 6-10 September 1999, 9pp.

APPLICATION OF RELIABILITY THEORY IN SERVICE LIFE PREDICTION OF INITIATION TIME

J Andrade

Lutheran University of Brasil

D Dal Molin

Federal University of Rio Grande do Sul

Brazil

ABSTRACT. This paper discusses a new form of project for reinforced concrete structures, considering aspects related to the durability in service life prediction. The concepts of limit state function and failure probability are showed, where a reliability analysis was conducted for to verify the influence of the several factors that influence the initiation of the corrosive process induced by chloride ions, taking in consideration to 2nd Fick's Law to model the chloride transport for the interior of the concrete. The results showed the importance of the accomplishment of an analysis that considers statistical aspects for the prevision of the project service life of the reinforced concrete structures.

Keywords: Service life, Chloride penetration, Concrete, Reliability.

Jairo Andrade, Civil Eng, M Sc, received his Dr Eng Degree from Federal University of Rio Grande do Sul, Brasil. He is a Professor at the Civil Engineering Department of Lutheran University of Brasil (ULBRA). His research interests are concrete durability and modelling.

Denise Carpena Coitinho Dal Molin received her D Sc from the University of São Paulo. She is the Professor at the Civil Engineering Department of the Federal University of Rio Grande do Sul. Her research interests are the development of new materials and concrete technology.

INTRODUCTION

The chloride-induced reinforcement corrosion is one of the most common durability problems associated with modern good quality reinforced concrete structures exposed to marine environments or to de-icing salts, being the main cause of decrease of service life of reinforced concrete structures. The service life is actually divided in two phases. The first is calling initiation stage, which is defined as the time from exposure until chloride ions have penetrated the concrete cover, and the chloride content at the surface rebars reaches a threshold level, calling critical value (C_{cr}), leading to corrosion of steel. The second phase is the propagation stage, being considered to the time from when the reinforcement starts to corrode until a critical limit of a material property has been reached. In present work, the service life term is used to mean only the initiation stage, which is to say that only the time necessary for the chloride penetrate into concrete is considered.

The current form of projecting concrete structures with relationship to the durability is reasonable, but extremely qualitative [1]. The performance criteria are not appropriately established, and the service life of the structures is not expressed into a quantitative way. In case certain project approaches are executed, as cover minimum, maximum w/c ratio and limitation of the crack width, among other, the structure can reach a satisfactory service life, but without the prescription of a numeric value of reference. That analysis form is considered inadequate in many aggressive environments, while it can be essentially rigorous in others. Besides, a big amount of parameters that have a great influence in deterioration processes (mainly the concrete characteristics and the environmental conditions) have a great variability in time. For to incorporated this variability in design process, the service life of a reinforced concrete structure can be considered as a stochastic (or random) quantity. This idea was proposed by the CEB [2], where a performance and reliability based service life design would be the basis for the future developments on concrete durability [3].

The project service life of a structure can be represented by a certain density probability function, as presented in Figure 1. Supposes that one wants to build a structure that for a service life of L years. However, it is necessary that a warranty is had that the same will reach the value foreseen with a certain occurrence probability. This way, an index is added to the value of the wanted service life, establishing like this a project life service (PSL), represented by μ this way, it is had that PSL would serve to guarantee that the structure reached, with an established probability of occurrence, the specified service life (L). In such proposition, it's admitted that the service life distribution of the reinforced concrete structure have a certain statistical distribution, where some authors [4,5] consider that the lognormal distribution can be used for to describe this property.

The definition of the value of reliability index (β) depends on some factors, where the performance levels that are defined for the structure have a significant importance. Such levels should be specified previously by the engineer, in order to separate clearly the limits among the failure state and safety state. In this way, the two main limit state that are used in this approach are:

- Ultimate limit state, that they refer to events that present irreversible consequences, generally associated with high financial damages and/or human losses.
- Serviceability limit state, that are related to events that restrict in some way the appropriate use of the structure.

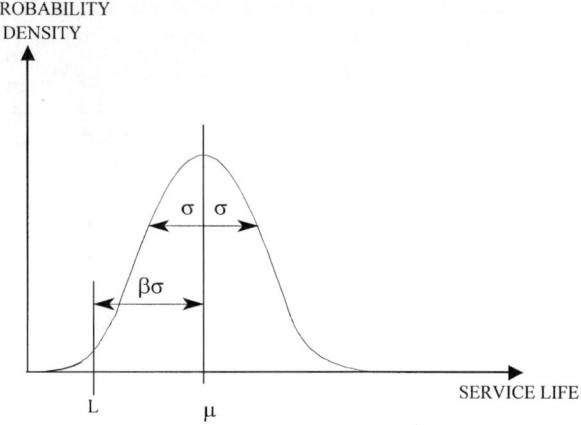

Figure 1   Example of a probability density function for the service life [1, 2]

The effective application of such concept was adopted by some researchers [1], where was presented some reliability indexes extracted of EUROCODE and of Dutch Code, associating them to certain failure probabilities (P_f), is showed in Table 1.

Table 1   Examples of some reliability indexes in structural codes [1]

TYPE OF PERFORMANCE	RELIABILITY INDEX (β) IN A PERIOD OF 50 YEARS (L)		APPROXIMATE FAILURE PROBABILITY IN 50 YEAR
	EUROCODE	DUTCH BUILDING CODE	
Ultimate limit state	3.8	3.6	10^{-4}
Serviceability limit state	1.5	1.8	10^{-2}

The application of this concept can be used for to predict the service life of a reinforced concrete structure located in a saline environment.   This structure is inserted in an environment with high amount of chloride ions, where such elements can penetrate into concrete, causing the corrosion of reinforcement.   Based in proposal presented, a certain failure probability can be adopted, that will be related at a performance limit previously established.   In a hypothetical situation of a structure where the same be built in saline area, a failure probability can be specified for a certain serviceability limit state (represented physically by the despassivation of the reinforcement by chloride ions), being extracted of there a reliability index to be specified for the situation.   The choice of a value for the reliability index it should take in consideration the consequences of the failure in terms of losses human and/or serious economic damages and the necessary effort to be reduced the failure probability in a specific situation [6].

This way, it is observed that the first step for the reliability analysis consists in knowing the mean values and the variability of each one of the factors that influence more significantly in project service life (PSL) of reinforced concrete structures, when inserted in a marine environment. Through the accomplishment of a reliability analysis an indicative of the PSL of a structure can be made, since it is had characteristics values about the environmental conditions and the concrete properties [7,8].

Thus, with base in the considerations presented, a reliability analysis was made for concrete test specimens. Was simulated a condition where some structures will be built with these concretes and inserted in a certain environmental condition, in order to predict the service lifetime (related at the initiation stage).

METHODOLOGY

Was chosen the model based in 2nd Fick's Law to represent the mechanism of transport of substances for the interior of the concrete in present case. The end of performance level was established as the moment when the chloride ion concentration, next at the reinforcement, will be equal at critical concentration for reinforcement despassivation, admitted as 0.4% in cement mass, being considered the end of the project service life (PSL).   Under this condition, a solution for 2nd Fick's Law in function of t is being represented by (1).

$$t = \frac{x^2}{4D_i}\left[erf^{-1}\left(\frac{C_s - C_{cr}}{C_s}\right)\right]^{-2}$$

(1)

The schematic representation of the limit state function for the example in analysis is showed in Figure 2.

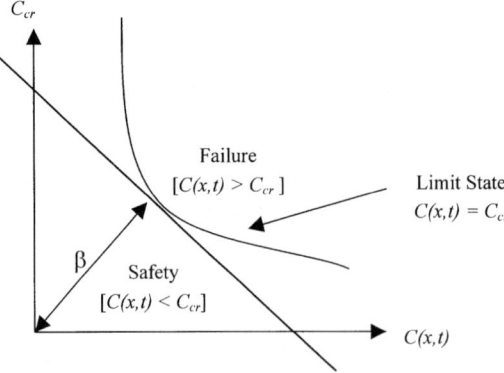

Figure 2   Graphic representation for limit state function for the proposed
case in two-dimensiona space

Defined the limit state function, was investigated the properties of some types of concrete. So, were prepared cylindrical test specimens (10 x 20 cm) for the determination of diffusion coefficient and compressive strength for each type of concrete. Were chosen 3 w/c ratios (0.28, 0.45 and 0.75) and the Brazilian cement type CP II F 32 was used in analyzed concretes.

The values of the chloride diffusion coefficients were obtained by an accelerated test method that consists of submitting a slice (2.5 cm) of the concrete sample among two cells with different chloride concentrations (the used solutions were distilled water, in positive cell, and a solution of NaCl 0,5 M in negative cell) and input a potential difference (10V) among the two cells [9]. The equation of Nernst-Plank was used (Equation 2) for to determine the value of the effective diffusion coefficient [10].

$$D_{ef} = \frac{JRTl}{zFC_{Cl}\Delta E} \qquad (2)$$

where:

D_{ef} = effective chloride diffusion coefficient (cm^2/s);
J = ions flow [mol/(s.cm^2)];
R = gas constant [1,9872 cal/(mol.K)];
T = temperature (Kelvin);
l = thickness of the test specimen (cm);
z = electric charge (for the case of chloride z=1);
F = Faraday constant [23063 cal/(volt eq)];
C_{Cl} = chloride concentration in negative cell; and
ΔE = potential difference applied (V).

Being determined the evolution of the chloride concentration in the positive cell, it could be built graphs that considers the increase of the chloride concentration in time. The ions flow (J) was calculated from graphic, where the D_{ef} values for each concrete analyzed was determined. In this work will be presented some values of the chloride diffusion coefficient, where the other results will be presented opportunely [9].

In order to apply the results from migration tests for predict the service life of concrete was used the equation proposed by some researchers [11], when was presented a solution for 2nd Fick's Law considering the migration and diffusion under a electrical field (Equation 3).

$$\left(\frac{C_l}{C_0}\right) = erfc\left\{\frac{\left[L - \left(\frac{zF\Delta UD_a}{RTL}\right)\right]}{2\sqrt{D_d t}}\right\} \qquad (3)$$

where D_a= apparent diffusion coefficient (cm^2/year);
             C_t = concentration of ions in positive cell at time t (mol/l);
             C_0 = initial concentration of ions in negative cell (mol/l);
             R = gas constant [1,9872 cal/(mol.K)];
             T = temperature (Kelvin);
             L = thickness of the test specimen (cm);
             t = time (years);
             $erfc()$ = error-function complement.
             z = electric charge (for chloride z=1);
             F = Faraday constant [23063 cal/(volt. eq.)]; and
             ΔU = applied potential difference.

The mathematical model developed provides a method of calculating the chloride diffusion coefficient for accelerated diffusion test which gives values comparable to those obtained by normal diffusion tests [11]. So the Equation 3 was applied for to calculate the apparent chloride diffusion for to insert in (1) for to predict the service life in concretes analyzed.

For simulation effect, it was admitted that would be built a element in a saline environment, where the surface chloride concentration (C_s) is equal at 2.5% in cement mass; the critical chloride concentration (C_{cr}) was considered equal at 0.4% in cement mass and the concrete cover (x) was admitted as being equal at 3.0 cm. Also was admitted that the values of the standard-desviation of x, D, C_s and C_{cr} would be all the same and equal to 10%, when all parameters have a lognormal distribution. Monte Carlo with Importance Sampling was employed for the accomplishment of the predictions [12], where was made 10000 simulations for each analyzed case. The simulations made reflect a worst exposure condition (eg elements immersed at sea water), when can be adopted which the end of project service life (PSL) was reached when to occur the despassivation in a structural element.

RESULTS

The values of D_{ef}, D_a and the compressive strength obtained in experiment are presented in Table 2. The values of D_a were used in simulation process, as explained previously.

Table 2   Characteristic of concretes at T = 25° C [9]

CONCRETE	W/C RATIO	EFFECTIVE DIFFUSION COEFFICIENT - D_{EF} (EQUATION 2) (CM^2/YEAR)	APPARENT DIFFUSION COEFFICIENT – D_A (EQUATION 3) (CM^2/YEAR)	COMPRESSIVE STRENGTH (28 DAYS) (MPA)
C1	0.28	0.094	0.059	66.1
C2	0.45	0.164	0.10	37.9
C3	0.75	0.206	0.142	14.9

In this way, was made a reliability analysis for each concrete presented in Table 2. For concrete C1, the value of the project service life (PSL) was calculated through the Equation 2, meeting the value of 39 year. The values of the reliability indexes (β) and the respective failure probabilities (P$_f$) were calculated through Monte Carlo's simulation, considering the values of L presented in Table 3.

Table 3   Values of L, β and P_f for concrete C1

L (YEARS)	20	25	30	35
β	2.70	1.36	1.01	0.40
P_f	0.003	0.038	0.154	0.346

It is verified that a structure built with a w/c ratio equal at 0.28, would have a probability of, at 20 years, to present only 0.3% of despassivation probability; at 35 years, such value would increase for 35%. In the simulations made it was observed that only few points had dropped in the failure region before 20 years. This means that a structure it was built with the concrete of characteristics presented and introduced in a environment specified, the chloride penetration would not probably reach the reinforcement before the structure reached 20 years of having built.

Was made the reliability analysis for the sample C2. Firstly the value of PSL was determined, being equal at 23 years. The values of the reliability indexes (β) and the failure probabilities (P_f) associated calculated through Monte Carlo's simulation are presented in Table 4.

Table 4   Values of L, β and P_f for concrete C2

L (YEARS)	10	15	20	22
β	3.35	1.70	0.52	0.14
P_f	0.004	0.045	0.30	0.44

The data show that a structure built with a w/c ratio equal at 0.45 have, at 10 years, probably 0.4% of it's elements in a despassivation state, and with approximately 22 years, 44% of elements that was inserted in environment it would be despassivated.

For the case of concretes with w/c ratio equal at 0.75, the value of PSL achieved was approximately 16 years. In Table 5 are showed the values of β and P_f associated to some points in time.

Table 5   Values of L, β e P_f for concrete C3

L (YEARS)	8	11	14	16
β	2.91	1.54	0.55	0.00
P_f	0.002	0.062	0.29	0.5

At 8 years, the structure will present despassivation in just 0.2% of its total; and the maximum value of failure probability (50%) was obtained at end of PSL (16 years).

The accomplished analysis showed the importance of staying in adapted levels the mean value of the chloride diffusion coefficient. As smaller its value, better the concrete quality, where the same presents an improvement of it's microstructural conditions, hindering the chloride penetration into concrete, increasing the project service life. Such effect is well evidenced in the concretes with lower w/c ratio, leading at high values of reliability indexes, when showed in Figure 3.

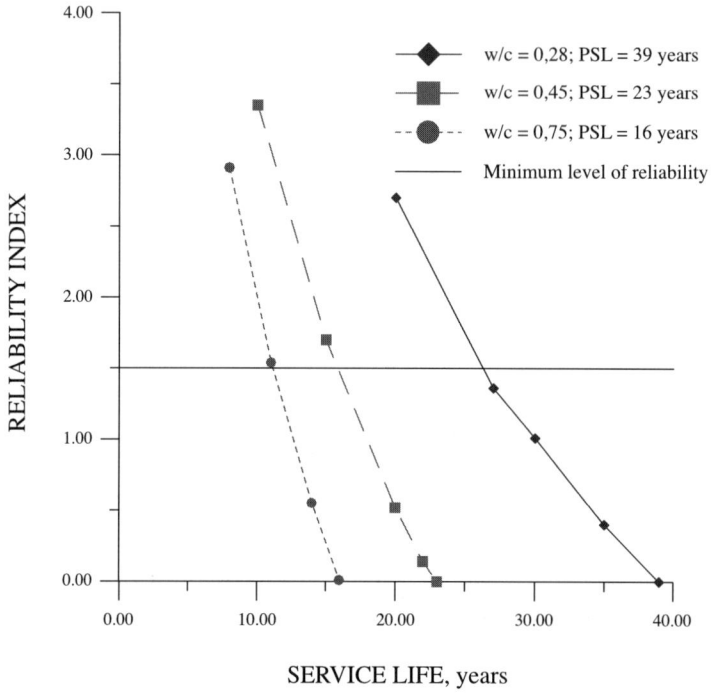

Figure 2   Reliability index (β) for all the analyzed concretes for initiation

time in corrosive process

First can be observed that any concrete evaluated reached the minimum value recommended by EUROCODE (Table 1) for the serviceability limit state (50 years). Besides, considering that the minimum level of reliability recommended by EUROCODE for 50 years is equal at 1.5, can be observed that the more durable concrete (w/c = 0.28) have this reliability index approximately at 26-27 years, for the environment specified. So, the use of protective measures for concrete protection is need for to decrease the chloride penetration into concrete element.

FINAL CONSIDERATIONS

When improving the characteristics of the concrete (resulting in a consequent decrease of the diffusion coefficient), can be obtained more durable structures, when submitted in a certain environmental condition. However, when exist an appropriate knowledge of the behavior of the variables that influence in chloride penetration (mean values and standard deviation), with a larger certainty degree, the project service life (PSL) of those constructions can be estimated. Thus, in the moment of the planning of a reinforced concrete structure that will be inserted in a chloride-laden environment, the engineer have considering durability parameters requested for maximize it's service life.

The results show that, without a probabilistic analysis, it is very difficult to establish adequated values of PSL for a structure that will be inserted in a certain environmental condition. Some studies should be made for to verify the methodology proposed, as the adapted definition of the failure probabilities associated with different types of structure, the study of another limit state functions in relation to the despassivation for chloride ions and the statistical characterization of the variables that influence in these models. The analysis here presented it is in development for the researchers of UFRGS (Federal University of Rio Grande do Sul) team, in order to find quantitative parameters for to guide the project professionals and maintenance engineers, taking in consideration the service life required by reinforced concrete structures [8, 13]. The study of the modelling techniques is being accomplished in a systematic way and it comes presenting satisfactory results for the analyzed cases.

REFERENCES

1. SIEMES, T et al, Design of Concrete Structures for Durability. Example: Chloride Penetration in the Lining of a Bored Tunnel. Heron, v 43, N° 4, 1998, p 227-244.

2. COMITE EURO-INTERNATIONAL DU BETON, New Approach to Durability Design: An Example for Carbonation Induced Corrosion. Bulletin D'Information N° 238, Suíça, 1997, 142 p.

3. SIEMES, T, History of Service Life Design of Concrete Structures. In: Workshop Design of Durability of Concrete. Proceedings. Berlin, 1999, 8 p.

4. MATSUSHIMA, M, A Study of the Application of Reliability Theory to the Design of Concrete Cover. Magazine of Concrete Research, v 50, N° 1, Mar 1998, p 5-16.

5. ENRIGHT, M; FRANGOPOL, D, Probabilistic Analysis of Resistance Degradation of Reinforced Concrete Bridge Beams Under Corrosion. Engineering Structures, v 20, N° 11, 1998, p 960-971.

6. SCHIESSL, P et al, Durability Aspects of Probabilistic Ultimate Limit State Design. Heron, v 44, N° 1, 1999, p 19-29.

7. POULSEN, E, Estimation of Chloride Ingress into Concrete and Prediction of Service Lifetime with Reference to Marine RC Structures In: Durability of Concrete in Saline Environment. Proceedings. P Sandberg (Ed) Lund, May 1996, p 113-126.

8. ANDRADE, J, Service Life of Reinforced Concrete Structures for the Corrosion of the Reinforcement: Initiation by Chloride Íons. D.Sc. Thesis. Federal University of Rio Grande do Sul, Porto Alegre, 2001. (In portuguese).

9. PEREIRA, V. Chloride Diffusion Coefficients in Concretes. M.Sc. Dissertation. Federal University of Rio Grande do Sul, Porto Alegre, 2001. (In portuguese).

10. ANDRADE, C. Calculation of Chloride Diffusion Coefficients in Concrete from Ionic Migration Measurements. Cement and Concrete Research, Vol 23, pp 724-742, 1993.

11. SHA'AT, A et al, Reliability of the Accelerated Chloride Migration Tests as a Measure of Chloride Diffusivity in Concrete. International Conference Concrete Repair, Rehabilitation and Protection, Dundee, 1996. Proccedings, p 245-255.

12. ANG, A, TANG, W, Probability Concepts in Engineering Planning and Design: Decision, Risk and Reliability, v 2, 1ª Ed, Ed John Wiley and Sons, 1984, 562 p.

13. CABRAL, A, ANDRADE, J, DAL MOLIN, D, Service Life of Repair Systems Used in Reinforced Concrete Structures. 2nd International Conference on the Behaviour of Damaged Structures. Proceedings CD-ROM. Rio de Janeiro, Jun 2000. (In portuguese).

FAILURE PROBABILITY ASSESSMENT FOR ADDITIONAL DESIGN LOAD OF AN EXISTING STRUCTURE

D Bandyopadhyay

S Saraswati

Jadavpur University

India

ABSTRACT. Condition assessment and failure probability assessment of old structure is significantly important particularly, for its safety, determining future usefulness and adopting effective restoration scheme for its life enhancement. The understanding of structural behavior will indicate the failure probability and future usefulness of the structure. In this particular study, the suitability of an existing industrial steel-concrete composite structure for the desired renovation and modernization has been examined. The installation of pollution control equipment at the roof of the coal-feeding bunker of an age old thermal power station has been essentially required for its sustainable development particularly from environmental consideration. Full-scale load test is incorporated for the additional design load to be subjected. The elastic property and the symmetry of the structure have been verified in this technique. The comparison of measured static responses and those obtained theoretically on idealised model has also been studied. It is inferred that, the condition can be assessed and the failure probability of the structure can be evaluated from the study. Based on the above observations the suitability of the structural for its further renovation can be recommended.

Keywords: Assessment, Load Test, Deflection, Recovery, Vibration, Non-linearity.

D Bandyopadhyay, born 1965 received his Civil Engineering degree from Jadavpur University, Calcutta, and did his Master of Technology, specialisation in Structural Engineering from Indian Institute of Technology, Kharagpur, India. He was Executive Engineer (Civil) in Damodar Valley Corporation, Government of India prior to joining the University. His research interest includes Damage Detection, Failure Investigation & Condition Assessment of structures, Repair and Rehabilitation of structures.

S Saraswati, born 1961 received his Civil Engineering degree from Bengal Engineering College, Calcutta, and did his Master of Technology, specialisation in Geotechnical Engineering from Indian Institute of Technology, Kharagpur, India. He was in Gammon India prior to joining the University. His research interest includes Soil investigation, Concrete Technology & Repair and Rehabilitation of structures.

INTRODUCTION

The importance of condition monitoring of structure in assessing its integrity in the context of safety or failure probability is immense. The assessment of an industrial building structure subjected to additional load due to modernization is significantly important both from safety consideration as well as from employing economical strengthening scheme. The particular structure New Cossipore Thermal Power Station is located at the east bank of the Ganges at Cossipore of Calcutta, West Bengal, India. The powerhouse was constructed more than 50 years back as per available information. The objective of the work is to assess the structural integrity under the additional load due to come from the pollution control equipment. The future use of the said structure and the mode of strengthening, if required, will be based on the present condition of the structure. Static load test was carried out to assess the present condition of the structure, particularly its capability of sustaining the additional load from the pollution control equipment to be installed at the roof of the bunker bay. The scope of the work consist of carrying out the load test at the selected portion of the roof made of RCC slab supported on steel beams at bunker bay of boiler house.

TEST DETAILS

The details of the load test for the condition assessment of the particular structure subjected to the additional load are governed by the equipment design data and the viable testing condition in the practical situation. The location of the test spot is at the northwest corner of the bunker roof of boiler house. The structural layout at this particular test location is shown in Figure 1.   The span of the main girders is 7.315 m.   Seven numbers secondary beams having a span of 5.476 meter and relatively smaller depth are located in the perpendicular direction to the main girder @ 1.240 m C/C.

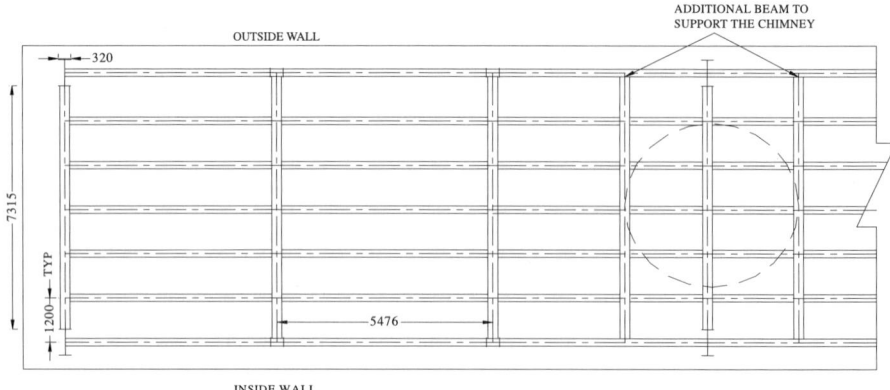

Figure 1   Structural layout of the test location at bunker bay roof

Loading Arrangement

The layout of the proposed equipment and its load distribution at the worst condition is calculated. Accordingly, the load area is marked and supporting arrangement is simulated close to the actual assumed condition. The full load of 1000 KN is distributed uniformly over an area of 5.125 meters by 12.0 meters platform supported on a grid of main and secondary beams as shown in Figure 2.

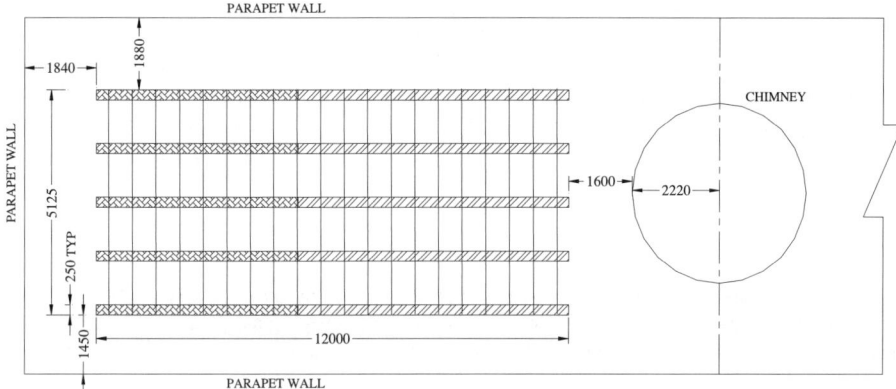

Figure 2   Loading arrangement at the test location of bunker bay roof

Loading is being made by pre-weighted sand bags uniformly spread over the loading platform. The loading was made in a controlled and smooth manner to avoid the effect of impact. Considering the rainy season the effort was taken to the possible extent to prevent the increase of weight of sand bags due to water absorption.

Deflection Measurement

The detail measurements of deflection are taken at central portion of the particular unit, however deflections at the center of edge girders are also noted. The deflection locations are selected at the soffit of girders and beams. The typical layout of dial gauge locations is given in Figure 3.

The deflections of all the dial gauges are noted for the different stages, such as at 0%, 25%, 50%, 75%, 100% of total loading and unloading. Monitoring of readings instantaneously and at its short term steady condition was also made during the test. Additional load up to the maximum extent of 25% of imposed load was also incorporated for a better understanding of the structural behavior and adequacy. The sequence of loading and unloading are maintained in the same manner for a continuous and symmetrical understanding of the structural behaviour.

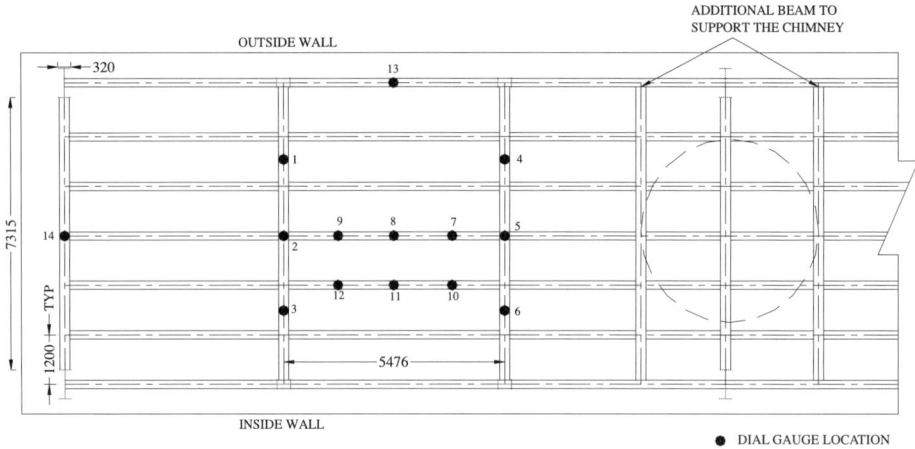

Figure 3   Layout of the dial gauge location at the test spot

Independent arrangement of dial gauge support for deflection measurement and reading platform are fabricated. It is observed that vibrations at the dial gauges are very small in case of main girders as the supports are from the main column whereas in case of secondary beam the vibration are quite high due to the operation of plant and the supporting arrangements are made from the bunker bay floor. The average vibrate readings are taken under this compelled situation.

Being an important industry and considering the emergency, it was not possible to stop the operation of the thermal power station throughout the test. Thus vibration of the structure was not able to avoid in the test. However the deflection readings are taken at different instances after closing the conveyor belt system, which cause the major vibrations.

It is assumed that the vibrations occurred during the loading are linear in nature and will not cause any permanent deflection. Secondly, the boiler, forced draft fan and the auxiliaries are in full swing operation during the test even during the measurement of the deflection. Thus the probability of errors due to this vibration could not be eliminated.

RESULT AND DISCUSSION

Considering the unavoidable hazard involved at the test location due to vibration etc, deflections are measured at 14 selected locations for six stages of loading and unloading conditions. The measured deflection data sheet for different percentage of loading is given in Table-I. Similarly, the measured recovery of deflection data for sequential unloading is given in Table-2.

Table 1   Dial gauge reading for different stages of loading

DIAL GAUGE NO.	LEAST COUNT (mm)	INITIAL READING	25 % LOADING	50 % LOADING	75 % LOADING	100 % LOADING	125 % LOADING	125% + CREEP (Min 24 Hrs)
1	0.002	1422	1422	1432	1684	1726	1822	1823
2	0.01	200	215	241	265	293	309	310
3	0.002	1592	1702	1925	1986	2101	2285	2318
4	0.002	918	1133	1358	1342	1450	1517	1443
5	0.01	905	929	1062	1098	1125	1138	1129
6	0.002	1300	1387	1505	1608	1735	1805	1805
7	0.002	1760	1741	2000	2290	2490	2500	2490
8	0.01	310	350	400	445	494	512	515
9	0.002	1150	1015	1250	1465	1672	1760	1790
10	0.01	642	660	705	740	773	790	776
11	0.01	492	500	552	596	637	662	667
12	0.002	1380	1398	1463	1521	1602	1696	1275
13	0.01	896	812	834	956	985	978	983
14	0.01	550	555	573	578	594	602	620

Table 2   Dial gauge reading for different stages of unloading

DIAL GAUGE NO.	LEAST COUNT (mm)	INITIAL READING (125%)	100 % LOADING	75 % LOADING	50 % LOADING	25 % LOADING	0 % LOADING	0% + CREEP (min 24 hrs)
1	0.002	1823	1823	1823	1823	1695	1676	1233
2	0.01	310	305	293	280	262	235	221
3	0.002	2318	2283	2194	2101	1994	1826	1826
4	0.002	1443	1443	1403	1272	1151	1007	958
5	0.01	1129	1123	1105	1071	943	904	901
6	0.002	1805	1731	1785	1715	1628	1468	1405
7	0.002	2490	2430	2396	2100	2080	1605	1585
8	0.01	515	503	470	420	373	305	240
9	0.002	2372	2281	2090	1515	1248	870	865
10	0.01	776	768	741	710	670	620	600
11	0.01	667	655	620	580	530	494	485
12	0.002	1275	1091	1000	996	907	805	609
13	0.01	983	985	973	959	947	822	800
14	0.01	620	615	605	595	578	562	523

The deflections at different dial gauge location for maximum loading and unloading condition with and without creep are given in Table 3 and Table 4.

Table 3   Calculation of deflection and recovery without creep

Dial Gauge No	Least Count. (mm)	DEFLECTION DUE TO LOADING			RECOVERY OF DEFLECTION			
		Initial Reading 0 %	Final Reading 125%	Deflection (mm)	Initial Reading 125% + Cr	Final Reading 0%	Recovery (mm)	% Recovery
1	0.002	1422	1822	0.8	1823	1676	0.294	36.75
2	0.01	200	309	1.09	310	235	0.75	68.8
3	0.002	1592	2285	1.386	2318	1826	0.984	71.0
4	0.002	918	1517	1.198	1443	1007	0.872	72.8
5	0.01	905	1138	2.33	1129	904	2.25	96.6
6	0.002	1300	1805	1.01	1805	1468	0.674	66.7
7	0.002	1760	2500	1.48	2490	1605	1.77	119.6
8	0.01	310	512	2.02	515	305	2.10	104.0
9	0.002	1150	1760	1.22	1790	870	0.974	79.8
10	0.01	642	790	1.48	776	620	1.09	73.6
11	0.01	492	662	1.70	667	494	1.73	101.7
12	0.002	1380	1696	0.632	1275	805	0.94	148.0
13	0.01	896	978	0.82	983	822	1.61	196.3
14	0.01	550	602	0.52	620	562	0.58	111.5

Table 4   Calculation of deflection and recovery with creep

Dial Gauge No	Least Count. (mm)	DEFLECTION DUE TO LOADING			RECOVERY OF DEFLECTION			
		Initial Reading 0%	Final Reading 125% + Cr	Deflection (mm)	Initial Reading 125% + Cr	Final Reading 0% + Cr	Recovery (mm)	% Recovery
1	0.002	1422	1823	0.802	1823	1233	1.18	147
2	0.01	200	310	1.10	310	221	0.89	80.9
3	0.002	1592	2318	1.452	2318	1826	0.984	67.8
4	0.002	918	1443	1.05	1443	958	0.970	92.3
5	0.01	905	1129	2.24	1129	901	2.28	101.8
6	0.002	1300	1805	1.01	1805	1405	0.80	79.2
7	0.002	1760	2490	1.46	2490	1585	1.81	124.0
8	0.01	310	515	2.05	515	240	2.75	134.1
9	0.002	1150	1790	1.28	1790	865	1.014	79.2
10	0.01	642	776	1.34	776	600	1.76	131.3
11	0.01	492	667	1.75	667	485	1.82	104.0
12	0.002	1380	1275	-	1275	609	1.332	-
13	0.01	896	983	0.87	983	800	1.83	210
14	0.01	550	620	0.70	620	525	1.0	142.9

It is noted that few readings are erratic and inconsistent, which may be due to
- The constant vibration results from the operation of the plant during the test,
- The random locking of the gauge spring by the coal dust accumulated in the dial gauges.

However, the maximum deflection observed at the mid span of the main girder and secondary beam for the 100% loading are tabulated below and compared with the theoretical values. The limiting deflection for the recovery calculation as stipulated in the Indian standard code for these locations is also compared in the following Table 5.

Table 5 Comparison of measured, theoretical and limiting deflections

SL NO	LOCATION (DIAL GAUGE NO)	DEFLECTION MEASURED (mm)	THEORETICAL DEFLECTION (mm)	LIMITING DEFLECTION FOR RECOVERY (mm)
1	Central span of main girder (5)	2.20	2.10	2.23
2	Central span of secondary beam (8)	1.84	1.75	2.53

From these above observation it is noted that at the particular location of roof, the deflection measured are more or less comparable with the theoretical one. However irregular and inconsistent data are also noted in few locations.

The maximum deflections observed are 2.33 mm & 2.02 mm at mid span of main, girder and secondary beam respectively for the maximum loading. The recovery of deflection for maximum loading and unloading are tabulated below in Table 6 at the location where a relative consistent behavior is observed.

Table 6 Recovery of deflections after unloading considering without and with creep

SL NO	LOCATION	RECOVERY OF DEFLECTION (%)		AVERAGE RECOVERY (%) (max 100%)	REMARK
		Without creep	With creep		* Recovery is limited to 100% ie the maximum value
1	2	68.8	80.9	75	
2	3	71.0	67.8	70	
3	4	72.8	92.3	83	
4	5	96.6	101.8	99	
5	11	101.7	104.0	100*	

In most of the cases the recovery of deflection are above the acceptable limit. However, few irregular behaviours are also observed.

The graphical representations of load deflection behavior in few important locations under loading and unloading condition are given in Figure 4a to Figure 4c. It is noted that in few cases little non-linearity have been observed at the higher loading condition particularly above 100% loading.

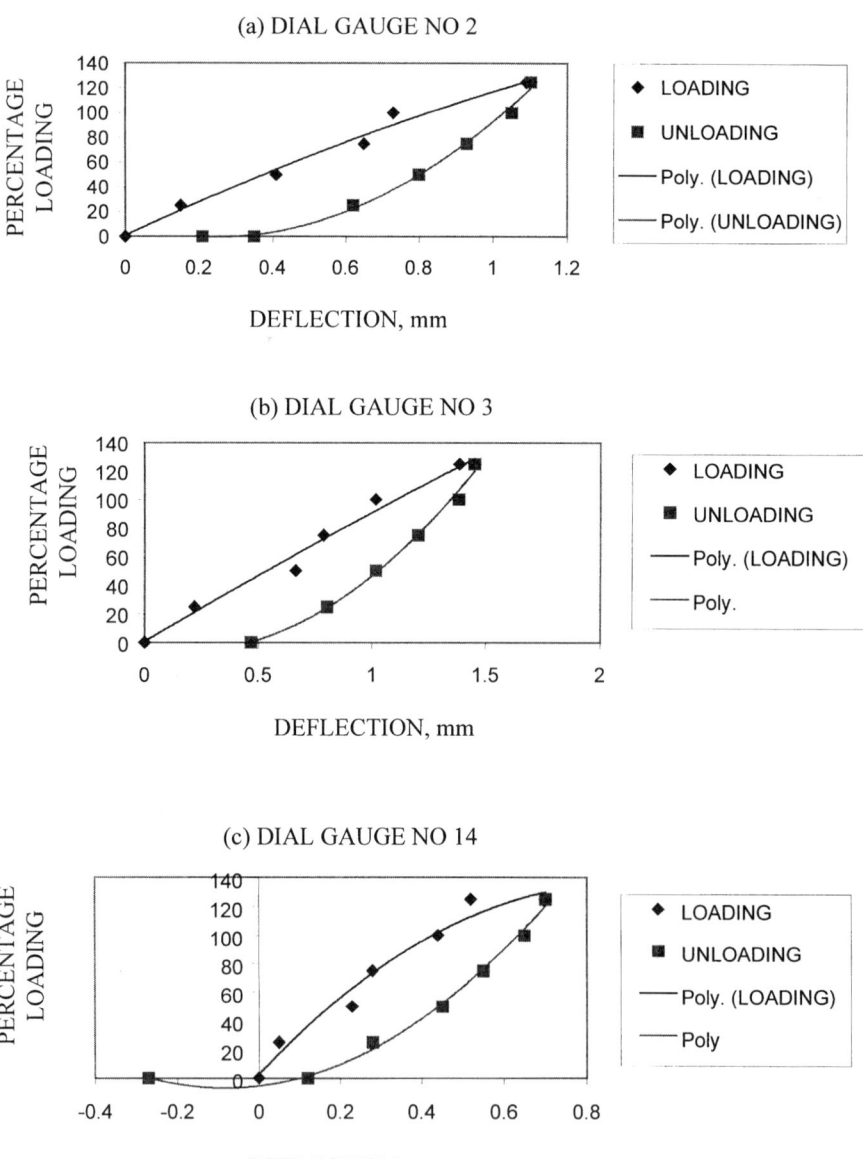

Figure 4   Load deflection curves

CONCLUSIONS

The acceptance criterion for this industrial building structure is governed by the stipulation of Indian Standard Code. The recovery of deflection should be at least 75% of deflection. In addition the study of load deflection behavior of the structure will able to indicate about the condition of the structure. It seems that over loading of structure beyond 1000 KN is not a safe proposition considering the non-linear behavior of the load deflection curve. Further the behavior of the structure in case of changed loading pattern demands due consideration. Again, keeping in view the age of the structure, proper care should be taken during construction and installation of the additional structure. The assessment of the dynamic behavior of the said structure is beyond the scope of this test.

Keeping in view the above stipulation and based on measured data and subsequent analysis it may be inferred that the structure is safe against the given loading condition at the particular location of testing. However inconsistent and irregular behavior of the structure is also observed at few locations, which may be due to the constant vibrating disturbances during the testing.

REFERENCES

1.   ACI 437R-91, Strength Evaluation of existing concrete buildings.

2.   STAAD - FEM based software for Structural Analysis And Design, developed by Research Engineers Incorporated.

3.   BANDYOPADHYAY, D., SARASWATI, S. Report on the load test on the roof of the bunker bay of New Cossipore thermal power station in connection with installation of pollution control equipment

4.   BANDYOPADHYAY, D. Structural condition assessment from simple field test data. IABSE Congress, Lucerne, Switzerland, 18-21 September 2000.

LABORATORY AND ON-SITE CARBONATION DATA CORRELATION IN ESTIMATING THE LIFE SPAN OF CONCRETE STRUCTURES

M F Nuruddin

K S Ali   H M Saman

Universiti Teknologi

A B M Diah

Universiti Sains

Malaysia

ABSTRACT. Deliberation on the influence of carbonation on covercrete is of paramount importance as it affects the life span of the concrete structure when corrosion of reinforcement takes place. This is because the diffusion of CO_2 into the concrete depletes the alkalinity level and causes the reinforcement to be exposed to the danger of corrosion. Subsequently corrosion causes the concrete to deteriorate and disintegrate. The epidermis of the concrete exposed to the environment experiences different effects, which depends on the concrete grade and type of exposures. This research elucidates the carbonation effect on concrete from different environments. Concrete samples of various grades were used. The measurement of carbonation depths of structures exposed to various environments was taken using phenolphthalein spray method. Two sets of samples were considered namely controlled sample in the laboratory and on-site samples. In the laboratory, the effect of carbonation was accelerated by using accelerated carbonation chamber (4% CO_2) and monitoring was done from the first week until 40th week. At the sites, samples of similar grades as the ones in the laboratory were taken from existing structures from different exposures. Based on 1 week in accelerated carbonation chamber as equivalent to 1 week on site, a correlation between the two sets of data was done. As a result an estimation chart was produced to analyze data obtained so that the laboratory data could explain the different on site carbonation characteristics via correlation Equations acquired. It was found that good correlation prevailed between laboratory data and the on-site data. From the correlation, the life span and the covercrete could be estimated or proposed for design purposes. These estimates were introduced via simple and user-friendly computer programs.

Keywords: Carbonation, Covercrete, Accelerated carbonation chamber, Correlation, Life span.

Dr M Fadh Nuruddin, Faculty of Civil Engineering, Universiti Teknologi Mara, Malaysia

Dr A B M Diah, School of Civil Engineering, Universiti Sains Malaysia.

Dr K Sh Ali, Faculty of Arch, Planning & Surveying, Universiti Teknologi Mara, Malaysia.

Dr H M Saman, Faculty of Civil Engineering, Universiti Teknologi Mara, Malaysia.

INTRODUCTION

Researches on the dependency of the life span of concrete structure to the level of resistance towards deteriorating onslaught have been in the limelight for decades. Indicators like chloride ingress and CO_2 diffusion have been taken as viable telltale to durability problems and subsequently its life span. Carbonation attack is known to take quite a long time to have any effect on the concrete but it was found that recently built concrete structures (less than 8 years old) experienced up to 15 mm carbonation depth [1]. This phenomenon causes corrosion to take place. Table 1 shows a list of repair projects that have to be done in Malaysia due to corrosion problems.

By and large concrete structures that is alkaline in nature (with a pH of about 11-13) has the ability of protecting the reinforcement from corrosion via a virtual passive coating or oxide film. Unless with the presence of chloride ion and CO_2, this virtual coating will remain intact as long as the alkaline environment within the concrete is maintained. For instance with the diffusion of CO_2 into the structure, carbonic acid is produced; as a result of reaction between CO_2 and $Ca(OH)_2$, that reduces the alkalinity level to a pH of about 8 [2]. Consequently depletion of passive coating occurs and corrosion begins with the presence of moisture and oxygen. Initiation of corrosion, if not treated, is also the initiation of the end of the life span of the structure. Therefore, remedial action need to be undertaken to ensure serviceability is upheld.

Table 1   Concrete structures repair project undertaken

PROJECT	YEAR	VALUE (RM)
Sultanah Aminah Hospital Johor Bahru	1990	20 million
Kota Kinabalu Port	1992	8 million
KCT Wharf	1993	6 million
Sandakan Wharf	1995	11 million
Klang Port	1995	6 million
TLDM Jetty at Lumut	1995	16 million
Penang Port Wharf Phase I	2001	28 million
Penang Port Wharf Phase II	2001	30 million

COMPRESSIVE STRENGTH DEVELOPMENT

It is commonplace that the compressive strength of concrete structures is taken as a measure of its quality. Non-conformance to its 28-day design strength indicates failure in serviceability [3]. It is accustomed that the 28-day cube strength is taken as the determining strength. Albeit the 28-day duration is the determining period, its strength development from day zero to 28 days needs to be looked critically. It is also a well-accepted fact that the compressive strength shall achieve 40% and 70% of the 28-day strength at 3 and 7 days respectively [4].

DURABILITY REQUIREMENTS

Durability can be defined as the ability of the concrete structures to withstand the damaging effect of the environment without maintenance work over a long period of time. Nevertheless there is always an end to almost everything; therefore concrete structures tend to deteriorate whenever its ability to resist against external and internal attacks ceases to exist. Figure 1 shows the relation between concrete integrity against time whereby three time-dependent stages can be drawn namely development stage, reliability stage, and finally deterioration stage.

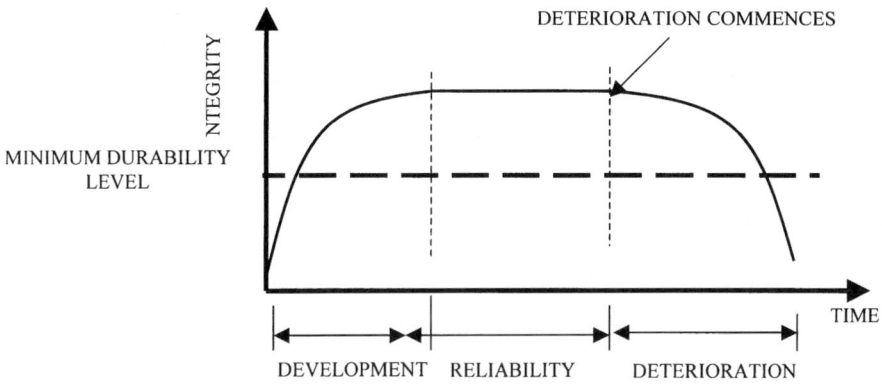

Figure 1   Graph of quality over time

Development stage is a stage where the concrete starts its hardening process and hydrates of calcium and aluminates start to take shape. Tricalcium silicates and tricalcium aluminates are the major compound compositions that contribute to this development stage. Subsequent stage is the reliability stage whereby the concrete illustrates a fully formed durable situation. In this stage the servicing concrete structure faces the environment, which can be categorized as mild moderate or severe [5]. Its ability to face the harmful agents from the environment depicts its level of reliability. After this stage, as when the concrete 'grows older', a threshold is reached whereby the concrete structures lose its battle against the environmental attack and give way for deterioration to take place. It is at this stage (or perhaps before) that maintenance or repair works need to be considered to uplift its service life span. This can be done via various methods and its major aim is to extend the serviceability capability.

The Malaysian Standards [MS 1195, 1991] on durability focuses on the covercrete for different types of exposures. Severe exposure needs thicker covercrete and concrete grade plays a role in determining the thickness of covercrete.

OBJECTIVE AND SCOPE OF RESEARCH

Durability is supposed to be a value proposition for concrete constructors. Nevertheless incessant onslaught originating from the environment has never fail to deteriorate concrete structures. The main objective of the research is to come up with a chart and a computer program that tells the life expectancy of concrete structures exposed to different environments by looking at the carbonation depth of the structure.

Two sets of data were taken; laboratory (controlled) data and on site (real) data.   Via correlation between the two, a life expectancy chart aided by a computer program is produced. The life expectancy chart can also be utilized to improvise the Malaysian Standard on covercrete requirements exposed to different exposures.

Concrete samples of grades 15, 20, 25, 30, 35, 40, and 50 were monitored [6].   The measurements of carbonation depths of structures exposed to various environments were taken using phenolphthalein spray method.

In the laboratory, using accelerated carbonation chamber (55% RH and 4% CO_2) enhanced the effect of carbonation and monitoring was done from the first week until the 40th week. This was done because the carbonation process is a slow process and takes many years (depending on the durability and concrete grade) to carbonate a few millimeters.

At the site, samples of similar grades as the ones in the laboratory were taken from existing structures from six different exposures.   The six exposures were L1 (rural - not exposed to rain), L2 (rural- exposed to rain), L3 (inside building), L4 (urban – not exposed to rain), L5 (urban – exposed to rain), and L6 (marine).   The age monitored for carbonation test was from 1 year up to 40 years.   Detail of the research undertaken is shown in Table 2 whilst Table 3 and 4 illustrate the type of curing and characteristics of different exposures respectively.

Table 2   Detail research works done (laboratory as well as on site)

TYPE OF TEST	EXP	AGE OF SAMPLE	SAMPLE	STANDARD USED
Carbonation	P(a/tp)	1,2,4, till 40 weeks	Cube 150mm	RILEM
(phenolphthalein)	L1 – L6	1,2,4, till 40 years	Core 100mm	CPC-18, 1984

Table 3   Type of curing and exposures of samples

ENV	CODE	CURING	REMARKS
LABORATORY	P(a/tp)	Curing in accelerated carbonation tank	Sample undergo 28 day water curing and 14 day normalization process and then placed in accelerated carbonation tank (4% CO_2 55% RH) until age of testing
ON–SITE	L1	rural; not exposed to rain	Samples taken from existing structures or secondary data from various parties
	L2	rural; exposed to rain inside	
	L3	building	
	L4	urban; not exposed to rain	
	L5	urban; exposed to rain marine	
	L6		

Table 4   Characteristics of different exposures

EXP	DEFINITION	TEMP(°C)	R.H (%)	CO_2 CONT (PPM)
P(a)	Water tank	28	100	-
L1	Rural; not exposed to rain	26-32	64-76	304-431
L2	Rural; exposed to rain	25-30	69-82	326-483
L3	Inside building	27-32	67-77	359-611
L4	Urban; not exposed to rain	27-34	55-68	418-766
L5	Urban; exposed to rain	26-35	67-81	481-784
L6	Marine	27-38	74-86	523-886

ANALYSIS OF RESULTS

The carbonation depth measurements for samples P(a/tp) and those exposed to 6 different types of exposures were plotted against age.   This was done for the various grades introduced.   Table 5 shows the coefficient of determination of individual data and it was found that the R^2 approaches nearly 1.   For instance values of R^2, for controlled samples, equal to 0.956, 0.932, 0.900 for low, medium, and high grade concrete respectively were obtained.   This illustrates that the ages of the samples could well foretell the carbonation depth given the grades and type of exposures.

Table 6 shows the relation between real and laboratory data assuming 1 week in accelerated carbonation chamber as equivalent to 1 year at site. The coefficients of determination, across the board R^2 obtained were 0.989, 0.977, 0.947 for low, medium, and high grade respectively, whilst the average value of all the three was 0.97. This concludes that the data on sample age can be utilized to represent the carbonation data in the laboratory as well as at the site.

Table 6 also shows that the two sets of data could be correlated well.   The correlation between real and laboratory data and the coefficient of correlation also nearly approaches the value of 1. It was found that the coefficient of correlation is between 0.945 and 0.998. This trend shows that the laboratory and the real data could be closely correlated via the Equations acquired and the assumption that a week in accelerated carbonation chamber as equivalent to one year at site conform findings by many researchers [7-11].

The numerical values of life span of the structure can be estimated with the aide of a computer program.   Table 7 shows the estimated life span for different types of exposures and grades. As an example assuming a cover of 35mm is adopted for concrete structures of grade 30 that being exposed to inside building (L3) condition, it was found that at the age of 25.92 years, carbonation depth has reached a stage whereby corrosion is prone to happen. At this stage maintenance work need to be done, without which the structure will deteriorate further.

CONCLUSIONS

The laboratory data correlates well with the on site data.   This implies that by using the laboratory carbonation data, the life expectancy of the concrete structures on site can be estimated via correlation Equations discovered. With this, planned maintenance work can be drawn up to ensure the concrete structures are well kept. If the carbonation depth is as deep as the concrete cover, depassivation of rebar occurs and corrosion is inevitable with the presence of oxygen and moisture.

Table 5   Equation of rate of carbonation and coefficient of determination

GRADE (N/mm²)	SALIENT FEATURES	P(a/tp)	L1	L2	L3	L4	L5	L6
					EXPOSURES			
				Low Strength				
15	Equation	$y = 7.348x$	$y = 6.187x$	$y = 6.797x$	$y = 7.169x$	$y = 8.130x$	$y = 8.846x$	$y = 9.151x$
	R^2	0.962	0.972	0.987	0.991	0.975	0.978	0.974
20	Equation	$y = 6.798x$	$y = 5.725x$	$y = 5.894x$	$y = 6.867x$	$y = 7.674x$	$y = 8.069x$	$y = 8.335x$
	R^2	0.949	0.935	0.977	0.993	0.975	0.969	0.967
				Medium Strength				
25	Equation	$y = 6.052x$	$y = 4.903x$	$y = 5.504x$	$y = 5.911x$	$y = 6.086x$	$y = 7.578x$	$y = 7.763x$
	R^2	0.939	0.945	0.959	0.988	0.964	0.972	0.976
30	Equation	$y = 5.456x$	$y = 4.542x$	$y = 4.950x$	$y = 5.691x$	$y = 5.774x$	$y = 7.142x$	$y = 7.389x$
	R^2	0.969	0.972	0.950	0.981	0.983	0.940	0.956
35	Equation	$y = 4.994x$	$y = 4.280x$	$y = 4.414x$	$y = 5.460x$	$y = 5.501x$	$y = 6.797x$	$y = 7.019x$
	R^2	0.887	0.947	0.963	0.973	0.971	0.937	0.955
				High Strength				
40	Equation	$y = 3.859x$	$y = 3.851x$	$y = 4.157x$	$y = 4.934x$	$y = 4.934x$	$y = 6.338x$	$y = 6.477x$
	R^2	0.933	0.897	0.920	0.958	0.958	0.914	0.913
50	Equation	$y = 3.075x$	$y = 2.852x$	$y = 3.295x$	$y = 4.659x$	$y = 4.659x$	$y = 5.321x$	$y = 5.641x$
	R^2	0.867	0.741	0.797	0.981	0.981	0.854	0.862

Note: In the Equation, y is the carbonation depth and x represents ⊕age.

Table 6 Correlation Equation, coefficient of determination, and coefficient of correlation

GRADE (N/mm²)	SALIENT FEATURE	EXPOSURES					
		L1-a/tp	L2-a/tp	L3-a/tp	L4-a/tp	L5-a/tp	L6-a/tp
		Low Strength					
15	Equation	y=1.060x+2.220	y=1.012x+0.732	y=0.996x-0.521	y=0.833x+1.195	y=0.777x+0.694	y=0.761x+0.319
	R^2	0.990	0.995	0.996	0.989	0.989	0.988
	Correlation	0.995	0.996	0.998	0.993	0.993	0.992
20	Equation	y=1.008x+4.339	y=1.108x+0.767	y=1.018x-1.372	y=0.863x+0.313	y=0.829x+0.025	y=0.809x-0.280
	R^2	0.975	0.989	0.995	0.989	0.985	0.984
	Correlation	0.988	0.994	0.998	0.994	0.992	0.992
		Medium Strength					
25	Equation	y=1.016x+3.112	y=0.937x+2.289	y=0.977x-0.580	y=1.010x-2.506	y=0.695x+1.819	y=0.698x+1.078
	R^2	0.959	0.970	0.991	0.982	0.975	0.976
	Correlation	0.979	0.985	0.995	0.991	0.988	0.988
30	Equation	y=1.038x+2.711	y=0.955x+2.592	y=0.912x+0.416	y=0.923x-0.362	y=0.648x+3.037	y=0.644x+2.443
	R^2	0.986	0.964	0.989	0.991	0.965	0.976
	Correlation	0.993	0.985	0.994	0.995	0.982	0.988
35	Equation	y=0.937x+3.771	y=0.962x+2.648	y=0.818x+1.595	y=0.815x+1.519	y=0.588x+3.780	y=0.597x+2.848
	R^2	0.978	0.980	0.984	0.982	0.964	0.967
	Correlation	0.989	0.990	0.992	0.991	0.982	0.983
		High Strength					
40	Equation	y=0.700x+4.017	y=0.672x+3.586	y=0.678x+0.983	y=0.677x+0.983	y=0.435x+3.777	y=0.431x+3.603
	R^2	0.947	0.954	0.953	0.953	0.961	0.959
	Correlation	0.973	0.977	0.976	0.976	0.980	0.980
50	Equation	y=0.665x+5.494	y=0.622x+4.787	y=0.676x-0.300	y=0.676x-0.300	y=0.449x+3.203	y=0.416x+3.444
	R^2	0.900	0.892	0.974	0.974	0.944	0.950
	Correlation	0.949	0.945	0.987	0.987	0.972	0.975

Note: In the Equation, y is the laboratory carbonation depth and x represents real carbonation depth

Table 7   Service life span expectancy from computer program

	GRADE (N/mm^2)	EXP (L)	COVER (mm)	L.SPAN (year)	COVER (mm)	L.SPAN (year)	COVER (mm)	L.SPAN (year)
LOW STRENGTH	15	1		21.44		67.84		108.16
	15	2		17.90		58.90		94.87
	15	3		15.96		54.53		91.38
	15	4		12.70		40.93		65.58
	15	5		10.67		34.93		56.20
	15	6		9.92		32.94		53.19
	20	1		25.87		77.33		121.39
	20	2		25.03		72.40		112.76
	20	3		18.41		64.55		105.69
	20	4		14.86		49.40		79.79
	20	5		13.41		45.03		72.93
	20	6		12.45		42.30		68.71
MEDIUM STRENGTH	25	1	thirty five	30.81	sixty	95.01	seventy	150.45
	25	2		25.23		79.10		125.80
	25	3		22.54		77.14		125.54
	25	4		21.09		76.82		116.97
	25	5		14.03		43.78		69.54
	25	6		13.24		42.53		68.09
	30	1		38.49		120.13		190.84
	30	2		32.79		102.06		161.99
	30	3		25.92		85.93		138.70
	30	4		25.09		85.34		138.67
	30	5		16.97		50.25		78.68
	30	6		15.91		48.16		75.87
	35	1		40.75		122.64		192.90
	35	2		39.81		113.76		186.40
	35	3		27.39		87.02		138.89
	35	4		27.04		86.12		137.54
	35	5		18.40		52.31		80.98
	35	6		17.28		51.05		79.89
HIGH STRENGTH	40	1		42.03		183.75		198.75
	40	2		37.86		110.39		172.11
	40	3		30.53		98.36		157.58
	40	4		30.45		98.36		157.13
	40	5		19.01		51.53		78.67
	40	6		18.35		50.08		76.59
	50	1		68.47		187.17		286.45
	50	2		58.14		160.83		247.00
	50	3		42.21		143.84		233.82
	50	4		42.21		143.31		233.82
	50	5		29.40		82.31		126.85
	50	6		26.82		73.28		112.15

REFERENCES

1. NURUDDIN, M F, DIAH, A B M, ATAN, I, ARBAI, S, Life expectancy of concrete structures due to carbonation, Proceedings of 7[Th] International Conference on Concrete.

2. NURUDDIN, M F, RIDZUAN, A R M, Factors affecting carbonation: some results based on accelerated tests, Journal IEM, Vol 60, No 2, 1999, p 53-61.

3. ENGINEERING AND TECHNOLOGY (CONCET 2001), 5th – 7th June 2001, p 298-305.

4. MALAYSIAN STANDARDS MS 1195: PART 1, Structural use of concrete – code of practice for design and construction, SIRIM, 1991.

5. JACKSON, N, DHIR, R K, Civil Engineering Materials 5th Edition, Published by MacMillam 1996, p 534.

6. GURUSAMY, K, Concrete durability in Malaysia: A much needed reassessment, 6th International Conference on Concrete Engineering and Technology CONCET, Special Lecture, 1999.

7. NURUDDIN, M F, DIAH, A B M, SAMAN, H M,, Covercrete estimation based on carbonation effect, Proceeding, 18th Conference of AFEO, Hanoi 22-24 November 2000, p 815-823.

8. DHIR, R K, JONES, M R, MUNDAY, J G L, HUBBARD, F H, A practical approach to studying carbonation of concrete, Concrete, Vol 19, No 10, October 1985, p 75-87.

9. JONES, M R, On physical characterization of PFA and its use in concrete, PhD Thesis, University of Dundee, 1986.

10. MCCARTHY, M J, Chloride and carbonation-induced reinforcement corrosion in PFA concrete, PhD Thesis, University of Dundee, 1991.

11. DIAH, A B M, Strength development and durability of GGBS concrete, PhD Thesis, University of Dundee, 1994.

REPAIR OF CONCRETE BEAM-COLUMN JOINTS USING FIBROUS COMPOSITES

M Shannag

S Barakat   M Kareem

Jordan University of Science & Technology

Jordan

ABSTRACT. Presented in this paper is a new technique proposed for repairing reinforced concrete beam-column joints that have been failed first under lateral cyclic loads; and then repaired using high performance fiber reinforced cementitious composite jackets all around the joint column regions; and retested up to failure. The repaired specimens have exhibited higher load levels, substantial energy dissipation, more ductile behavior and slower stiffness degradation compared to reference specimens. The preliminary results indicate that the cementitious composite designed exhibit high strength, high toughness, and excellent crack control, and thus may be considered as a promising material for design-maintain-rehabilitate program of concrete structures.

Keywords: Reinforced concrete, Repair, Joints, Fibers.

M J Shannag is an assistant professor of Civil Engineering at Jordan University of Science & Technology. He received his PhD in civil engineering from the University of Michigan, Ann Arbor, in 1995. His research interests include high-performance fiber reinforced cement-based composites, concrete materials and repair.

S A Barakat is an associate professor of civil engineering at Jordan University of Science & Technology. He received his PhD from the University of Colorado, Boulder, in 1994. His research interests include reliability-based structural optimization, earthquake structural resistance, and composite materials.

M A Kareem is a graduate student in the Department of Civil Engineering at Jordan University of Science & Technology.

INTRODUCTION

In the past, a large number of reinforced concrete structures have been damaged by severe earthquakes, and some of these structures have been repaired or strengthened [1-3]. Several researchers [4-6] have indicated that the optimal use of fiber reinforced cementitious composites (FRC) will improve seismic performance and increase durability of the structure. Substantial improvement in the seismic behavior of a beam-column sub-assemblage can be achieved by using even low fiber volume fractions. Hence, the use of FRC in the critical regions of seismic resistant buildings will yield safer and more economical design.

Extensive experimental studies have previously been performed to investigate the seismic behavior of beam-column joint details in reinforced concrete buildings, and to investigate the joint behavior under simulated seismic loading [7-9]. However, systematic studies to determine the behavior of the repaired or strengthened members under lateral cyclic loading are still very limited. The main objective of this study is to propose a new technique for strengthening non-ductile reinforced concrete frames that are subjected to lateral seismic loads, using a newly developed high performance cementitions composites (HPFRC) that contain small amounts of brass-coated steel fibers for improving the seismic performance of the beam-column joints. Use of HPFRC may be considered as a potential alternative for providing more cost–effective ductile beam–column joints for structures constructed in active seismic zones.

EXPERIMENTAL PROGRAM

In this investigation the structural seismic behavior of beam-column joints is studied, monitored, and evaluated by simulating seismic loading on the most critical regions in the structure (joints) with lateral cyclic loads. A number of reinforced concrete beam-column joints with different reinforcement details will be prepared and tested up to failure. The failed specimens will be repaired with HPFRC jackets and retested to determine the effects of the repairing process on their overall behavior. These as-built specimens represent the beam-column joints that are designed only for gravity loads according to the ACI code, and detailed according to the existing practice in Jordan; representing the case of our old buildings.

Materials

The materials needed for this investigation included the concrete used for casting the reinforced concrete beam-column joints and the HPFRC matrix used for preparing the jacket.

Concrete the concrete mix proportions were (1:3:2.6:0.55) by weight of cement, limestone coarse aggregates, limestone fine aggregate, and water respectively, with a compressive strength of about 27 MPa. The yield strengths of the deformed bars used as the longitudinal reinforcement was 310 MPa.

HPFRC matrix the cementitious matrix was designed and optimized to provide very high strength while maintaining high workability. The mix proportions used were (1:0.60:0.15:0.35:0.02) by weight of commercially produced portland pozzolana cement, silica sand, silica fume, water, and superplasticizer respectively. The mix made contained 2%

by volume of short, brass coated steel fibers of 0.15 mm diameter, 6 mm length and tensile yield strength of 2950 MPa. The HPFRC matrix provided an average strength in tension and compression of (8.5 MPa) and (75 MPa) respectively.

Specimen geometry and reinforcement details

One third scale reinforced concrete interior beam-column joint specimens were prepared in this study.   A schematic sketch of the specimen used is shown in Figure 1 and the reinforcement details are listed in Table 1 and demonstrated in Figure 1.

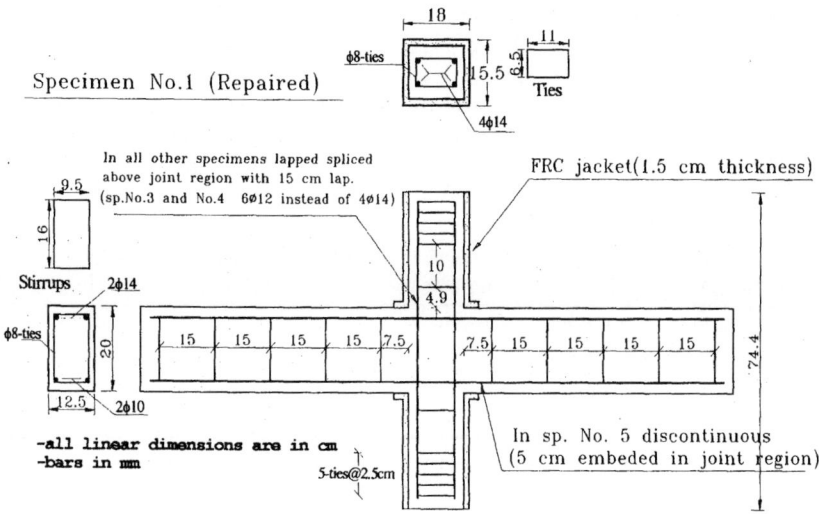

Figure 1   Sketch showing reinforcement details and jacketed regions

Preparation of HPFRC Jacket

The entire joint and the maximum moment column regions were covered using a 15 mm thick jacket of high performance HPFRC mix. The HPFRC mix was then poured in a specially designed wooden formwork to cover the full joint and the maximum moment column regions, and left to dry for 48 hours at laboratory temperature, and cured for one week by wrapping the specimens with wet burlap.

Test setup

The test setup consists of a hydraulic actuator of a 150 kN capacity in tension and 250 kN in compression, that provides the cyclic loading on the top column of the beam column joint; roller supports at both beam-ends and hinged support at the bottom of the lower column; the point of load at the upper column end is free to rotate in the loading plane. The supports were made from hardened steel plates, cut and formed in the engineering workshop of our university with a suitable thickness to sustain the applied load without any deformation that may affect the test results.   The layout of the specimen with its boundary conditions (supports), and the applied loads is shown in Figure 2. A convenient cyclic loading history (simulating earthquake loading) was applied in testing the specimens, as shown in Figure 3. All specimens were tested under stroke control at a rate of 0.02 mm/sec using a hydraulic actuator jack with a cyclic loading applied at the top of the column. Four LVDTs were also used to measure the curvature in two beams-joint faces. The LVDTs and the corresponding load and top column displacement were connected to a data acquisition system that recorded the data every 10 seconds.

Table 1   The reinforcement details of the specimens investigated

NO.	BOTTOM	TOP	COLUMN	JOINT	COLUMN TIES	BEAM	$\rho^{(1)}$
1	continuous (2Ø10)	continuous (2Ø14)	continuous (4Ø14)	No	Ø8@ 12.5cm	Ø8 @15 cm	0.0072
2	continuous (2Ø10)	continuous (2Ø14)	lap splices (4Ø14)	No	Ø8@ 12.5cm	Ø8 @15 cm	0.0072
3	continuous (2Ø10)	continuous (2Ø14)	lap splices (6Ø12)	No	Ø8@ 12.5cm	Ø8 @15 cm	0.0072
4	continuous (2Ø10)	continuous (2Ø14)	lap splices (6Ø12)	Yes	Ø8@ 12.5cm	Ø8 @15 cm	0.0072
5	discontin. (2Ø10)	continuous (2Ø14)	lap splices (4Ø14)	No	Ø8@ 12.5cm	Ø8 @15 cm	0.0072

$\rho^{(1)}$ (beam-bottom reinforcement ratio)

TEST RESULTS AND DISCUSSION

During testing the reference specimens, the test was continued until the joint region was exposed to extensive diagonal tension cracks on both diagonals as shown in Figure 4, when these cracks propagated and extended from both sides, the load carried started to decrease. The test was stopped at this stage so that the specimens could be repaired and re-tested. For all the reference specimens tested, the cracks started at the diagonal of the joint during the push part of the cycle, at approximately 5 mm net applied displacement, this crack started to initiate at the maximum tension point at the two opposite corners. When the applied peak net displacement reaches 5 mm during the pull part of the cycle; the cracks started at the opposite diagonal at the maximum tension point at the two opposite corners.

During the repeated cycles of the same peak, the cracks appeared clearly and their width kept on increasing. With the increase in the peak displacement of the push part of the cycle, the cracks that appeared previously at the opposite corners started to extend; at the same time another crack at this diagonal in the middle part of the joint started to propagate. The same happened for the other diagonal during the pull part of the same cycle.

Figure 2   Layout of the specimen with its steel supports and loading actuator

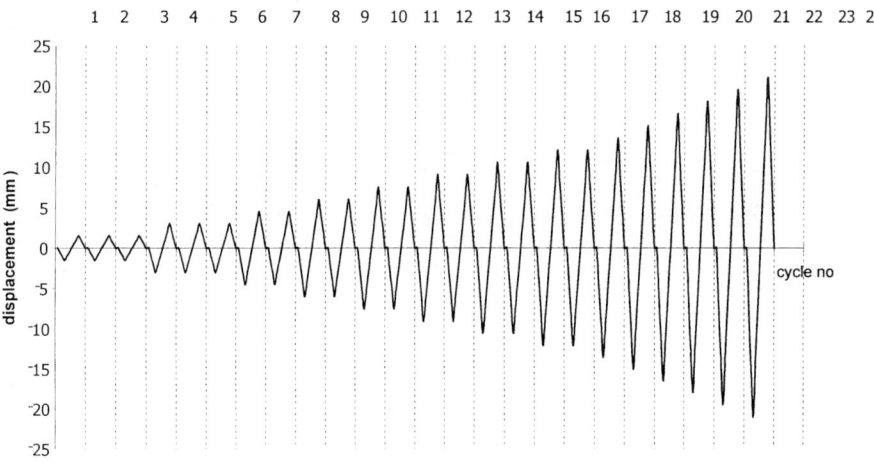

Figure 3   Cyclic-loading history used in this study

After repairing, the specimens performed in a different manner. The initiation of cracks started at a higher load levels in pulling, and in pushing. For all specimens, the cracks started to initiate during the pushing part of the cycle of 8 mm-applied displacement.

During this part of the cycle, the cracks started at the two opposite corners of the joint diagonal along with flexural cracks which were noted at both beams, on their tension side of the section. At final cycles, a large rotation was noticed at both rollers and hinge supports as shown in Figure 5. This is a verification of the good simulation of the sidesway movement during seismic loading. When the load carrying capacity of the specimens started to decrease with increasing the applied cyclic displacement, extensive cracks appeared and propagated at the joint and beams sections, with a maximum beam flexural crack width of about 4 mm.

Load-displacement histories

The load-displacement loops were plotted from the recorded data of the applied tip displacement and the corresponding lateral load. These loops were used to evaluate the behavior of the specimens under cyclic lateral loads such as load carrying capacity, capability of specimens to resist displacement through ductility and stiffness degradation; and to provide a quantitive idea about the stability of the hysteresis loops throughout the test. The load-displacement histories of reference and repaired specimens are shown in Figure 6 through Figure 10.

Load carrying capacity

For all reference specimens, the behavior was very weak, they have a maximum load carrying capacity of (11.4 kN) in pushing and (10.7 kN) in pulling. This is because these specimens are designed only to resist gravity loads, and are detailed to represent the critical steel details under seismic loading. The load carrying capacities in pushing was slightly larger than that in pulling, this is due to unequal top and bottom steel reinforcement, and due to starting the pushing part of the cycle before the pulling part.

Figure 4   Failure of the beam-column joint region of reference specimen

Figure 5   Failure of the beam-column joint region of repaired specimen
(note the large rotation at both rollers and hinge supports at final cycles of the test)

The repairing technique used in this study was very effective in increasing the cyclic load carrying capacity of all specimens. Figure 11 shows about three times increase in load carrying capacity of repaired compared to reference specimens. This is due to the capability of repaired specimens to carry and displace largely without extensive cracking (i.e. behave in a ductile manner) compared to reference specimens.

Energy dissipation

As a measure of the dissipated energy of reference and repaired specimens, the area under the full load-displacement envelopes was computed and defined as the energy that could be dissipated by the specimens before the system looses its stability. For reference specimens, the energy dissipation was relatively low. This is clear from the small areas enclosed by the hysteresis loops seen in part a of Figure 6 through 10.

The energy dissipation for repaired specimens was much larger than that for reference specimens. This is clear from the large areas enclosed by the hysteresis loops seen in part b of Figure 6 through 10. For the full response, the energy dissipation increased by about twenty times over that of reference specimens as shown in Figure 12.

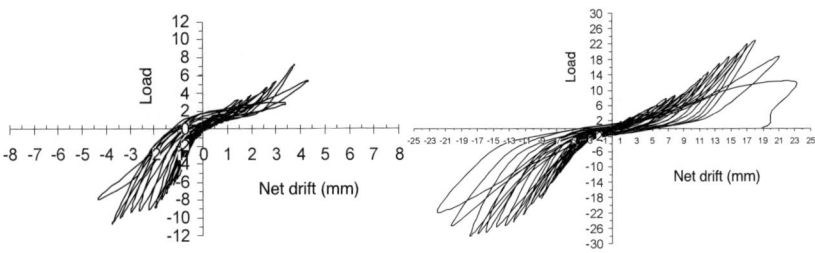

Figure 6   Load-displacement histories of specimen number 1, (a) reference and (b) repaired specimen

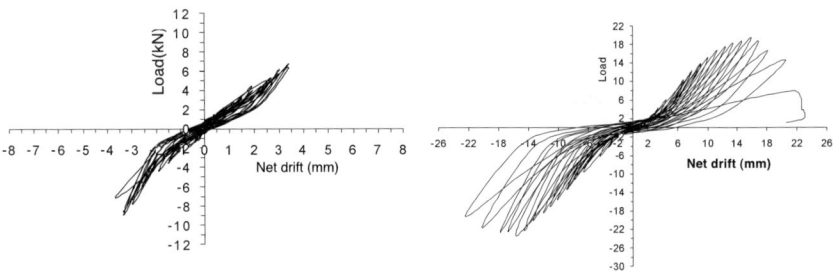

Figure 7   Load-displacement histories of specimen number 2, (a) reference and (b) repaired specimen

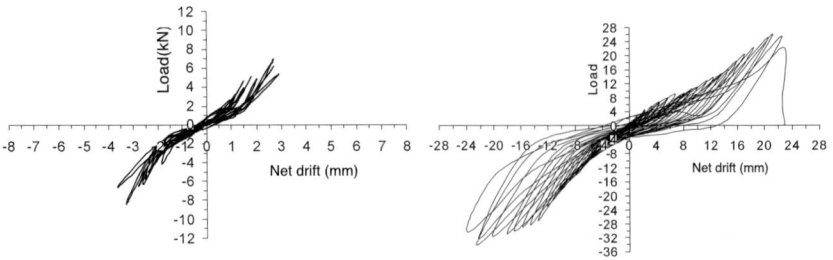

Figure 8   Load-displacement histories of specimen number 3, (a) reference and (b) repaired specimen

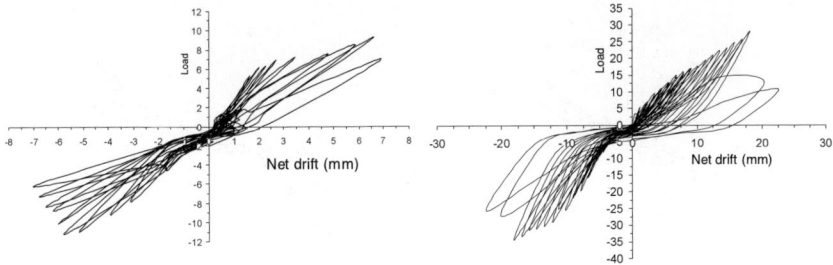

Figure 9   Load-displacement histories of specimen number 4, (a) reference and (b) repaired specimen

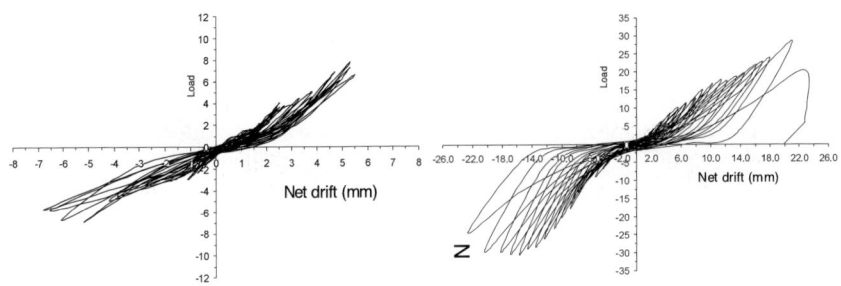

Figure 10   Load-displacement histories of specimen number 5, (a) reference and (b) repaired specimen

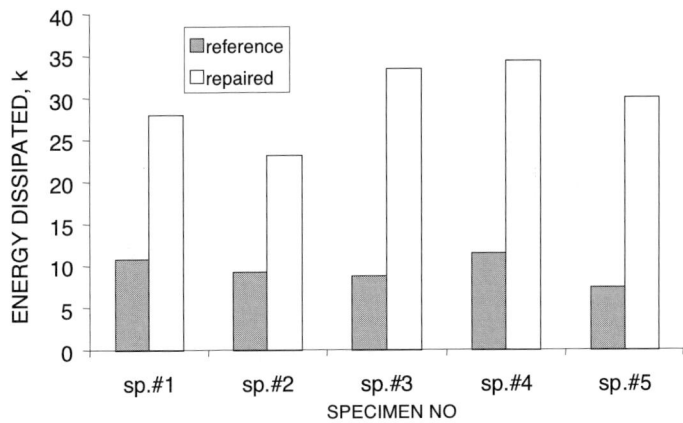

Figure 11   Comparison of load carrying capacity of reference and repaired specimens

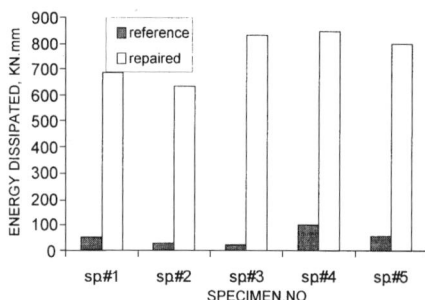

Figure 12 Comparison of energy dissipation of reference and repaired specimens

CONCLUSIONS

Obtained information clearly shows that the repair of concrete beam-column joints that exhibit significant deterioration due to lateral cyclic loading, using a 15 mm thick HPFRC jacket, served to transform the joint shear failure to a more ductile flexural one. Compared to reference specimens, much improved behavior was observed for all repaired specimens, higher load levels were attained, and substantial energy dissipation values were recorded during each loading cycle. The repair technique presented can thus be implemented and easily introduced to the field in particular for non-seismically designed concrete structures that are built in seismically active regions.

ACKNOWLEDGMENTS

The authors acknowledge the technical and financial support provided by Jordan University of Science and Technology. The steel fibers used in this investigation were provided by Bekaert Company, Belgium.

REFERENCES

1. TSONOS, A G, Lateral load response of strengthened reinforced concrete beam-to-column joints, ACI Structural Journal, 96, 1, 1999, pp 46-56.

2. RODRIGUEZ, M, PARK, R, Seismic load tests on reinforced concrete columns strengthened by jacketing, ACI Structural Journal, 91, 2, 1994, pp 150-159.

3. BRACCI, J, REINHORN, A, MANDER, J, Seismic retrofit of reinforced concrete building designed for gravity loads: performance of structural model, ACI Structural Journal, 92, 6, 1995, pp 711-723.

4. SOUBRA, K, WIGHT, J, NAAMAN, A, Cyclic response of fibrous cast in place connections in precast beam-column subassemblages, ACI Structural Journal, 90, 3, 1993, pp 316-323.

5. FILIATRAULT, A, PINEAU, S, HOUDE, J, Seismic behavior of steel-fiber reinforced concrete interior beam-column joints, ACI Structural Journal, 92, 5, 1995, pp 543-552.

6. 6. FILIATRAULT, A, LADICANI, K, MASSICOTTE, B, Seismic performance of code-designed fiber reinforced concrete joints, ACI Structural Journal, 91, 5, 1993,   pp 564-571.

7. PESSIKI, S P, CONLEY, C, GERGELEY, P, WHITE, R N, Seismic behavior of lightly-reinforced concrete column and beam-column joint details, Technical Report NCEER-90-0014, Department of Structural Engineering, School of Civil and Environmental Engineering, Cornell University, Ithaca, NY, 1990.

8. BERES, A, WHITE, R N, GERGELY, P, PESSIKI, S P, EL-ATTAR, A, Behavior of existing non-seismically detailed reinforced concrete frames, Earthquake Engineering Tenth World Conference, Madrid, 1992, pp 3359-3363.

9. KRAMER, D, SHAHROOZ, B, Seismic response of beam-columns knee connection, ACI Structural Journal, 91, 3, 1994, pp 251-260.

CARBONATION INDUCED CORROSION OF REINFORCEMENT

G C Bouquet
ENCI BV
Netherlands

ABSTRACT. The Dutch Cement Industry (ENCI) has developed a physical model, which simulates the carbonation induced corrosion of reinforcement. Important input variables for the model are parameters describing the climate conditions. For outdoor concrete the duration of contact with water (rain) and the frequency are determining factors for the occurrence of carbonation-induced corrosion of the reinforcement. For the extreme climate conditions: no wetting (indoor concrete) and no drying (under water concrete) corrosion do not occur. The reinforcement will only corrode when the concrete directly in contact with the reinforcement bar is intermitted dry and wet. The results of the calculations show that the value of the diffusion coefficient for CO_2 or the density of the concrete has little influence on the corrosion of the reinforcement. The model clearly illustrates that simple laboratory experiments on measuring the carbonation depth of indoor exposed concrete with or without an increased partial CO_2 pressure are irrelevant for determining the onset of corrosion of the reinforcement.

Keywords: Reinforcement corrosion, Concrete carbonation, Concrete cover, Durability, Permeability, Service life.

Mr G C Bouquet, is currently Manager, Applied Concrete Research in the R&D Department of ENCI B.V., M.Sc. degree in Civil Engineering, Delft University of Technology and Member of the Institute of Concrete Technology (MICT).

INTRODUCTION

The durability of concrete is principally related to the ability of water (without or with aggressive ions), oxygen and carbon dioxide to penetrate the concrete pore structure. In this paper the approach, developed by Dr R F M Bakker, is presented for the simulation of one of the most common detoriation processes: carbonation induced corrosion of reinforcement [1].

The rate of carbonation and corrosion strongly depend on the climate conditions. A realistic outdoor climate simulation, with wet and dry periods, is incorporated in the model.

Service life of the reinforced concrete is determined by two distinct periods: an initiation period (the carbonation period) and a propagation period (the corrosion period), as was first described by Tuutti [2].

In case of carbonation induced corrosion the initiation time is the time needed for the carbonation front to reach the reinforcement. Only after the initiation time corrosion will start with a rate depending on the climate conditions. In general these conditions will be equal to the conditions during the initiation period. A realistic 'West European climate' with wetting and drying periods can lead to corrosion of reinforcement. Drying is needed to let the concrete carbonate and water is needed to let the reinforcement corrode. The model takes into account the influence of wetting and drying on the rate of carbonation.

INITIATION PERIOD

The model is based on the following starting points:

- When the concrete is wet the rate of carbonation is negligible.
- Justification: the rate of diffusion of carbon dioxide through water is about 10^4 smaller than the diffusion of carbon dioxide through air.
- The carbonation stops as soon as the concrete is wetted.
- Justification: The absorption of water by concrete is instantaneous because of the capillary suction of the small pores.

Drying Depth

Because at the starting point (poring of the concrete) all concretes are more or less water saturated it follows from these assumptions that concrete can only carbonate when it dries out. From the assumptions mentioned above it also follows that the limit of the carbonation depth is determined by the drying depth.

Based on Fick's law the drying depth is defined by [3]:

$$x_d = B \sqrt{t} \tag{1}$$

With the coefficient B:

$$B = [2D_v (c_3 - c_4)/b \,]^{0,5} \tag{2}$$

With:

x_d = drying depth at time t [m].
t = drying time [s].
D_v = effective diffusion coëfficiënt for water vapour by a given moisture distribution in the pores [m^2/s].
b = amount of water to be evaporated from the concrete [kg/m^3].
$(c_3 - c_4)$ = moisture difference between air and the evaporation front [kg/m^3].

Based on the estimation of mean values [1]: $D_v = 4,8.10^{-7}$ m^2/s; b = 132 kg/m^3; $c_3 = 17,6.10^{-3}$ kg/m^3 and $c_4 = 11.4.10^{-3}$ kg/m^3. From this it follows B = 67.10^{-7} m/$\sqrt{s}$

Carbonation Depth During Drying

At the same time concrete starts drying the carbonation starts. Based on Fick's law and in analogy with the drying depth, the carbonation depth is defined by [3]:

$$x_c = A \sqrt{t} \tag{3}$$

With the coefficient A:

$$A = [2D_c (c_1 - c_2)/a]^{0,5} \tag{4}$$

With:

x_c = carbonation depth at time t [m].
t = drying time [s].
D_c = effective diffusion coëfficiënt for CO_2 by a given moisture distribution in the pores [m^2/s].
a = amount of alkaline compounds in de concrete to be neutralised by CO_2 expressed in kg CO_2 [kg/m^3]. This value depends on the kind of cement used in the concrete and the cement content.
$(c_1 - c_2)$ = concentration difference of CO_2 between air and the carbonation front [kg/m^3].

Based on the estimation of mean values [1]: $D_c = 4.10^{-8}$ m^2/s; a = 164 kg/m^3; $c_1 = 0,66.10^{-3}$ kg/m^3 and $c_2 = 0$ kg/m^3. From this it follows A = $5,7.10^{-7}$ m/$\sqrt{s}$

Carbonation Depth under Drying and Wetting

The climate is schematised in dry and wet periods. The time [s] after n cycles of drying and wetting is:

$$t_n = n.t_d + (n-1) t_w \tag{5}$$

in which t_d [s] is the length of the dry period and t_w [s] is the length of the wet period.

After the first dry period the carbonation depth has reach a depth of:

$$x_{c,1} = A \sqrt{t_{d,1}} \tag{6}$$

During the next wet period the carbonation depth ($x_{c,1}$) remains constant. In the next (second) dry period, first the concrete must dry till the depth $x_{c,1}$ before carbonation can start again. With Formula 1 the effective time for carbonation during the second drying period can expressed as:

$$t_{eff,2} = t_{d,2} - (x_{c,1} / B)^2 \qquad (7)$$

The effective carbonation time after n cycles of drying and wetting is:

$$t_{eff,n} = t_{d,1} + t_{d,2} - (x_{c,1} / B)^2 + t_{d,3} - (x_{c,2} / B)^2 \quad \text{etc}$$

The effective carbonation time after n cycles of drying and wetting can be expressed as:

$$t_{eff,n} = t_{d,1} + \sum_{n=2} [t_{d,n} - (x_{c,n-1}/B)^2] \qquad (8)$$

With $x_{c,n-1} = A \sqrt{t_{eff,n-1}}$ $\qquad (9)$

The carbonation depth after n cycles is:

$$x_{c,n} = A \sqrt{t_{eff,n}} \qquad (10)$$

The initiation time (t_i) is the time needed for the carbonation front to reach the reinforcement, thus when after n cycles $x_{c,n} = c$ then:

$$t_i = n.t_d + (n-1) t_w \qquad (11)$$

The model clearly predicts the influence of the length of a wetting/drying cycle on the depth of the carbonation. Long dry and long wet periods are worse than short cycles even if the average time of wetting and drying are the same. In Figure 1 an example for two different artificial climates are given.

CARBONATION OF CONCRETE IN AN ARTIFICIAL ENVIRONMENT

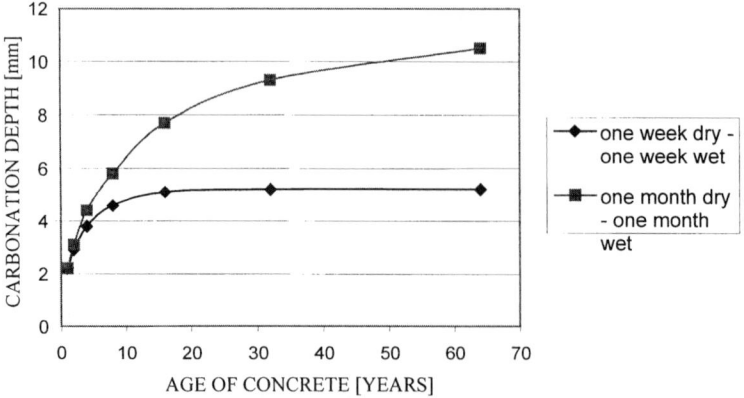

Figure 1   The carbonation depth of blast furnace slag concrete in an artificial climate

Experimental Verification

Verification of the model has been carried out on well-defined concretes under different artificial and natural climate conditions [6, 7, 9].

Concrete samples (500 x 100 x 100 mm) have been made with the following composition:

- 330 kg Blastfurnace Slag Cement (CEM III/B 42,5 N), CaO-content is 44% (alkali content: a = 330 x 0,44 = 145 kg/m^3).

- 165 l water.

- 1868 kg aggregate.

After casting and compaction the samples were kept covered with plastic in the mould during 3 days then placed in the test environments. After start of the test programme (January 1993), the carbonation depth of the samples is measured after 4, 12, 24, 52, 105, 208 and 345 weeks. In Figure 2 the carbonation depth under the five climate conditions is given.

CARBONATION OF CONCRETE
Blast Furnace Slag Cement CEM III/B 42,5 N
Water/cement ration 0,50

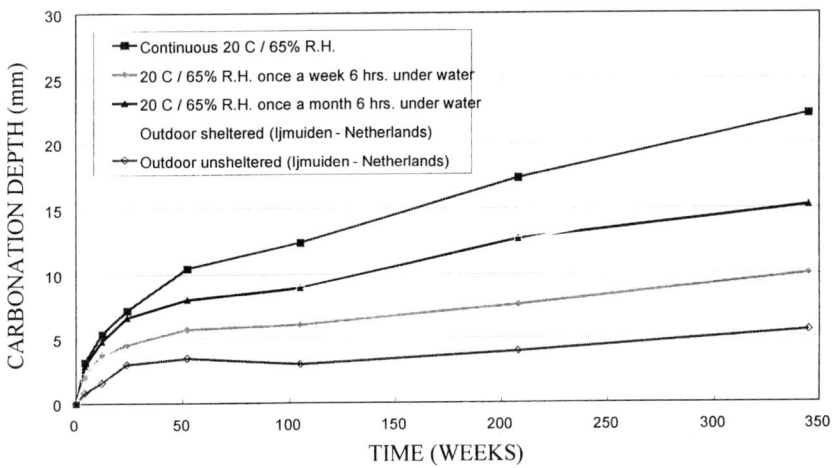

Figure 2 measured carbonation depths under five different climate conditions

These results show that there is a fundamental difference between the concrete wetted or not wetted. The non wetted concrete's follow the square root function. The wetted concrete's tend to reach a limit. The frequency of the cycles influences the limit. More cycles give a lower limit. Compare Figure 1 (predicted) with Figure 2 (measured). For the non wetted concrete's rate of carbonation is determined by the average R.H. The explanation for the diminishing carbonation depth of the 'outdoor sheltered' concrete after 104 weeks is that the hydration does not stop after the curing period but starts again if the concrete is rewetted.

If the amount of new alkalinity is greater than the amount that is neutralised by carbonation there will be some realkalisation. The model parameters (D_c, D_v and a) are calibrated on the artificial climate conditions 'once a week 6 hrs under water' and 'once a month 6 hrs under water. See dataset III in Table 1 and the results of the calculated service life in Table 4.

PROPAGATION PERIOD

The end of the propagation or corrosion period is by Tuutti defined as the moment when visible cracks appear along the reinforcement. The propagation time will than be [2]:

$$t_p = P_{lim} / W.I_{corr} \qquad (12)$$

in which P_{lim} is the corrosion depth [mm] before cracking, W is the percentage ($0 \leq W \leq 1$) of the year with high humidity at the reinforcement and I_{corr} the corrosion rate [mm/year]. See also Andrade [4] and Parrott [5].

The Time During Which the Reinforcement Corrodes

A wet environment of the reinforcement with respect to corrosion of the reinforcement can be defined as a relative humidity (R.H.) of 80%. Only above this R.H. corrosion will take place at a relevant rate [4]. The factor W can now be defined as the part of the lifetime in which the reinforcement lays in concrete with a R.H. above 80%:

$$W = t_{wr} / (t_w + t_d) \qquad (13)$$

In which t_{wr} is the time the reinforcement is in a wet condition (R.H. > 80 %).

During a wet period the time the reinforcement is in wet condition is:

$$t_{wr} = t_w - \Delta t_d + \Delta t_w \qquad (14)$$

With:

Δt_d = time delay, after start of a wet period, before the wetting front has reached the reinforcement;
Δt_w = time delay, after start of a dry period, before the drying front has reached the reinforcement.

In Figure 3 the Δt_d and Δt_w are illustrated.

The wetting depth is defined by:

$$x_w = A_w \sqrt{t} \qquad (15)$$

With:

x_w = wetting depth (moisture front) [m].
A_w = water absorption coefficient [m/√s].
t = time before the reaching a wetting depth x_w [s].

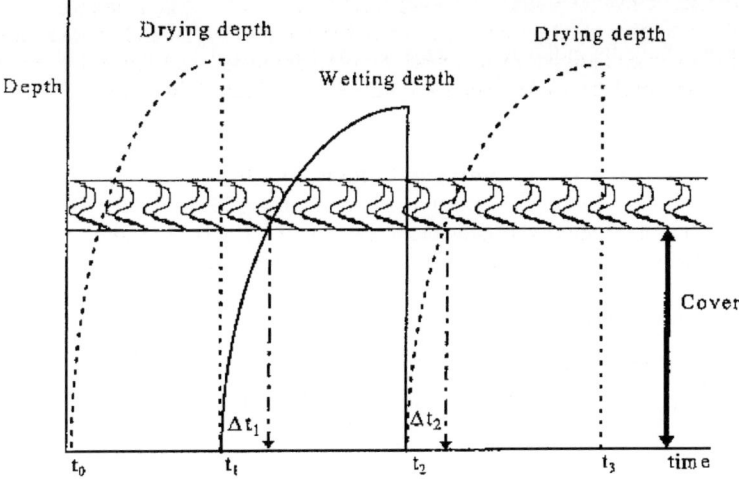

Figure 3   Schematic sketch of drying and wetting depth
$\Delta t_1 = \Delta t_d$ and $\Delta t_2 = \Delta t_w$, wet period $t_w = t_2 - t_1$

From the beginning of the wet period there is a delay Δt_d before the wetting front reaches the reinforcement.   With Formula 15 and the wetting depth equal to the cover on the reinforcement ($x_w = c$), the additional time before the moisture front reach the reinforcement can be expressed as:

$$\Delta t_d = (c / A_w)^2 \tag{16}$$

With Formula 1 and the drying depth equal to the cover on the reinforcement ($x_d = c$), the additional time, at the end of the wet period, before the drying depth is equal to the cover on the reinforcement can be expressed as:

$$\Delta t_w = (c / B)^2 \tag{17}$$

CALCULATION OF SERVICE LIFE

In the model the following boundary conditions are used for the calculation of the service life (SL). If the carbonation depth after n cycles $x_{c,n}$ (Formula 10) is smaller than the concrete cover: $x_{c,n} < c$, then the service life $t_{SL} = \infty$.   When the carbonation depth can reach the reinforcement: $x_{c,n} \geq c$, then the service life is limited to $t_{SL} = t_i + t_p$ (Formula 11 and 12).

Before starting the calculations we have to estimate the mean values of the parameters used in the model.

Parameter a: Alkali Content

The amount of alkalis in the concrete that can be neutralised by carbonation depends mainly on the amount of CaO in the cement. The CaO content in clinker is ≈ 65% and in blast furnace slag ≈ 38%.

In a normal concrete mix with 300 kg/m^3 cement, the CaO content may vary between ≈ 200 kg/m^3 for Ordinary Portland Cement (CEM I) and ≈ 120 kg/m^3 for Blast Furnace Slag Cement with a minimum slag content of 65% (CEM III/B). For the 'reference' concrete calculation 160 kg/m^3 is taken.

Parameter b: Evaporable (Free) Water

The amount of evaporable water in the concrete will depend on the water content of the mix minus the water that is chemically and physically bound. Depending on the degree of hydration and the cement content the value will be between ≈ 150 and 50 kg/m^3. For the 'reference' concrete calculation 120 kg/m^3 is taken.

Parameter D_c: Effective Diffusion Coefficient for CO_2

According to research carried out by Tuutti the diffusion coefficient for oxygen in concrete with 50 % R.H. will vary between ≈ 2.10^{-8} and 10.10^{-8} m^2/s. In air the rate of diffusion of carbon dioxide is 0,78 times the diffusion of oxygen. If we assume that this factor is also valid within concrete, we can estimate the diffusion coefficient for carbon dioxide: D_c will be in the range between $1,5.10^{-8}$ and 8.10^{-8} m^2/s. For the 'reference' concrete calculation 4.10^{-8} m^2/s is taken.

Parameter D_v: Effective Diffusion Coefficient for Water Vapour

The diffusion coëfficiënt for water vapour can be calculated by dividing the free diffusion coefficient in air (≈ 24.10^{-6} m^2/s) by the water vapour resistance factor (≈ 50). For the 'reference' concrete calculation 24.10^{-6} / 50 = $4,8.10^{-7}$ m^2/s is taken.

Parameter (c_1-c_2): Concentration Difference of CO_2 Between Air and the Carbonation Front

(c_1-c_2) is the driving force being the difference in concentration of carbon dioxide at the carbonation front and the carbon dioxide concentration outside the concrete. The concentration in normal air (c_1) is ≈ 0,033 vol. %. The concentration at the carbonation front (c_2) is zero. The weight of 1 m^3 CO_2 is ≈ 2 kg. In 1 m^3 air there is $0,033.2.0,01 = 0,66.10^{-3}$ kg/m^3 CO_2. This value is taken for the 'reference' concrete.

Parameter (c_3-c_4): Moisture Difference Between Air and the Evaporation Front

The moisture difference between air and the evaporation front (c_3-c_4) is the driving force being the difference in water vapour concentration at the drying front and the water vapour concentration outside the concrete. At the drying front the concentration can be derived from the vapour pressure at 20°C and 100% R.H. The water vapour concentration outside is 65% if the R.H. is 65%.

Under these conditions the values are: $c_3 = 17,6.10^{-3}$ kg/m^3 and $c_4 = 11,4.10^{-3}$ kg/m^3. This result in a value for (c_3-c_4) of $6,2.10^{-3}$ kg/m^3 for 'reference' concrete.

Parameter P_{lim}: Limit Amount of Corrosion Products Before Cracking

For the 'reference' concrete calculation the value of 100 µm/year, proposed by Wierig [8], is taken.

Parameter I_{corr}: Rate of Corrosion in the Wetting Period (Carbonated Cover)

For the 'reference' concrete calculation a value of 50 µm/year is assumed.

Parameter A_w: Water Absorption Coefficient

Based on mean water absorption of 20 mm in a week, the water absorption coëfficiënt 25.10^{-6} m/√s is taken for the 'reference' concrete.

Calculation of the Service Life

The calculations of the service life are made for the 'reference' concrete (dataset I) and for a concrete with a bad performance (dataset II) for which the value of the diffusion coefficient for CO_2 (D_c) is twice the value of the reference concrete. The diffusion coefficient for H_2O (D_v) is about four times greater than the value of the reference concrete. Dataset III represents the blastfurnace slag cement concrete used in the carbonation tests (Figure 2).

Table 1   Datasets for service life calculation

PARAMETERS	DATASET I	DATASET II	DATASET III
Concrete:			
D_c [m^2/s]	$4,0.10^{-8}$	$8,0.10^{-8}$	$4,0.10^{-8}$
D_v [m^2/s]	$4,8.10^{-7}$	$2,0.10^{-6}$	$1,0.10^{-6}$
a [kg/m^2]	160	160	145
b [kg/m^2]	100	100	100
A_w [m/√s]	$2,5.10^{-5}$	$2,5.10^{-5}$	$2,5.10^{-5}$
Steel:			
I_{corr} [µm/year]	50	50	50
P_{lim} [µm]	100	100	100
Environment:			
(c_1-c_2) [kg/m^3]	$0,66.10^{-3}$	$0,66.10^{-3}$	$0,66.10^{-3}$
(c_3-c_4) [kg/m^3]	$6,2.10^{-3}$	$6,2.10^{-3}$	$6,2.10^{-3}$

The results of the calculations are summarised in Table 2, 3 and 4.

Table 2   Calculated service life of concrete with 'reference' concrete (dataset I)

ENVIRONMENT		SERVICE LIFE [YEAR]				
t_{dry} [days]	t_{wet} [days]	c = 5 mm	c = 10 mm	c = 15 mm	c = 20 mm	c = 25 mm
1	7	no carb.	-	-	-	-
7	1	8	no carb.	-	-	-
7	7	11	no carb.	-	-	-
7	28	23	no carb.	-	-	-
28	7	10	23	no carb.	-	-
28	28	9	35	no carb.	-	-
56	7	14	18	50	no carb.	-
56	56	9	27	86	no carb.	-

Table 3   Calculated service life of concrete with 'bad' concrete (dataset II)

ENVIRONMENT		SERVICE LIFE [YEAR]				
t_{dry} [days]	t_{wet} [days]	c = 5 mm	c = 10 mm	c = 15 mm	c = 20 mm	c = 25 mm
1	7	no carb.	-	-	-	-
7	1	11	no corr.	-	-	-
7	7	6	19	no carb.	-	-
7	28	9	42	no carb.	-	-
28	7	11	14	22	no corr.	-
28	28	6	14	30	66	no carb.
56	7	18	19	23	no corr.	-
56	56	6	14	28	50	88

Remark: with $t_{dry} = t_{wet} = 56$ days the service life is 163 years for c = 30 mm, maximum carbonation depth is 35 mm.

Table 4   Calculated service life of concrete with 'Blastfurnace Slag Cement' concrete (dataset III)

ENVIRONMENT		SERVICE LIFE [YEAR]				
	t_{wet} [days]	c = 5 mm	c = 10 mm	c = 15 mm	c = 20 mm	c = 25 mm
1	7	no carb.	-	-	-	-
7	1	9	no corr.	-	-	-
7	7	8	no carb.	-	-	-
7	28	16	no carb.	-	-	-
28	7	11	18	48	no carb.	-
28	28	8	24	75	no carb.	-
56	7	17	19	33	no corr.	-
56	56	9	23	52	117	no carb.

The results are consistent with the experience in practice that a cover of about 20 to 25 mm under the Dutch climate conditions are sufficient to prevent carbonation induced corrosion. Comparing Table 2 with Table 3 we can conclude that the effect of climate conditions is far greater than the effect of the change of diffusion coefficients.

CONCLUSIONS

A refined model for the calculation of the service life of reinforcement in concrete under different climate conditions is developed. The model clearly shows that the dominating factor for carbonation induced corrosion protection is the thickness of the cover. The cover needed is mainly determined by the climate conditions. The test results under artificial and natural climate conditions (Figure 2) clearly show the fundamental difference between the carbonation depth and the frequency of rewetting. Important input variables for the model are parameters describing the climate conditions. For outdoor concrete the duration of contact with water (rain) and the frequency are determining factors for the occurrence of carbonation-induced corrosion of the reinforcement. For the extreme climate conditions: no wetting (indoor concrete) and no drying (under water concrete) corrosion do not occur. The model clearly illustrates that simple laboratory experiments on measuring the carbonation depth of indoor exposed concrete with or without an increased partial CO_2 pressure are irrelevant for determining the onset of corrosion of the reinforcement. The results of the carbonation tests (Figure 2) show that a good simulation of the natural climate 'outdoor sheltered' can be made with the artificial climate 'once a month 6 hrs. under water' under the standard laboratory conditions (20°C / 65% R.H.).

REFERENCES

1. BAKKER, R F M, Prediction of service life of reinforcement in concrete under different climatic conditions at given cover, Report ENCI, Ijmuiden.

2. TUUTTI, K, Corrosion of steel in concrete. CBI-Forskning: fo 4:82, Cemnet-Och Betonginstitutet, Stockholm, 1982.

3. SCHIEβL, P, Corrosion of Steel in Concrete, RILEM report of the Technical Committee 60-CSC, 1988.

4. ANDRADE, C, ALFONSO, C, Life time of rebars in carbonated concrete, 10th European Corrosion Congress, Paper 165.

5. PARROT, L, Contribution to CEN/TC 104/WG 1/TG 1, Paper No 133.

6. BAKKER, R F M, ROESSINK, G, Carbonation, corrosion and Moisture, CUR report 90-3, Stichting CUR, Gouda, 1990.

7. BAKKER, R F M, The Significance of Carbonation Measurements to the Corrosion Risk of Reinforcement, Betonwerk + Fertigteil-Technik BFT, Heft 12/1992.

8. WIERIG, H J, Longtime studies on the carbonation of concrete under normal outdoor exposure, RILEM Seminar Hannover, 1984.

9. CEN-report CR 12793 (Part II), 1997.

CLOSING PAPER

CONCRETE MAINTENANCE AND REPAIR: THE LESSONS AND THE FUTURE

P C Robery
FaberMaunsell
United Kingdom

ABSTRACT. Although more than 50% of UK construction expenditure is now attributable to refurbishment projects, maintenance and repair is still the poor relation in the construction sector. Yet proper maintenance and lasting repair of an asset over its life is one of the most difficult technical challenges facing the industry today. Increasingly, industry is being asked to predict whole-life expenditure profiles for an asset, such as for private financing. In short, tomorrow's asset managers need to be "deteriorologists", who understand why infrastructure assets of various kinds fail over the service life. The future can unlock many of the difficult issues through training, research and development; but we will first have to overcome the negative attitudes about maintenance and repair that pervade the industry.

Keywords: Concrete repair, Maintenance, Strategy, Deterioration modelling, Monitoring, Research and development, Education.

Professor P Robery, is Business Stream Leader for Infrastructure Maintenance in FaberMaunsell and Head of the Birmingham Office. He is also a visiting Professor at the University of Leeds, from where he graduated and now lectures on concrete repair and maintenance strategy. He also represents the UK's interests for BSI on the CEN Standards Working Group TC104/SC8/WG2, covering mortars and concretes for structural and non-structural concrete repair.

INTRODUCTION

Over the past 100 years there has been considerable changes in the nature of construction work, particularly reinforced concrete structures. For reasons that are discussed later, the use and abuse of reinforced concrete as a construction material has to a backlog of deteriorating structures, particularly those in the most severe exposure environments found with bridge decks. But despite the fact that more than 50% of UK construction expenditure is now attributable to refurbishment projects, maintenance and repair is still the poor relation in the construction sector. Project managers effuse about their new buildings or bridges that rise out of the ground like gleaming diamonds, whereas traditionally their planned maintenance regime for their structures include little more than an occasional coat of paint (if you are lucky). Maintenance and repair has not generally been perceived as an attractive discipline, more "bore-ology" than technology; but this attitude is changing and must change if we are to make construction a truly sustainable discipline.

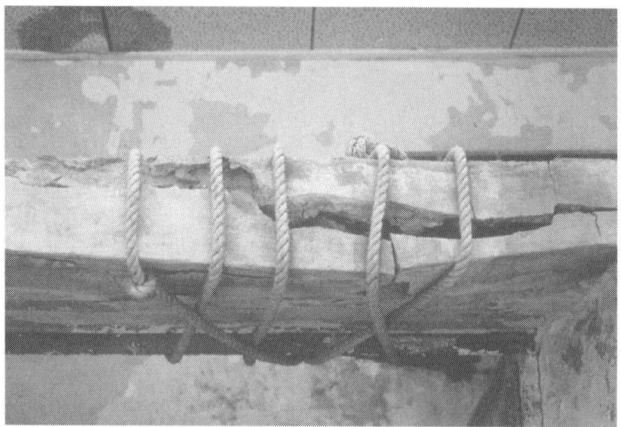

Figure 1   If we can't stop the corrosion process, then at least we can catch the bits!

This poor perception about concrete repair and maintenance is not helped by the failures that have occurred in the past. A lack of understanding of the concept of durability and exposure classifications, led to a multitude of structures built in the 1960s to 1970s that required patch repairs or even demolition and replacement, well short of their intended service life. Worse still, the defects were treated without any real understanding of the causes of the problem, leading to a catalogue of unsuccessful repairs. These failures fostered an attitude among engineers that concrete repair products were "no good", because the patches always fell off. Not that the formulators were completely innocent, but more that there was a general lack of understanding of the needs for lasting concrete repair work. Some good examples are included in the following list:

- Concrete repair products bonded with materials that are attacked by the alkali and moisture naturally present in concrete.
- Use of materials with additives that speed up both the setting of the mix and the corrosion rate of the embedded steel onto which they are applied (e.g. calcium chloride).
- Use of ultra fast setting materials that cure by exothermic reaction, leading to cracking as they cool down (e.g. polyester resin mortars).

Some engineers have learned from the mistakes of the past and have taken to challenge the claims for every new product and system; others lack detailed technical knowledge and put their trust in extended warranties and case histories, often for structures with very different diurnal temperature, precipitation and humidity exposure than the current candidate for repair.

In reality the proper maintenance and lasting repair of an asset over its life is one of the most difficult technical challenges facing the industry today. Increasingly, private organisations need to predict whole-life expenditure profiles for an asset, to facilitate the private financing of new or renovated developments, and then manage the expenditure, often on behalf of the traditional public sector. Before the future life and maintenance costs can be predicted, the design life of the asset needs to be defined and the required performance of each element over its life needs to be established. Then, the technology is that of deterioration modelling, considering local macroclimates and understanding the properties of the concrete materials used around the original structure. Finally the future performance of the structure needs to be monitored, using corrosion detection and control systems to give assurance that the asset continues to perform as planned.

It is important therefore for engineers of tomorrow to understand the problems of the past, challenge the "common knowledge" advice of the present and think about providing lasting reinforced concrete structures that will be both a credit to the industry in 100 years and be still standing!

THE LESSONS FROM THE PAST

To understand the future for concrete repair and maintenance, an appreciation is needed as to how the industry has ended up in the current position of having a multitude of failing structures. The first topic has to be the original concrete itself.

Concrete Durability Factors

It remains an unfortunate truth about the history of reinforced concrete design and specification, that had we learned our lessons from the early research into the performance of reinforced concrete, many of the problems facing us today would not have occurred [1]. In the 1920s, research into the effects of exposing reinforced concrete to chloride ions, such as are present in seawater, led to very firm conclusions about keeping the two apart.

Where concrete structures were designed for use in seawater, extreme care was taken. In fact, some of the most durable reinforced concrete maritime structures were built in the period from 1910 to 1950. A good example is the concrete used in the Mulberry Harbour units, which were conceived by Guy Maunsell and designed by Oscar Faber (among others). These floating structures were built during 1939-1945 using no more sophisticated concrete technology than a rich, water-tight concrete (1:1:2 by weight of cement : sand : 10mm coarse aggregate) [2]. The units have survived exposure in seawater for over 60-years [3]. With a cement content equivalent to over 550kg/m^3, far too high by today's "modern" standards, and with high cover to the reinforcement, recent surveys have shown that chloride ions from the seawater have hardly penetrated into the concrete at all and are certainly well away from the deep-set reinforcement.

A lesson from this 1940s structure is that by using large quantities of coarse-ground cement, necessary to get sensible early strengths, a highly durable structure would result. Yet the trends of the 1950s to 1970s very much changed the design concepts and hence the durability of structures. Driven by the need to build quickly, industry required that concrete for both precast and insitu works should be fast setting, which resulted in changes to the cement chemistry and an increase in the fineness and hence reactivity of the cement. With the use of the high reactivity cement, it was found that less cement was needed to get the same compressive strength at 28-days, this well-known time horizon of obscure origin that takes no heed of the materials being used or their rate of strength gain. Figure 2 illustrates that by using a higher cement content of say $550kg/m^3$, a coarse ground cement can provide similar 28-day strength to $350kg/m^3$ of a "modern" cement of normal fineness, but that the long term strength gain and imperviousness is far superior. Equally, by using the same cement content, but with a blend of PFA (fly ash) and cement, similar strengths are achieved at 90 days, through the 28 day strengths are different. So why do we need to stick rigidly to our 28-day strength requirement?

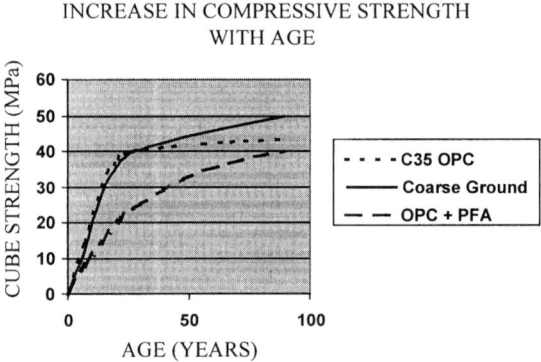

Figure 2   Typical strength gain relationship for concrete made with
different types/fineness of cement

Architectural pressures dictated the need for an increasingly slender form of construction and reduced concrete cover. With pressure to shorten the construction programme, a concrete "anti-freeze" began to be introduced, based on calcium chloride, which was added to the mix both to accelerate the set and allow concrete to be cast under low temperature conditions. Allied to these issues were a workforce and a supervisory team that generally did not under-stand the importance of cover, compaction and curing. Therefore, structures of this era were commonly built using low cement contents (some barely over $300kg/m^3$) low cover (anywhere from 0 mm upwards) and poor quality concrete, with some containing added calcium chloride.

The result was a large number of structures suffering from premature corrosion of the reinforcement, resulting in cracking, spalling and loss of section of the bar, caused by one or more corrosion initiators:

- Chloride added to the mix, from contaminated marine sourced aggregates or calcium chloride accelerator.

- A weak and porous concrete matrix, due to low cement content, high water/cement ratio and poor compaction and curing of the concrete, prone to atmospheric carbonation, coupled with low cover protection to the reinforcement.

Added to these in-built problems, there was also a lack of appreciation of the exposure environment. Many structures, such as car parks, were built for the building sector and would have a low strength requirement and therefore poor inherent durability. Yet these structures were exposed to a chloride ion build-up that was as severe as that found for bridge decks. Bridges were designed to take much higher stress levels than car parks and therefore used a higher strength of concrete and had better durability performance. Was it therefore just a quirk of fate, in an era when structures were designed for strength alone, that highly stressed structures such as bridge decks would have more resistance to chloride ions?

Concrete Repairs

To combat the problem of reinforcement corrosion, the "ubiquitous" patch repair was born. The name itself gives some idea as to how concrete repair works were generally specified and used – "patch it up".

While in theory patching up an area of spalled concrete with new cementitious material appears to be the correct approach, variations on this theme soon led to problems, which are generalised for emphasis in the following list:

- Mortar mixes of cement, sand and water (CC) were found to crack and fall off, due to high shrinkage and lack of bond, leading to the development of polymer-modifier dispersions for adding to the mix (PCC) to improve bond and reduce shrinkage.
- High flow concrete mixes were specified for repairs to beam soffits, based on plasticising admixtures, but these tended to fail by collection of air and bleed water at the upper concrete interface (Figure 3). Concrete technologists had to learn about the inter-relationship between cement fineness, water/cement ratio and the rate and duration of bleed – one T.C Powers in 1939 had explained the process perfectly [4]!

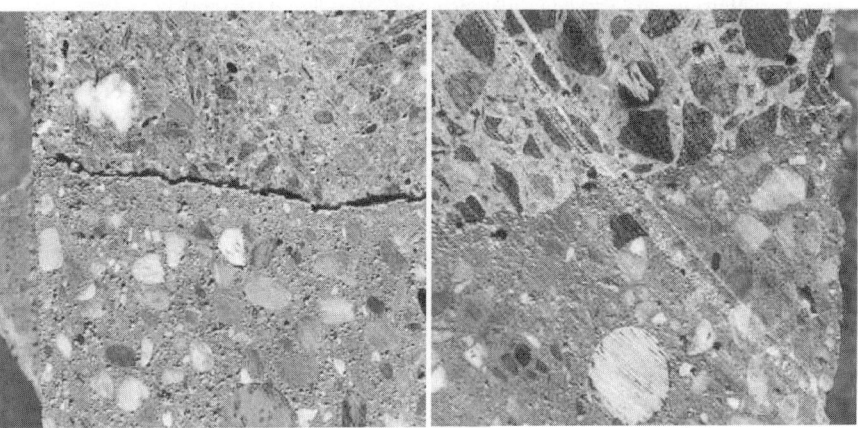

Figure 3   Repair to a beam soffit using a high bleed, cohesive mix (left) and a low bleed, superplasticised mix (right)

- Polymer-based mortars, using epoxy or polyester resin as the binder (PC), began to crack and fall off, as the thermal, elastic modulus and tensile strength differences created high tensile strains in the concrete around the patch.
- Certain types of PC and PCC mortars began to fail, because the products were either affected by the continued presence of moisture in concrete (typically between 3% and 5%) or the strong alkali in the pore water (upwards of pH 13).
- Special fast-setting mortars were developed, but with the side effect of having a high temperature rise during cure, and in the case of polyester resins a high chemical shrinkage during cure as well, leading to failure of the repairs when they cooled and contracted.

However, even with the best materials and workmanship, concrete repairs began to fail in some cases, but not in others, and industry began to understand the destructive power of the chloride ion in concrete. Terms like incipient anode, electrochemical corrosion currents and chloride ion corrosion threshold began to be used, in an attempt to explain why patch repairs to damaged areas of chloride-contaminated concrete lasted only a few years, before either the patch repair failed again, or new spalling appeared alongside the existing repair. The corrosion cell was born (Figure 4) [5].

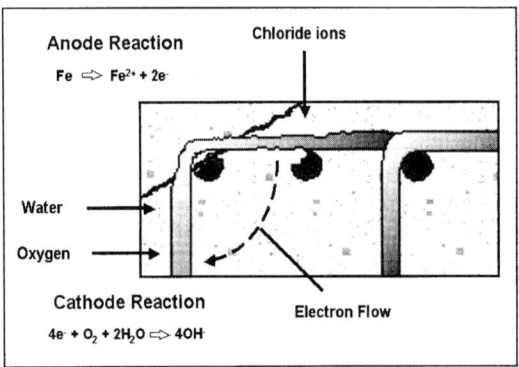

Figure 4   Typical anode-cathode relationship in chloride-contaminated concrete, showing the corrosion site (Anode) sacrificially protecting the cathode (until it is repaired)

Over the past 20 years, methods for electrochemical corrosion control have taxed the minds of many researchers, leading to a wide range of imaginative solutions for the repair of reinforced concrete. These include: impressed current cath dic protection, electrochemical chloride extraction, anodic and cathodic corrosion inhibitors, high resistivity repair products, protective coatings to keep the concrete dry. The message here is that we have got better at understanding the deterioration processes at work and the best methods of combating them.

All of these techniques, and others besides, have now been incorporated in a new standard for concrete repair (BS EN 1504 series), with Part 9 of this series [6] setting out the principles to be used for repairing concrete that is either damaged, under strength or insufficiently durable for its conditions of exposure. The scope of the series of documents is summarised in Table 1.

The main requirement from this Standard is to ensure that the mistakes of yesterday, such as using incompatible, untried, or "wishful" solutions are eliminated, by requiring that that all construction products sold for concrete repair works meet a series of minimum performance standards and are therefore "CE-marked" as fit for purpose.

Table 1 Summary of repair principles and methods for concrete repair to BS ENV 1504-9 [5]

REPAIR PRINCIPLE		REPAIR METHOD	
1 PI	Protection against ingress	1.1	Impregnation
		1.2	Surface coating with and without crack bridging ability
		1.3	*Locally bandaged cracks*
		1.4	*Filling cracks*
		1.5	*Transferring cracks into joints*
		1.6	*Erecting external panels*
		1.7	*Applying membranes*
2 MC	Moisture Control	2.1	Hydrophobic impregnation
		2.2	Surface coating
		2.3	*Sheltering or over cladding*
		2.4	*Electrochemical treatment*
3 CR	Concrete Restoration	3.1	Applying mortar by hand
		3.2	Recasting with concrete
		3.3	Spraying concrete or mortar
		3.4	*Replacing elements*
4 SS	Structural Strengthening	4.1	Adding or replacing embedded or external reinforcing steel bars
		4.2	Installing bonded rebars in pre-formed or drilled holes in the concrete
		4.3	Plate bonding
		4.4	Adding mortar or concrete
		4.5	Injecting cracks, voids or interstices
		4.6	Filling cracks, voids or interstices
		4.7	*Prestressing - (post tensioning)*
5 PR	Physical Resistance	5.1	Overlays or coatings
		5.2	Impregnation
6 RC	Resistance to chemicals	6.1	Overlays or coatings
		6.2	Impregnation
7 RP	Preserving or restoring passivity	7.1	Increasing cover to reinforcement with additional cementitious mortar or concrete
		7.2	Replacing contaminated or carbonated concrete
		7.3	*Electrochemical realkalisation of carbonated concrete*
		7.4	*Realkalisation of carbonated concrete by diffusion*
		7.5	*Electrochemical chloride extraction*
8 IR	Increasing Resistivity	8.1	Limiting moisture content by surface treatments, coatings or sheltering
9 CC	Cathodic Control	9.1	Limiting oxygen content (at the cathode) by saturation or surface coating
10 CP	Cathodic Protection	10.1	Applying electrical potential
11 CA	Control of Anodic Areas	11.1	Painting reinforcement with coatings containing active pigments
		11.2	Painting reinforcement with barrier coatings
		11.3	Applying inhibitors to the concrete

Note: Methods in italics may make use of products and systems that are outside the scope of the EN 1504 series.

The BS EN 1504 series greatly assists engineers in Europe, by ensuring that products and systems sold for concrete repair meet a series of minimum performance criteria and that single methods of test exist for assessing compliance with those criteria (as opposed to national Standards that vary between countries).   However, the BS EN 1504 series specifically excludes the key areas of: investigation, testing, residual life prediction and whole life costing as the basis for option selection.   These remain areas for future development and agreement [7].

WHAT THE FUTURE BRINGS

In practice, the proper maintenance and lasting repair of an asset over its life is one of the most difficult technical challenges facing the industry today.   Increasingly, whole-life expenditure profiles need to be determined for an asset, such as for the 20-year concessions commonplace in the private financing of new or renovated developments. If engineers are uncertain about the rate of future deterioration of structural elements, then the expenditure required over the life of the concession will remain uncertain and financially weighted accordingly, possibly making the opportunity unviable.

Before the future life and maintenance costs can be predicted, the design life of the asset needs to be defined and the required performance of each element over its life needs to be established. The necessary technologies include the following [8]:

- **Deterioration modelling** that considers local macroclimates and takes account of the properties of the concrete materials used around the original structure (Figure 5). Considerable research work is needed in this area if industry is to move forward with deterioration and life prediction, using statistical approaches such as reliability analysis, as current techniques rely heavily on establishing actual time-performance curves

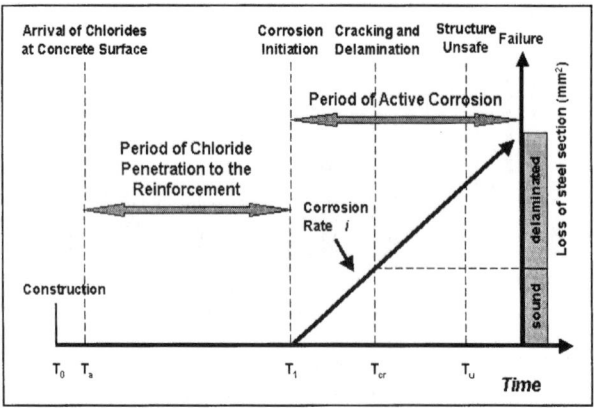

Figure 5   Typical T0-T1 curve for deteriorating reinforced concrete structures subject to chloride ion exposure

- **Whole life costing** of the various approaches to deal with future deterioration, based on the intended service life (Figure 6).

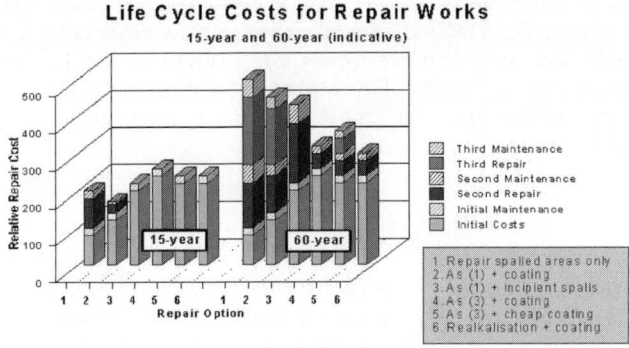

Figure 6   Typical life cycle costing of different repair options to combat carbonation of a reinforced concrete building

- **Predictive Planned repairs and maintenance** regimes, targeted only to the key areas of the structure that are critical to the future performance, rather than just following an arbitrary list of works and "cosmetic" repairs, while missing the unplanned major events. Options include a variety of measures that suppress the rate of corrosion, thereby delaying any necessary repair works (Figure 7).

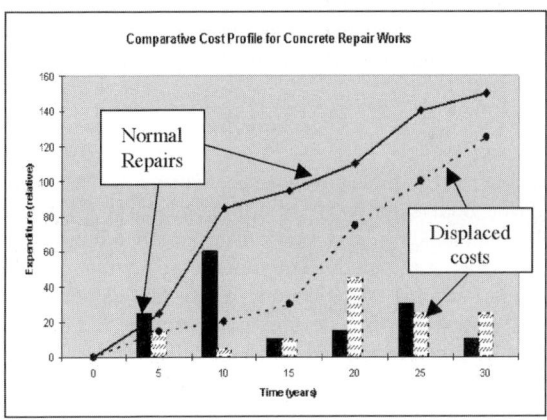

Figure 7   Schematic representation of delayed repair works displacing the expenditure profile

- **Performance assurance,** using advanced monitoring and control techniques to provide feedback that all is well in the critical areas of a structure, or early warning of impending problems.   Figure 8 shows an advanced bridge monitoring system giving real time information on the incidence of a bridge strike and its severity [9].

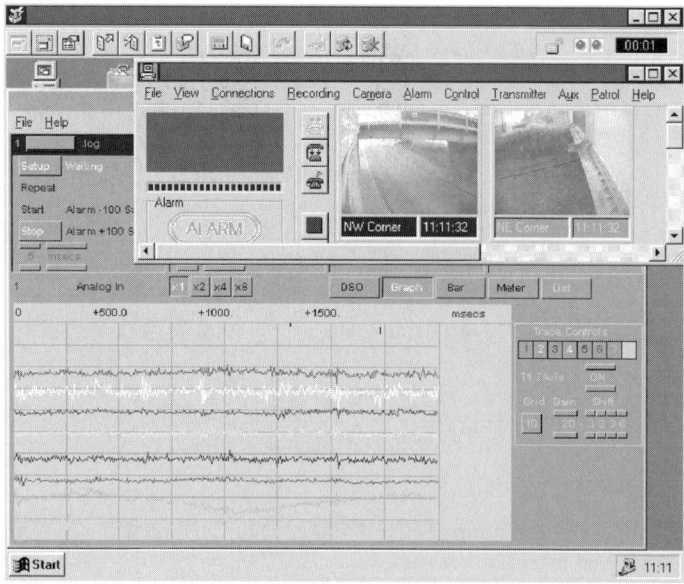

Figure 8   Advances in structure monitoring provide real-time information to the desktop

Only with significant research and development in the above areas can the expenditure for an asset be effectively predicted and managed, giving confidence to clients and financiers alike and recognising real cost savings over the life of the structure.

CONCLUSIONS

To tackle the problems of tomorrow, asset managers need to be (or to employ) "deteriorologists", who understand why infrastructure assets of various kinds fail over the service life. Such understanding goes beyond general building and civil engineering and delves into specialist areas of materials, deterioration modelling, monitoring, assessment and strengthening and repair scheme specification. The future can unlock many of the difficult issues through training, research and development; but we will first have to overcome the negative attitudes about maintenance and repair that pervade the industry. This will require close collaboration between industry, Universities and funding organisations to ensure the repair and maintenance industry is prepared to tackle the problems that will arise in the future.

ACKNOWLEDGEMENT

The author would like to express his thanks to the Infrastructure Maintenance team in FaberMaunsell for helping to contribute to the concepts and conclusions in this paper, born of many years of field research into reinforced concrete deterioration. Acknowledgements also to Grace Construction Chemicals for providing images used in this paper.

REFERENCES

1. LOOV, R E, Reinforced Concrete at the Turn of the Century, Concrete International, December 1991, p 67-73.

2. HARTCUP, G, Code Name Mulberry – The planning, building and operation of the Normandy Harbours, 1968.

3. CIRIA UEG TN.5/1, Concrete in the Oceans: Marine durability survey of the tongue sands tower, Report 5652, CIRIA Underwater Engineering Group, London, 1979.

4. POWERS, T C, "Bleeding of Cement Pastes", PCI, Stokoe, 1939.

5. CURRIE, R J, ROBERY, P C, Repair and Maintenance of Reinforced Concrete, Building Research Establishment Report No BR 254, April 1994, p 34.

6. BS ENV 1504-9, Products and systems for the protection and repair of concrete structures - Definitions, requirements, quality control and evaluation of conformity - Part 9 : General

7. principles for use of products and systems, British Standards Institution.

8. ROBERY, P C, Standards for Concrete Repair and Protection, Proc 4th South African Conference on Polymers in Concrete, 20 – 23 June 2000, South Africa.

9. ROBERY, P C, Maintenance strategies for highway structures, journal of the Institute of Highways, October 1997, p 14-16.

10. ROBERY, P C, Remote monitoring and control systems for steel reinforced concrete structures, ICRI Spring Convention, Seattle, April 1997.

LATE
PAPER

MOTORWAYS: CEMENT OR ASPHALT

T Chrzan

Technical University of Zielona Gora

Poland

ABSTRACT. In this paper presented are the plans of the construction of the paid motorways in Poland. Given is the way of funding of their construction. Discussed are the benefits of the asphalt and concrete motorways. Presented are the conclusions resulting from the operation of the concrete roads built on the cement, asphalt and boulder foundations: (i) after 10 years of the operation the concrete roads have less fractures than the asphalt ones, (ii) the concrete roads built on the existing asphalt roads have the most of the damages to their surfaces.

Keywords: Asphalt, Concrete motorways, Motorways.

Professor T Chrzan, is a Research/Teaching Fellow in the Technical University of Zielona Gora, Environmental Inginiering Institute. His research focuses on the construction of asphalt and concrete motorways.

INTRODUCTION

The plans (Chrzan T 1996) of construction of motorways in Poland have been existing for decades already. Presently there are 257 km of motorways in Poland in five sections: Wrzesnia - Konin, Tuszyn – Piotrkow Trybunalski, Katowice – Krakow, Wroclaw – Golnice and a several kilometres long section from Kolbaskowo to Szczecin. Outside that there are 342 km of so called express roads such as the ones from Warsaw to Katowice, from Poznan to Konin or the tri-city ring road, Tczew – Gdańsk – Gdynia.

The program of motorways' construction adopted by Sejm, Polish parliament, in 1994 assumes the construction of five motorways within 15 years. It is planned that by 2007 there will be four main routes of the combined length of some 2400 km [1]:

A-1:   Gdansk-Torun-Łodz-Katowice-Gorzyce near Rybnik (597 km);
A-2:   Swiecko-Poznan-Strykow-Warszawa-Terespol (626 km);
A-3:   Szczecin-Gorzow Wielkopolski-Zielona Gora-Legnica-Lubawka on the Polish-Czech border (440 km);
A-4:   Zgorzelec-Wroclaw-Gliwice-Katowice-Krakow-Medyka (738 km);

In the future the following motorways will be build:

A-12:   Olszyna-Golnice near Legnica;
A-8:   Wroclaw-Łodz.

In order to implement the program of construction of the paid motorways several companies and consortiums were founded: Stalexport S. A., Autostrada Wielkopolska S. A., Opolskie Konsorcjum Budowy Autostrad, Autostrada Sląsk S.A., Autostrada Wschod S. A., Dogus-Holding, Przedsiębiorstwo Budowy i Eksploatacji Autostrad Automedon S. A., Gdansk Transport Company S. A., Sudecko-Pomorskie Towarzystwo Drogowe S. A., Konsorcjum Drogowo-Mostowe Autostrada.

It is assumed that the cost of 1 kilometre of a motorway will be around $ 3 million.

THE WAY OF FUNDING OF MOTORWAYS

The implementation of the paid motorways' building programme will be based on the licence system. The state will grant a business entity (consortium) the licence for the construction and operation of a motorway or its section for a precisely specified time (20-30 years).

The licensee has to draw up technical design of the motorway, built it using own funds and loans that will be paid back during the operation of the motorway using the income from the tolls. The state will purchase and lease to the licensee the land, on which the motorway will be built, for the period of the license. After its expiry the licensee will transfer the rights to the motorway to the state free of charge.

The system of motorway licences enables the following (Chrzan T 1996):

  – Lowering of the state budget input to 50%,
  – Reliving the state from the duty to built the motorways and manage them,
  – Sharing of the risks related to the programme of the construction of motorways.

Because of that the license system is the target system of the organisation of construction and operation of the paid motorways in Poland.

The selection of a licensee takes place during a two-step tender procedures organised by the Agency of Construction and Operation of Motorways (ABiEA by its Polish acronym):

a) initial selection,
b) actual tender, limited to the selected, offering parties.

Only limited companies and share companies based in Poland with the share or initial capital equal to a minimum of Euro 10 million may take part in the tender.   The initial documentation and submitted offers are assessed by the tender commission appointed by the Minister of Transport and Maritime Economy. As the result of tender procedures the said minister grants a licence in the form of an administrative decision to that company whose offer is considered as the most beneficial one.   Such licence may be granted for the construction and operation or only for the operation of the whole motorway or its section. The licensee builds and operates the motorway based on the conditions set in the negotiations with ABiEA.   The funds for the construction of the motorway (outside the costs of the purchase of the land for the construction of a motorway) the licensee gathers from various sources. Those can be:

– licensee's own funds,
– loans from commercial banks,
– loans from international banks (EBI, EBOR).

During the operation of a motorway the licensee, from the annual income, pays the following:

– taxes,
– maintenance and operational costs of a motorway in accordance with the standards set in the agreement with ABiEA,
– loan instalments,
– recovery of the invested capital together with the profit rate specified in the licence.

TYPES OF THE SURFACES

The surfaces of the motorways may be bituminous or concrete ones (Chrzan T 2000). In the construction of bituminous surfaces interested is the oil industry and in concrete ones the concrete industry. The experience with bituminous surfaces shows that they are not very durable (there are ruts on almost all roads in Poland). Positive results with the usage of concrete surfaces caused that in the Czech Republic the concrete surfaces have 70% of the new roads and in former East Germany, where the motorways are being built, their sections with big traffic loads are made of concrete. The concrete surfaces are considered better that the asphalt ones. The concrete does not deform as much as asphalt does. The concrete of higher strength has the grindability comparable to that of an intrusive rock. The concrete surface is safer because of its porosity better is the tractive adhesion, and it is also bright coloured making the night driving easier. The durability of an asphalt surface is assumed to be 18 years and the concrete plate one, 26 years. The construction of a concrete road is more expensive but its operation is much cheaper. In Poland both cement and asphalt are locally made. The road asphalt is the product of processing of the heavy ends after the distillation of oil.   Polish road building companies have been using asphalt for years and they have necessary experience and equipment. To prevent the deformation of the surface (ruts) needed are better asphalts and the roads should be designed to withstand heavier loads. The new technologies of the bituminous surfaces that are resistant to a deformation are started to be used – thanks to the Research Institute of Roads and Bridges – by several road building companies. They already built a few hundred of sections of roads with the usage of such technologies in several regions of Poland [1].

Polish cement industry has the capacity that is used in about 70%. The cement production [1] in the years 1990-1996 was from 11.9 million Mg/year to 13.9 million Mg/year. To built 2600 km of motorways together with their infrastructure needed is from 6 million Mg to 10 million Mg of the cement (bituminous surfaces) and 13 million Mg to 17 million Mg for concrete surfaces [1]. Taking into account the maximum values, for 1 km of a motorway, one needs about 3850 Mg of the cement (bituminous surface) and about 6540 Mg of the concrete for the concrete surface.

The type of the technology to be used depends on the results of the analysis of the transport costs, accessibility of the needed raw materials and technological abilities of the contractors. In the areas where there is no crushed aggregate the costs of its transportation on the distance of 80 km from the construction site are equal to the costs of is purchase. That's why it would be cheaper to bring there the cement and use a local, natural aggregate to built a concrete road foundation than to make the road foundation using the crushed aggregate. Similarly, one can make a concrete surface instead of an asphalt one.

CONCRETE ROADS CONSTRUCTION EXPERIENCE

The experimental sections of the concrete roads of the total length of 3.5 km were built in the years 1983-1986, and based on them, in 1987, the Research Institute of Roads and Bridges drawn up *The initial guidelines for the construction of the concrete surfaces on the roads with the traffic loads lower than average.* In accordance with those guidelines, in the years 1988-1991, in former Zamość province, 20 km of the roads were built (Chrzan T 2000). The concrete for the road construction was produced at a plant of the capacity of 60 m^3/h. The process of the production of a class B-25 concrete was programmed and the control of that process was automatic. The components of the concrete mix were proportioned by weight: aggregate with the accuracy of 2%, cement with the accuracy of 1% and chemical additions were apportioned by volume. The concrete was spread using a caterpillared spreader that would make the strips of the pavement of the width of up to 3.5 m. The thickness of such layer was 150 mm.

To cut the hardened concrete and to cut out the contraction joints every 5 m (50 mm in depth and 4 mm in width) used was a cutting machine powered by a combustion engine and with a disk of diameter of 350 mm and width of 3 mm. The expansion gaps of the width of 20 mm and depth of 150 mm were made every 100 m.

For the surfaces used was B-25 class concrete of a flexible consistence, content of air of 4% to 4.5% and resistance to freezing (after 150 cycles) measured by the loss of mass smaller than 5% and the decrease in strength lower than 20%. The weight absorbability did not exceed 5% and the tensile strength at bending was, on average, 3.5 MPa. To make the cement concrete used were: a P-35 and P-45 pure clinker cements from Chełm cement plant, an aeration addition, Roksol B-3A, in the amount of 0.08% relative to the cement and a plasticisation compound, SK-1, in the amount of 0.8% relative to the cement.

The cement surfaces were spread on the following road foundations:

a) existing boulder pavements widened, using the broken stone, to a standard width, profiled using a mineral and bituminous mix,
b) existing bituminous surfaces on the road foundations made out of a soil stabilised with cement.
c) road foundations made of the soil stabilised with the cement of the thickness of 140 mm – 160 mm, made using an agricultural equipment.

After 10 years of operation the sections of roads with concrete surface were in good condition. It was discovered that the intensity of damages depends on the type of the road foundation. Within the section of the provincial road of the length of 4.0 km, built on the road foundation made of an existing asphalt pavement – between Mircze-Borsuk-Stara Wies - there were 97 cracks in the plates. There were also local, longitudinal and oblique cracks and numerous edge cracks.

Within the sections on the foundations made of the soil stabilised with cement and others made of boulders the number of cracks ranged only from a few to several on 1 km of the road. Within the whole 10-year period of the operation of roads within the sections with the concrete surface there was no need of any repairs because there were no ruts, bumps or fissures. The road was coarse, it had a high wheel load capacity and adhesion, and it was well visible at night.

So, the concrete roads should not be built on existing bituminous surfaces. As per present prices the cost of making a layer of the thickness of 1 cm of asphalt is more or less equal to the cost of making a layer of the thickness of 1 cm of concrete. According to the catalogue of typical flexible and semi-stiff pavements of 1997 the thickness of the layer of the asphalt pavement is from 100 mm to 140 mm, and for the concrete (plate) pavements it is 150 mm. And that favours the building of the roads with a concrete pavement in Poland.

CONCLUSIONS

Building the concrete or asphalt motorways depends, most often, on their costs, and they are usually lower for the asphalt roads.

REFERENCES

1. CHRZAN, T., Autostrady I materiały do ich budowy (Motorways and materials for their construction) Edited by Wrocław Polytechnic.

2. CHRZAN, T., Jakość i wystarczalność zasobów złóż surowców skalnych dla potrzeb produkcji kruszyw drogowych. Konferencja: Jakość i wystarczalność zasobów złóż w Polsce. Centrum Podstaw Problemów Gospodarki Surowcami Mineralnymi. Polska Akademia Nauk, Kraków, 1996, s. 225-265. (The quality and availability of the rock materials used for the production of the road aggregates. The Conference: The quality and availability of the rock materials in Poland. The Centre of the Mineral Resources Management. Polish Academy of Science, Cracow, 1996, pages 225-265).

CONGRESS
CLOSING
PAPER

CONCRETE: VADE MECUM

P C Hewlett

British Board of Agrément

United Kingdom

ABSTRACT. In putting together this Congress review from the many papers submitted, I have been looking for significant trends that can give direction to the way forward. Both the Seminars and Conferences have been taken into account but it has been written in advance. As a consequence the views expressed and conclusions drawn may well change as a result of the Congress itself and the exchanges that will occur during the event. There will be opportunity to cover such developments during the Closing Address Ceremony.

A number of ongoing challenges for concrete have been identified that suggest a way forward. The intent to change is serious but the consequences of not changing are even more so. Quo Vardis?

Keywords: Cement, Environment, Durability, Composite materials, Toughness, Pathology, Sustainability, Aesthetics, Waste, Deconstruction.

P C Hewlett is a chartered chemist and Chief Executive of the British Board of Agrément. He is visiting Industrial Professor in the Department of Civil Engineering at the University of Dundee and an active member of the Concrete Technology Unit. He is Chairman of the Technical Executive Committee of the UK Concrete Society and Chairman of the Editorial Board of the Magazine of Concrete Research.

Particular research interests cover durability, surface and bulk characteristics of concrete modified using chemical additions.

INTRODUCTION

A Congress is defined as a formal meeting of delegates with the purpose of discussing between those present issues such as special studies. In this regard, a Congress about concrete is both timely and needed and since concrete is a global material an International Congress seems very appropriate. This is now the 5th event held in Dundee on a triennial basis, the others being:

'Protection of Concrete' 1990
'Concrete 2000' 1993
'Concrete in the Service of Mankind' 1996
'Creating with Concrete' 1999

For those not involved with concrete it may seem strange after so many years of use, thousands of published papers and many books on the subject, all that could be known about concrete was known. So why does the debate and development continue?

Firstly, concrete is the most widely used construction material globally, it has become established in technically advanced countries and it has been estimated that some 1,200-2,400 kg per head of population are made per year [1]

Secondly, because of its adaptability it can be used for almost all construction situations, both structural and non-structural.

Thirdly, it is capable of being developed further in response to environmental concerns, energy considerations and new functional demands extending the material's performance in answer to engineering needs.

In other words, concrete and concreting are dynamic and will never reach a static position, unless of course our imaginations stagnate and we run out of ideas. For all of these reasons concrete is a material of opportunity and the debate will continue.

The theme of the present Congress is 'Challenges of Concrete Construction' embracing the weather, fire, seismic and marine situations and all that interlinks them. However, concrete faces other challenges from alternative materials such as steel, wood, glass, plastics and natural masonry. Alternatives do not readily have their justification in being straight material's replacements so much as finding their own niche in the design and functional requirements of buildings.

Other challenges covered at this Congress are,

1. Exploitation of the planet – Awakening of conscience in our own self interest
                             – Environmental concerns/auditing/marking

2. Adoption of cleaner technologies

3. Sustainability – Design aspects
                    – Brownfield development
                    – Role of taxing and charging (landfill and aggregate tax)
                    – Standardisation of components to assist reuse.

4. Elimination of waste

5. Whole life costing – life and death

6. Design for deconstruction

7. Alternatives to Portland cement

8. Wide/mandated use of cement replacement materials

9. High volume use of by-products

10. Energy conservation

11. Education for such involvements

Notwithstanding such challenges, concrete remains the most widely used construction material globally and that situation is likely to continue.

There is a constant probing for new developments reflecting drivers for change both direct and indirect and some suggested drivers are noted below.

- functional
- decorative
- competitiveness
- opportunity
- serviceability
- environmental issues
- safety
- fashion

These matters are dealt with in the Congress that comprises three Seminars and three Conferences.

CONGRESS REVIEW

Seminar 1 is concerned with 'Composite Materials used Internally and Externally', both organic and in-organic. These new materials offer better durability, lower weight and higher strength, ease of transportation, low thermal conductivity and reduced energy to make. For all of these reasons, adoption of composite materials technology should be welcomed but as with so many innovations in our industry, exploitation is guarded. It is clear that questions are being raised over slowness to adopt new ideas and innovations within concrete construction.

Conceptually fibre reinforced composites are not new and the ground rules for design and selecting them are established [2]. However, reinforced concrete is still conceptually large steel fibres (rebar) as reinforcement in an inorganic matrix rather than organic.

Proven performance based on unequivocal data are key to acceptance by specifiers and users alike to give confidence in the adoption of such new technologies.

Ironically, reinforced concrete itself was originally not backed up by a great deal of pre-use data as are fibre reinforced composites today and yet it was adopted with commitment. Why was this? Perhaps we are both set in our ways as well as worried about litigation. In this respect the composite beam example of Van Elp, Cattel and Heldt [3] is a good one that contradicts the apparent trend. The logical and prospective replacement of the established convention by optimum use of new materials resulting in a hybrid beam that outperforms the traditional reinforced concrete beam. The new types of beam have high load capacity, excellent fatigue resistance, outstanding durability, although detailed reaction to fire and comparable costs need to be addressed.

Are such radical developments welcomed and do we express the functional benefits well enough to persuade clients and specifiers and designers to adopt the new options? When considering new possibilities there is also the issue of appropriate technology. It is sometimes tacitly assumed that technologies rooted in advanced industrialised countries, may be used with equal effect in less developed countries where materials and skills may be different. Since concrete in one form or another is global, it should be possible to evolve appropriate applications for different locations. We talk readily of buildability but in the area of new ideas we should also consider adoptability.

There is merit in simplicity, both in concept, manufacture and use and a shift in Europe to prefabrication, giving better control over the process and a reduction in the need for established site skills.

Composite materials should respond in a ductile manner, eg be tough rather than simply strong. The issue of toughness is repeated throughout the Congress papers. The same trend applies to the development of lightweight structural materials, probably based on waste products, reflecting the emergence of conscience as well as functional and cost requirements. The phrase 'priorities for change' appears in Lowe's keynote paper [4].

It is suggested that value should replace cost in selecting options. This is compatible with the concept of whole life costing, another strand repeated throughout the Congress.

Do regulations and standardisation stimulate or impede innovation? Regulation can stimulate by demanding new and more exacting performance levels. The means of showing compliance with the regulation, eg conforming to a standard or established technical specification may not assist change. Performance based not on prescriptive specifications are an answer but to implement such an approach requires a will to adopt and change.

Mention is made to high temperature and fire effects on organic binders and one paper by Ballaguru [5] introduced an inorganic adhesive for bonding carbon fibre sheets to concrete beams. The water-based adhesive is composed of an alumna-silicate and is stable up to 1000°C. Despite a very complicated application procedure it is indicative of what can be achieved.

One stimulant to development of fibre-reinforced plastics has been seismic performance. Typical fibres are carbon, glass and aramids maintaining cohesion beyond the point of failure.

The pursuit of lightweight and toughness is highly desirable and whilst carbon fibres have a role to play in achieving this objective it needs to be remembered that carbon can act as a noble metal in the galvanic series and whilst it will not corrode it may cause less noble metals relative to it to do so. The carbon itself is conductive and given the right combination of conditions may exacerbate the process.

Seminar 2 is concerned with 'Floors and Slabs' with an emphasis on flooring that seems to represent a typical case of 'we seem to know what to do but do not always do it'. Self imposed inadequacy!

Flooring is undoubtedly a substantial activity in the distribution depot and monolithic construction sectors that use some $1.5m^3$ of concrete a year of concrete in the UK alone.

However, what in principle could be simpler than a slab? What could be simpler than a slab made of concrete? The physical principles of which are known – or are they? When you take regard of the conditions under which concrete for flooring is laid that assumption might be questioned.

There would appear to be too many homespun practices observes Harvey [7] but the problems with floors are global and commonplace. Notwithstanding all of this there would appear to be no substitute for concrete.

There is still much emphasis on Concrete Society Report TR34 [8]

Seidler [9] challenged that notwithstanding progress in concrete generally over the last 150-200 years, concrete flooring has not advanced significantly.

Incipient cheapness whilst a determining factor in this apparent lack of progress at some 27 Euros per square metre (approximately £17.50 per square metre) the equivalent of a medium quality fitted carpet! I cannot image the average fitted carpet functioning like an industrial floor. The requirements of industrial floors are clearly stated, namely,

- non shrink concrete
- monolithic construction
- non dusting
- adequate strength and surface resistance
- complex serviceability requirements

Chemical modifications can help overcome the known problems but the addition of polymers increases the cost substantially.

An industrial floor is a good example of a multi faceted performance requirement from something that is basically very simple. Why has this topic received so little attention? Value, longevity and robust performance rather than cost might be a better approach. In this sector cost seems to dominate.

Fast track construction leading to long strip and large bays are a trend to be acknowledged, together with joint-free slabs with speciality surface finishing and finishes. These are added demands to what superficially is a simple functional element.

Floor quality appears to have reached a plateau in terms of economically obtained quality. Again a plea for whole life costing.

The paper by Watanabe et al [10], suggests a form of flooring categorisation in the range 1-7, linked to the use to which the floor may be put and the development of a so-called U-scale as a measure of anti-static performance.

Seminar 3 deals with 'Repair, Rejuvenation and Enhancement of Concrete'.

Concrete repair and rehabilitation still dominate the concrete scene and yet concrete's failures are a relatively small proportion of concrete's use. Failure is also small, relative to new applications of concrete, eg, self compacting and ultra-high strength concrete and in relation to major projects such as the second Severn Crossing. However, in money and nuisance value terms, the profile of rehabilitation and repair is high and in that sense it attracts attention. It is the longer term inadequacies such as poor appearance, sulphate and chloride attack, sulphation and carbonation, that need to be addressed. Have these problems resulted from pursuing cost containment, lower cement contents, reduced cover rather than long term value?

The entire topic of concrete's pathology and degradation processes is worthy of our attention, how do we extend with confidence, the lifespan of concrete structures?

I refer to the paper of Tuutti [11] that quotes some telling figures relating capital values of buildings and structures to the value of all stocks as they apply to Sweden but may well reflect global trends in industrialised countries. Much of this stock has to be repaired and/or replaced on a 50 year cycle and that represents vast sums of money. If that is so, do we simply use concrete and design buildings, towns and cities as we have always done? Is such an approach sustainable?

Exploitation of new materials options that will have a longer service life and perhaps even be 'smart' will become the norm. Autogenous healing of cracked concrete is an example of a 'smart' material. It is difficult to predict where the initiative for change will come from. Will it be radical design, reflecting efficient function or will it be ad hoc picking up on personal preferences and available options? Will the drive be regulation or market opportunism? Did the development of self-compacting concrete start with an engineering need or more the availability of such a material's option, driven in turn by dispersant technology applied to cementitious suspensions rather than sound market research? In a capitalist economy the ultimate drive is financial well-being and planning and opportunism will live close together resulting in a somewhat volatile cocktail. So will concrete's future development be ad hoc and random or will sustainability, efficiency and environmental concerns dominate? Will such concerns only be responded to by wealthy economies with the less endowed creating their own appropriate technologies?

The existence of historical structures built from concrete is tangible evidence of the material's good latent durability. However, what remains is the best and a great deal has not lasted and for many reasons. Therefore the historical legacy has lessons to teach us and they are worthy of study.

In determining the effect of challenges to concrete, the diagnosis stage is very important. A comparison with forensic science is justified and the diagnostician has many techniques at his disposal. I am of course referring to the paper by Sims [12].

Threats, such as thaumasite sulphate attack (TSA), alkali-silica reaction (ASR), alkali-carbonate reaction (ACR) and delayed ettringite formation (DEF), all challenge concrete but we have to keep the potential problems in context. Notwithstanding this the avoidance of alkali-silica and alkali-carbonate reaction by selecting suitable aggregates remains a global issue.

Physico-chemical techniques have resulted in preventative and remedial measures such as cathodic protection, electro chemical realkalisation and chloride removal. All have a place and the last two may well have moved on from being something of a curiosity to full-scale practical application [13].

The themes developed in the Seminars enlarge and extend into the three Conferences. Conference 1 is concerned with 'Innovations and Developments in Concrete Materials and Construction' and Shah [14] sees the following targets for concrete in the 21st Century,

- more durable
- more constructable
- more predictable
- greener

I would add, more sustainable and more competitive.

By judicious use of investigative and monitoring techniques, all these objectives are attainable. However, since concrete is a global material, a case could be made for such developments to be globally supported with the results available to all. At the present time research cultures are very national and even regional, resulting in considerable duplication of effort on the one hand but also inventive variety on the other. The basic principle of such issues as rheology, hydration, corrosion and loading characteristics could be conducted in one or two locations whereas much effort is dispersed at numerous locations around the world, competing for funds and technical recognition. It is only when we draw people together in a Congress such as this that we identify the common features. It depends upon whether a global material such as concrete requires global development in a global economy with appropriate planning and prioritising or whether the more parochial approach is more beneficial if somewhat wasteful.

For instance, the use of electronic speckle pattern interferometry (ESPI) to study cracking in fibre reinforced concrete [14] compared with the K and F functions to describe fibre distributions resulting in stress intensity factors that govern crack propagation, assists in defining the capability limits of such reinforcements. Do such studies need to be duplicated? Is the study itself not definitive?

Shah et al's work[14] has shown that extrusion can improve fibre distribution and result in a stronger and tougher composite. Prefabricated components might well lend themselves to such techniques.

In the area of self compacting concrete, rheological studies assisted the optimisation of concrete mixes. A balance has to be struck between high flow and no segregation with yield values permitting deformability of the concrete to accommodate awkward shapes. The absolute rheological terms of yield value and viscosity do not relate directly to deformability, placeability and segregation that describe what the material has to do in practice. However, such an approach permits judgements and selection to be made.

Rossi [15] has developed ultra high strength performance fibre reinforced concrete (UHPFRC) to the point that concrete without conventional reinforcement might be a prospect. This concept is represented by the MSCC (multi scale concept concrete) that consists of short fibres (6mm) and long fibres (13mm) mixed together at 7% of the cement content.

Perhaps we have too many options e.g.

- MDF (macro defect free) concrete
- DSP (densified small particle concrete)
- CRC (compact reinforced composites) and its BPR variant (similar to CRC but with longer fibres)
- RPC (reactive powder concrete)
- MSCC (multi-scale cement composite)
- SIFCON (slurry infiltrated fibre concrete)
- ECC (engineered cementitous composite)

to mention but a few.

Having developed new materials, techniques of placing and finishing them have to be considered. These are not concretes as we know them. With approximately 1000 kg/m^3 of cement and cement:aggregate ratios of 4:1 and tensile strengths of 40-50 MPa being attainable!

Costs of these developments, at this stage, are not given but clearly it is cost effectiveness and value that matters – they are not like for like replacements of normal concretes. The importance of such extreme developments as this is that it shows what can be done and, like radical fashions, they set a pattern within which there is general advancement.

It is encouraging to note in a paper by Garshol and Constantiner [16] that at long last the concept of incorporating chemicals to control and modify the plastic and harden states of concretes and mortars has consolidated into normal practice. Indeed without certain admixtures and reactive additives these new concepts would remain only a wish.

Admixtures are now being tailored to the known chemistry of cements and the required physical and chemical characteristics of resulting mortars and concretes. Science is replacing intuitive flair and materials engineering replacing a 'try and see' approach. Integrating disciplines in this way will bring construction in line with other process engineered activities, eg, aircraft construction. Such trends are predicted in the Egan Report – 'Rethinking Construction' [17].

Having made and placed the concrete its maintained appearance does not receive the status it deserves. The visual impact of concrete is not popular with the community at large and Kronlof [18] is concerned with the aesthetics of concrete, as we all should be. The phrase 'concrete is beautiful when it makes the designer and user happy' can hardly be argued with.

Concrete's form and appearance should be predictable – no unpleasant surprises. Natural ageing should be taken into account and neatness should be a prime aim for concrete development but again a low cost outlook does not aid aesthetic development.

General expectations of society and designers may differ from those of the industry and users. Kukko [19] considers that designers and society may be concerned with image and public benefit, but industry is more concerned with economic and technical profits. Is this division true? If it is how can it ever be reconciled?

Production of high quality surfaces, edges and joints is a priority with high quality materials being required, resulting in ready for finishing details, eg, painting and wallpapering.

There are opportunities for higher strengths and toughness – thinner and slender structures, shells, lattices, profiled beams and columns, all are attainable. Kukko identifies some aims, namely,

- environmental friendliness
- improving the quality of life
- competitiveness
- improved employment prospects
- improved working conditions, in particular, safety

Concrete is 'a material for all reasons' and, if its permanency could be assured, we might invest more in some of these value added options.

Conference 2 deals with 'Sustainable Concrete Construction' and Nixon's paper [20] is both forthright, to the point and very relevant. He makes some telling criticisms of man's exploitation of the planet. He contends we should be concerned with,

- adaptable buildings
- minimum waste
- design for deconstruction
- low energy cements
- reduced energy in use by using concrete intelligently

In attaining these aims concrete is the premier construction material.

Firstly, some disturbing facts.

In the UK about 25% of the energy used in industry is accounted for by the manufacture and transportation of building materials.

- It is estimated that 8% of global CO_2 emissions result from concrete production.
- One ton of CO_2 is produced per ton of Portland cement.
- Cement production is growing, particularly in developing countries and what we gain by going in one direction to save the environment may be offset by trends in the opposite direction.
- In summary we need an alternative to Portland cement – such options are coming into play already.

Cements that require reduced energy for production (less by 16%) and in turn produce less carbon dioxide (less by 10%) are seemingly attainable. Cements based upon belite with properties comparable to Portland cements and with some evidence to indicate that durability of resulting concretes might even be improved have been produced on a commercial scale in China. There is also the recent TecEco development based on magnesium oxide.[21] What is the future for these alternatives?

Nixon also makes the point that we should use concrete innovatively, taking regard of its high thermal mass resulting in substantial economies in the running of buildings.

Jensen and Glavind [22] remind us that to make $1m^3$ of office space requires something in the region of 500 MJ whereas for the same office space it requires 15,000 MJ to heat and light. How can concrete assist in the use and running of buildings is an issue that needs to be addressed.

Jensen and Glavind continue a similar theme noting that 2-6% of worldwide CO_2 stems from cement production and cement manufacture is increasing at 5% per year, equivalent to an increase of 10 million tons of CO_2 per year.

Perhaps we will only take the environmental issues seriously when failure to achieve set aims is legislated for. The Eco Management and Audit Scheme (EMAS) coupled with a statutory instrument against which a company can be registered might be a way forward. Various tools are available to engendering an environmental culture but what creates the will to do so? The principles of life cycle assessment (LCA) and life cycle inventory (LCI) can have aims and set targets such as,

- CO_2 reduction by at least 30%
- 20% of all concrete should use residual products as aggregates
- use concrete industry's own residual products
- CO_2 neutral waste derived fuel being used at a rate of at least 10% of all fuel used in cement production

In the USA some 5 billion tons of non hazardous by-product materials are produced annually (NAIK [23]). Major inputs from agriculture (2.1 billion tons) and mineral sources (1.8 billion tons).The UK construction industry produces 20% of all UK waste [24].

The use of fly ash and bottom ash and clean coal ash in cement production with new energy generating technologies yielding different coal derived ashes assists the quest for energy containment. There are many new end uses for waste materials, for instance, sewage sludge for lightweight aggregates and for making clay bricks.

The task of carrying the environmental banner often falls to the lot of the manufacturer but contractors have a role to play as well (Goring [25]). Greater integration and co-operation emphasising quality and safety and less so cost, as has been the habit to date. These principles are set out in Egan's 'Rethinking Construction' [25]. How do these attitudes impinge upon concreting activities? Firstly, starting with design concepts – concrete can play a role by way of its thermal mass in providing better air quality and natural ventilation. To effect radical change we need an integrated approach involving concrete design and function and increasing the overlap between environmental concerns and how we build.

Torring and Lauritzen [26] estimate a potential of 400 million tons of reusable concrete, stone and brick from industrialised countries. We have to consider the means of deconstruction and reclaiming the materials used. – Joined up construction underwritten by a joined up sense of public conscience.

Pocklington and Glass [24] believe that the energy performance of buildings are a key to sustainability in which case there is good reason for using concrete. Phrases such as 'burn and bury', 'dilute and disperse' and 'end of pipe' are no longer acceptable. The term anthropogenic was used – greenhouse gas emissions caused by man. We need some form of fiscal encouragement to create a culture of change. We also need to plan for longer life and adaptability of buildings.

Further telling statistics supplied by Glavind and Munch-Petersen [27]. Some 5 km^3 of concrete are used per year globally and whilst some would contend that CO_2 produced per ton of cement is small in itself, it becomes large due to the amount of concrete produced. The prospect of quantified benchmarking of attainable objectives for CO_2 reduction say by 30% and recycled concrete used as aggregate, making an energy reduction of 20% is tantalising. The authors also consider that waste derived fuels should replace 10% of fossil fuels and their paper results in a specification for 'green' concrete types – some 14 in number. The authors were very conscious of solving one problem but creating a second order unwelcome legacy, eg, kiln dust containing zinc, vanadium, lead and copper as well as phosphorus pentoxide. We have to maintain a sense of proportion and perspective. The various phases of materials production and construction activity cannot be dealt with in isolation, one from the other.

Conference 3 deals with 'Concrete for Extreme Conditions' and the paper by De Vries [28] endeavours to put the problems of durability into perspective. On balance the performance of concrete is not as bad as many would contend. However, there are problems of poor workmanship and with new materials. Codes and specifications are not particular enough and matters of maintenance and repair not covered sufficiently well. De Vries would like to see performance and reliability based service life design and makes reference to the European Brite/Euram research project – 'Duracrete'. A plea is made to involve the client and give contractors a number of options.

Durability design should get as much attention as structural design. The Eastern Schedlt barrier has a service life-span of some 200 years! However, a period of 85 years was settled for the concrete when it was accepted that the cover will have to be replaced. An example of integrating maintenance with design life. We have to quantify the anticipated functional life-span. To do this a knowledge of the durability of materials is required and the effect of workmanship on achieving these properties needs to be addressed.

De Vries is also concerned with the interaction of structure and the environment and uses a probabilistic technique to determine the likelihood of failure and target service life. There are problems of defining the limit state requirements eg the onset of corrosion. Models exist for defining degradation conditions and the design can be modified to offset adverse predictions of service life based on such models – a preemptive approach.

Slater [29] concentrates on marine structures and these represent a severe exposure condition but relates the data so that it is relevant to all types of construction. The emphasis is on buildability and durability. Buildability covering such issues as safety and economy and durability fixed by design and exposure (environment). Reassuring to see a preference for low w:c ratios and a useful quoted rule of thumb 'a reduction in w:c ratio of .05 is the equivalent of an increase in a cover of 5mm'. Slater concludes with an excellent series of pragmatic recommendations.

Over recent times there has been much discussion on the existence of threshold chloride levels below which passivation is maintained and above which active corrosion commences. The paper by Paramasivam et al [30] is a good example of data obtained on an actual structure, in this case a 32 year old wharf from which 6 marine piles (driven) were recovered and analysed for chloride content and penetration. Comparisons were made of the actual with predicted ingress levels of chloride and the relationship between such levels and loss in mass of reinforcing bars. Reference is made to threshold chloride levels (Reference 2 of [30])again and the establishment of critical chloride levels in the range .03 - .1% weight/weight concrete, covering in the first instance the splash zone and the latter submerged (Reference 6 [30]). Alternatively, Browne (Reference 7 of [3]) again showed such values to be in the range .2 - .49% w/w cement. Therefore values of .034 - .068% w/w concrete were considered appropriate.

These figures are to be doubted where a plentiful supply of oxygen is available. There was broad agreement between predicted and actual values confirming the various models used. As an extension to the problems caused by chloride ingress and contamination, Masuda [31] considered salt damage to reinforced concrete buildings resulting from both seawater and airborne salt. Some 4,363 buildings were investigated, covering everything from domestic to industrial, schools, offices, hospitals, etc resulting in 60% or so being perfectly satisfactory in the age range 7-46 years old. The distance from the coast was a significant factor in determining the residual chloride levels. Insufficient cover was the cause of deterioration in most cases but the examination was primarily visual. The threshold levels in this study were broadly corroborated.

CONCLUSIONS

This Congress has brought together a great deal of data and experience that within itself has trends that indicate the way ahead and help to establish attitudes and create priorities. Some noted trends are given below.

1.   Concrete is capable of considerable further performance-based development and should not posture as a low technology stereotype .

2.   Sustainability will remain a motivator for regulators, designers and concrete material providers. Is there a sustainable alternative to Portland cement?

3.   Adopted technology should be in proportion to prevailing local conditions.

4.   Creating a concrete culture at operative level with recognition of skill status will help to exploit new developments and make the aim of best practice a reality.

5.   Laboratory-based data must reconcile with what happens in practice — transfer of micro mechanisms to macro fact. Methods of diagnosis should be accurate and unambiguous, performed by those qualified to do so and interpretation should be subject to severe scrutiny.

6.   The role of water needs more committed study. It is necessary for cement to transform into masonry but is also responsible for much of concrete's degradation.

7.   Development of tough concrete rather than high strength but brittle concretes.

8.   The visual appearance of concrete with time is of concern. Concrete should remain pristine and not take on a patina of industrial downgrading.

9.   Coating and sealing may not be sufficient. We have to understand how concrete behaves to fluctuations in its surroundings at a micro mechanistic level.

10. Does a true threshold chloride level exist before steel corrosion occurs?

11. Can we consider structures that do not contain normal reinforcement but rely solely on metal fibres and a reconstituted matrix?

12. The use of recycled and waste materials should be encouraged using legislation and tax incentives for those that comply – a stick and carrot approach.

ACKNOWLEDGEMENTS

I am indebted to all the authors of those Congress papers to which I have made reference and indeed those that I have not, I also thank colleagues with whom I have endless discussions about concrete, without which opinions and viewpoints could not be formed and conclusions drawn.

REFERENCES

With the exception of References 1, 8, 17 and 21, all are drawn from papers given at the Dundee Congress 'Challenges of Concrete Construction' 5-11 September 2002. To simplify the listing only the relevant Seminar or Conference is identified.

1.   GLASSER F. P., Private Communication

2.   All Authors, Composite Materials in Concrete Construction, Seminar 1.

3.   VAN ERP G., CATTEL C., HELDT T., Fibre Composites in Civil Engineering: An Opportunity for a Novel Approach to Traditional Reinforced Concrete Concepts. Proceedings of International Congress: Challenges of Concrete Construction, Seminar 1 – Composite Materials in Concrete Construction, Dundee, Scotland, 5-6 September, 2002, pp 1-16

4.   LOWE P., Composite Materials in Concrete Construction, Proceedings of International Congress: Challenges of Concrete Construction, Seminar 1 – Composite Materials in Concrete Construction, Dundee, Scotland, 5-6 September, 2002, pp 17-30

5.   BALAGURU P.N., Inorganic Polymer Composites in Concrete Construction: Properties, Opportunities and Challenges, Proceedings of International Congress: Challenges of Concrete Construction, Seminar 1 – Composite Materials in Concrete Construction, Dundee, Scotland, 5-6 September, 2002, pp 109-126

6.   VAN GEMERT D., BROSENS K., Non-Metallic Reinforcements for Concrete Construction, Proceedings of International Congress: Challenges of Concrete Construction, Seminar 1 – Composite Materials in Concrete Construction, Dundee, Scotland, 5-6 September, 2002, pp 225-236

7.   HARVEY D.J.J., Developing a Greater Understanding of the Nature and Usability of Concrete in Industrial Floor Applications, Proceedings of International Congress: Challenges of Concrete Construction, Seminar 2 – Concrete Floors and Slabs, Dundee, Scotland, 5-6 September, 2002, pp 183-194

8.   CONCRETE SOCIETY, Concrete Industrial Ground Floors – A Guide to their Design and Construction, Technical Report No 34, 1994, pp 170.

9.   SEIDLER P., How Polymers Improve Floors – Possibilities Today and Prospects for the Future, Proceedings of International Congress: Challenges of Concrete Construction, Seminar 2 – Concrete Floors and Slabs, Dundee, Scotland, 5-6 September, 2002, pp 1-14

10.   WATANABE H., ONO H., KAIZU H., Performance Evaluation of Floors for Static Charge on Human Body – Proposal of a New Method, Proceedings of International Congress: Challenges of Concrete Construction, Seminar 2 –Concrete Floors and Slabs, Dundee, Scotland, 5-6 September, 2002, pp 233-244

11.   TUUTTI K., Repair, Rejuvenation and Enhancement of Concrete – A Fast Growing Market, Proceedings of International Congress: Challenges of Concrete Construction, Seminar 3 – Repair, Rejuvenation and Enhancement of Concrete, Dundee, Scotland, 5-6 September, 2002, pp 1-10

12.   SIMS I., Diagnosing and Avoiding the Causes of Concrete Degradation, Proceedings of International Congress: Challenges of Concrete Construction, Seminar 3 – Repair, Rejuvenation and Enhancement of Concrete, Dundee, Scotland, 5-6 September, 2002, pp 11-24

13.   VENNESLAND O., Documentation of Electrochemical Maintenance Methods, Proceedings of International Congress: Challenges of Concrete Construction, Seminar 3 – Repair, Rejuvenation and Enhancement of Concrete, Dundee, Scotland, 5-6 September, 2002, pp 191-198

14.   SHAH S.P., AKKAYA Y., BUI V.K., Innovations in Microstructure, Processing and Properties, Proceedings of International Congress: Challenges of Concrete Construction, Conference 1 - Innovations and Developments in Concrete Materials and Construction, Dundee, Scotland, 9-11 September, 2002, pp 1-16

15.   ROSSI P., Developments of New Cement Composite Materials for Construction, Proceedings of International Congress: Challenges of Concrete Construction, Conference 1 - Innovations and Developments in Concrete Materials and Construction, Dundee, Scotland, 9-11 September, 2002, pp 17-30

16.   GARSHOL K.F., CONSTANTINER D., Super-Concrete Examples: Complete Rheology Control and Passive Fire Protection, Proceedings of International Congress: Challenges of Concrete Construction, Conference 1 - Innovations and Developments in Concrete Materials and Construction, Dundee, Scotland, 9-11 September, 2002, pp 411-422

17. EGAN J., Rethinking Construction, July 1998, pp 39.

18. KRONLÖF A., Concrete Aesthetics: Flexible or Stiff – Humble or Arrogant, Proceedings of International Congress: Challenges of Concrete Construction, Conference 1 - Innovations and Developments in Concrete Materials and Construction, Dundee, Scotland, 9-11 September, 2002, pp 751-762

19. KUKKO H., Requirements for Advanced Concrete Materials, Proceedings of International Congress: Challenges of Concrete Construction, Conference 1 - Innovations and Developments in Concrete Materials and Construction, Dundee, Scotland, 9-11 September, 2002, pp 949-956

20. NIXON P.J., More Sustainable Construction: The Role of Concrete, Proceedings of International Congress: Challenges of Concrete Construction, Conference 2 – Sustainable Concrete Construction, Dundee, Scotland, 9-11 September, 2002, pp 1-12

21. GLASSER F.P., TecEco: Cements Based on Magnesium Oxide, A Private Communication, 2001, pp 9.

22. JENSEN B.L., GLAVIND M., Consider the Environment – Why and How, Proceedings of International Congress: Challenges of Concrete Construction, Conference 2 – Sustainable Concrete Construction, Dundee, Scotland, 9-11 September, 2002, pp 13-22

23. NAIK T.R., The Role of Combustion By-Products in Sustainable Construction Materials, Proceedings of International Congress: Challenges of Concrete Construction, Conference 2 – Sustainable Concrete Construction, Dundee, Scotland, 9-11 September, 2002, pp 117-130

24. POCKLINGTON D., GLASS J., Economics, Sustainability and Concrete, Proceedings of International Congress: Challenges of Concrete Construction, Conference 2 – Sustainable Concrete Construction, Dundee, Scotland, 9-11 September, 2002, pp 683-694

25. GORING P.G., Rethinking Sustainable Concrete Construction, Proceedings of International Congress: Challenges of Concrete Construction, Conference 2 – Sustainable Concrete Construction, Dundee, Scotland, 9-11 September, 2002, pp 439-456

26. TORRING M., LAURITZEN E., Total Recycling Opportunities – Tasting the Topics for the Conference Session, Proceedings of International Congress: Challenges of Concrete Construction, Conference 2 – Sustainable Concrete Construction, Dundee, Scotland, 9-11 September, 2002, pp 501-510

27. GLAVIND M., MUNCH-PETERSEN C., Green Concrete – A Life Cycle Approach, Proceedings of International Congress: Challenges of Concrete Construction, Conference 2 – Sustainable Concrete Construction, Dundee, Scotland, 9-11 September, 2002, pp 771-786

28. DE VRIES H., Durability of Concrete : A Major Concern to Owners of Reinforced Concrete Structures, Proceedings of International Congress: Challenges of Concrete Construction, Conference 3 – Concrete for Extreme Conditions, Dundee, Scotland, 9-11 September, 2002, pp 1-16

29. SLATER D., Marine and Underwater Concrete – Buildability and Durability, Proceedings of International Congress: Challenges of Concrete Construction, Conference 3 – Concrete for Extreme Conditions, Dundee, Scotland, 9-11 September, 2002, pp 189-204

30. PARAMASIVAM P., LIM C.T.E., ONG K.C.G., Performance of Reinforced Concrete Piles Exposed to Marine Environment, Proceedings of International Congress: Challenges of Concrete Construction, Conference 3 – Concrete for Extreme Conditions, Dundee, Scotland, 9-11 September, 2002, pp 525-536

31. MASUDA M.Y., Condition Survey of Salt Damage to Reinforced Concrete Buildings in Japan. Proceedings of International Congress: Challenges of Concrete Construction, Conference 3 – Concrete for Extreme Conditions, Dundee, Scotland, 9-11 September, 2002, pp 823-836

INDEX OF AUTHORS

Abe, M	329-338	Kimberley, D A	317-328
Abu-tair, A	139-148	Kobuliev, Z	49-54
Ackerman, C E	149-158	Kuchikulla, S R	103-110
Ahwazi, B B N	247-256	Kumar, S	83-92
Ali, K S	443-452	Kurshpel, A V	199-206
Al-Mattarneh, H M A	277-288	Kwan, A K H	237-246
Amirov, O H	49-54	Law, D W	339-348
Amleh, L	247-256	Lee, P K K	237-246
Andrade, J	423-432	Lindvall, A	169-178
Aoyama, T	329-338	Lyness, J F	139-148
Bahar, R	289-296		227-236
Bandyopadhyay, D	433-442	Maier, M	349-360
Barakat, S	453-464	Majid, W M bin W A	277-288
Bocca, G	267-276	Mantegazza, G	391-402
Bosunia, S Z	179-190	Mays, G C	361-370
Bouquet, G	465-476	McFarland, B J	227-236
Brosens, K	371-380	McParland, C	139-148
Browne, T M	403-414	Mezzi, M	391-402
Bulteel, M D	93-102	Mirza, M S	247-256
Castro, P	159-168	Molloy, D	103-110
Cervenka, V	217-226	Moriwake, A	37-48
Chernyavskiy, A V	199-206	Muhamadiev, M S	49-54
Chrzan, T	489-494	Mukhametshin, A M	199-206
Cousins, W	227-236	Nadjai, A	139-148
Crotti, M	267-276	Nikiforov, J V	111-118
Dal Molin, D	423-432	Nuruddin, M F	443-452
Daly, A F	381-390	Olague, C C	159-168
Davison, N	307-316	Owolawi, O T	207-216
De Corte, W	119-126	Pasynkov, B P	199-206
De Leersnyder, D	119-126	Penny, J E T	207-216
De Winne, E	119-126	Pukl, R	217-226
Diah, A B M	443-452	Purkiss, J A	207-216
Edwards, P J	65-74	Radicchia, R	391-402
Efimov, S N	111-118	Riche, M J	93-102
Eperjesi, L	75-82	Roberts, A C	307-316
Fedner, L A	111-118	Robery, P C	477-488
Ferreyra Hirschi, E	75-82	Rusina, R	217-226
Fiorior, B	25-36	Sadka, B	381-390
Fraaij, A L A	297-306	Sadovich, M A	55-64
Fried, A N	339-348	Safarov, M M	49-54
Garcia-Diaz, M E	93-102	Saman, H M	443-452
Gatti, A	391-402	Samohvalov, A B	111-118
Ghodgaonkar, D K	277-288	Saraswati, S	433-442
Giovambasttista, A	75-82	Saxena, A	103-110
Hayes, F O	103-110	Saxena, D S	103-110
Hewlett, P C	495-510	Seki, H	329-338
Hoque, A M	179-190	Shannag, M	453-464
Igawa, K	329-338	Short, N R	207-216
Ignoul, S	371-380	Siegwart, M	227-236
Kareem, M	453-464	Sims, I	11-24
Kenai, S	289-296	Siwak, M J M	93-102

511

Skorobogatov, S M	199-206
	415-422
Slater, D	127-138
Sokolvskaya, AA	55-64
Tanabe, T	37-48
Taylor, J M	307-316
Tuutti, K	1-10
Van Bruegel, K	297-306
Van Gemert, D	371-380
Vennesland, O	191-198
Vernet, MC	93-102
Vesely, V	217-226
Vorechovsky, M	217-226
Wang, S	257-266
Watry, P M	403-414
Wenglas, G	159-168
Wimpenny, D	127-138
Wood, M G	207-216
Yaogfarov, A K	199-206
Ye, G	297-306
Yokota, H	37-48
Zaripova, M A	49-54
Zheng, L	257-266
Zheng, W	237-246

SUBJECT INDEX

This index has been compiled from the keywords assigned to the papers, edited and extended as appropriate. The page references are to the first page of the relevant paper.

Accelerated carbonation chamber 443
    corrosion 247
Accidental damage 83
    impact 403
Additives 111
Aesthetics 495
Aggregate porosity 93
Aggregates 159
Airport pavement 179
Alkali silica reaction 93, 317
Alkali-reactivity 11
Alkalis 159
Anode materials 329
ASR 103
Asphalt 489
Assessment 403, 433
    methods 1
Aviation 103

Balustrade 149
Bicarbonation 11
Bond behaviour 247
Bridge 149
    foundations 127
    volume 297
Buildings 83

Capillary absorption 75
Capital destruction 1
Carbon fibre reinforced plastic 391
Carbonation 11, 75, 443
Catastrophe 415
Cathodic protection 329
Cement 495
    paste 297
CFRP bonding 381
Chemical and mineral composition of cement 111
Chloride 307, 339
    content 37
    extraction 317
    penetration 169, 423
Coastal structures 37
Composites 495

Component removal 55
Concrete 25, 207, 247, 267, 289, 297, 339, 361, 423
    bridges 139
    carbonation 465
    corrosion 55
    cover 465
    endurance 55
    motorways 489
    repair 317, 403, 477
    resistivity 227
    strength 199
    strengthening 381
Condition assessment 349
Contact 25
Cores 289
Correlation 443
Corrosion 329
    products 247
Covercrete 443
Crack 257
    hierarchy 415
Cracking behaviour 247
Cracks 199

Dam 55
Damage 257
Database 139
Deconstruction 495
Defect history 139
Deflection 433
Degradation of concrete 83
Degree of deterioration 37
De-icing salt 169
Density 75
Desalination 339
Deterioration 207, 317
    mechanisms 1
    modelling 477
Diagnosis 11, 127
Dielectric properties 277
Diffusion coefficient 37
Discrete anode 307
Documentation 191

Durability   11, 83, 111, 465, 495
        breakdown   65
        design   75

Early age   297
Education   477
Effective porosity   75
Elastic modulus   267
Electrochemical   307, 339
        chloride extraction   227
        maintenance methods   191
Engineering properties   111
Environmental actions   169, 495
Ettringite   11
Europe   361
Evaluation   149
Experimental   49
Externally bonded reinforcement   381

Failure pattern   257
Fibers   453
        concrete   277
        reinforced mortar   391
Filtration seepage   55
Finite element calculation   119
Fire damage   83
        proofing   83
        resistant structures   83
        safety   83
Fixed assets   1
Flexural strength   103
Fly ash   247
Framework agreement   65
Free-space   277
Freeze/thaw resistance   111
Freeze-thaw   11
Friction   25

Galvanic   307

Heterogeneity of properties   199
Historic structures   349
Historical buildings   391
Honeycomb   289
Hybrid solutions   381

ICCP   307
Ice   25
Incipient anode   307
Information entropy   415
Inlay technique   119
Inspection   37

Integrity   207
Interaction   25
Ion redistribution   227

Joints   453

Latin hypercube sampling   217
Leaching   55
Life span   443
Load Test   349, 433
Longitudinal cracks   415
Long-standing capacity   199
Low strength   289

Maintenance   477
        management system   37
Marine environment   307, 403
Microstructure   297
Microwave   277
Migration   227
Model   257
Modelling   139, 297
Models   49
Monitoring   477
Mortar   93
Motorway bridges   169

NDE   207
New materials   1
Non-destructive   277
        testing   289
Nonlinear fracture mechanics   217
Non-linearity   433

On site   267
Overload   403
Oversize   415

Pathology   495
PC strands   329
PCI   103
Performance   149, 179
Permeability   465
Petrography   11
Pore size   227
Pratinum's of hydroelectric power station   49
Prediction   37
Pressure front   55
Pretensioned PC members   329
Protection   361
Pull-out   267

Quality control testing   103
    of the covercrete   75

Reaction degrees   93
Reactivity   159
Rebar corrosion   37
Recovery   433
Refurbishment strategy   65
Rehabilitation   339, 391
Reinforced concrete   169, 391, 453
Reinforcement   339
    corrosion   465
Reinforcing steel   247
    corrosion   191
Reliability   423
Renovation   179
Repair   127, 179, 289, 361, 453
Research   1
    development   477
Retrofitting   391
Roughness   25
Runways   103

Sacrificial zinc   307
Safety   403
Sea environments   329
Segregation   289
Seismometry profiling   199
Seismotomography   199
Service life   1, 139, 423, 465
    prediction   75
Serviceability   415
Silanol   93
Silica   159
Size effect   237
Software   217
Sprayed mortar   289
Standardisation   361
Statistics   217
Steel bonding   381
    corrosion   75
Steel/CFRP   381
Strategy   477
Strengthening   391
    systems   1
Stress-strain   257
Structural alteration   349
    fire engineering   83
Structure   25, 207
Sulfate   11
    attack   127
Sustainability   495

Swelling mechanism   93

Tensile strength   237
Tension test   237, 247
Testing   207
    material   267
Thaumasite   11, 127
Theory   49
Thermophysical properties   49
Toughness   495
Transportation construction   111
Trial   317

Ultra thin whitetopping   119
Ultrasonic pulse velocity   297

Vandalism   149
Vessel / Ship collisions   403
Vibration   207, 433

Waste   495
Water filtration   55
    retaining structures   65
Wear   25
Weathering   11